"Knowing is not enough; we must apply.
Willing is not enough; we must do."
—Goethe

INSTITUTE OF MEDICINE

Shaping the Future for Health

THE NATIONAL ACADEMIES

National Academy of Sciences
National Academy of Engineering
Institute of Medicine
National Research Council

The **National Academy of Sciences** is a private, nonprofit, self-perpetuating society of distinguished scholars engaged in scientific and engineering research, dedicated to the furtherance of science and technology and to their use for the general welfare. Upon the authority of the charter granted to it by the Congress in 1863, the Academy has a mandate that requires it to advise the federal government on scientific and technical matters. Dr. Bruce M. Alberts is president of the National Academy of Sciences.

The **National Academy of Engineering** was established in 1964, under the charter of the National Academy of Sciences, as a parallel organization of outstanding engineers. It is autonomous in its administration and in the selection of its members, sharing with the National Academy of Sciences the responsibility for advising the federal government. The National Academy of Engineering also sponsors engineering programs aimed at meeting national needs, encourages education and research, and recognizes the superior achievements of engineers. Dr. Wm. A. Wulf is president of the National Academy of Engineering.

The **Institute of Medicine** was established in 1970 by the National Academy of Sciences to secure the services of eminent members of appropriate professions in the examination of policy matters pertaining to the health of the public. The Institute acts under the responsibility given to the National Academy of Sciences by its congressional charter to be an adviser to the federal government and, upon its own initiative, to identify issues of medical care, research, and education. Dr. Kenneth I. Shine is president of the Institute of Medicine.

The **National Research Council** was organized by the National Academy of Sciences in 1916 to associate the broad community of science and technology with the Academy's purposes of furthering knowledge and advising the federal government. Functioning in accordance with general policies determined by the Academy, the Council has become the principal operating agency of both the National Academy of Sciences and the National Academy of Engineering in providing services to the government, the public, and the scientific and engineering communities. The Council is administered jointly by both Academies and the Institute of Medicine. Dr. Bruce M. Alberts and Dr. Wm. A. Wulf are chairman and vice chairman, respectively, of the National Research Council.

Consultants

NEAL BENOWITZ, M.D., Professor and Chief of the Division of Clinical Pharmacology, Departments of Medicine, Biopharmaceutical Sciences, and Psychiatry, University of California, San Francisco
CHRISTOPHER LOFFREDO, Ph.D., M.S., Assistant Professor, Lombardi Cancer Center, Georgetown University Medical Center

Staff

KATHLEEN STRATTON, Ph.D., Study Director
PADMA SHETTY, M.D., Program Officer
ANN W. ST. CLAIRE, Senior Project Assistant
ROSE MARIE MARTINEZ, Sc.D., Director, Board on Health Promotion and Disease Prevention
TERRY PELLMAR, Ph.D., Director, Board on Neuroscience and Behavioral Health

Preface

Tobacco has been used by humans for at least a millennium, and its harmful effects have been suspected for at least 200 years. In the last 50 years, convincing and generally accepted evidence has established the fact that exposure to tobacco products is the major single cause of early human mortality and morbidity in developed nations and in many developing nations as well.

Even nonsmokers suffer morbidity and excess mortality from the toxic effects of inhalation of sidestream smoke. Both smokers and their nonsmoking associates are more likely to be injured in fires and automobile accidents. The personal and social price we pay for establishing and sustaining nicotine addiction through exposure to tobacco smoke is our greatest controllable health cost and one of our greatest social burdens.

It has been scientifically established that reduced exposure to tobacco smoke by lifelong abstinence and avoidance of smoke eliminates the added risk and harm and that cessation, even after many years of smoking, reduces risk and harm both immediately and in the long term for many tobacco-related conditions.

Several smoking cessation programs, some aimed at individuals and some at communities, have been shown to be modestly effective in assisting smokers to quit smoking. These programs have been shown to be more effective with the added use of nicotine replacement by patches for absorption through the skin, by nicotine-containing chewing gum or sprays for absorption through oral or nasal mucous membranes, or by the administration of psychotropic drugs to reduce the desire for nicotine.

However, with the most intensive application of the most effective known programs for prevention and cessation, approximately 10-15% of the adults in the United States are expected to be regular users of tobacco in 2010, and they will continue to suffer the increased incidence of harmful and lethal consequences. Among this group are many who cannot or will not stop using tobacco, and it is to this group that effective programs and products of harm reduction should be directed.

New tobacco products and nicotine replacement products are being marketed frequently and, along with products now on the market, often have associated direct or implied health claims. Some of the new products differ from traditional products in ways that appear minor, whereas others involve substantial changes in types of tobacco, in additives, or in curing, blending, or processing of the tobacco. New products may also change the composition of the aerosol the consumer inhales compared to cigarette smoke by changing the burning temperature of the tobacco by new methods of combustion, by limiting the release of smoke into the atmosphere, by dilution of the smoke with air, and/or by adding unnatural carriers for smoke particles.

Although many components of tobacco are known to be toxic, little is known of the specific dose-response relations of the individual toxins as they occur in cigarette smoke or of the interactions between the constituents of tobacco smoke. There is little direct evidence that removal of specific substances from tobacco smoke or from tobacco actually reduces risk or harm to human health. In considering the health effects of modified tobacco products it is important to remember that the health consequences of the use of any such product are determined not by the toxic agents removed from the product but by the actual exposure to the toxins that remain. Harm reduction is the net difference in harm between the products as actually used.

There is strong evidence that in the range of exposures involved in smoking, there is a quantitative relationship between the magnitude of exposure and the incidence of cancer, coronary vascular disease, pulmonary disease, and several other tobacco-related illnesses. Rarely if ever is there impartial and thorough assessment of the risk associated with new tobacco products relative to the risk of abstinence or the risk of other tobacco products prior to marketing. Unlike new tobacco products, nicotine replacement products are subject to full disclosure of content, rigorous testing, and the regulation of marketing claims by the of the Food and Drug Administration.

In addition to cigarette smoke, other forms of tobacco such as cigars, chewing tobacco, and snuff are also vectors of nicotine addiction and often have their own sets of serious toxic consequences.

The latent period between beginning exposure to tobacco and the development of most of the major adverse consequences is so long that empirical, direct evidence (assessment of immediate and long-term toxicity of individual tobacco products in humans) that one tobacco product is less harmful than another will rarely be available in time to be a basis for informing users. In the absence of direct evidence, conflicting claims of the degree of harm reduction are likely and informed usage decisions by smokers and nonsmokers will be difficult.

No one knows the dose-response relations, the specific toxins, the pathogenic mechanisms, or the interrelationship between the many components of tobacco smoke with enough precision to make scientifically reliable quantitative judgments about the risk or actual harm reduction associated with use of any tobacco product. Since we do not know which of many toxins may be the cause of specific harmful effects, we can only infer but we cannot know the health effects of the elimination of any one or several tobacco components. Further, we are just beginning to identify and understand the genetic basis and other causes of the differences in susceptibility to toxic effects among groups or individuals that largely determine the response of an individual to a toxin.

Nonetheless, it is reasonable to expect that some of the new products will reduce exposure to tobacco toxins and possibly reduce harm to some users and to others who are exposed to them. It is, therefore, urgent and important that the assessment of exposure to tobacco toxins resulting from the use of modified tobacco products or drugs be based on the best available evidence, made by the most qualified judges, and communicated to policy makers and the public completely and honestly.

There is little direct evidence available to serve as a basis for judgment as to the potential for harm reduction of specific new tobacco and pharmaceutical products. Therefore, any conclusions as to the relative harm of these products must necessarily be inferred from a base of indirect knowledge. The continuing introduction of new tobacco products with implicit or explicit claims of risk or harm reduction makes it important and urgent that the capacity for the best possible scientific assessment of these claims be put in place.

Since even the availability of harm reduction products may deter some from following the healthier course of abstinence or cessation, assessment of health claims should be based on an estimate of the effect of the product on the prevalence of smoking in the population, as well as the effect on the health risk to the individual smoker.

The most reliable scientific interpretation of necessarily incomplete indirect evidence comes when individuals who are experts in the related fields are not biased and are free of conflict of interest form a consensual judgment. Such a judgment based on evidence of high quality should be a requirement for a conclusion that the use of a product is in fact associated with decreased exposure to toxins and that the decreased exposure is likely to be associated with less harmful outcomes.

Further, since these judgments of risk will necessarily be inferential because they are based on indirect and inconclusive evidence, some form of postmarketing surveillance of each product is important.

The charge to the committee is to address the science base for harm reduction from tobacco. The committee concluded early in its deliberations that the science base for harm reduction will evolve over time. There will inevitably be important interactions between the types of products that are developed and the science base. There will also be interactions between any regulatory process and the science base (the science base will influence regulation, and regulation will focus pertinent science) and, obviously, between regulations and products. For these reasons, the committee realized that the science base for harm reduction can be usefully considered only in the context of some sense of the types of specific products and of the consequences of regulation. Accordingly, portions of this report address both general categories of potential harm reduction products and regulatory considerations.

It is the strong sense of the committee that claims of less harm or risk associated with the use of tobacco products or drugs should be available—but only if four conditions are met: (1) There should be strong and widely available programs designed to avoid initiation and to achieve abstinence; (2) There should be premarketing evidence satisfactory to a group of disinterested experts that—as the product will actually be used by consumers—there is less exposure to toxic agents without coincidental increase in harm to the individual from other smoke components or to the population from encouraging initiation or continuation of smoking, the burden of proof of assertions of harm reduction should rest entirely with those making the assertion; (3) The public should be fully informed of the strength of the claims as assessed by an independent panel of experts; and (4) There should be an effective surveillance system in place to determine short-term behavioral and the long-term health consequences of the use of the new products.

The committee wishes to express its great appreciation to the many individuals, listed in Appendix B, who contributed generously and substantially to its deliberations. Representatives of many health agencies as well as tobacco interests responded thoughtfully and extensively to the committee's questions.

Dr. Kathleen Stratton contributed perspective, insight, meticulous attention to detail, and essential oversight to the work and report. This report would not be possible without her substantial and important contributions.

Dr. Padma Shetty assumed responsibility for blocks of the report, and both the full report and many specific parts are testimony to her analytic, organizational, and expressive proficiency. Ann St. Claire organized the arrangements for the work of the committee with great finesse and also made useful contributions to the analytical work of the committee. Every member of the committee is deeply appreciative of the work of Dr. Stratton, Dr. Shetty, and Ms. St. Claire.

Stuart Bondurant
Chair

REVIEWERS

The report was reviewed by individuals chosen for their diverse perspectives and technical expertise in accordance with procedures approved by the National Research Council's Report Review Committee. The purpose of this independent review is to provide candid and critical comments to assist the authors and the Institute of Medicine in making the published report as sound as possible and to ensure that the report meets institutional standards for objectivity, evidence, and responsiveness to the study charge. The content of the review comments and the draft manuscript remain confidential to protect the integrity of the deliberative process. The committee wishes to thank the following individuals for their participation in the report review process:

Daniel Azarnoff, D.L. Azarnoff Associates
Alfred Fishman, University of Pennsylvania
Margaret Gilhooley, Seton Hall University
Jack Henningfield, Pinney Associates
Peter Barton Hutt, Covington and Burling
Roger McClellan, Chemical Industry Institute of Toxicology, President Emeritus
Patricia Dolan Mullen, University of Texas
Peter Nowell, University of Pennsylvania
William Parmley, University of California, San Francisco
Thomas Pearson, University of Rochester
Frederica Perera, Columbia University
John Pierce, University of California, San Diego
Stephen Rennard, University of Nebraska
Joseph Rodricks, The Life Sciences Consultancy
Jonathan Samet, Johns Hopkins University
Thomas Schelling, University of Maryland
John Slade, University of Medicine and Dentistry of New Jersey-Robert Wood Johnson Medical School
Judith Wilkenfeld, Campaign for Tobacco Free Kids
Gerald Wogan, Massachusetts Institute of Technology
Raymond Woosley, Georgetown University

Although the reviewers listed above have provided many constructive comments and suggestions, they were not asked to endorse the conclusions or recommendations nor did they see the final draft of the report before its release. The review of this report was overseen by **David Challoner**, (review monitor) University of Florida and **Hugh Tilson**, (review coordinator) University of North Carolina. Appointed by the

National Research Council and Institute of Medicine, they were responsible for making certain that an independent examination of this report was carried out in accordance with institutional procedures and that all review comments were carefully considered. Responsibility for the final content of this report rests entirely with the authoring committee and the institution.

Contents

SECTION II: EVIDENCE FOR THE SCIENCE BASE

APPENDIXES

Executive Summary

BACKGROUND

Tobacco smoke is the cause of the most deadly epidemic of modern times. Smoking causes cancer (e.g., lung, oral cavity, esophagus, larynx, pancreas, bladder, kidney), chronic obstructive pulmonary disease (COPD), myocardial infarction, and stroke. The continuing toll of tobacco use has prompted the search for means of harm reduction for those who cannot or will not stop using tobacco. Numerous products that make implied or explicit claims to reduce the burden of smoking while allowing continued nicotine consumption are now entering the market. This report is concerned with the evaluation of these products.

Nearly one-quarter of adult Americans—an estimated 47 million people—smoke cigarettes (CDC, 2000a). Although this is far lower than the 42% recorded in 1965, the decline in the rates of smoking among adults appears to have leveled off during much of the 1990s (PHS, 2000). In a recent survey, 12.8% of middle school children and 34.8% of high school students reported some form of tobacco use during the month prior to their being interviewed (CDC, 2000b). The vast majority of smokers begin tobacco use during adolescence (IOM, 1994). However, 70% of smokers say they want to quit (CDC, 1994), and 34% of smokers make an attempt to quit each year. Thus, many but not all tobacco users find it very difficult to quit and continually expose themselves to known toxic agents.

DEFINITION OF HARM REDUCTION

For the purposes of this report, a product is harm-reducing if it lowers total tobacco-related mortality and morbidity even though use of that product may involve continued exposure to tobacco-related toxicants. Many different policy strategies may contribute to harm reduction. However, this report focuses on tobacco products that may be less harmful or on pharmaceutical preparations that may be used alone or concomitantly with decreased use of conventional tobacco. The committee does not use the term "safer cigarette," in particular, in order to avoid leaving the impression that any product currently known is "safe." Every known tobacco-containing product exposes the user to toxic agents; every pharmaceutical product can have adverse effects.

HISTORY OF EFFORTS TO REDUCE HARM FROM CIGARETTES

There have been many efforts in the past to develop less harmful cigarettes, none of which has proved to be successful. One of the first innovations with the promise of harm reduction was the development of cigarettes with filters. Filters attempt to reduce the amount of toxicants that go into the smoke inhaled by the smoker. The next major modification of cigarettes with safety implications was "low-yield" cigarettes. These products emit lower tar, carbon monoxide (CO), and nicotine than other products as measured by the Federal Trade Commission (FTC) assay (the "smoking machine"). Many consumers believed, and still do, that these products pose less risk to health than other cigarettes.

However, data on the health impact of low-yield products are conflicting, in part due to a lack of systematic study early in the introduction of the products. Most current assessments of morbidity and mortality suggest that low-yield products are associated with far less health benefit, if any, than would be predicted based on estimates of reduced toxic exposure using FTC yields. In order to maintain the desired intake of nicotine, many smokers who changed to low-yield products also changed the way they smoked (e.g, compensated by inhaling more deeply than when smoking higher-yield products). Thus, their exposure to tobacco toxicants is higher than would have been predicted by standardized assays and people who have continued to use these products have not significantly reduced their disease risk by switching to them. Moreover, widespread use of these products might have increased harm to the population in the aggregate if tobacco users who might otherwise have quit did not, if former tobacco users resumed use, or if some people who would otherwise not have used tobacco did so because of perceptions that the risk with low-yield products was minimal.

TYPES OF EXPOSURE REDUCTION PRODUCTS

Tobacco and cigarette-like products have been introduced recently that, under measurement systems such as the FTC smoking machine, result in decreased emission of some toxicants compared to conventional tobacco products. Currently available products include tobacco with reportedly reduced levels of some carcinogens and cigarette-like products that deliver nicotine with less combustion than cigarettes. Two classes of pharmaceutical products approved by the Food and Drug Administration (FDA) for short-term use in smoking cessation might also be used for harm reduction. These include nicotine products, such as in patch, gum, inhaler, and nasal spray preparations, and a nonnicotine product that reduces the craving for tobacco. These cessation drugs could be used long term to maintain cessation or concomitantly with continued but decreased use of conventional tobacco products (see Table 1).

These tobacco and pharmaceutical products could *potentially* result in reduced *exposure* to toxicants. The committee uses *potentially*, because whether exposure to tobacco toxicants is reduced depends on the user's behavior, such as frequency and intensity of use. Reduced exposure, however, does not necessarily assure reduced risk to the user or reduced harm to the population. Therefore, and in order to avoid misinterpretation, the committee uses the generic phrase "potential reduced-exposure products," or PREPs, when discussing the modified tobacco products, cigarette-like products (whether tobacco containing or not), or pharmaceutical products and medical devices (whether nicotine containing or not) used for their tobacco harm reduction potential. More such products are likely to be introduced in the near future, perhaps accompanied by claims that they are less harmful than conventional products.

THE COMMITTEE CHARGE AND ASSUMPTIONS

The Institute of Medicine (IOM) convened a committee of experts to formulate scientific methods and standards by which PREPs (pharmaceutical or tobacco-related) could be assessed. Four questions were imbedded within the charge given to the committee by the Food and Drug Administration in December 1999. Where there are not yet answers, the committee was asked to outline the broad strategy by which the knowledge base might be assembled.

1. Does use of the product decrease exposure to the harmful substances in tobacco?
2. Is this decreased exposure associated with decreased harm to health?

TABLE 1 Potential Reduced-Exposure Products

Category	Descriptors	Examples
Modified tobacco	Reduced yield of selected toxicants	Advance™, low-nitrosamine tobacco cigarettes, Snus, reduced nitrosamine smokeless tobacco
Cigarette-like products	Less combustion than cigarettes	Premier™ (off market) Eclipse™ Accord™
Pharmaceutical products	Nicotine replacement	Nicotine gum, patches, inhaler, nasal spray
	Antidepressants that reduce nicotine craving	Bupropion SR, nortriptyline
	Other medications	Nicotine antagonists, clonidine

3. Are there surrogate indicators of this effect on health that could be measured in a time frame sufficient for product evaluation?
4. What are the public health implications of tobacco harm reduction products?

The first three questions deal with the adequacy of current scientific methods to determine whether and to what extent these products reduce the risk of morbidity and mortality and the nature of the advice to give to citizens, health professionals, and others. The fourth question is important because it addresses the population impact of these products. That is, although a product might be risk-reducing for an individual's health compared to conventional tobacco products, its use might not be harm-reducing for the population as a whole. The fourth question is also important because the answer lays the groundwork for educational, policy, and regulatory actions.

The committee reviewed the literature and assessed the nature and availability of the data needed to evaluate the feasibility of tobacco harm reduction. Its review encompassed the major disease categories linked by scientific evidence to tobacco consumption, including cancer, cardiovascular disease, respiratory disease, reproductive and developmental disorders, and others. The report is offered to relevant federal and state regulatory and policy bodies, Congress, scientists and health care professionals, tobacco and pharmaceutical industries, and—most importantly—the public, who will have to decide whether or not to use these products.

The committee began with fundamental operating precepts, reiterating and reaffirming overwhelming scientific evidence and the conclusions of many scientific and policy advisory bodies:

Precept 1. Tobacco use causes serious harm to human health.

Precept 2. Nicotine is addictive.

Precept 3. The best means to protect individual and public health from tobacco harms are to achieve abstinence, prevent initiation and relapse, and eliminate environmental tobacco smoke exposure.

Precept 4. A comprehensive and authoritative national tobacco control program, with harm reduction as one component, is necessary to minimize adverse effects of tobacco.

PRINCIPAL CONCLUSIONS

The committee does not evaluate specific PREPs in this report, since the currently available tobacco-related PREPs in particular are most likely prototypes of limited life span. Under present regulatory conditions, tobacco-related PREPs can be changed with little assessment and without disclosure of their contents. Therefore, the committee considered the types of PREPs currently or likely to become available in the foreseeable future. After reviewing a large body of scientific documents and data, hearing presentations from many scientific, regulatory and industrial interests, and publicly soliciting comments on the issues at hand, the committee reaches the following principal conclusions regarding the questions posed by the charge:

Conclusion 1. *For many diseases attributable to tobacco use, reducing risk of disease by reducing exposure to tobacco toxicants is feasible.* This conclusion is based on studies demonstrating that for many diseases, reducing tobacco smoke exposure can result in decreased disease incidence with complete abstinence providing the greatest benefit.

Conclusion 2. *PREPs have not yet been evaluated comprehensively enough (including for a sufficient time) to provide a scientific basis for concluding that they are associated with a reduced risk of disease compared to conventional tobacco use.* One narrow exception is the use of nicotine gum in one study for maintenance of cessation, described in Chapters 8, 13, and 14. Carefully and appropriately conducted clinical and epidemiological studies could demonstrate an effect on health. However, the impact of *PREP*s on the incidence of most tobacco-related diseases will not be directly or conclusively demonstrated for many years.

Conclusion 3. *Surrogate biological markers that are associated with tobacco-related diseases could be used to offer guidance as to whether or not PREPs are likely to be risk-reducing.* However, these markers must be validated as robust predictors of disease occurrence and should be able to predict the range of important and common conditions associated with conventional tobacco products. Furthermore, the efficacy of PREPS will likely depend on user population characteristics (e.g., those defined by gender, genetic susceptibility, ethnicity, tobacco history, and medical history).

Conclusion 4. *Currently available PREPs have been or could be demonstrated to reduce exposure to some of the toxicants in most conventional tobacco products.* Many techniques exist to assess exposure reduction, but the report contains many caveats about the use of all of them, including usually an unknown predictive power for harm.

Conclusion 5. *Regulation of all tobacco products, including conventional ones as recommended in IOM, 1994, as well as all other PREPs is a necessary precondition for assuring a scientific basis for judging the effects of using PREPs and for assuring that the health of the public is protected.* Regulation is needed to assure that adequate research (on everything from smoke chemistry and toxicology to long-term epidemiology) is conducted and to assure that the public has current, reliable information as to the risks and benefits of PREPs. Careful regulation of claims is needed to reduce misperception and misuse of the products. If a PREP is marketed with a claim that it reduces (or could reduce) the risk of a specific disease(s) compared to the risk of the product for which it substitutes, regulation is needed to assure that the claim is supported by scientifically sound evidence and that pertinent epidemiological data are collected to verify that claim.

Conclusion 6. *The public health impact of PREPs is unknown. They are potentially beneficial, but the net impact on population health could, in fact, be negative.* The effect on public health will depend upon the biological harm caused by these products and the individual and community behaviors with respect to their use. Regulation cannot assure that the availability of risk-reducing PREPs will lead to reduced tobacco-related harm in the population as a whole. However, a regulatory agency can assure that data are gathered that would permit the population effects to be monitored. If tobacco use increases or tobacco-related disease increases, these data would serve as a basis for developing and implementing appropriate public health interventions.

PRINCIPAL RECOMMENDATIONS

The committee believes that harm reduction is a feasible and justifiable public health policy—but only if it is implemented carefully to achieve the following objectives:

- Manufacturers have the necessary *incentive* to develop and market products that reduce exposure to tobacco toxicants and that have a reasonable prospect of reducing the risk of tobacco-related disease;
- Consumers are fully and accurately *informed* of all of the known, likely, and potential consequences of using these products;
- Promotion, advertising, and labeling of these products are firmly *regulated* to prevent false or misleading claims, explicit or implicit;
- Health and behavioral effects of using PREPs are *monitored* on a continuing basis;
- Basic, clinical, and epidemiological *research* is conducted to establish their potential for harm reduction for individuals and populations; and
- Harm reduction is implemented as a *component* of a comprehensive national tobacco control program that emphasizes abstinence-oriented prevention and treatment.

Recommendations about future research needs are based on Principal Conclusions 1-4 and can be found in the following section. They flow primarily from material presented in Section II of the report. Progress in these areas will permit the application of the principles of risk assessment to the evaluation of PREPs in the future. At present, judgement informed by incomplete science will be used to evaluate PREPs. However, immediate actions are required. Therefore, the committee makes recommendations that address Principal Conclusions 5 and 6. These conclusions and recommendations are particularly intertwined, requiring immediate attention and swift action.

The effect of PREPs could be to increase or decrease tobacco-related disease in the population. Assessing a positive public health impact will be difficult and will require extensive surveillance and research to ensure that the impact is positive. Even the strongest surveillance system could not alone provide minimal assurance of safety or protection of the public. Currently there is little public authority over tobacco products of any type. Whatever the current legal or regulatory posture with respect to these products, the committee realized that in order to obtain the best available scientific evaluation of emerging tobacco-related PREPs and to provide the best advice on use of all PREPs to the public, some national authority over these PREPs is needed. Only a comprehensive program of

regulation and assessment including extensive premarket testing and surveillance offers a reasonable possibility of net gain in health from use of
PREPs instead of conventional tobacco product use.

**Therefore, the committee recommends development of a surveil-
lance system to assess the impact of promotion and use of PREPs on the
health of the public.** A national comprehensive surveillance system is
urgently needed to collect information on a broad range of elements necessary to understand the population impact of tobacco products and
PREPs, including attitudes, beliefs, product characteristics, product distribution and usage patterns, marketing messages such as harm reduction
claims and advertising, the incidence of initiation and quitting, and nontobacco risk factors for tobacco-related conditions. There should be surveillance of major smoking-related diseases as well as construction of
aggregate population health measures of the net impact of conventional
products and PREPs.

The surveillance system should consist of mandatory, industry-
furnished data on tobacco product constituents and population distribution and sales. Resources should be made available for a program of
epidemiological studies that specifically address the health outcomes of
PREPs and conventional tobacco products, built on a robust surveillance
system and using all available basic and clinical scientific findings.

**The committee further recommends strengthened federal regula-
tion of all modified tobacco products with risk reduction or exposure
reduction claims, explicit or implicit, and any other products offered to
the public to promote reduction in or cessation of tobacco use. The
committee outlines 11 principles to govern the regulation of PREPs.** The
regulation proposed by this committee is narrowly focused on assuring
that the products reduce risk of disease to the user and accumulating data
that would indicate whether or not the products are harm-reducing for
the population in the aggregate. Other potential regulatory approaches to
tobacco control are not addressed within this report.

The recommended regulatory structure builds on the foundation of
existing food and drug law, with appropriate adaptations to take into
account the unique history and toxicity of tobacco products. These principles envision testing and reporting for all tobacco products, approval of
claims regarding reduced exposure and reduced risk regarding tobacco
or cigarette-like products, and retention of current FDA regulation of
pharmaceutical PREPs. Manufacturers of tobacco products and pharmaceuticals should be encouraged to develop and introduce new products
that will reduce the burden of tobacco-related disease. The regulatory
system proposed in this report is not to be viewed in isolation. It is

proposed as an essential component of a package of public health initiatives (including research, education, and surveillance) that this committee believes is necessary to realize whatever benefit tobacco and pharmaceutical product innovation can offer in reducing the nation's burden of tobacco-related illness and death. (See Box 1.)

Research Conclusions and Recommendations

Many fruitful research directions should be explored to strengthen the scientific basis for assessing harm reduction. In reviewing the range of scientific disciplines and disease areas, the committee specifically noted five general scientific issues: (1) description of the dose-response relationship between tobacco smoke and/or constituent exposure and health outcomes, (2) identification and development of surrogate markers for disease, (3) the utility of preclinical research, (4) utility of short-term clinical and epidemiological studies, and (5) the role of long-term epidemiological studies and surveillance. The committee has reviewed the evidence available regarding these points and has described a research agenda to facilitate evaluation of the harm reduction potential of these products. This section summarizes the committee's conclusions and recommendations for future research, which are elaborated in detail in Section II of this report.

1. *Currently available data allow estimation, albeit imprecise, of a dose-response relationship between exposure to whole tobacco smoke and major diseases that can be monitored for evaluation of harm reduction potential.*

There are more than 4,000 different chemicals in tobacco smoke; many of these are known to be toxic. Many of the mechanisms of pathogenesis attributed to tobacco use have been explicated, and in a few cases, causative tobacco constituents have been identified. In order to effectively evaluate the toxic effects of tobacco smoke and identify the primary causal agents, the toxic components of PREPs and comparison products must be identified and be disclosed. For the most part, the data are insufficient to accurately describe the relationship of tobacco use and disease formation at the level of detail that would establish all causal agents involved or the exact dose-response relationship. The characteristics of this relationship vary among diseases and are affected by differences in compensation and actual exposure and by interindividual or population differences. Consequently, the confidence with which the adverse effects or harm reduction potential of PREPs can be extrapolated, especially at low doses, is uncertain. Also, there is currently no evidence to support a threshold level of tobacco exposure below which no risk exists for any of the reviewed health outcomes.

BOX 1
Regulatory Principles

Regulatory Principle 1. Manufacturers of tobacco products, whether conventional or modified, should be required to obtain quantitative analytical data on the ingredients of each of their products and to disclose such information to the regulatory agency.

Regulatory Principle 2. All tobacco products should be assessed for yields of nicotine and other tobacco toxicants according to a method that reflects actual circumstances of human consumption; when necessary to support claims, human exposure to various tobacco smoke constituents should be assessed using appropriate biomarkers. Accurate information regarding yield range and human exposure should be communicated to consumers in terms that are understandable and not misleading.

Regulatory Principle 3. Manufacturers of all PREPs should be required to conduct appropriate toxicological testing in preclinical laboratory and animal models as well as appropriate clinical testing in humans to support the health-related claims associated with each product and to disclose the results of such testing to the regulatory agency.

Regulatory Principle 4. Manufacturers should be permitted to market tobacco-related products with exposure-reduction or risk-reduction claims only after prior agency approval based on scientific evidence (a) that the product substantially reduces exposure to one or more tobacco toxicants and (b) if a risk reduction claim is made, that the product can reasonably be expected to reduce the risk of one or more specific diseases or other adverse health effects, as compared with whatever benchmark product the agency requires to be stated in the labeling. The "substantial reduction" in exposure should be sufficiently large that measurable reduction in morbidity and/or mortality (in subsequent clinical or epidemiological studies) would be anticipated, as judged by independent scientific experts.

Regulatory Principle 5. The labeling, advertising, and promotion of all tobacco-related products with exposure-reduction or risk-reduction claims must be carefully regulated under a "not false or misleading" standard with the burden of proof on the manufacturer, not the government. The agency should have the authority and resources to conduct its own surveys of consumer perceptions relating to these claims.

Regulatory Principle 6. The regulatory agency should be empowered to require manufacturers of all products marketed with claims of reduced risk of tobacco-related disease to conduct post-marketing surveillance and epidemiological studies as necessary to determine the short-term behavioral and long-term health consequences of using their products and to permit continuing review of the accuracy of their claims.

Regulatory Principle 7. In the absence of any claim of reduced exposure or reduced risk, manufacturers of tobacco products should be permitted to market new products or modify existing products without prior approval of the regulatory agency after informing the agency of the composition of the product and certifying that the product could not reasonably be expected to *increase* the risk of cancer, heart disease, pulmonary disease, adverse reproductive effects or other adverse

continues

BOX 1 Continued

health effects, compared to similar conventional tobacco products, as judged on the basis of the most current toxicological and epidemiological information.

Regulatory Principle 8. All added ingredients in tobacco products, including those already on the market, should be reported to the agency and subject to a comprehensive toxicological review.

Regulatory Principle 9. The regulatory agency should be empowered to set performance standards (e.g., maximum levels of contaminants; definitions of terms such as "low tar") for all tobacco products, whether conventional or modified, or for classes of products.

Regulatory Principle 10. The regulatory agency should have enforcement powers commensurate with its mission, including power to issue subpoenas.

Regulatory Principle 11. Exposure reduction and risk reduction claims for drugs that are supported by appropriate scientific and clinical evidence should be allowed by the FDA.

In summary, current knowledge of the dose-response relationships is sufficient to support risk reduction through exposure reduction as a goal for the individual through the use of these various products. To date, these relationships are not defined well enough in terms of specific components of smoke to serve as a predictive tool for the effect a particular product will have on most health outcomes. However, a strong quantitative relationship between maternal tobacco exposure and the incidence of spontaneous abortions and intrauterine growth retardation leading to low infant birthweight has been documented extensively. This population is one in which the actual health effects of PREPs and potential for harm reduction may be most directly evaluated. Further discussion regarding dose-response can be found in the disease-specific chapters in Section II (Chapters 12–16).

2. Although candidate disease-specific surrogate markers are currently available, further validation of these markers is needed. In addition, other biomarkers that accurately reflect mechanisms of disease must be developed to serve as intermediate indicators of disease and disease risk.

Biomarkers are measurements of any tobacco constituent, tobacco smoke constituent, or effect of such a compound in a body fluid (including exhaled air) or organ. Although some biomarkers currently exist, these require further validation and more must be developed that have adequate sensitivity, specificity, and limited complexity and that quantitatively link biological exposure of tobacco smoke or specific constituents to disease induction or progression prior to the advent of clinically apparent disease. Validation and development of biomarkers will provide a stronger foundation by which to make scientific evaluations and regulatory decisions regarding PREPs.

Although no panel of markers can be utilized currently to evaluate the health effects of PREP use, potential biomarkers have been and are being developed for many of the relevant disease categories. The committee recommends further study of biomarkers for various disease categories that may potentially be determined to be intermediate indicators of disease and disease risk. For example, possible measures include markers of platelet and vascular activation, lipid peroxidation, and inflammation, which have the potential to be related to measures of cardiovascular physiology and, ultimately, reflect the risk of clinical cardiovascular disease as well as markers of inflammation in lung disease. Also, biomarkers of cancer that may indicate early carcinogenic processes and risk of cancer development include but are not limited to markers of genetic damage in blood, sputum, urine, and internal organs. Another potential marker is the measurement of bone density as a direct reflection of the severity and risk of osteoporosis.

Ideally, a set of behavioral markers is needed to monitor product use patterns, thereby enabling clinicians and researchers to measure substitution of PREP use for cessation. Although the committee realizes the difficulty of developing a set of such behavioral markers, they are needed for a comprehensive evaluation of PREPs. A further detailed research strategy regarding the development of biomarkers can be found in the disease-specific chapters (Chapters 12–16) and the chapter on exposure and biomarkers assessment (Chapter 11).

3. *The evaluation of PREPs will be facilitated by the development of appropriate animal models and in vitro assays of the pathogenesis of tobacco-attributable diseases.*

Animal models and in vitro testing can contribute to the evaluation of individual PREPs and to the development of a scientific basis for designing and evaluating harm reduction products. Such studies could include cell culture, animal studies, and molecular studies to document specific toxicants as the most likely causative agents, to better define pathogenic effects of tobacco smoke exposure, to better explain the relationship of disease risk regression and exposure regression (dose-response relationships), and to validate biomarkers of exposure and biological effect.

The new technologies of genomics and proteomics have the potential for evaluating and comparing the effects of tobacco exposure and PREP use on gene translation and expression in neoplastic and nonneoplastic disease.

The committee recommends specific applications of pre-clinical models for specific tobacco-attributable disease. For example, the committee recommends the utilization of genomic and proteomic technologies to investigate the effect on gene translation and expression of tobacco smoke exposure and its relevance for pulmonary, cardiovascular, and neoplastic health outcomes. Also, accurate models are needed for smoke or tobacco constituent exposure (including nicotine) and exposure to PREPs and their effects on the development of COPD, cardiovascular disease, neoplasia, and in utero injury. Again, a more detailed pre-clinical research agenda can be found in the disease-specific chapters in Section II (Chapter 12–16).

4. *Short-term clinical and epidemiological studies in humans are required for the comprehensive evaluation of PREPs.*

Some effects of PREPs in humans could be evaluated by epidemiological studies, by measurement of intermediate disease markers or, in some cases, by clinical studies of smokers who are unwilling or unable to quit but are willing to use PREPs (compared to a control group of conventional tobacco product users). The committee recommends the utilization of validated intermediate biomarkers of disease effect in these studies in order to assess potential harm reduction within a practical time frame for diseases that occur only after prolonged exposure. Examples of potential measures include the use of lung function tests or inflammatory changes, evaluated through bronchoalveolar lavage, as intermediate markers for COPD in interventional studies.

A few smoking-attributable diseases develop after relatively brief exposure (weeks to months) and provide an opportunity for strong short-term clinical and epidemiological studies. These diseases include intrauterine growth retardation leading to low infant birthweight, slowed wound or ulcer healing, and perhaps acute myocardial infarction. Human studies are also required for evaluating the relationship of individual smoking history, environment, gender, race, and other factors (e.g., diet) to disease risk and susceptibility. Further discussion regarding the utilization of clinical studies can be found in Section II (Chapters 12–16).

5. *Long-term epidemiological studies of populations and/or pilot groups of users should monitor the incidence of disease or other adverse effects.*

Most tobacco-related diseases develop clinically over many years, and the only direct and definitive way to evaluate the harm reduction value of PREPs is to monitor the health outcomes of users compared to appropriate control groups over an extended period of time. Such surveillance could be an add-on to other epidemiological studies and should include ongoing reports of smoking behavior, types of products used, and health outcomes, as well as intermittent collection of biological samples for biomarker assessment in a segment of users. Further discussion can be found in Chapter 6 and in the disease-specific chapters in Section II (Chapters 12–16).

Risk Assessment

A report published in 1983 by a committee of the National Research Council (NRC) outlined important steps and considerations in risk assessment (NRC, 1983). The "Red Book" identified important steps: hazard identification (Does the toxicant cause the adverse effect?), dose-response assessment (What is the relationship between dose and incidence in humans?), exposure assessment (What exposures are currently experienced or anticipated under difference circumstances?), and risk characterization (What is the estimated incidence of the adverse effect in a given population?). A risk characterization provides important information for risk management, which also includes public health, social, economic, and political considerations.

The committee did not do a formal risk assessment of PREPs. The knowledge base is inadequate to do so at this time. However, the "Red Book" framework has great utility in presenting the committee's work. Table 2 uses it to summarize material discussed in Chapters 1, 5, 6, 7 and 8. Even though the committee has concluded that harm reduction through the use of PREPs is not yet convincingly demonstrated, Table 2 illustrates how the committee's conclusions and recommendations are key to gathering important data. This new knowledge base will permit a more definitive evaluation of harm reduction as a strategy and of PREPs as tools for reducing tobacco-related morbidity and mortality.

Based on an extensive review of the scientific and medical literature, the committee concludes that although harm reduction is feasible, no currently available PREPs have been shown to be associated with biologically relevant exposure reduction or with decreased tobacco-related harm. One narrow exception is the use of nicotine gum in one study for maintenance of cessation, described in Chapters 8, 13, and 14. Without a comprehensive program of scientific research, surveillance, and regulation, the potential benefit of harm reduction will go unrealized. Furthermore, without such a comprehensive program PREPs could, in fact, be detrimental to both individual and public health.

TABLE 2 Use of Risk Assessment Framework in Assessing Tobacco Harm Reduction

	Hazard Identification	Dose Response	Exposure Assessment	Risk Characterization	Risk Management
Information required as described in 1983 "Red Book"	Epidemiology Animal bioassay Short-term studies Comparisons of molecular structure	Epidemiology Low-dose extrapolation Animal to human extrapolation	Dose to which humans are exposed Dose of special populations Estimation of size of population potentially exposed	Estimate of the magnitude of the public health problem	A risk-assessment (qualitative or quantitative) may be one of the bases of risk management
Challenges in risk assessment of conventional tobacco products	Complex mixture Animal models are limited Constituents and additives are proprietary information	Dose changes for an individual over time Dose of individual toxicants varies over time Exposure at time of disease progression	Changes in smoking topography Complex mixture	For which disease? At which point in smoking history?	FTC regarding advertising
Additional challenges of PREP risk assessment	Tobacco-related products will change rapidly with time	Assessing effect of moving backwards on a dose-exposure curve, assuming long-time previous higher exposure	Changing exposure after long-term higher dose exposure Some toxicants could increase	Need models to consider effects on initiation, cessation, and relapse	FDA authority currently exerted only over pharmaceutical PREPs

continues

TABLE 2 Continued

	Hazard Identification	Dose Response	Exposure Assessment	Risk Characterization	Risk Management
Committee charge	1. Does product decrease exposure to the **harmful substances** in or produced during use of tobacco?	2. Is decreased exposure associated with decreased harm to health? 3. Are there useful surrogate indicators of disease that could be used?	1. Does product **decrease exposure**?	4. What are the public health implications?	4. What are the public health implications?
Disease-specific summary data (Chapter 5; Section II)	3. Utility of preclinical research to judge feasibility	1. Dose-response data for conventional tobacco products 2. Validation and development of biomarkers 4. Short-term clinical and epidemiological studies	2. Validation and development of biomarkers 4. Short-term clinical and epidemiological studies	5. Long-term epidemiological studies and surveillance	
Principal conclusions	1. Risk reduction is feasible 4. Exposure reduction can be demonstrated.	3. Surrogate measures could be used to predict risk reduction	4. Exposure reduction can be demonstrated	1. Risk reduction is feasible 2. Risk reduction not yet demonstrated 6. Public health impact is unknown	5. Regulation is a necessary precondition for assuring a science base and for assuring protection of the health of the public

Elements of surveillance system	Specific tobacco constituents of both the products and the smoke they generate	Disease outcomes	Consumption of tobacco products and of PREPs Biomarkers of exposure to tobacco products Personal tobacco product use and related behavioral patterns	Disease outcomes	Tobacco product marketing, including PREPs
Regulatory principles (all refer to tobacco-related PREPS, except for 11)	1. Ingredient disclosure 3. Preclinical testing required to support health-related claims 7. Evidence for no increased risk 8. Added ingredient review 9. Performance Standards	6. Products with claims would require post-marketing surveillance and epidemiological studies	2. Yield Assessment 4. With specific claims, no increased exposure to unclaimed compounds 9. Performance Standards 11. Exposure reduction claims for pharmaceutical PREPs	5. Labeling for products with claims cannot be false or misleading	10. Enforcement power
Recommendations	3. Develop appropriate animal models and in vitro assays of pathogenesis	1. Sufficient data to allow estimation of dose–response 2. Need to develop validated biomarkers of disease	4. Clinical and epidemiological studies in human are required	Comprehensive surveillance is recommended	Regulation is recommended

REFERENCES

CDC (Centers for Disease Control and Prevention). 1994. Health objective for the nation; cigarette smoking among adults–United States, 1993. *Morbidity and Mortality Weekly Report* 43(50):925-930.

CDC (Centers for Disease Control and Prevention). 2000a. cigarette smoking among adults–United States, 1998. *Morbidity and Mortality Weekly Report* 49(39):881-884.

CDC (Centers for Disease Control and Prevention). 2000b. Tobacco use among middle and high school students–United States, 1999. *Morbidity and Mortality Weekly Report* 49(3): 49-53.

IOM (Institute of Medicine). 1994. *Growing Up Tobacco Free.* Lynch, BS, and RJ Bonnie, Eds. Washington, DC: National Academy Press.

NRC (National Research Council). 1983. *Risk Assessment in the Federal Government: Managing the Process.* Washington, DC: National Academy Press.

PHS (Public Health Service). 2000. *Health, United States, 2000 With Adolescent Health Chartbook.* Hyattsville, MD: National Center for Health Statistics.

Section I

Introduction,
Background,
and Conclusions

1

Introduction

"...tobacco use, particularly among children and adolescents, poses perhaps the single most significant threat to health in the United States."

Justice Sandra Day O'Connor,
FDA v. Brown & Williamson Tobacco Corp. et al. 2000.

"...as a practical matter, it is important to appreciate that a virtually harmless cigarette smoked by only 1% of the population will have a lesser impact on the reduction of tobacco-related diseases than a somewhat more harmful cigarette smoked by 80% of the total smoking population. Research on the less harmful cigarette should therefore be directed toward developing a cigarette containing the lowest possible amount of harmful elements for all tobacco-related diseases, but one that has sufficient acceptability for the largest segment of smokers."

Ernst Wynder, Banbury Conference of Safer Cigarettes, 1979.

"...the use of tobacco, especially cigarette smoking, has been causally linked to several diseases. Such use has been associated with increased deaths from lung cancer and other disease, notably coronary artery disease, chronic bronchitis, and emphysema. These widely reported findings, which have been the cause of much public concern over the past decade have been accepted in many countries by official health agencies, medical associations, and voluntary health organizations."

Smoking and Health, Report of the Advisory Committee to the
Surgeon General of the Public Health Service, 1964.

Knowledge of the devastating consequences of tobacco use to human health has burgeoned in the 37 years since release of the first U.S. Surgeon General's report on smoking and health (U.S. Public Health Service, 1964). As scientific evidence has steadily accumulated, policy actions and the political will to reduce this toll have waxed and waned. Tobacco-caused death, illness, personal suffering, and costs are a major scourge of our time. If the toll due to tobacco were instead due to an infectious agent for example, most societies would move vigorously to stem the losses with every resource. Public health, medical, and tobacco control professionals, however, continue their campaign against a major threat to the public health. Numerous products that make implied or explicit claims to reduce the burden of smoking while allowing continued nicotine consumption or smoking are now entering the market. This report is concerned with the evaluation of these products.

Nearly one-quarter of adult Americans—an estimated 47 million people—smoke cigarettes (CDC, 2000a). Although this is far lower than the 42% recorded in 1965, the decline in the rates of smoking among adults appears to have leveled off during much of the 1990s (PHS, 2000). A slightly higher percentage of American men (26%) than women (22%) currently smokes (CDC, 2000a). Smoking rates vary greatly among racial and ethnic groups within the United States (CDC, 2000b). American Indians or Alaskan Natives have much higher smoking rates than the national average, and Asians or Pacific Islanders have much lower rates (CDC, 2000a).

The prevalence of adolescents in the United States who report current smoking is high. In a recent survey, 12.8% of middle school children and 34.8% of high school students reported some form of tobacco use during the month prior to their being interviewed (CDC, 2000c). The vast majority of smokers begin tobacco use during adolescence (IOM, 1994). The important role of adolescent smoking in lifetime addiction and smoking-caused disease is an important justification for many tobacco control policies in the United States.

These high rates of tobacco use continue despite statistics showing that 70% of smokers say they want to quit (CDC, 1994) and 34% of smokers make an attempt to quit each year. However, 2.5% of smokers (or less than 10% of those who try) quit smoking (CDC, 1993b). Approximately half of high school seniors who smoked reported that they expect not to be smoking five years from the time of the survey (1976–1984), but 80% of those who smoked more than a half-pack per day were still smoking at follow-up (CDC, 1994). A recent study of high school students reported that almost three-fourths of those who currently smoked had tried to quit (CDC, 1998).

The reasons for great national concern over tobacco use are well known. The harm to human health from tobacco use is well documented and has been for decades. Large-scale epidemiological studies conducted in the 1950s supported the causal relationship between cigarette smoking and lung cancer and other diseases. These studies detailed the dose-response relationship between number of cigarettes smoked and individual risk for lung cancer. The 1964 Surgeon General's report stated that cigarette smoking was causally associated with lung cancer in men. The evidence for a causal relation in women was suggestive in 1964 but not considered established until 1971. It is now known that smoking also causes cancers of the oral cavity, esophagus, and larynx and is a contributing cause of cancers of the pancreas, bladder, kidney, and cervix. Smokers are at increased risk of lung cancer (a sixteenfold increase), chronic obstructive pulmonary disease (COPD; twelvefold), and a myocardial infarction (twofold) (Fielding et al. 1998). Pregnant women who smoke have twice the risk of nonsmokers of delivering low-birthweight infants. The Surgeon General of the United States has continued to publish reports on smoking and health. Some of the most significant publications are listed in Box 1-1.

Cigarette smoking is often called the single leading preventable cause of death in the United States (U.S. DHHS, 2000). Smoking results in more deaths each year in the United States than AIDS, alcohol, cocaine, heroin, homicide, suicide, motor vehicle crashes, and fires combined. Primary causes of death from tobacco are cardiovascular disease (approximately 906,600 deaths between 1990-1994), cancer (approximately 778,700 deaths between 1990-1994), and respiratory disease (approximately 454,800 deaths between 1990-1994) (CDC, 1993a). For 1990, smoking was estimated to be the cause of 20% of deaths from ischemic heart disease, 29% of all cancer deaths, 83% of lung cancer deaths, and 79% of COPD deaths. The average reduction in life expectancy for smokers is 6.6 years (Lew and Garfinkel, 1987). Detailed descriptions of the epidemiology of tobacco health effects can be found in Chapters 12–16 of this report.

The prevalence of tobacco use varies greatly outside the United States. A recent study from the World Bank reports that 59% of males in the regions of East Asia, the Pacific, Eastern Europe, and Central Asia smoke (World Bank, 1999). The prevalence of female smoking in these regions ranges from 4 to 26%. Worldwide, tobacco use is the sixth leading cause of lost disability-adjusted life-years, a measurement that combines the effects of morbidity and mortality for comparative purposes between and across health outcomes (Murray and Lopez, 1997). The World Health Organization (WHO) estimates that by 2030, tobacco will kill 10 million people per year worldwide (WHO, 1999), making it the leading cause of

BOX 1-1
Select Surgeon General's Reports Regarding Tobacco

1964 Report concluded that smoking causes cancer and other serious diseases.

1967 Report concluded that "cigarette smoking is the most important of the causes of chronic non-neoplastic bronchiopulmonary diseases in the United States." The report also identified measures of morbidity associated with smoking.

1969 Report made solid conclusions regarding the relationship between maternal smoking and low infant birthweight. It also identified evidence of increased incidence of prematurity, spontaneous abortion, stillbirth, and neonatal death.

1972 Report associated smoking with cancers of the oral cavity and esophagus.

1973 Report studied immunological effects of tobacco and tobacco smoke, and identified carbon monoxide, nicotine, and tar as the smoke constituents most likely to produce health hazards from smoking.

1977- Report focused on health effects of smoking on women, noting in particular
1978 the effects of oral contraceptives and smoking on the cardiovascular system.

1980 Report addressed women and smoking projecting that lung cancer in women will surpass breast cancer as the leading cause of cancer mortality.

1981 Report examined the health consequences of lower-tar and nicotine cigarettes.

1983 Report evaluated health consequences of smoking for cardiovascular disease, declaring cigarette smoking one of the three primary causes of coronary heart disease.

1984 Report examined the health effects of smoking on chronic obstructive lung disease (COLD). Smoking accounted for 80–90% of COLD deaths in the United States.

1986 Report concluded that involuntary smoking is a cause of disease in healthy nonsmokers.

1988 Report stated that nicotine is addicting.

1989 Report reported that cigarette smoking is a major cause of cerebrovascular diseases (stroke). The report addressed the future of nicotine addiction in light of new nicotine delivery systems test marketed in 1988 and associated smoking with cancer of the uterine cervix.

1990 Report presented data on the benefits of smoking cessation for most smoking-attributable diseases. Also, presented association of smoking with bladder and cervical cancers.

1994 Report looked at preventing tobacco use among young people including initiation, cessation, advertising influences, and school-based programing.

1998 Report examined tobacco use among U.S. racial and ethnic minority groups showing that patterns of use vary among these groups.

2000 Report analyzes approaches to reducing tobacco use and the future of tobacco control.

avoidable premature death in the developing world, as it is in developed countries today. In recognition of the difficulties of cessation, the WHO Framework Convention on Tobacco Control has been reviewing scientific information on harm reduction as part of its overall strategies.

However, despite the morbidity and mortality caused by tobacco and widespread knowledge by adults and adolescents of its adverse health effects, tobacco use continues. The biologically active component in to-bacco that is primarily responsible for this is nicotine. Nicotine acts on several organs, including the brain. Nicotine is pleasurable to the user, and it is addictive. Thus, many but not all tobacco users find it very difficult to break their addiction and reduce the risk to their health. The continuing toll of tobacco use has prompted the public health community to consider anew harm reduction approaches for tobacco.

Tobacco harm reduction refers simply to the goal of reducing harm to health from tobacco use, including environmental tobacco smoke (ETS). Harm avoidance is achieved by never using tobacco products or having contact with ETS. Harm is minimized by quitting tobacco use and reduc-ing exposure to ETS. *For the purposes of this report—and as the term is com-monly used in other disciplines—harm reduction refers to minimizing harms and decreasing total morbidity and mortality, without completely eliminating to-bacco and nicotine use.* This definition acknowledges that a significant pro-portion of individuals will continue to use tobacco for the foreseeable future. Harm reduction can be accomplished by decreasing the risk of an act (e.g., tobacco use), by decreasing the intensity per user, or by decreas-ing the prevalence. Chapter 2 includes a detailed discussion of harm re-duction in other areas of public health concern and sets forth some gen-eral principles relevant to tobacco.

A multitude of policy strategies, such as increased taxes, contribute to the goal of harm reduction. However, this report focuses on substituting conventional tobacco use with either newly developed so called less harm-ful tobacco products or with pharmaceutical preparations used alone or concomitantly with decreased use of conventional tobacco products. The committee uses the terms harm-reducing or risk-reducing. The term "safer cigarette" has often been used historically and colloquially. The commit-tee avoids the term "safer" in particular in order to avoid any impression that such products are safe. They are not. The U.S. cigarette manufacturers have recently publicly echoed the public health community's assertion that that there is no safe cigarette (Philip Morris USA, 2000; R.J. Reynolds Tobacco Company, 2000). Despite any promised harm reduction, the use of tobacco harm reduction products poses greater risks than no tobacco exposure.

HISTORY OF HARM REDUCTION

Attempts to reduce the known or suspected risks to health from tobacco use by modifying tobacco or cigarettes predate the first U.S. Surgeon General's report on smoking and health. One of the first product innovations introduced with the potential and promise of harm reduction was filters. Filters reduce the amount of toxicants that go into the smoke inhaled by a smoker. The sale of filter cigarettes went from 1% of the market in 1952 to more than half of the market by 1960 (U.S. DHHS, 1989). In 1998, 98% of cigarettes sold in the United States contained filters (FTC, 2000).

The next major product modification with safety implications was the introduction of "low-yield" cigarettes. These products emit lower tar, carbon monoxide (CO), and nicotine than other products as measured by the Federal Trade Commission (FTC) assay (the "smoking machine"). This is achieved through blending different types of tobacco, ventilation, addition of accelerants, and filtration, as discussed in Chapter 4. The utility, purpose, and inferences made of the FTC assay yields are discussed in Chapters 4 and 11.

Consumers believed, and still do, that these products pose less risk to health than other cigarettes. Typical advertising campaigns for low-yield products stressed the softer side of smoking: "For smokers who prefer the lighter taste of a low-tar cigarette" (Kluger, 1997). An advertisement in the mid-1970s for a low-yield product, True, stated, "Considering all I'd heard, I decided to either quit or smoke True. I smoke True" (Pollay and Dewhirst, 2000). The market share of products yielding 15 mg of tar or less (as measured by the FTC assay) increased from 4% in 1970 to more than 50% in 1981. These products commanded approximately 80% of the U.S. cigarette market in 1998 (FTC, 2000).

Data on the health impact of low-yield products are conflicting, in part due to a lack of systematic and comprehensive study early in the introduction of these products. Most current assessments of the epidemiological and toxicological data suggest, however, that low-yield products are associated with far less health benefit than predicted based on FTC assay-generated tar, CO, and nicotine levels. The sales-weighted average of tar and nicotine yields of cigarettes in the United States has decreased by approximately half since the 1950s (Health Canada, 1998), without a significant or proportional change in the harm or prevalence of specific smoking-related diseases. Some of this is explained by changes in smoking behavior, known as compensatory smoking. In an effort to maintain adequate exposure to nicotine, smokers who use low-yield products smoke differently (e.g, inhale more deeply) than those who smoke higher-yield products. Thus, exposure to tobacco toxicants from low-yield products is greater by an unknown amount than predicted by FTC assay values. In addition to the disappointing impact of low-yield products on an individual smoker's health, of concern is the health burden of smokers who might have quit smoking altogether had they not had the opportunity to switch to a product they assumed was less harmful.

DEFINITIONS

Definitions of tobacco products are important for legal and regulatory purposes. Definitions are important also as the reader considers the harm reduction strategies discussed in this report. The Food and Drug Administration (FDA), as part of its rule-making process in 1995–1996, defined a cigarette as "any product which contains nicotine, is intended to be burned under ordinary conditions of use, and consists of any roll of tobacco wrapped in paper or in any substance not containing tobacco; or any roll of tobacco wrapped in any substance containing tobacco which, because of its appearance, the type of tobacco used in the filler, or its packaging and labeling, is likely to be offered to, or purchased by, consumers as a cigarette" (21 CFR 897.3). The Bureau of Alcohol, Tobacco and Firearms uses a very similar definition of a cigarette for purposes of taxation (27 CFR 290). Some potential harm reduction strategies involve specific modifications of conventional cigarettes, as defined above. The committee struggled with terms for other products, which currently exist or could be developed, that are similar to, but not exactly, what is commonly recognized as a cigarette. When strict adherence to the FDA definition of a cigarette (which seems to require the presence of both nicotine and tobacco and to be dependent on high-temperature burning) is not intended, the committee uses the term "cigarette-like." These products are basically paper-covered cylinders of approximately 90 mm length that, when lit, heated or burned (usually controlled in a very rigorous manner and at lower temperatures than conventional cigarettes), and puffed, allow smoke or vapors with what is recognized as the flavor of tobacco or cigarettes to be inhaled, leading to the absorption of nicotine into the body. When referring both to modifications of conventional cigarettes and to cigarette-like products, the committee uses "tobacco-related." When referring specifically to pharmaceutical preparations, it will be noted.

There is no evidence currently that use of any product, other than those that lead to cessation, can achieve harm reduction from tobacco. Many tobacco and cigarette-like products have been introduced in the distant and recent past that do, under measurement systems such as the smoking machine, result in decreased emission of some toxicants compared to conventional tobacco products. These products could, therefore, *potentially* result in reduced *exposure* to toxicants. The committee uses "potentially," because whether exposure to tobacco toxicants is reduced depends on the user's behavior, such as frequency and intensity of use. Reduced exposure, however, does not necessarily assure reduced risk to the user or reduced harm to the population. Therefore, and in order to avoid misinterpretation, the committee will use the generic phrase "potential reduced-exposure products," or PREPs, when discussing modified tobacco products, cigarette-like products (whether tobacco containing or not), or pharmaceutical products and medical devices (whether nicotine containing or not) developed for their tobacco harm reduction potential. Demonstration of exposure reduction is possible but at this time, demonstration of harm reduction is not. This conclusion is reiterated and supported in subsequent chapters.

No tobacco-based PREPs other than conventional low-yield products have been used by enough consumers to assess health impact. The recent or forthcoming expansion of the test market for several new products, which are described in Chapter 4, and public statements by tobacco company executives suggest a new and critical opportunity for assessing harm reduction. The next few years may see an explosion of available tobacco or cigarette-like products that make some claim of harm reduction based on reduced tobacco or smoke content of one or more toxicants.

Uncertainty and skepticism remains about the potential health benefits of PREPs. Key to the skepticism, in addition to the lessons learned from low-yield products, is concern that such products will discourage quitting in smokers who might otherwise have stopped using tobacco or will encourage new tobacco use. Evidence to support this concern is limited, however. Historical data suggest that people "switched down" to low-yield products due in part to health concerns, but the studies were not designed to determine whether those smokers would have quit tobacco use altogether had only high-yield products been available. Most PREPs will maintain nicotine addiction and there is little agreement in the tobacco control field that public policy should encourage products that maintain nicotine addiction. Worrisome as harm reduction might be to those who have studied the history and disappointments of low-yield cigarettes, PREPs are currently available and likely are here to stay.

NICOTINE

There is disagreement among tobacco control experts about the optimal content of nicotine in tobacco products. Whether nicotine addiction is harmful beyond supporting tobacco use and should, therefore, be the target of public health efforts is not a simple question. Nicotine has toxic properties (see Chapter 9), but they are far fewer and less serious than those of other tobacco constituents. Some have proposed that nicotine should be removed from tobacco products in order to prevent addiction (Henningfield et al., 1998), decrease tobacco use, and thereby decrease exposure to the most toxic constituents in tobacco. The marketability of a nonaddicting tobacco product, however, is thought to be low. Nicotine has pleasurable or rewarding effects, in addition to its addictive properties. However, societal views of addictions per se surround the controversy.

Retaining nicotine at pleasurable or addictive levels while reducing the more toxic components of tobacco is another general strategy for harm reduction (for a recent review of this issue, see Russell, in Ferrence et al., 2000). The tobacco industry reportedly would support some FDA regulation of cigarette products (Schwartz and Kaufman, 2000). Key to its acceptance is that there be no upper level for nicotine that is set so low as to effectively ban cigarettes (Schwartz and Kaufman, 2000). Experience with NEXT, a cigarette with extremely low nicotine levels that did not succeed in the marketplace, suggests that nicotine is one of the factors crucial to the success of a tobacco product.

PHARMACEUTICAL PRODUCTS

Modified tobacco and cigarette-like products are not the only potential strategies for harm reduction. Two classes of pharmaceutical products might provide an alternative to the less harmful cigarette for harm reduction. These two classes of drugs are nicotine products, of which there are several such as patch, gum, inhaler, and nasal spray preparations, and nonnicotine products. To date, only one product that does not contain nicotine has been approved by the FDA for tobacco cessation and is on the market—a slow-release bupropion preparation, Zyban. This has been approved by the FDA for short-term (up to six months) use for tobacco cessation and was subject to standard FDA review and approval. Chapter 4 includes a detailed description of such products and the FDA approval process.

Early speculation that these products might be used on a long-term or continuing basis to reduce exposure to tobacco toxicants was supported by the observation that some people use them far longer than indicated in the FDA labeling and continue to smoke as well. It is possible that these smokers might decrease their tobacco use to a level that is less harmful than their prior tobacco use. Some smokers might cease using tobacco but use the pharmaceutical products on a long-term basis to help ensure abstinence. Claims by the manufacturers of usefulness for harm reduction short of cessation or for long-term maintenance would require FDA review and approval for either new indications or as new products. To date, no drugs have been approved for use with such harm reduction claims. The health effects of pharmaceuticals used as PREPs include both the effects of the drug itself (both benefits and risks) and of any concomitant tobacco use.

REGULATORY ISSUES AND AUTHORITIES

This strong federal regulatory authority for nontobacco, pharmaceutical PREPs contrasts with that for conventional tobacco products and, possibly, tobacco-related PREPs. The FDA asserted in 1996 that conventional cigarettes and smokeless tobacco were nicotine delivery devices intended to affect the structure and function of the human body and thus fell under its jurisdiction. On March 21, 2000, the Supreme Court of the United States denied FDA jurisdiction over these products (Legal Information Institute, 2000). The Supreme Court majority opinion stated that it was never congressional intent for FDA to regulate tobacco and that the products did not fall under the Food, Drug, and Cosmetic Act or its amendments. FDA does retain the authority to regulate health claims regarding tobacco products. However, there is no precise definition of the term "health claim."

The Federal Trade Commission has regulatory authority over the advertising and marketing of tobacco products. The purpose of FTC authority, however, is to ensure consumers have opportunities to exercise informed choice. The FTC enforces a variety of federal antitrust and consumer protection laws. They seek to ensure that the nation's markets function competitively and are vigorous, efficient, and free of undue restrictions. They work to eliminate acts or practices that are unfair or deceptive (FTC, 1999). Thus, the FTC has responsibility under various federal laws to ensure the proper display of health warnings in advertising and on packaging of tobacco products sold in the United States. Further, the agency collects and reports to Congress information concerning cigarette and smokeless tobacco advertising, sales expenditures, and the tar, nicotine, and carbon monoxide content of cigarettes (FTC, 1992).

Other federal agencies have jurisdiction over tobacco (e.g., the Bureau of Alcohol, Tobacco, and Firearms and the Department of Agriculture) but not regarding health issues. No agency is responsible for ensuring that any standards are met in the manufacturing and composition of tobacco products. This disparity between strict regulation of nicotine replacement products (and Zyban) compared to tobacco products puts the safest form of nicotine administration at serious marketing disadvantage. The so-called unequal regulatory playing field has led to suggestions to raise regulatory standards for tobacco or to lower the regulatory standards in terms of marketing, packaging, and over-the-counter availability of tobacco cessation and harm reduction pharmaceutical products.

COMMITTEE CHARGE AND PROCESS

Because of the staggering morbidity and mortality associated with tobacco use and because fewer than 10% of smokers who attempt to quit each year succeed, public health considerations suggest a need to study alternatives to cessation for some smokers. Purported harm reduction tobacco products have been introduced into the U.S. marketplace periodically over the last 50 years. There is a strong likelihood of increased marketing and more products in the near future (Philip Morris Incorporated, 2000). These products are often associated with marketing statements or claims interpreted by some as indicating less health risk than to conventional products. It is also possible that nicotine or other pharmaceutical products could be used on a long-term basis for harm reduction, either concomitant with decreased use of conventional tobacco products or by themselves for maintenance of tobacco abstinence.

In anticipation of these issues, in 1999 the FDA asked the Institute of Medicine (IOM) to convene a committee of experts to formulate scientific methods and standards by which PREPs (pharmaceutical or tobacco related) could be assessed. Specifically, the committee was asked to answer several questions about harm reduction products. Where there are not yet answers, the committee was to determine the broad strategy by which the knowledge base should be assembled. For each product or class of product, four questions were asked:

1. Does use of such a product decrease exposure to the harmful substances in tobacco?
2. Is this decreased exposure associated with decreased harm to health?
3. Are there surrogate indicators of this harm to health that could be measured in a time frame sufficient for product evaluation?
4. What are the public health implications of tobacco harm reduction products?

The first three questions are obviously necessary for regulatory review of PREPs for their ability to reduce the risk to an individual of tobacco-caused disease and for decision making by an individual about tobacco use. The fourth question is important to ensure the health of the public. That is, although a PREP might be risk-reducing for an individual's health compared to conventional tobacco products, the availability of

PREPs might not be harm-reducing for the population. This could occur if tobacco users who might otherwise have quit do not, if former tobacco users resume use, or if some people who would not have otherwise initiated tobacco use do so because of perceptions that the risk with the "new" products is minimal and therefore acceptable. The committee was not asked to evaluate, and so did not debate, the merits of pursuing harm reduction.

The committee was drawn by its charge necessarily and inevitably into considering the regulatory framework for products purported to reduce harm from tobacco use. First, regulation is a necessary precondition for advancing scientific knowledge on the effects of using these products, especially their impact on public health. Second, whether the public health impact will be beneficial is substantially dependent on the implementation of a comprehensive and carefully designed regulatory framework for these products.

An expert committee was convened by the IOM and is the author of this report. The report is offered to the FDA, FTC, other relevant federal and state regulatory and policy bodies, Congress, scientists and health care professionals, tobacco and pharmaceutical industries, and—most importantly—the public, who will have to consider advertising and marketing information and make decisions about whether to use these products. The committee hopes the report will also be useful to similar groups in other countries, some of which are examining the harm reduction issue at present and to the global tobacco control community.

Committee expertise includes clinical medicine, epidemiology, toxicology, pharmacology, behavioral sciences, regulatory policy, and public health. The committee met five times between December 1999 and August 2000. Three of the committee meetings involved open sessions, during which invited researchers, public health advocates, clinicians, and representatives of tobacco and pharmaceutical manufacturers presented relevant data and engaged in scientific discussions. Written submissions were encouraged and reviewed by the committee (see Appendix A for a listing of these presentations and submissions to the committee). The committee in part or as a whole reviewed thousands of articles from the scientific peer-reviewed literature, including original research, review articles, and reports of advisory bodies such as the office of the U.S. Surgeon General. All materials sent to the committee and those used in support of this report are listed in a public access file maintained by the National Research Council's (NRC) Public Access and Records Office.

This report, which contains a Preface, Executive Summary, and 16 chapters including this introduction, is divided into two sections. Section I contains background material, conclusions, and recommendations on surveillance implementation and general conclusions regarding the committee's charge. Chapter 2 reviews general principles of harm reduction. Chapter 3 discusses the historical record of harm reduction attempts with an eye toward lessons learned for the future. Chapter 4 discusses current products and strategies for tobacco harm reduction. Chapter 5 contains a summary of the technical evidentiary base, conclusions, and recommendations regarding the first three questions of the charge. Chapter 6 reviews the public health considerations of harm reduction, including conclusions and recommendations regarding surveillance and post marketing research. Chapter 7 provides an overview of conclusions regarding the implementation of a public health approach to tobacco harm reduction, including principles for regulating PREPs. Chapter 8 sums up section one with principal conclusions. Section II contains the technical evidentiary base summarized in Chapter 5. Chapters 9 and 10 describe the pharmacology of nicotine and the toxicology of tobacco. Chapter 11 discusses principles of exposure assessment. Chapters 12—16 discuss the evidence that harm reduction strategies might impact the major diseases associated with tobacco and nicotine use.

APPLICATION OF RISK ASSESSMENT TO TOBACCO HARM REDUCTION

Predicting human health risks from toxic exposures, such as those associated with occupational and environmental toxicants, frequently involves a risk assessment process such as that outlined in "Risk Assessment in the Federal Government" (NRC, 1983), a report produced at the request of the U.S. Food and Drug Administration.

The risk assessment process outlined in the NRC report, sometimes referred to as "The Red Book," includes four basic elements:

- hazard identification (Does the agent cause an adverse health effect?),
- dose-response assessment (What is the relationship between dose and incidence in humans?),
- exposure assessment (What exposures are currently experienced or anticipated under different conditions?), and
- risk characterization (What is the estimated incidence of the adverse health effect in a given population?).

Hazard identification includes laboratory and field observations of adverse health effects resulting from exposures to specific agents. For *dose-response assessment*, if clinical or epidemiological data are not available, extrapolation from high-to-low-dose exposures and from animal to human doses in dose-response assessment is required. To determine *exposures*, research based on field measurements, estimated exposures, and characterization of populations is required. The data required for the first three elements of the risk assessment process usually include in vitro and in vivo toxicology studies in animals, including dose-response studies; mathematical modeling; toxicokinetic studies in animals and human; other clinical studies in humans; and, if available, epidemiology studies in human populations.

A synthesis of those three components leads to a risk characterization. This can be either qualitative or quantitative. Considerations that can affect the risk characterization are the statistical and biological uncertainties in estimating the health effects, the choice of dose-response or exposure assessments used, and a determination of targeted population for protection. The risk characterization is used to guide regulatory or other action for the purposes of risk management. This entails consideration of political, social, economic, and other technical (e.g. engineering) information with risk-related information. Value judgments are involved, such as the acceptability of risk and the reasonableness of the costs of control.

Risk assessment is the link between research and decision making (risk management). Complex and conflicting scientific information must be presented in an organized manner such that its meaning and limits are clear to the risk manager. The more complex the science base and the more controversial the policy implications, the more a rational and explicit framework is needed. Tobacco harm reduction presents complex technical issues and controversial policy options.

The committee has used the Red Book risk assessment paradigm in its approach to assessing the science base for tobacco harm reduction. Chapter 5 includes a discussion of the special challenges tobacco (and PREPs in particular) poses to risk assessment. Chapter 5 (and the chapters from Section II summarized in it) also contains the evidence that PREPS could affect each major tobacco-related disease. Each type of evidence is linked to the Red Book paradigm. The data gaps that currently prevent a quantitative risk characterization for PREPs are also described. Chapter 8, Conclusions, returns to the Red Book paradigm to organize the conclusions and recommendations of the report.

This organization and synthesis demonstrates that each of the four steps in risk assessment and in risk management are informed by:

- the four specific questions posed by the committee (described above) as a means to answer its charge in order to provide regulatory, scientific, and medical guidance,
- the evidence reviewed by the committee in each of the disease-

specific chapters (contained in Section II and summarized in Chapter 5),
- the principal conclusions of the committee (found in Chapter 8) related to the questions contained within the charge,
- specific recommendations of the committee related to a comprehensive surveillance system (Chapter 6) and regulatory principles (Chapter 7), and
- the research gaps identified (Chapter 5 and Section II).

OPERATING PRECEPTS

The committee introduces the rest of the report by laying out four fundamental operating precepts. The committee reiterates and reaffirms decades of overwhelming scientific evidence and the conclusions and recommendations of advisory bodies such as those that authored the Surgeon General's reports, previous IOM and NRC committees, and international health experts such as the World Health Organization.

1. Tobacco use causes serious harm to human health.
2. Nicotine is addictive.
3. The best means to protect individual and public health from tobacco harms are to achieve cessation, prevent initiation and relapse, and eliminate ETS exposure.
4. A comprehensive and authoritative national tobacco control program, with harm reduction as one component, is necessary to minimize adverse effects of tobacco.

REFERENCES

21 CFR 897.3 (1997) Cigarettes and smokeless tobacco (eff. 8-28-97): Subpart A-General Provisions.
27 CFR 290.11 (1997) Exportation of tobacco products and cigarette papers and tubes, without payment: Subpart B-Definitions.
CDC (Centers for Disease Control and Prevention). 1993a. Cigarette smoking-attributable mortality and years of potential life lost-United States, 1990. *Morbidity and Mortality Weekly Report* 42(33):645-649.
CDC (Centers for Disease Control and Prevention). 1993b. Smoking cessation during previous year among adults-United States, 1990 and 1991. *Morbidity and Mortality Weekly Report* 42(26):504-507.
CDC (Centers for Disease Control and Prevention). 1994. Health objectives for the nation cigarette smoking among adults-United States, 1993. *Morbidity and Mortality Weekly Report* 43(50):925-930.
CDC (Centers for Disease Control and Prevention). 1998. Selected cigarette smoking initiation and quitting behaviors among high school students-United States, 1997. *Morbidity and Mortality Weekly Report* 47(19):386-389.

CDC (Centers for Disease Control and Prevention). 2000a. Cigarette smoking among adults-United States, 1998. *Morbidity and Mortality Weekly Reports* 49(39):881-884.

CDC (Centers for Disease Control and Prevention). 2000b. Trends in cigarette smoking among high school students-United States, 1991-1999. *Morbidity and Mortality Weekly Report* 49(33):755-758.

CDC (Centers for Disease Control and Prevention). 2000c. Tobacco use among middle and high school students-United States, 1999. *Morbidity and Mortality Weekly Report* 49(3):49-53.

Ferrence R, Slade J, Room R, Pope M, eds. 2000. *Nicotine and Public Health.* Washington, DC: American Public Health Association.

Fielding JE, Husten CG, Eriksen MP. 1998. Tobacco: health effects and control. Wallace RB, Doebbeling BN, ed. *Public Health & Preventative Medicine.* 14th ed. Samford, Conn.: Appleton & Lange. Pp. 817-846.

FTC (Federal Trade Commission). 2000. Federal Trade Commission Report to Congress for 1998. [Online]. Available: http://www.ftc.gov/os/2000/06/index.html [accessed September 6, 2000].

FTC (Federal Trade Commission). November, 1992. Tobacco products. [Online]. Available: http://www.ftc.gov/bcp/conline/pubs/products/baccy.html [accessed September 7, 2000].

FTC (Federal Trade Commission). June 17, 1999. Vision, mission, goals. [Online]. Available: http://www.ftc.gov/ftc/mission.htm [accessed September 7, 2000].

Health Canada. 1998. Toronto, Ontario: Health Canada.

Henningfield JE, Benowitz NL, Slade J, Houston TP, Davis RM, Deitchman SD. 1998. Reducing the addictiveness of cigarettes. Council on Scientific Affairs, American Medical Association [see comments]. *Tob Control* 7(3):281-293.

IOM (Institute of Medicine). 1994. Lynch BS, Bonnie RJ. *Growing Up Tobacco Free.* Washington, DC: National Academy Press.

Kluger R. 1997. *Ashes to Ashes.* New York, NY: Vintage Books.

Legal Information Institute. March 21, 2000. Food and Drug Administration, et al., Petitioners v. Brown & Williamson Tobacco Corporation et al. [Online]. Available: http://supct.law.cornell.edu/supct/html/98-1152.ZO.html [accessed March 22, 2000].

Lew EA, Garfinkel L. 1987. Differences in mortality and longevity by sex, smoking habits, and health status. *J. Soc Actuaries* 39:107-130.

Murray C, Lopez A. 1997. Global mortality, disability, and the contribution of risk factors; Global Burden of disease study. *The Lancet* 349:1436-1442.

NRC (National Research Council). 1983. *Risk Assessment in the Federal Government: Managing the Process.* Washington, DC: National Academy Press.

Philip Morris Incorporated. 2000. *Request for Applications: 2000 Research Focus.* Linthicum Heights, MD: Philip Morris.

Philip Morris USA. 2000. Cigarette smoking: health issues for smokers. [Online]. Available: http://www.PhilipMorrisUSA.com/DisplayPageWithTopic.asp?ID=60 [accessed September 7, 2000].

PHS (Public Health Service). 2000. *Health, United States, 2000; With Adolescent Health Chartbook.* Hyattsville, MD: National Center for Health Statistics.

Pollay RW, Dewhirst T. 2000. *Successful Images and Failed Fact: The Dark Side of Marketing Seemingly Light Cigarettes.* Vancouver, Canada: History of Advertising Archives, Faculty of Commerce, University of British Columbia.

R. J. Reynolds Tobacco Company. 2000. Risk reduction efforts. [Online]. Available: http://www.rjrt.com/TI/Pages/TIrisk_reduct_cover.asp [accessed September 7, 2000].

Schwartz J, Kaufman M. 2000. A tobacco giant backs FDA rule. *Washington Post.* February 29, 2000; A 01.

U.S. DHHS (U.S. Department of Health and Human Services). 1989. *Reducing the Health Consequences of Smoking: 25 Years of Progress; A Report of the Surgeon General.* Atlanta, GA: U.S. Department of Health and Human Services, Centers for Disease Control and Prevention.

U.S. DHHS (U.S. Department of Health and Human Services). January, 2000. Healthy People 2010. [Online]. Available: http://www.health.gov/healthypeople/Document/tableofcontents.htm [accessed September 11, 2000].

U.S. Public Health Service. 1964. *Smoking and Health: Report of the Advisory Committee to the Surgeon General of the Public Health Service.* U.S. Department of Health, Education, and Welfare, Public Health Service, Center for Disease Control. PHS Publication No. 1103.

World Bank. 1999. *Curbing the Epidemic: Governments and the Economics of Tobacco Control.* Washington, DC: The World Bank.

World Health Organization (WHO). 1999. *The World Health Report 1999: Making a Difference.* France: World Health Organization.

2

Principles of Harm Reduction

A broadly shared goal of public policy toward cigarettes and other tobacco products is to reduce their health burden (IOM 1994, 1998). That health burden is minimized if no individual begins smoking and those who are currently smoking quit promptly (U.S. DHHS, 1988). However, quitting is difficult for most smokers and many adolescents will experiment with smoking; experimentation predictably leads a substantial fraction to become regular smokers (U.S. DHHS, 1994). Thus in addition to interventions aimed at prevention and at promoting immediate quitting, it is appropriate to consider interventions that aim to reduce the harm that the remaining population of smokers cause themselves and others by continued smoking. This is the underlying concept of harm reduction or harm minimization.

The term "harm reduction" has a variety of applications. It can refer to a policy or strategy (a set of laws and programs) or to specific interventions (e.g., an individual product innovation or dissemination effort). A harm reduction policy or intervention (a) explicitly assumes continuation of the undesired behavior as a possibility and (b) aims to lower the total adverse consequences, including those arising from continuation. In this use, the term describes an assumption and a goal rather than a result. It can also be used as a criterion for evaluating results; an intervention or policy is harm-reducing if it does in fact reduce the total adverse consequences. Finally, harm reduction can also be viewed as a framework, a way of thinking about dealing with a harmful behavior, since it requires analysis of a broader set of outcome measures than would otherwise be considered. One cannot usually determine in advance, on theoretical grounds, whether a particular policy or intervention is harm-reducing. For example, it may turn out that a policy which aims to minimize prevalence (i.e., addresses only abstinence) reduces total harm as compared to any other policy. But the framework allows consideration of alternatives to reduction in the number of users as a complement to abstinence.

The concept of harm reduction has application in a number of policy areas apart from tobacco, including automobile safety, sex education for children, alcohol control and policy toward illicit drugs. In some instances the harm reduction considerations are only implicit, providing an ex post rationalization of decisions already made (e.g., automobile safety, MacCoun, 1998). In others (e.g., needle exchange programs) it is a very prominent element of policy discussions. While none of the harm reduction interventions in these other policy areas are exactly analogous to those in the tobacco field, they will be used to illustrate the potential strengths and weaknesses of this approach for tobacco.

The next section elaborates the basic framework of harm reduction and briefly relates it to risk assessment. The third section also shows some of harm reduction's applications in related areas and compares these applications to some potential smoking interventions, though a much more extended discussion of those interventions is provided later in the report. The chapter concludes with some observations on the difficulties of applying the harm reduction framework, in particular the problems of developing measures to establish whether harm has in fact been reduced, and the need to give greater weight to mistaken acceptance of a product as harm-reducing than to mistakenly rejecting a harm-reducing product.

CONCEPTUAL FRAMEWORK FOR HARM REDUCTION

Basic Concepts

Harm reduction accepts that interventions focused on reducing the harmfulness of a substance or behavior, even if they increase the extent of substance use or the frequency of the targeted behavior, may be able to reduce the aggregate of adverse consequences for society, including both users and nonusers. For example, referring to the alcohol field, a prominent group of researchers stated: "Unlike 'abstinence-only' or 'zero-tolerance' approaches, the harm reduction model supports any behavior change, from moderation to abstinence, that reduces the harm of problems due to alcohol" (Marlatt et al., 1993). For illicit drugs "the central defining characteristic of harm reduction is that it focuses on the reduction of *harm* as its *primary goal* rather than reduction of *use per se*, secondly that strategies are included to reduce the harms for those who continue to use drugs, and thirdly that strategies are included which demonstrate that, on the balance of probabilities, it is likely to result in a net reduction in drug-related harm" (Lenton and Single, 1998). Frequently definitions assess alternatives not in aggregate health measures but in terms of the harms associated with a single act or product or with the individual user; for example, "the term harm reduction originally referred to only those policies and programmes which attempted to reduce the risk of harm *among people who continued to use drugs*" (Lenton and Single, 1998, citing Single and Rohl, 1997).

The harms consist of all the adverse consequences borne by members of society. These include increased morbidity and mortality (among both users and nonusers) from all sources; addiction itself; expenditures on regulation or enforcement, since these are costs borne by taxpayers; the increased intrusiveness of the state; and crime that might be generated by regulation or enforcement or by the behavior itself.

These adverse consequences are borne by many different groups: users themselves; intimates of the user, particularly children and spouses; nonusers directly (e.g., through crime, in the case of illicit drugs, or traffic accidents in the case of alcohol); and nonusers indirectly or society generally (e.g., through taxation). The value society gives in considering interventions to the interests of these groups may vary (MacCoun et al., 1996); typically a greater consideration is given to the welfare of children or of neonates, since they are the most vulnerable victims, with very limited capacity to undertake actions in their own interests.

Analytics

Total harm can be expressed as a function of the number of individuals engaged in the behavior and the damage each causes. In turn, the damage caused by an individual is a function of intensity of use (or frequency of behavior) and of the harmfulness of each episode of use or behavior. MacCoun and Reuter (2001) suggest that total harm can then be expressed as

Total Harm = Harmfulness (per use) × Intensity (per user) × Prevalence (of use)

It can be reduced through declines in any of the three components individually, including intensity.

However the three components may not be independently determined. In particular, prevalence may be affected by harmfulness through three distinct, though related, paths; initiation, nondesistance, and relapse. The lower harmfulness may reduce *perceived* harmfulness and encourage someone to begin using drugs, to drive a car too fast, or engage in sex at too early an age. Perceived dangers may be influenced both by the actual dangers and by information about those dangers. Where the substance or behavior is addictive or habit-forming, this may generate a long-term increase in the number of users. In addition, lower perceived harmfulness may reduce incentives to quit or desist. Relapse may also be encouraged by the perception of less dangerous means of continuing the desired behavior. Lower perceived harmfulness may also increase intensity of use.

Changing the riskiness of an act is known to alter the behavior of the population. Generally the change is in the form of compensation, i.e., higher risk will reduce the prevalence of the behavior while lower risk will increase that prevalence. Engineers tend to overestimate the benefits from safety devices, since they ignore that behavioral adaptation. On auto safety, Evans (1991) has noted: "If the safety change affects vehicle performance, it is likely to be used to increase mobility. Thus, improved braking or handling characteristics are likely to lead to increased speeds, closer following, and faster cornering. Safety may also increase, but by less than if there had been no behavioral response." In its most extreme form, this kind of risk compensation has been labeled *risk homeostasis*—a term that implies implicit or explicit efforts to maintain a constant level of risk (Wilde, 1982).

These changes are logical possibilities. How substantial they are is an empirical matter. So is the extent of generalizability across product domains and populations. The introduction of a safer automobile may, for example, have negligible effect on the driving behavior of older drivers but sharply increase speeding by younger drivers, while condom availability may increase sexual behavior of older adolescents more than it affects that of younger adolescents. The psychological mechanisms generating alcohol and cigarette dependence may be different enough that harm reduction interventions in general are more effective in one field than in the other.

This lack of generalizability raises a question as to the relevance of examples from other fields to tobacco interventions. But tobacco harm reduction involves a large variety of potential interventions. They differ in some important dimensions, just as do automobile seat belts and needle exchange programs. The examples can help identify the dimensions that influence the outcomes of harm reduction interventions.

APPLICATIONS

The harm reduction framework can be applied not only in a number of different policy areas but to a variety of forms of interventions:

(a) Lowering the inherent harmfulness of a broad class of products (automobile safety regulation); this is the approach used for tobacco harm reduction products that rely on removal of some toxicants from tobacco.

(b) Shift to less toxic mode of ingestion. Needle exchange programs attempt to reduce the harmfulness of the act of injecting drugs, without requiring abstinence. That is the approach embodied by products such as Eclipse, with heated tobacco or tobacco-like materials providing nicotine by a similar mechanism that allows continuation of the act of smoking but attempts to make it less harmful.

(c) Behavioral change therapies (controlled drinking); many tobacco harm reduction strategies will require behavior change as a complement to product innovation.

(d) Adding a less harmful but dependency-creating product to the available mix of dangerous products (methadone for heroin addicts); this is the rationale for nicotine replacement therapy for long-term use.

This section provides a brief description of the nature of each of the non-tobacco interventions listed and how they have fared in the harm reduction framework. It also describes the extent to which they have had the effects predicted for them when introduced. The examples presented are, by nature, imperfect analogies of tobacco harm reduction but are offered to highlight the positive and negative implications of harm reduction interventions.

Automobile Safety Regulation

Automobiles are a source of numerous injuries and fatalities; in 1998 there were 41,200 deaths in the United States associated with automobiles (National Safety Council, 1999). Some of these, but not all, are a consequence of unsafe operation of vehicles, in particular driving while intoxicated or driving at high speed. The National Highway Traffic Safety Administration (NHTSA) estimates that speeding was a contributing factor in 30% of such deaths (NHTSA, 1999a). It was also estimated that in 1994, 16,600 traffic fatalities were alcohol-related (CDC, 1995). A series of product innovations, including seat belts, anti-lock braking systems, and air bags have led to large improvements in the crash-worthiness of vehicles. Lap/shoulder safety belts, when used, reduce the risk of moderate to fatal injury to front seat passenger occupants by approximately half (NHTSA, 1999b). Federal law now requires their installation in new vehicles and all states mandate that they be used.

From the earliest days of these innovations, there has been a research interest in behavioral adjustments that might reduce the effectiveness of these innovations. Given that cars are safer, drivers may be more inclined to exceed the legal speed limit as well as exercise reduced care with respect to driving while alcohol impaired. This is analogous to increased intensity of smoking when given access to low tar and nicotine cigarettes.

Peltzmann (1975) found that changes in driver behavior more than offset product improvement. However, since then a substantial literature has refuted that finding. The research generally concludes that there is in fact more speeding but that the net result is a reduced burden of automobile injury and mortality. For example, Chirinko and Harper (1993) found that the introduction of air bags reduced automobile fatalities by between 13.8 and 26.1%. They also observed a shift (as did Peltzmann) in the composition of those fatalities; non-occupant fatalities (i.e., deaths of pedestrians, cyclists, etc.) increased while occupant fatalities decreased. This latter finding is consistent with the hypothesis that the technology encourages less safe driving.

Other engineering improvements on roads (stronger guard rails, brighter lights) have also reduced the frequency of accidents or severity of injuries associated with unsafe driving (Ross, 1992). Evans (1991) notes that: "When safety measures are largely invisible to the user, there is no evidence of any measurable human behavioral feedback. Likewise, when measures affect only the outcome of crashes, rather than their probability, no user responses have been measured." These are instances of pure harm reduction interventions; no behavioral response diminishes their design effect.

Seat belts can be disabled; like many harm reduction interventions they require compliance for their effect. One factor explaining the relative modest impact of mandatory seat belt laws on traffic fatalities lies in substantial rates of non-compliance (Dee, 1998). Moreover the population is not homogeneous: "unsafe drivers may be the least likely to adjust their belt use after the introduction of the law" (Dee, 1998). Thus the safety effects of automobile innovations are less than expected due to both increased speeding and selective noncompliance regarding seat belts.

Though there is strong and increasing social disapproval of unsafe driving, as expressed in congressional passage of legislation in 2000, urging states to reduce the maximum allowable Blood Alcohol Content (BAC) to 0.08%, there is no public discussion that these mandated product changes might encourage faster or less safe driving or redistribute the damage toward pedestrians and other innocent parties. This lack of debate may reflect, *inter alia*, the compelling nature of the intervention, reducing the risks associated with being the driver of a car, an almost universal experience of American adults. What is immediately discernible is the reduction for the driver, not the potential increase for other parties. Moreover, the targeted behavior (unsafe driving) is so common that, even though there is support for increasingly stringent laws and enforcement, there is an acceptance of its inevitability; there are no calls for a "speeding free society." Speeding, though not driving while intoxicated, also confers a benefit—namely, reduced travel time.

These benefits of reduced mortality and injury have been obtained at considerable cost. The estimated cost per life-year saved varies widely for different devices and mandates; for example, the figure for driver airbag and manual lap belts (as compared to manual lap belts alone) is $6,700 while for the same devices for passengers *and* drivers, the figure is $62,000 (Tengs et al, 1995). Some interventions are rated as essentially costless; for example, driver automatic (vs. manual) seat belts. These require investments only of government authority, not financial resources.

Automobile safety also illustrates the potential conflict among social goals that may be ignored even in a harm reduction framework. To reduce gasoline consumption and air pollution, the corporate average fuel economy standard (CAFE) has led to lighter cars; these cars are correspondingly less safe (Crandall and Graham, 1989). The trade-off between the two goals (pollution abatement and reducing injury and death from automobile accidents) has not been evaluated and is the subject of little public discussion. Indeed, the National Highway Traffic Safety Administration has been accused of obfuscating precisely this issue (Kazman, 1991). Harm reduction, despite its ostensible breadth, does not necessarily cover all potential adverse consequences.

Automobile safety represents an instance of a successful harm reduction intervention. There is indeed an increase in harmful behavior but not so much as to overcome the reduction in adverse consequences of that behavior.

Teen Sexual Behavior

Births to teenagers have been identified as a major societal health and behavioral problem (Ventura et al., 1997). For example, teenage mothers are less likely to complete their own education or provide adequate parental supervision and more likely to give birth to a low-birthweight infant. Over one third of teenage pregnancies end in abortions (Henshaw, 1999). Unwanted children are particularly at risk of neglect from teenage mothers (Federal Interagency Forum on Child and Family Statistics, 1997). There is a societal interest in reducing the number of unwanted infants born to teenage females. In addition, teen sexual activity, which frequently occurs outside of monogamous relationships, also facilitates the spread of sexually transmitted diseases (STDs), including AIDS.

The severity of problems related to premature sexual intercourse is a function of the prevalence of the acts and their average safety—mostly the probability of pregnancy and of STDs. Until recently many schools chose to emphasize the value of abstinence before a certain age; not only does that represent community values but abstinence, if achieved, eliminates the risky acts (Kirby, 1992). However, the rise in sexual activity among adolescents and the emergence of the AIDS epidemic have led to a concern that "abstinence only" messages may be ineffective and that the adverse consequences of unprotected sex may be greater than previously estimated. There has been exploration of the effects of harm reduction interventions which accept the high probability of sexual activity at an inappropriately young age and aim to reduce the probability of pregnancy or disease (Kirby, 1997).

Such interventions can take a number of forms. One is curricula aimed at teaching adolescents about responsible behavior, if they choose to have sexual relations. Such a curriculum may reduce the adverse consequences of early sexual activity. Opponents of these programs believe that they may encourage early sexual activity, in itself socially undesirable. However, a recent review of evaluations of a number of curricula concluded that "none of the 11 studies that examined the impact of programs on the frequency of intercourse found a significant increase" (Kirby and Coyle, 1997).

Another possible method of reducing harm is to facilitate access to adequate birth control technologies to adolescents, in particular to school children. Kirby and Brown (1996) estimate that by the mid-1990s nearly 400 schools made condoms available to students. Four recent studies found little evidence that the programs raised the percentage of high school students who engage in sexual intercourse, either by increasing teen awareness of how frequent such activity is among their peers or by reducing its perceived risks (Kirby and Coyle, 1997). On the other hand, evaluations of these programs also provide mixed support as to an effect on utilization of condoms, or any other birth control device. For example, a study of nine Philadelphia schools which provided reproductive health information, condoms, and general health referrals found that these schools showed no significant increases or decreases in condom use over time as compared to schools which did not install these programs (Furstenberg et al., 1997). A study of Seattle schools which made condoms available in vending machines or in baskets at school health clinics concluded that students use a relatively large number of condoms distributed in this fashion, but this "did not lead to increases in either sexual activity or condom use" (Kirby et al. 1999). The condoms distributed through the schools may have substituted for others obtained through other channels. There was one discrepant study: a condom distribution program that was part of a comprehensive AIDS prevention program increased rates of condom use in a sample of Chicago schools compared to a set of matched controls (Guttmacher et al., 1997).

These interventions illustrate the effects of increasing the availability of a potentially harm-reducing product to a specific high-risk population. It appears that this one has slight impact on behavior, on either the frequency of the acts or their average safety. In the absence of a strong scientific base, the debate is largely in terms of values and impressions, a common characteristic of social policy fields in which harm reduction has been applied. Making condoms available to school students may "send the wrong signal." It involves the state in apparently facilitating acts of which society disapproves. Since the evidence to date is that these interventions have at most a modest effect on the frequency and damage of the targeted behavior, the harm reduction framework has not been explicit.

Alcohol

Alcohol policy raises many harm reduction issues, reflecting the mixed social message with respect to alcohol's health consequences. Light drinking is a socially acceptable behavior, with apparently health promoting consequences. Heavy drinking, particularly chronic heavy drinking, is the source of a huge burden of morbidity and mortality, and is acknowledged as dangerous both to the drinker and others.

Harm reduction enters alcohol control in a number of ways. Alcohol consumption is characterized, even for most light drinkers, by episodes of excessive drinking. Though each light drinker has only a small to moderate probability of an alcohol-related automobile accident or other kind of injury, their numbers are large enough that, as a group, light drinkers account almost half the damage associated with alcohol consumption (Kreitman, 1986). Duncan (1997) found that driving while intoxicated (DWI) rates across states were associated with binge drinking but not with chronic heavy drinking.

As a result, a central debate is whether alcohol policy should focus on heavy drinkers as a group, on the total amount of alcohol, consumed or primarily on intoxication as a behavior. For example, some programs in the last group emphasize how the potential harm of drinking a given amount can be reduced by consuming it over longer periods of time or eating food during the drinking session. That implies an acceptance of heavy drinking, itself an unhealthy behavior and one that is a risk factor for numerous diseases. In contrast, high alcohol taxes reduce aggregate consumption, including that which is nonharmful; these taxes can be seen as "punish[ing] the many for the sins of the few" (Stockwell et al, 1996). If all drinking is seen to generate some probability of adverse effects, as the total consumption model suggests, then measures that reduce overall drinking are likely to be harm-reducing. Single (1997) suggests that "[I]ncreased attention is likely to be given to prevention measures which focus on preventing problems associated with drinking rather than restricting access to alcohol." This is only partly rooted in public health; it also represents the political realities of decreasing public support for restricting alcohol availability and the dissemination of data showing that moderate levels of alcohol consumption appear to reduce all-cause mortality (Edwards, 1995; NIAAA, 2000).

Harm reduction shows up in other aspects of alcohol policy; examples include "measures to reduce nonbeverage alcohol by 'Skid Row' inebriates, measures to reduce intake of alcohol by drinkers (e.g., promotion of low-alcohol beverages, server training programs) and measures to reduce the consequences of intoxication" (Single, 1997). For example, impoverished, single men who have chronic alcohol problems are at risk of drinking various liquids that contain alcohol but are not fit for human consumption (e.g., methylated spirits). Since these men are unable (both for financial reasons and because of poor self-control) to maintain stocks of alcohol, they are likely to consume these more dangerous substances if they cannot readily obtain alcoholic beverages, as happens early in the morning when liquor stores are closed. One method of alleviating this problem is to expand opening hours for stores operating in Bowery-like areas. Early opening may reduce the extent of drinking of nonbeverage alcohol and thus alcohol-related morbidity and mortality. However, it also signals society's willingness to facilitate drinking by individuals with serious drinking problems who will remain untreated and some of whose problems may be exacerbated or ameliorated by allowing easier access to alcohol.

Some harm reduction interventions in the alcohol field are indeed aimed directly at harms and seem unlikely to induce behavioral responses that would ameliorate their effects. For example, intoxication leads to numerous violent fights in pubs and bars. In Scotland, one intervention involved serving drinks in glass which crystallized rather than shattered when it broke, thus reducing the damage caused by such fights (Plant et al., 1994). That seems highly unlikely to increase the extent of fighting (the proximate source of harm) or heavy drinking (the more distal source of the harm); indeed, if the purpose of fighting is to cause the maximum injury to others, it may actually reduce the prevalence of fighting.

Harm reduction also operates at the clinical level. A long-standing belief among treatment specialists is that any message for individuals with drinking problems other than abstinence imposes unacceptable risk of relapse to dangerous drinking behavior. Over the last quarter century, however, a number of studies have found that controlled drinking may be a better goal for at least some problem drinkers (e.g., Miller and Caddy, 1977; Sobell and Sobell, 1978), though the results have not been consistent (Foy et al., 1984; Vaillant, 1996). Its appropriateness is in part a function of targeting. Can one identify subpopulations of problem drinkers with higher probability of benefiting from goals other than abstinence?

Illicit Drugs

Harm reduction has been most explicitly discussed in terms of policy toward illicit drugs. A distinctive aspect of drug policy is that the policies themselves have direct adverse consequences. Toughly enforced prohibitions aimed at use reduction can lead to drug overdose deaths due to drugs of unknown purity and poor quality. That does not imply that the prohibitions should be relaxed. These current policies may in fact be harm-minimizing, since the high prices and limited accessibility reduce the use and volume of addiction- or intoxication-related harms. This presents an empirical issue that remains unexamined (Caulkins and Reuter, 1997). However the very prominence of the harms directly related to prohibition, such as the violence and disorder around drug markets has generated an interest in the possibility that society would benefit from less punitive policies, *even* if they increase prevalence of drug use.

Two important interventions have been the subject of harm reduction debates: needle exchange programs (NEPs) to reduce the spread of AIDS and the provision of methadone to ameliorate the adverse consequences of opiate dependence.

The spread of HIV among intravenous drug users (IVDUs) and their sex partners is primarily a function of needle sharing, not of the drugs consumed. NEPs aim to reduce harm caused by IV drugs by reducing the risky practice. Opponents of NEPs argue that these programs facilitate a dangerous and illegal behavior: IV drug use. The proponents of needle exchange programs argue that whatever its symbolism, both public health and considerations of humane treatment of drug addicts require NEP.

There is a base of research demonstrating the positive public health effects. As summarized by a National Research Council panel, "NEPs increase the availability of sterile injection equipment. For the participants in a (NEP) . . . this amounts to a reduction in an important risk factor for HIV transmission. . . . There is no credible evidence to date that drug use is increased among participants as a result of programs that provide legal access to sterile equipment. The available scientific literature provides evidence based on self-reports that needle exchange programs do not increase the frequency of injection among program participants and do not increase the number of new initiates to injection drug use" (Normand et al., 1995). Several recent policy reports have upheld those conclusions (IOM, 2000; U.S. DHHS, 1998). There is also strong popular support as revealed in survey research (Henry J. Kaiser Family Foundation, 1996). Congress has been unpersuaded by either the research or the moral arguments. The belief that it will "send the wrong signal" and thus increase drug use by undermining the abstinence message, both to current and potential users, does not seem to be responsive to empirical findings.

Perhaps more relevant to the current concerns about harm reduction in the tobacco field is the introduction of methadone as a maintenance drug for heroin addicts. Methadone is a long-acting agonist that reduces craving for other opiates; see Rettig and Yarmolinsky (1995) for a review. Provision of low-cost methadone has enabled hundreds of thousands of opiate addicts in a dozen countries (North America, Western Europe, and Australia) to lead substantially better lives; they are able to avoid the humiliation of searching for an expensive prohibited drug, achieve modest levels of workplace functioning, and mitigate major threats to their own health and the health of others. It has helped limit the spread of HIV (Longshore et al., 1993). While it is an abusable drug, addicts on methadone have a much lower mortality rate than untreated addicts (Ball and Ross, 1991).

Methadone dependence seems to be at least as difficult to end as heroin dependence. Patients who discontinue methadone use relapse to opioid dependence at a high rate (Ward et al., 1992). The perception of heightened dependence is common amongst patients. Some addicts use methadone when their heroin use has become particularly problematic, with the expectation of returning to heroin use when they are past this particular crisis.

Methadone was a major ideological battlefront in the 1970s and 1980s (Rosenbaum, 1997; White House Conference for a Drug Free America, 1988). It has been noted that "the controversies over methadone treatment stem almost entirely from philosophical differences—objections to the substitution of one drug for another—and not from doubts about the pharmacological safety and efficacy of methadone" (Newman and Peyser, 1991). Methadone is a dependency-creating opiate, as is heroin. Methadone dependence is more acceptable than heroin dependence because it improves the user's function as a member of society. That the methadone patient gets less pleasure from the substitute drug is probably not critical in itself, though the lessened intensity of pleasure may allow for greater engagement with others and less self-centeredness and thus helps generate popular acceptance.

For some policy makers, however, the reduction in the burden of disease and other social dysfunction is not enough to justify government funding and provision of an addictive drug. In this case, in contrast to the battle over NEP, the proponents have prevailed. Methadone maintenance, though inadequately funded and poorly delivered, is the mainstay of the U.S. treatment for heroin addicts (Rettig and Yarmolinsky, 1995).

There is no evidence that methadone has increased initiation into opiate addiction where it has been made available. However, it is difficult to develop a powerful design for testing the hypothesis, particularly given the paucity of local indicators of heroin use. There has been no association between the number of persons in methadone programs and the number of reported new users in the National Household Survey on Drug Abuse, but that is at best a weak test. Assessing whether methadone prolongs a career of opiate dependence is difficult because of the very different characteristics of the methadone and heroin initiate populations. Even among addicts, many desist for long periods without treatment; for example, Anglin et al. (1986) found that 56.4% of a sample of 406 heroin addicts were able to desist for three years or more without formal assistance. However, those who initiate methadone use for addiction treatment are those who were unable to quit heroin; hence the difficulty of comparison.

Summary of Applications

Harm reduction has been controversial wherever it has been applied explicitly; moreover it often has a weak base in terms of assessed outcomes. Automobile safety is one instance for which there is good evidence of both compensatory behavior and net harm reduction. It is also the instance in which harm reduction issues have been least clearly articulated in public debate, though widely discussed in the traffic safety community.

Interventions to reduce the danger of adolescent sexual behavior have shown no adverse effect, in terms of increased sexual activity, but also little indication that the interventions have reduced the average harmfulness of the acts. For alcohol, harm reduction is gathering momentum but with only a modest scientific basis at either the population or clinical level. For heroin, research on methadone and needle exchange programs provide evidence that they do reduce total harms resulting from use by currently dependent users; there is a weaker evidentiary base for concluding that initiation is unaffected.

COMPARING OTHER HARM REDUCTION INTERVENTIONS TO THOSE FOR SMOKING

These examples are offered for the insights they may provide as to the likely consequences of tobacco-related potential reduced-exposure products (PREPs). However, harm reduction can work through a number of mechanisms and has consequences in a number of dimensions. Assessing the relevance of these harm reduction interventions to PREPs requires consideration of those mechanisms and consequences.

Table 2-1 compares three types of tobacco-related PREPs with five other harm reduction interventions discussed above. It aims to show in what ways the various PREPs are similar to or different from the other interventions. It is intended to be illustrative rather than conclusionary; a number of entries are conjectural. The comparisons are in terms of:

(a) Theory of how the product/policy might reduce harm (Presumed Mechanism). Each intervention posits a specific mechanism as to how it might diminish adverse consequences at the individual level. For example, modified tobacco attempts to present a less toxic version of a familiar product, in contrast to nicotine replacement therapy (NRT), which involve a substitute product with a very different mode of consumption and action. Methadone is similar to NRT in that respect, while NEPs are not. This is not a judgment of actual success but simply of the theoretical basis for believing that it is possible the intervention reduces adverse consequences.

(b) Effect on prevalence of the undesired behavior. These are crude summaries of the empirical literature; where the result is particularly conjectural, this is indicated by a question mark. For modified tobacco products the entry reflects the almost certain effect of allowing their availability, with claims. The undesired behavior in the case of PREPs is smoking; for others it is fast driving, underage sex, injecting of powerful illegal opiates, and drinking by problem drinkers. This is an estimate of the effect on the number of undesired acts, not of the total harms themselves.

(c) Effect on intensity of use. As for prevalence, these are crude summaries of complex empirical literatures. Harms are a function of both prevalence and intensity of use. Low tar and filter cigarettes not only led more individuals to smoke but also on average led them to higher daily consumption. Early opening of bars on the Bowery may lead to higher alcohol consumption by chronic alcoholics, while needle exchange programs generally either reduce injecting frequency or leave it unchanged.

(d) Effects on others. Again, these are summaries of the empirical literature. In the case of mandated seat belts, the increase in speeding may have negative consequences for pedestrians, while reducing mortality of vehicle passengers or even total traffic-related mortality. Nicotine replacement products, even if they lead to more nicotine consumption, may reduce ETS sufficiently to lower harms to others.

(e) Whether the intervention conveys symbolic approval of the undesired behavior. Allowing cigarette manufacturers to market cigarettes with the claim of lower carcinogens requires the government to approve the act of smoking cigarettes, even if accompanied by warning signs, just as do condom programs for kids (underage sex) and needle exchange for addicts (injecting drug use). Other interventions have no such effect. NRT meets a physiological need through such different mechanisms that they

TABLE 2-1 Characteristics of Eight Interventions With Harm Reduction Rationales

Product	Presumed Mechanism	Effect on Prevalence	Harms to Others	Symbolic Approval?	Effect on Intensity of Use	Potential Threats to Reducing Harm
Light & filter cigarettes	Less dangerous product	Increased cigarette consumption	Increased	No (govt. action not required)	Increase	Adaptive behavior negates technology/raises prevalence
Modified tobacco	Less dangerous product	Increased cigarette consumption	Increased	Yes (if claims allowed)	Increase	Exposure reductions not realized/prevalence rises too much
Nicotine replacement therapy	Substitute product	Decreased smoking?	Reduced	No	Reduce	Prolongs smoking careers/incomplete compliance
Oral methadone	Substitute product	Reduced heroin/raise opiate use	Reduced	No	Lower heroin/possibly higher opiate	Prolongs opiate use
Needle exchange programs	Reducing danger of act	None	Reduced	Yes	Reduce/unchanged	Incomplete compliance
Early opening of bars on the Bowery	Shifting consumption from dangerous forms	None	Unchanged	Yes	Increase	Behavioral model incorrect
Adolescent condom distribution	Reducing danger of act	Unclear	Unclear	Yes	Unclear	Increased sexual activity among kids
Mandated seat belts	Safer product	Increase	Unclear	No	Increase	Faster driving/selective compliance

constitute no approval of smoking. Oral methadone also represents no endorsement of injecting powerful opiates, it may prolong opiate use but in a form that permits social functioning. This dimension may be important for establishing popular and professional acceptability.

(f) An assessment of why a harm reduction intervention or PREP might fail to reduce total harm. There are many paths to failure. For example, even if the mechanism of modified tobacco products is correct and they are less toxic than conventional products, the population changes may be so great as to lead not to harm reduction but to greater total adverse consequences. In the case of NEPs, it may be that compliance is so incomplete, or that the HIV epidemic is so far advanced, that needle exchange programs fail to have any detectable effect. The Skid Row intervention may underestimate the effect of early opening on other heavy drinkers.

The table indicates that no two interventions are identical in all dimensions. If one accepts the utility of distinguishing between making an existing product safer and offering a very distinctive substitute, only mandated seat belts parallel modified tobacco products. However the seat belts involve no endorsement of unsafe driving, while the modified tobacco products, if the government allows regulated claims of reduced harms, does provide endorsement of the very act of smoking tobacco. This heterogeneity complicates projections from the other harm reduction experiences to PREPs. However it indicates where one might turn for insight into likely effects.

CONCLUSIONS

Harm reduction is a viable approach to government interventions. It has informed policy debates in a number of areas, and there is a modest research base indicating that some interventions with a harm reduction focus may indeed be harm-reducing. However, the framework is not yet well developed in either theory or application and continues to encounter both popular and professional skepticism. That skepticism will form part of the backdrop to decisions about implementation of the framework in the area of reducing tobacco-related harms.

Social Values

There are well-documented health effects of tobacco exposure on the nonuser. Environmental tobacco smoke has just been included on the National Toxicology Program's list of known-carcinogens and has long been linked to respiratory diseases and cancers in nontobacco users exposed to the smoke of others. Chapter 15 includes well-documented evidence of effects of tobacco exposure in utero. One could argue that these involuntarily exposed people should be the touchstone for harm reduction demonstration. However it is difficult to identify other applications of the harm reduction framework in which this kind of reasoning has been used; for example, automobile safety modifications may have shifted the burden of traffic accidents to non-occupants, but this has not been given prominence.

Another important consideration in decision making about harm reduction is the social value of the product. Many judgments about regulating risks include such considerations. For example, some risks of pharmaceutical products are acceptable because the benefit outweighs the risk, with benefit measured as decreased mortality, decreased disease prevalence or severity, and in some cases as quality-of-life and social well being. As a society, we accept certain levels of air pollution and attendant harm to health because we value the right to drive a car.

Freedom of choice (to smoke or not) and the tobacco industry's right to exist are values to the American public that preclude tobacco prohibition. However, the American public overwhelmingly supports restrictions or bans on tobacco use in indoor public places (The Gallup Organization, 2000) and restriction of youth access to tobacco products (American Heart Association, 1998). Moreover, most tobacco users themselves would like to quit and there are very few health benefits associated with tobacco use (See Chapter 16). Therefore, acceptable restrictions on potential exposure or harm reduction products might be stringent, although not severe enough to put the industry out of business. However, just as social values assure the continued availability of tobacco, social values can influence the limits of harm reduction strategies. It is unclear how much actual reduction in harm should be required for approval and marketing of a harm reduction product. Should a new product be allowed and encouraged if the gain is incremental only? Because regulation of novel tobacco or cigarette-like products involves a change in philosophy about tobacco, and because this is fundamentally a value judgment absent specific scientific guidance, one gauge of how much harm reduction should be required of a novel tobacco or cigarette-like product could be an assessment of how the public values increased health or reduced harm from tobacco. Such information is obtainable in a scientific way and could guide policymakers as they set the "bar" for regulatory action.

Analytic Problems

Though conceptually simple, harm reduction is difficult to apply when assessing alternative interventions. Following MacCoun and Reuter (2001), three basic problems emerge:

1. Aggregating harms. For illicit drugs in particular, a very heterogeneous set of adverse consequences need to be considered, such as loss of family cohesion, the spread of HIV and crime. Even if each effect can be measured, itself a major undertaking for any nonmarginal intervention (Reuter, 1999), there is little agreement about how to value them in monetary terms. The range of adverse consequences is narrower for the other cases, but even for alcohol, the range is wide enough (including crime and health) that aggregate measures in monetary terms are very approximate (e.g., Harwood et al, 1998). Some of these consequences are difficult to estimate and may consequently have been set at zero in quantitative studies because there is no systematic empirical base for an estimate. The harms derived from tobacco-related products are predominantly health related but the debate around economic interests of farmers in tobacco policy is a reminder that some other outcomes need to be taken into account.

2. Weighting the interests of different parties. The interests of users versus non-users has already been mentioned but there may also be other considerations of relative culpability and vulnerability. For example, the health burden of the poor may be given greater weight or consideration than the burden borne by the nonpoor, simply because the former can do less to ameliorate their own problems. It is also important to note that harms are not uniformly distributed; any particular option may benefit one group at the expense of another.

3. Harm reduction as a hazardous policy choice. An unsuccessful harm reduction intervention may lead to long-lasting and broadly distributed adverse consequences. For example, methadone maintenance might have turned out to increase heroin initiation and to prolong opiate addiction; it would be difficult to reverse those consequences or to be able to predict them in the early years of the intervention, let alone in advance of the decision. That suggests that harm reduction interventions may have to be held to a higher standard of proof and that government should be particularly careful, adopting one step at a time and closely monitoring the results. The fact that it will take decades to be certain about harm reduction effects of tobacco-related PREPs is a reason for particular caution in this case; this holds for none of the other applications. MacCoun (1998) summarizes the research on risk compensation: "[t]o date, research on compensatory responses to risk reduction provides little evidence that behavioral responses produce net increases in harm, or even the constant level of harm predicted by the "homeostatic" version of the theory. Instead, most studies find that when programs reduce the probability of harm given unsafe conduct, any increases in the probability of that conduct are slight, reducing, but not eliminating, the gains in safety" (Chirinko and Harper, 1993; Stetzer and Hofman, 1996). However, the

applicability of this finding to specific tobacco interventions remains to be established.

The committee concludes with two other observations:

Harm reduction presents major problems of measurement. Harm is much more difficult to measure than is prevalence or even quantity consumed, the conventional targets of control. As Single (1997) notes with respect to illegal drugs: "[A]s a practical matter, it is very difficult to determine whether specific policies involve a net reduction in drug-related harm." It requires estimation not only of numerous disparate outcomes but also assessment of how much of their change can be attributed to the intervention. For example, vehicle fatalities are determined by many factors; it is a complex research task to identify the contribution of seat belt laws or more stringent Blood Alcohol Content laws. Harm reduction in the tobacco control field will require the development of complex surveillance programs and very difficult issues of attribution.

Public health advocates opine that tobacco is a "special case," because tobacco is the only consumer product that when used exactly as intended is lethal. Further, they posit that it is unconscionable to market an addictive product to youth who are not competent to make informed judgments about long-term risks in the face of perceived short-term benefits. Finally, an undeniable history of suppression of information about the health risks of tobacco and tobacco product design changes leads many to seriously question any assessment of harm reduction potential by the manufacturers of the products. Just as harm reduction with respect to illicit drugs has been hurt by its association with the legalization movement, so too has the tobacco companies' use of false messages about the benefits of light and filter-tipped cigarettes created suspicion in the field of tobacco control. Therefore, the burden of proof for a benefit of novel, potential exposure or harm reduction tobacco products entails special considerations beyond that required of many other scientific questions.

REFERENCES

American Heart Association. 1998. Tobacco industry's political and economic influence. [Online]. Available: http://www.americanheart.org/Heart_and_Stroke_A_Zguide/tobec.html [accessed 2001].

Anglin MD, Brecht ML, Woodward JA, et al. 1986. An empirical study of maturing out: conditional factors. *International Journal of Addiction* 21:233-246.

Ball JC, Ross A. 1991. *The Effectiveness of Methadone Maintenance: Patients, Programs, Services and Outcomes.* New York: Praeger.

Caulkins JP, Reuter P. 1997. Setting goals for drug policy: harm reduction or use reduction? *Addiction* 92(9):1143-1150.

CDC (Centers for Disease Control and Prevention). 1995. Update: alcohol-related traffic crashes and fatalities among youth and young adults-United States, 1982-1994. *Morbidity and Mortality Weekly Report* 44(47):869-874.

Chirinko RS, Harper EP. 1993. Buckle up or slow down? New estimates of offsetting behavior and their implications for automobile safety regulation. *Journal of Policy Analysis and Management* 12(2):270-296.

Crandall RW, Graham JD. 1989. The effect of fuel economy standards on automobile safety. *Journal of Law and Economics* 32(1):97-118.

Dee TS. 1998. Reconsidering the effects of seat belt laws and their enforcement status. *Accid Anal Prev* 30(1):1-10.

Duncan DF. 1997. Chronic drinking, binge drinking and drunk driving. *Psychological Reports* 80:681-682.

Edwards G. 1995. *Alcohol Policy and the Public Good*. Oxford: Oxford University Press.

Evans L. 1991. *Traffic Safety and the Driver*. New York: Van Nostrand Reinhold.

Federal Interagency Forum on Child and Family Statistics. 1997. *America's Children: Key National Indicators of Well-Being*. Washington, DC: U.S. Government Printing Office and the U.S. Department of Health and Human Services.

Foy DW, Nunn LB, Rychtarik RG. 1984. Broad-spectrum behavioral treatment for chronic alcoholics: effects of trained controlled drinking skills. *J Consult Clin Psychol* 52(2):218-230.

Furstenberg FF Jr, Geitz LM, Teitler JO, Weiss CC. 1997. Does condom availability make a difference? An evaluation of Philadelphia's health resource centers. *Fam Plann Perspect* 29(3):123-7.

Guttmacher S, Lieberman L, Ward D, Freudenberg N, Radosh A, Des Jarlais D. 1997. Condom availability in New York City public high schools: relationships to condom use and sexual behavior. *Am J Public Health* 87(9):1427-33.

Harwood H, Fountain D, Livermore G. 1998. *The Economic Costs of Alcohol and Drug Abuse in the United States, 1992*. Rockville, MD: National Institute on Drug Abuse.

Henry J. Kaiser Family Foundation. 1996. *The Kaiser Survey on Americans and AIDS/HIV*. Menlo Park, CA: Kaiser Family Foundation.

Henshaw SK. 1999. U.S. *Teen Pregnancy Statistics: With Comparative Statistics for Women Aged 20-24*. New York: Alan Guttmacher Institute.

IOM (Institute of Medicine). 1994. Lynch BS, Bonnie RJ, eds. *Growing Up Tobacco Free*. Washington, DC: National Academy Press.

IOM (Institute of Medicine). 1998. *Taking Action to Reduce Tobacco Use*. Washington, DC: National Academy Press.

IOM (Institute of Medicine). 2000. *No Time to Lose: Getting More From HIV Prevention*. Washington, DC: National Academy Press.

Kazman S. 1991. Death by regulation. *Regulation* 14(4):18-22.

Kirby D. 1992. School-based programs to reduce sexual risk-taking behaviors. *J Sch Health* 62(7):280-7.

Kirby D. 1997. *No Easy Answers: Research Findings on Programs to Reduce Teen Pregnancy*. Washington, DC: National Campaign to Prevent Teen Pregnancy.

Kirby DB, Brown NL. 1996. Condom availability programs in U.S. schools. *Fam Plann Perspect* 28(5):196-202.

Kirby D, Brener ND, Brown NL, Peterfreund N, Hillard P, Harrist R. 1999. The impact of condom availability [correction of distribution] in Seattle schools on sexual behavior and condom use [published erratum appears in Am J Public Health. 1999; 89(3):422]. *Am J Public Health* 89(2):182-7.

Kirby D, Coyle K. 1997. School-based programs to reduce sexual risk-taking behavior. *Children and Youth Services Review* 19(5-6):415-436.

Kreitman N. 1986. Alcohol consumption and the preventive paradox [see comments]. *Br J Addict* 81(3):353-63.

Lenton S, Single E. 1998. The definition of harm reduction. *Drug and Alcohol Review* 17(2):213.

Longshore D, Hsieh S, Danila B, Anglin MD. 1993. Methadone maintenance and needle/syringe sharing. *Int J Addict* 28(10):983-96.

MacCoun RJ. 1998. Toward a psychology of harm reduction. *Am Psychol* 53(11):1199-208.

MacCoun RJ, Reuter P. 2001. *Drug War Heresies: Learning From Other Vices, Times and Places.* Cambridge, England: Cambridge University Press.

MacCoun R, Reuter P, Schelling T. 1996. Assessing alternative drug control regimes. *J Policy Anal Manage* 15(3):330-52.

Marlatt GA, Larimer ME, Baer JS, Quigley LA. 1993. Harm reduction for alcohol problems: Moving beyond the controlled drinking controversy. *Behavior Therapy* 24(4):461-504.

Miller WR, Caddy GR. 1977. Abstinence and controlled drinking in the treatment of problem drinkers. *Journal of Studies on Alcohol* 38:986-1003.

National Safety Council. 1999. Report on injuries in America. [Online]. Available: http://www.nsc.org/lrs/statinfo/99report.htm [accessed October 19, 2000].

Newman RG, Peyser N. 1991. Methadone treatment: experiment and experience. *Journal of Psycoactive Drugs* 23:115-121.

NHTSA (National Highway Traffic Safety Administration). 1999a. Traffic safety facts 1998: speeding. [Online]. Available: http://www.nhtsa.dot.gov/people/ncsa/pdf/Speeding98.pdf [accessed October 19, 2000].

NHTSA (National Highway Traffic Safety Administration). 1999b. Traffic safety facts 1998: occupant protection. [Online]. Available: http://www.nhtsa.dot.gov/people/ncsa/pdf/OccPrt98.pdf [accessed October 19, 2000].

NIAAA (National Institute on Alcohol Abuse and Alcoholism). 2000. *Tenth Special Report to the U.S. Congress on Alcohol and Health.* Washington, DC: U.S. Department of Health and Human Services and National Institute on Alcohol Abuse and Alcoholism.

Normand J, Vlahov D, Moses L. 1995. *Preventing HIV Transmission: The Role of Sterile Needles and Bleach.* Washington, DC: National Academy Press.

Peltzmann S. 1975. The effects of automobile safety regulation. *Journal of Political Economy* 83(4):677-725.

Plant M, Miller P, Nichol P. 1994. Preventing injuries from bar glasses-no such thing as safe glass. *British Medical Journal* 308(6938):1237-1238.

Rettig R, Yarmolinsky A. 1995. *Federal Regulation of Methadone Treatment.* Washington, DC: National Academy Press.

Reuter P. 1999. Are calculations of the economic costs of drug abuse either possible or useful? [comment]. *Addiction* 94(5):635-8.

Rosenbaum M. 1997. The De-medicalization of methadone. Erickson PG, Riley DM, Cheung YW, O'Hare PA, ed. *Harm Reduction: A New Direction for Drug Policies and Programs.* Toronto, Canada: University of Toronto Press. Pp. 69-79.

Ross HL. 1992. *Controlling Drug Driving: Social Policy for Saving Lives.* New Haven: Yale University Press.

Single E. 1997. The concept of harm reduction and its application to alcohol: The 6th Dorothy Black lecture. *Drugs: Education, Prevention and Policy* 4(1):7-22.

Single E, Rohl T. 1997. *The National Drug Strategy: Mapping the Future.* Canberra, AGPS: Report Commissioned by the Ministril Council on Drug Stategy.

Sobell MB, Sobell LC. 1978. *Behavioral Treatment of Alcohol Problems: Individualized Therapy and Controlled Drinking.* New York: Plenum Press.

Stetzer A, Hofman DA. 1996. Risk compensation: implications for safety interventions. *Organizational Behavior and Human Decision Processes* 66(1):73-88.

Stockwell T, Hawks D, Lang E, Rydon P. 1996. Unravelling the preventive paradox for acute alcohol problems. *Drug and Alcohol Review* 15(1):7-15.

Tengs TO, Adams ME, Pliskin JS, et al. 1995. Five-hundred life-saving interventions and their cost-effectiveness [see comments]. *Risk Anal* 15(3):369-390.

The Gallup Organization. November 29, 2000. Smoking in restaurants frowned on by many Americans. [Online]. Available: http://www.gallup.com/poll/releases/pr001129.asp [accessed 2001].

U.S. DHHS (U.S. Department of Health and Human Services). April 20, 1998. Press Release: Research shows needle exchange programs reduce HIV infections without increasing drug use. [Online]. Available: http://www.hhs.gov/news/press/1998pres/980420a.html [accessed October 19, 2000].

U.S. DHHS (U.S. Department of Health and Human Services). 1994. *Preventing Tobacco Use Among Young People; A Report of the Surgeon General.* Atlanta, GA: U.S. Department of Health and Human Services, Centers for Disease Control and Prevention.

U.S. DHHS (U.S. Department of Health and Human Services). 1988. *The Health Consequences of Smoking; Nicotine Addiction; A Report of the Surgeon General.* Atlanta, GA: U.S. Department of Health and Human Services, Centers for Disease Control and Prevention.

Vaillant G. 1996. A long-term follow-up of male alcohol abuse. *Archives of General Psychiatry* 53:243-249.

Ventura SJ, Curtin SC, Mathews TJ. 1997. *Teenage Births in the United States: National and State Trends, 1990-96.* Atlanta, GA: Centers for Disease Control and Prevention, National Center for Health Statistics.

Ward J, Mattick R, Hall W. 1992. *Key Issues in Methadone Maintenance Treatment.* Australia: University of New South Wales Press.

White House Conerence for a Drug Free America (1988).

Wilde GJS. 1982. The theory of risk homeostasis: implications for safety and health. *Risk Analysis* 2:209-255.

3

Historical Perspective and Lessons Learned

When filtered and low-yield cigarettes were introduced into U.S. markets, they were heavily promoted and marketed with both explicit and implicit claims of reducing the risk of smoking. Even as data accumulated, albeit slowly, that these products did not result in much—if any—decrease in risk, consumers have continued to believe otherwise. The population continues to misunderstand the meaning of numbers that purport to describe yields of tar and nicotine. Consumer misunderstanding is explained in part by the ways in which these products are marketed and in part by general theories of risk perception. This chapter reviews some key evidence regarding tobacco marketing, risk perception, knowledge about toxicity, and reasons for using low-yield products in hopes of illuminating the possible effects, including the promise of benefit and the risk of increased harm, of potential reduced-exposure products (PREPs) on tobacco use and on the health of the public as harm reduction is pursued in the near future. This material provides some of the evidentiary base for several of the conclusions and recommendations found in Chapters 6 and 7 of this report.

TOBACCO MARKETING: EARLY HEALTH CLAIMS

Tobacco companies have promoted the manufactured cigarette for more than a century (Kluger, 1996). The history of cigarette marketing has been discussed in several sources (e.g., Altman et al., 1987; Cohen, 1996;

Kluger, 1996; Kozlowski et al., 2000b; Pollay 1989, Pollay and Dewhirst, 2000; Ringold and Calfee, 1989; Swedrock et al., 1999; U.S. DHHS, 1989, 1994, 1998; Warner, 1985). As shown in Table 3-1, the tobacco companies have appealed to health concerns of smokers at least since 1927. Claims about tar and nicotine levels appeared as early as 1942.

The federal government, primarily via the Federal Trade Commission (FTC), has been involved in regulating tobacco marketing. The FTC is empowered by Congress to "prevent persons, partnerships, or corporations . . . from using unfair or deceptive acts or practices in commerce" (McAuliffe, 1988; OSH, 2000; U.S. DHHS, 1989). Using this authority, the FTC has undertaken several legal actions on health or medical claims made by tobacco companies. In 1942, for example, the FTC pressed Brown & Williamson Tobacco Corporation to stop making claims that Kool cigarettes, among other things, protected against colds and were soothing to the throat, and the company entered into a stipulation that it would refrain from doing so (FTC, 1942). As seen in Table 3-1, however, claims about Kools continued for several years after the 1942 stipulation.

The FTC also took legal action in the 1940s against claims that lower-tar cigarettes were beneficial (e.g., produced less throat irritation) (Cohen, 1996). These claims subsided. However, early in the 1950s, *Consumer Reports* published brand-specific tar and nicotine ratings. When claims increased, the FTC filed more suits against companies making claims about tar and nicotine levels. In September 1955, the FTC published "Cigarette Advertising Guides" that prohibited health claims (Cohen, 1996; Kozlowski, 2000b). Specifically, these guides prohibited claims about tar and nicotine that were not supported by "competent scientific proof" of a substantial (physiological) difference between brands (Cohen, 1996; McAuliffe, 1988).

The FTC explicitly stated that the guides would not prohibit statements about taste, flavor, aroma, or enjoyment (Kozlowski, 2000b). This led to the use of terms such as "mild," "light," and "smooth," which likely were better advertising techniques than direct health claims. The health-protecting messages brought to mind the fact that protection was needed, raising anxiety in the smoker. However, terms such as mild, light, and smooth only suggested the concept of safety and did not bring to mind the health-compromising properties of the product. The industry received counsel in the 1950s from motivational researchers who advised companies against using verbally explicit health appeals and suggested more visual images (Pollay, 1989). A comment by Martineau (1957), cited by Pollay (1989), is highly relevant. Martineau stated that direct claims "may offer some reassurance for the inveterate smokers, but they do utterly nothing to widen the market . . . to make smoking seem reasonable, justifiable, and highly desirable" (see Box 3-1).

TABLE 3-1 Selected Advertising Text Messages for Cigarettes and PREPs-United States, 1927-2000

1927: "OLD GOLD cigarettes. Better . . . smoother . . . not a cough in a carload."

1928: "It's toasted." "No Throat Irritation—No cough." (Lucky Strike)

1929: "20,679 physicians have confirmed the fact that Lucky Strike is less irritating in the throat than other cigarettes."

"Many prominent athletes smoke Luckies all day long with no harmful effects to wind or physical condition." (Lucky Strike)

"MILD . . . and yet THEY SATISFY." (Chesterfield)

1930: "It's toasted." "Your throat protection—against irritation—against cough." (Lucky Strike)

1932: "Do you inhale? What's there to be afraid of? Seven out of 10 inhale knowingly; the other 3 do so unknowingly. Do you inhale? Lucky Strike meets the vital issue fairly and squarely . . . for it has solved the vital problem. Its famous purifying process removes certain impurities that are concealed in even the choicest, mildest tobacco leaves. Lucky's created that process. Only Lucky's have it!" "IT'S TOASTED! Your protection against irritation, against cough."

1933: "Does winter make your head feel stuffy? Steam-heated rooms parch your throat? Heavy smoking 'brown' your taste? Then you've three extra reasons for changing to KOOLS. They're mildly mentholated. Light up and feel that instant refreshment. Smoke deep; the choice Turkish-Domestic tobacco flavor is all there. Smoke long; your throat and tongue stay cool and smooth, your mouth clean and fresh. Change to KOOLS. It's a change for the better."

"Give your throat a Kool vacation! Like a week by the sea, this mild menthol is a tonic to hot, tired throats."

1935: "They don't get your wind!" (Camel)

1936: "Ask your doctor about a light smoke." (Lucky Strike)

"The truth about irritation of the nose and throat due to smoking. Philip Morris & Company do not claim that Philip Morris cigarettes cure irritation. But they do say that an ingredient—a source of irritation in other cigarettes—is not used in the manufacture of Philip Morris . . . Their [group of doctors] tests proved conclusively that changing to Philip Morris, every case of irritation due to smoking cleared completely or definitely improved."

1937: "They're so mild and never make my throat harsh or rough." (Camel)

1938: ". . . so smooth and mellow you can smoke them in any number *without cigarette hangover*." (Old Gold)

"Viceroy's filter neatly checks the throat-irritants in tobacco . . . Safer smoke for any throat. Inhale without discomfort."

1939: "Your throat will like the change. The mild menthol is definitely refreshing." (Kool)

continues

TABLE 3-1 Continued

1940s-1950s: "Outstanding . . . and they are mild!" (Pall Mall)

"Pall Mall's greater length of fine tobaccos travels the smoke further on the way to your throat—filters the smoke and makes it mild."

1942: "*Reader's Digest* exposes cigarette claims . . . Impartial tests find Old Gold lowest in nicotine and throat-irritating tars . . .".

1943: ". . . filtering the flavor and aroma of the world's finest tobaccos into the smoothest of blends and checking OUT resins, tar, and throat irritants that can spoil the EVENNESS for smoking enjoyment!" (Viceroy)

1944: "Try Camels as your own T-Zone. T for taste. T for throat. The true proving ground for a cigarette."

1946: "More Doctors Smoke Camels Than Any Other Cigarette."

"Pasteurized for your protection." (Philip Morris)

"HEAD STOPPED UP? GOT THE SNEEZES? SWITCH TO KOOLS. . . THE FLAVOR PLEASES!"

1947: "When you have a cold and can't taste a thing, always smoke KOOLS to get back in the swing!"

1949: "Not one single case of throat irritation due to smoking Camels."

". . . Remember: Less irritation means more pleasure. And Philip Morris is the ONE cigarette proved less irritating—definitely milder than any other leading brand."

"Got a COLD? Smoke KOOLS as your *steady* smoke for that clean, KOOL taste!"

1951: "Notice that Philip Morris is definitely less irritating, definitely milder!"

"Filtered cigarette smoke is better for your health." (Viceroy)

1952: "No other cigarette approaches such a degree of health protection and taste satisfaction." (Kent)

"Because this filter is exclusive with KENT, it is possible to say that no other cigarette offers smokers such a degree of health protection and taste satisfaction."

". . . like millions today, you are turning to filter cigarettes for pleasure plus protection . . . it's important that you know the Parliament Story."

1953: "First cigarette ever to give you black and white proof of greatest health protection . . . with full smoking pleasure!" (Kent)

"The American Medical Association voluntarily conducted in their own laboratory a series of independent tests of filters and filter cigarettes. As reported in the *Journal of the American Medical Association*, these tests proved that of all the filter cigarettes tested, one type was the **most effective** for removing tars and nicotine. This type **filter** is used by Kent . . . and only

continues

TABLE 3-1 Continued

Kent!" "KENT. For the greatest protection of any filter cigarette with exclusive MICRONITE filter."

". . .Alpha Cellulose. Exclusive to L&M Filters, and entirely pure and harmless to health."

"Parliament filters 100% of the smoke—recessed filter keeps trapped tars and nicotine from touching lips or mouth!"

"New King-Size Viceroy gives Double-Barreled Health Protection . . . is safer for throat, safer for lungs than any other king-size cigarette."

"The nicotine and tars trapped by Viceroy's Double-Filtering action cannot reach your throat or lungs!"

"*FILTERED* CIGARETTE SMOKE IS BETTER FOR YOUR HEALTH. The nicotine and tars trapped by this Viceroy filter cannot reach your mouth, throat, or lungs!" *Reader's Digest*, January 1950.

1954: "L&M Filters are Just What the Doctor Ordered!"

"The cigarette that takes the FEAR out of smoking!" (Philip Morris)

1956: "Good news for ALL smokers . . . **Salem** filter cigarettes. THE FIRST TRULY NEW SMOKING ADVANCE IN OVER 40 YEARS! Menthol fresh. Most modern filter. Rich tobacco taste."

1958: "The *first* filter cigarette in the world that meets the standards of United States Testing Co. New Hi-Fi Filter Parliament." (note: the parenthetical phrase "high filtration" was printed, in small type, "Hi-Fi")

1966: "The truth is out: The wire services recently released a new report that revealed that, new *TRUE* Filter Cigarettes delivered less tar and nicotine than other brands tested . . . It's TRUE . . . without our knowledge or permission, these tests were conducted and TRUE Filter Cigarettes were found to be 'most effective in removing tar and nicotine.'"

1971: "You don't cop out. We don't cop out. You demand good taste, But want low 'tar' and nicotine. Only Vantage gives you both."

1976: "Considering all I'd heard, I decided to either quit or smoke True. I smoke True."

"If you are a smoker: There are many reasons to smoke Now. If you are a smoker who has been thinking about 'tar' and nicotine, these are the reasons to smoke Now: Reason. Now has the lowest 'tar' and nicotine levels available to you in a cigarette, king-size or longer. 2 mg. 'tar,' 2 mg. nicotine."

1977: "Vantage is changing a lot of my feelings about smoking. I like to smoke, and what I like is a cigarette that is not limited on taste. But I am not living in an ivory tower. I hear the things being said about high tar smoking as much as the next guy. So, I started looking for a low tar smoke that had some honest-to-goodness taste."

continues

TABLE 3-1 Continued

1978: *"National Smoker Study:* Merit Science Works! Low tar MERIT with 'Enriched Flavor' tobacco delivers taste equal to—or better than—leading high tar brands."

"I'm realistic. I only smoke Facts. We have smoke scrubbers in our filter."

1980s-1990s: "Alive with pleasure." (Newport)

1981: "The pleasure is back. BARCLAY. 99% tar free."

1985: "Latest U.S. Gov't Laboratory test confirms, of all cigarettes: Carlton is lowest. Box King—lowest of *all* brands—*less* than 0.01 mg. tar, 0.002 mg. nic."

1987: "Ultra taste in ultra low tar. Test a pack today." (Vantage)

1997: "No Additives" (Winston)

1998: "New Marlboro ULTRA LIGHTS. Famous Marlboro flavor now in an ULTRA LIGHT."

1999: "1 mg. Isn't it time you started thinking about number one?" (Carlton)

"Discover the rewards of thinking light." (Merit Ultra Lights)

2000: "The best choice for smokers who are worried about their health is to quit. Here's the next best choice." (Eclipse™)

"...a smooth satisfying taste with less smoke around you, virtually no lingering odor, and no ashes." (Accord™)

SOURCES: Anonymous, 1946; Arnett, 1999; Caples, 1947; Chickenhead Productions, 2000; Glantz et al., 1996; Harris, 1978; Kluger, 1996; Kozlowski, 2000; Lewine, 1970; Mullen, 1979; Pollay and Dewhirst, 2000; R.J. Reynolds Tobacco Company, 2000; Sobel, 1978; Swedrock et al., 1999.

BOX 3-1
1977 British American Tobacco Company Document
Describing Advertising and Communication Strategies to
Reassure Consumers About Safety

"All work in this area [communications] should be directed towards providing *consumer reassurance* about cigarettes and the smoking habit . . . by claimed low deliveries, by the perception of low deliveries and by the perception of "mildness." Furthermore, advertising for low delivery or traditional brands should be constructed in ways so as not to provoke anxiety about health, but to alleviate it, and to enable the smoker to feel assured about the habit and confident in maintaining it over time" [emphasis in original].

SOURCE: 030, Minnesota Litigation.

BOX 3-2
Summary Analysis of Cigarette Industry Advertising

"The cigarette industry has not voluntarily employed its advertising to inform consumers in a consistent and meaningful way about any of the following: (1) the technologies employed in fabricating the products, (2) the constituents added in the manufacturing processes, (3) the residues and contaminants that may be present in the combustible column, (4) the constituents of smoke that may be hazardous, (5) the addictiveness of nicotine, or (6) the health risks to which its regular customers and their families are inevitably exposed. Their advertising for low-yield products, instead, has relied on pictures of health and images of intelligence, and has misled consumers into believing filtered products in general and low tar products in specific to be safe(r) than other forms without knowing exactly why."

SOURCE: Pollay and Dewhirst, 2000.

Business Week publicly criticized the industry on the issue of direct health claims. McAuliffe (1988) cites the following quote from the December 5, 1953, issue: "Why has the industry persisted in this negative form of advertising even when, as tobacco growers and others complain, it hurts the trade by making people conscious that cigarettes can be harmful?" (Anonymous, 1953).

Advertisements using health claims in *Time* and *Life* magazines, which had increased substantially in 1952 and 1953 (after the first cancer scare), dropped to pre-1952 levels in 1954 (Swedrock et al., 1999). There has been a steady increase since the early 1960s in the percentage of magazine advertisements using visual images of bold and lively behaviors in pristine environments (Pollay, 1989; Pollay and Dewhirst, 2000; see Box 3-2).

HEALTH IMPACT OF LOW-YIELD PRODUCTS

In a recent review of epidemiologic data on the disease risks associated with the changing cigarette, Samet (1996) concluded that low-tar and nicotine cigarettes, when compared with relatively higher-tar and nicotine cigarettes, were associated with modest decreases in lung cancer risk and similar cardiovascular risk. The data on chronic obstructive pulmonary disease (COPD) were inconclusive. For all diseases, smokers of lower-tar and nicotine cigarettes experienced substantially higher disease risks than persons who did not smoke. These findings were similar to those in a 1981 report of the U.S. Surgeon General (U.S. DHHS, 1981).

The nature of these relationships may be influenced by misclassification bias, in that smokers may not be able to accurately recall lifetime brand use patterns. In addition, selection bias may occur if symptoms or diseases cause smokers to switch to lower-yield brands and if persons who switch to lower-yield brands exhibit different lifestyle characteristics that may influence disease risk (Samet, 1996). For example, persons who switch to lower-tar products appear to be more likely to eat more fruits and vegetables (Haddock et al., 1999). Although the magnitude of such biases may eventually prove to be small, research to assess their potential impact is warranted.

In related analyses, Thun and colleagues (1997a) compared the relative risks of lung cancer and coronary heart disease in Cancer Prevention Study I (from 1959 to 1965) with those observed in Cancer Prevention Study II (from 1982 to 1988). The risks did not decrease across studies when the authors stratified for gender, duration of smoking, age, and number of cigarettes smoked daily, even though the average tar yield of cigarettes consumed decreased substantially during that time. The results of this analysis could be influenced at least in part by unmeasured differences in lifetime smoking patterns (e.g., number of cigarettes smoked daily during adolescence) and by lifestyle factors (e.g., fruit and vegetable consumption). The weight of the evidence indicates that lower-tar and nicotine yield cigarettes have not reduced the risk of disease proportional to their FTC yields, in part because smokers compensate to obtain more nicotine (Burns, 2000; Kozlowski and Pillitteri, 1996) and in part because the products themselves contain higher concentrations of selected carcinogens (Hoffman et al., 1996).

Increased prevalence of the use of lower-tar and nicotine cigarettes has been associated with an increase in the percentage of lung cancers that are adenocarcinomas and a decrease in squamous cell carcinomas (Levy et al., 1997; Thun et al., 1997b). This changing histological pattern may be influenced by increased levels, over time, of nitrosamines in cigarette smoke and by increased inhalation of lower-nicotine-yield cigarettes, as smokers attempt to compensate to reduced nicotine yields by inhaling more deeply (Hoffman et al., 1996; Kozlowski and Pillitteri, 1996).

RISK PERCEPTION

The Gallup Organization has polled Americans about their perceptions of cigarette smoking since 1949 (Moore, 1999). For example, in 1999, 95% of Americans considered smoking to be harmful, up from 60% in 1949. In 1999, 92% of Americans considered cigarette smoking to be one of the causes of lung cancer, compared to only about 40% in 1954. Additionally, about 80% of smokers considered smoking to be one of the causes of heart disease in 1999, compared to about 37% in 1957. For the purposes of this analysis, a focus on smokers' perceptions is crucial to understanding the possible ways in which various harm reduction approaches, including the marketing of PREPs, may be understood or interpreted and, as a result, may affect consumer behavior.

Perceptions by Smokers of the Risk of Cigarette Smoking

It is now well established that perceptions of the harmfulness of smoking affect behavior. Much might be learned about the possible consequences of introducing PREPs from analysis of studies of smokers' knowledge of the consequences of various types of tobacco products. Although Viscusi (1992, 1998) argues that smokers tend to overestimate their risks, the vast majority of research conducted to date supports the opposite conclusion. Weinstein (1999) has reviewed much of this research. In general, the work of Weinstein and others (Ayanian and Cleary, 1999; Cohn et al., 1995; Hahn and Renner, 1998; Slovic 1998, 2000) suggests the following:

- Although smokers acknowledge that smokers have higher risks for various health problems than persons who do not smoke, most smokers did not view themselves as having higher risks of heart disease or cancer compared to other adults their age.
- Smokers' estimates of the number of years of smoking that are needed to produce health consequences increase with the number of years they have been smoking.
- Many young smokers perceive themselves to be at minimal risk from each cigarette they smoke, because they intend to stop smoking before any damage to their health occurs.
- Adolescents and adults believe that they are less likely than their peers to become addicted to cigarettes.

These findings highlight the frequently observed phenomenon of "optimism bias" or "unrealistic optimism" (Weinstein, 1999). For most hazards—and regardless of how they perceive the risks for people in general—individuals tend to perceive their own risks as less than those of other people (Weinstein, 1999; see Box 3-3).

Perceptions of the Risk from Low-Yield Cigarettes

In this report, cigarettes with tar yields on the FTC method of 15 mg or less are classified as low-yield cigarettes. In 1998, 81.9% of cigarettes consumed in the United States were low yield, up from 2.0% in 1967 (FTC

BOX 3-3
Industry Document from 1974 Reporting on a Study of Young Smokers' Health Concerns

"Health concerns exist among younger smokers. . . . One type of smoker rationalized smoking as a pleasure that outweighed the risks. Another felt that they didn't smoke enough to be dangerous. A third type rationalized his use of cigarettes by feeling he would quit before it was 'too late.' A final smoker group said that science would come to his rescue."

SOURCE: 018, K0028.

2000). Cigarette companies generally use "light" to refer to cigarettes with 6 through 15 mg tar and "ultralight" to refer to cigarettes with 1 to 5 mg of tar (Davis, 1987; Townsend, 1996). The terms are often used in surveys (e.g., Cohen, 1996), although the cut points used to classify brands may vary slightly.

The various studies described below have assessed perceptions of risks for smoking high- and low-tar cigarettes. These assessments have been conducted for smokers overall and, at various times, by tar level of the smoker's usual brand and by whether the smoker has switched to a low-yield brand.

Data on perceptions of risks from types of cigarettes from the 1966 and 1975 Adult Use of Tobacco Surveys (AUTSs) are summarized in the 1981 Surgeon General's report (U.S. DHHS, 1981). The percentage of smokers who felt that "some cigarettes [are] more hazardous than others" increased from 29.9% in 1966 to 49.1% in 1975.

Perceptions Held by Smokers of Low-Yield Products Regarding Risk

Cohen (1992) reports on the results of a 1980 industry-sponsored Roper survey that he learned about during preparation for testimony in a court case. In 1980, more than one-third (36%) of smokers of low-yield cigarettes reported that smoking their brand of cigarette did not significantly increase a person's risk of disease over that of nonsmokers; another 32% stated that they weren't sure if this was the case.

More recently, data from the 1986 AUTS and the 1987 National Health Interview Survey (NHIS) were used to assess attitudes, knowledge, and beliefs about low-yield cigarettes among adults and adolescents (Giovino et al., 1996). Both low-tar smokers and persons who've ever switched to lower-tar brands were (compared, respectively, to persons who smoke higher-tar brands and those who've never switched) more likely to (1) acknowledge the dangers of smoking, (2) be concerned about the health effects of smoking, (3) say that their health has been affected by smoking, and (4) believe that their cigarettes are safer. In 1987, 44% of smokers reported that they had ever switched to a low-tar brand to reduce their health risks. Among persons 10- to 20-years old in 1993 who smoked light or ultralight cigarettes, 33% said that they smoked these brands because they taste better, 20% because they are less irritating, and 21% because they thought these cigarettes were healthier than other brands.

In another study, a 1995-1996 survey of individuals who enlisted in the Air Force, which was administered during the first week of basic military training (Haddock et al., 1999), 32% of current smokers reported that they had switched during the previous 12 months to a lower-tar and nicotine cigarette brand just to reduce their health risk.

Shiffman and colleagues (In Press) analyzed 1999 national survey data to assess health perceptions, tar and nicotine delivery characteristics, and smoking sensations among cigarette smokers. Most smokers, particularly smokers of light and ultralight (L/UL) cigarettes, believed L/UL cigarettes were less harsh and that they delivered less tar and nicotine than regular cigarettes. The beliefs that L/UL cigarettes delivered less tar and nicotine and that they were less harsh than regulars each independently contributed to predicting the belief that L/UL cigarettes were safer than regulars.

Knowledge and Perceptions About Yields and Yield "Terms"

A Gallup Organization poll conducted in 1993 asked respondents the following question: Besides selling the product, what message do you think cigarette advertising is trying to get across when it uses terms like low-tar, low-nicotine, or low-yield? (Gallup Organization, 1993). Of the respondents, 56% of smokers and 60% of nonsmokers stated that the terms indicated a positive health benefit, with specific meanings that included being safer, less harmful, healthier, not as bad for you, or less cancerous.

Bolling and colleagues (2000) reported that English smokers did not find tar numbers (printed on cigarette packages) meaningful, claiming instead that the numbers were "scientific" and that they weren't sure if the numbers indicated yield per cigarette or per pack. Instead, smokers tended to rely on terms such as light, mild, and ultralight and on pack color to make assessments about various brands.

Based on results from a survey conducted in 1994, Cohen (1996) reported that 28% of smokers thought that switching from a 20-mg-tar cigarette to a 16-mg-tar cigarette would substantially reduce health risks.

About one-quarter felt that a smoker could smoke 10 cigarettes yielding 1 mg of tar and take in the same amount of tar as smoking one cigarette yielding 10 mg of tar.

In Cohen's survey (1996), only 14% of smokers overall reported that they used the actual tar numbers to make judgments regarding the relative safety of different brands. However, 56% of smokers of cigarettes yielding 1 to 5 mg of tar reported that they used tar numbers when assessing health risks of various cigarettes. Tar numbers, which are printed on less than 10% of cigarette packages sold in the United States, are more likely to be printed on the packages of the lowest-yielding cigarettes (FTC, 2000).

Kozlowski and colleagues conducted several informative surveys on the topic of light and ultralight cigarettes (Kozlowski et al., 1996, 1998a, 1998b, 1999, 2000a, 2000b). In one study, about two-thirds of smokers either reported not having seen or heard that there were "rings of small holes on the filters of some cigarettes" or did not know that blocking these holes would increase tar yields (Kozlowski et al., 1996). Less than 20% of smokers of ventilated brands knew that their cigarettes had filter vents (Kozlowski et al., 1998b). These data show that although intense marketing by tobacco companies about low-yield products has influenced consumers' perception of product safety, a lack of marketing or communication about ventilation holes has led to less informed consumers.

Another analysis indicated that the majority of smokers in a national sample responded either "don't know" or "two" in response to the question, How many LIGHT cigarettes would someone have to smoke to get the same amount of tar as from one REGULAR cigarette? (Kozlowski et al., 1998b). Less than 10% knew that one light cigarette could give the same amount of tar as one regular cigarette.

Reasons for Switching to Low-Yield Products

Smokers of light or ultralight cigarettes were asked if they smoked for each of the following reasons (presented in random order): "as a step toward quitting smoking completely," "to reduce the risks of smoking without having to give up smoking," "to reduce the tar you get from smoking," "to reduce the nicotine you get from smoking," and "because you prefer the taste compared to Regular cigarettes" (Kozlowski et al., 1998b). This study showed that 80% of smokers of light and 69% of ultralights preferred the taste of their brands; 50% of smokers of light and 72% of ultralight endorsed less nicotine as a reason; 57% of smokers of light and 73% of ultralights endorsed less tar; 39% of smokers of light and 58% of ultralights endorsed reducing risks without having to give up smoking; and 30% of lights and 49% of ultralights endorsed smoking their brand as a step toward quitting completely.

Knowledge Regarding Constituents, Additives, and Toxicity

Bolling and colleagues (2000) conducted a survey of 1,036 smokers in England in October 1998 to examine consumers' reactions to cigarette yield and product information. They also conducted focus groups and in-depth interviews in February 2000. While most smokers knew that tar (92%) and nicotine (98%) were present in tobacco smoke, fewer (29%) knew that carbon monoxide was present and almost none (≤5% for all chemicals) knew about the presence of toxic chemicals such as arsenic, lead, and cyanide. Bolling and colleagues also reported that smokers were shocked to learn about the presence of dangerous chemicals in their cigarettes, in part because such information undermined their beliefs that cigarettes were "natural" products.

In 1997, R.J. Reynolds Tobacco Company re-positioned the Winston brand with an advertising campaign claiming its product was made with "100% tobacco," containing "no additives" (see table 3-1, Arnett, 1999). In mall intercept interviews, 400 adolescents were surveyed about the Winston advertisements in Arizona and Washington, and 203 adults were surveyed in Washington. The two most common responses to the question, What do you think the Winston ads mean by saying that Winstons have "no additives"? were that Winston cigarettes contained only tobacco and that Winston cigarettes have no added chemicals. However, 36% of adolescents and 18% of adults also perceived them as meaning that Winston cigarettes were healthier than other cigarettes. Furthermore, 39% of adolescents and 20% of adults perceived the advertisements as claiming that Winstons were "less likely than other cigarettes to harm your health." Additionally, 42% of adolescents and 14% of adults stated that the advertisements meant that Winstons are "less likely than other cigarettes to be addictive." Overall, about two-thirds of adolescents and one-quarter of adults believed that the no additives claim meant at least one of the above implied health claims.

Evidence also suggests that consumers perceive menthol-containing cigarettes to be less harmful because they seem less harsh on the throat. Menthol cigarettes appear to be at least as dangerous as nonmenthol cigarettes (U.S. DHHS, 1998). About 26% of all cigarettes sold in the United States are mentholated (FTC, 2000). The vast majority of African-American smokers smoke menthol brands (U.S. DHHS, 1998). Epidemiological studies of mentholated cigarettes suggest that these products are at least as dangerous as nonmentholated cigarettes (Herbert and Kabat, 1989; Kabat and Herbert, 1991, 1994; Sidney et al., 1995). Additionally, the committee is aware of no evidence in support of the assertion that cigarettes without additives are less hazardous than those with additives.

Given the persistent health claims that were made about these products earlier in the century (Table 3-1), surprisingly little research has been conducted on people's perceptions of them. In a recent study, Hymowitz and colleagues (1995) questioned 213 adult smokers of menthol cigarettes who participated in a stop-smoking study. Among 174 African Americans, the main reasons for smoking menthols included the following: menthol cigarettes tasted better than nonmenthol cigarettes (83%); they had always smoked menthol cigarettes (63%); menthol cigarettes were less harsh on the throat than nonmenthol cigarettes (52%); inhalation was easier with menthol cigarettes (48%); and menthol cigarettes could be inhaled more deeply (33%). Among 39 white smokers of menthol cigarettes, reasons for their choice of menthols included menthol cigarettes tasted better than nonmenthol cigarettes (74%); menthol cigarettes were more soothing to the throat (51%); they had always smoked menthol cigarettes (39%); and inhalation was easier with menthol cigarettes (21%).

An industry document (Tibor Koeves Associates, 1968) reports the results of in-depth interviews (most likely conducted in 1968) of 10 African-American smokers of menthol cigarettes. The authors of the report concluded that two underlying factors "generated the great enthusiasm for menthol cigarettes." The preference for menthols seemed "based both on dynamic sensory and on psychological gratifications." The taste of menthol, which reminded many of candy, was a major attraction. The fact that the smoke wasn't hot or burning was also important. Psychologically, menthols were perceived to be modern and youthful. More relevant to this discussion, they were "considered as generally 'better for one's health.'" Most respondents viewed menthols as "less strong" than regular cigarettes, with the understanding among interviewees that cigarettes that were less strong were less dangerous to one's health.

POTENTIAL INFLUENCE OF PREPS
ON TOBACCO USE BEHAVIORS

The introduction of products to reduce harm in a population can result in both intended and unintended consequences. Both Pauly and colleagues (1995) and Hughes (1998) raise the possibility that the introduction of PREPs and their promotion as less harmful ways to smoke could lead to increased initiation. Behavioral adaptation can occur in ways that diminish the possible beneficial consequences of potentially harm-reducing products. In this section, consideration is given to studies relating to the possible influence of low-yield products on initiation and quitting.

Initiation

Arnett (1999) raised concerns that adolescents' beliefs about claims made in advertisements for the no-additive products described earlier could influence susceptible nonsmoking adolescents by inflating whatever optimism bias they already posses. Marketing strategies for the products could influence perceptions of product safety and the overall acceptability of smoking. In addition, such information might undermine the advice of parents, teachers, and health professionals. Unfortunately, there is little evidence one way or the other on whether the introduction of filters or light cigarettes has affected rates of initiation.

Silverstein and colleagues (1980) observed that smoking prevalence increased substantially for young girls in the 1970s. Their research had indicated that high school females experienced greater societal pressure to smoke (exhibited by a higher prevalence of trying smoking than for males) but a greater physiological pressure not to smoke (exhibited by a higher sensitivity to nicotine). However, the pressures to smoke may also have been due to marketing for women's brands, which increased markedly in the late 1960s and early 1970s (Pierce et al., 1994). Silverstein and colleagues (1980) concluded that females resolve the competing pressures by smoking fewer cigarettes per day and using low-yield cigarettes. The authors argued that if low-nicotine cigarettes were less available, many females would choose not to smoke rather than experiencing the unpleasant effects of nicotine reactions.

Quitting Smoking

The introduction of PREPs into a population may increase, decrease, or have no effect on the rate of quitting smoking in that population. As with initiation, effects on quitting could be positive or negative. The direction and magnitude of these effects, if real, could influence the population impact of PREPs.

Effects on Motivation

PREPs may influence quitting by changing people's motivation to quit (Russell, 1978) in either direction. It is possible that switching to low-yield cigarettes has facilitated quitting by some people, because successful switching might increase smokers' confidence in their ability to control their smoking behavior and thereby encourage a future quit attempt (Hughes, 1995). However, switching to low-yield cigarettes could also reduce the motivation to quit. The 1980 finding that about 36% of smokers of light cigarettes reported that smoking light did not increase risk compared to not smoking provides indirect evidence that they could have decreased motivation to quit (Cohen, 1992).

Users of low-yield cigarettes are generally more interested in quitting than those who smoke regular cigarettes (Giovino et al., 1996; Jarvis et al.,

1989), although Shiffman and colleagues (In Press) found this to be true only for smokers of light (but not ultralight) cigarettes. They may also be more confident in their ability to quit (Haddock et al., 1999; Jarvis et al., 1989). In one recent national survey, 30% of light smokers and 49% of ultralight smokers reported that they used light cigarettes as a step toward quitting (Kozlowski et al., 1998a). In another study, 5.2% of persons who tried to quit during the previous 10 years but relapsed reported using low-tar and nicotine cigarettes as a quitting strategy during their most recent quit attempt (Fiore et al., 1990). However, a similar (4.6%) percentage of persons who tried to quit within the previous 10 years and were abstinent for at least 1 year at the time of being surveyed reported using low-tar and nicotine cigarettes as a quitting strategy during their most recent quit attempt. This finding suggests that even though many people use low-yield cigarettes as a quitting strategy, the efficacy of this strategy is doubtful.

In a recent national survey, 32% of smokers of lights and 26% of ultralights reported that they would be likely to quit smoking if they learned that one light or one ultralight cigarette could provide as much tar as one regular cigarette (Kozlowski et al., 1998a). Another study (Kozlowski et al., 1999) used a simulated radio message to inform smokers that one light cigarette could provide as much tar as one regular cigarette. More than half (55%) stated that it made them think more about quitting, and nearly half (46%) said that the message increased the amount they wanted to quit. These data suggest that correcting misperceptions can motivate health-promoting intentions. Whether it can affect behavior remains unknown.

Effects on Quitting

Only a few large studies have assessed the association between tar levels or switching to lower-tar brands and actual quitting. In 1959, the American Cancer Society fielded the Cancer Prevention Study (CPS) I of approximately 1,078,000 adults. Hammond (1980) used the data to study quit rates as a function of tar levels. Persons who smoked cigarettes with ≤17.6 mg tar were classified as smoking low-tar cigarettes. Persons who smoked cigarettes with ≥25.8 mg tar were classified as smoking high-tar cigarettes. Those who smoked cigarettes with intermediate tar yields were classified as smoking medium-tar cigarettes.

Participants enrolled in 1959 were followed to 1965. Those who were still smoking in 1965 were followed until 1972. There was an inverse relationship between tar yield and the probability of quitting. Low-tar smokers at baseline and in 1965 were the most likely to be abstinent at each respective follow-up observation.

The 1986 AUTS was used to assess quitting among a nationally representative sample of persons who had ever smoked regularly (i.e., ≥100 lifetime cigarettes) (Giovino et al., 1996). Tar levels were ≤6 mg tar, 7-15 mg tar, and ≥16 mg tar. The prevalence of cessation was higher for persons who had never switched brands to reduce their level of tar and nicotine. The relationship was observed overall and in the three age groups examined (17 through 34 years, 35 through 64 years, and ≥65 years). The prevalence of cessation was directly related to tar yield (of the current smoker's brand and the former smoker's last brand) overall and for persons age 35 through 64 and ≥65. More recent multivariate analysis of data for persons who had smoked within 15 years of the survey confirmed the direct relationship between tar level and abstinence (Giovino, 2000). Additionally, abstinence was more common among persons who had reported that they had never switched brands to reduce their tar and nicotine levels.

The differences in the direction of the finding between the CPS and the 1986 AUTS data may be due simply to methodological differences between the studies. Additionally, one must consider that the overall tar level in 1959 was substantially higher than in 1986, making comparisons unrealistic. However, the differences may also represent a real phenomenon in which patterns of quitting changed over a period of nearly three decades. Low-tar smokers of 1959 may have been more motivated to quit than higher-tar smokers. In addition, some may have switched to lower-tar brands as a step to quitting. Around the mid-1970s, various advertising campaigns were run (see Table 3-1) introducing the notion that people could switch to lower-tar brands instead of quitting. It is possible that these advertisements decreased switchers' motivations to quit (see Box 3-4).

Most recently, Haddock and colleagues (1999) conducted a study of smokers who were forced to abstain during Air Force basic training. About 32% of smokers reported upon entering basic military training, that they had switched brands during the previous 12 months to reduce health risks. At one year follow-up, the percentage of switchers who reported being abstinent was slightly higher than nonswitchers (12.5% vs. 11.1%). However, this difference was not statistically significant in a multivariate analysis that controlled for demographic and smoking history variables. Further, the intervention represents involuntary deprivation and the sample is not generalizable to the U.S. population.

SUMMARY AND RELEVANCE TO PREPS

Studies on risk perception indicate that smokers tend to underestimate their overall risks from smoking. A large proportion of smokers considers smoking low-yield cigarettes to be safer than smoking regular

---

BOX 3-4
British American Tobacco Memo Describing Use of Low-Yield Products by Smokers Trying to Quit

"Smokers *needed* light brands for tangible, practical, understandable reasons" [emphasis in original]. "It is useful to consider lights more as a third alternative to quitting and cutting down—a branded hybrid of smokers' unsuccessful attempts to modify their habit on their own."

SOURCE: 081, PSC 60.

---

cigarettes, even though evidence does not support such a conclusion. Menthol and additive-free cigarettes, which likely pose as much risk, respectively, as nonmenthol and additive-containing cigarettes, also seem to be perceived as somewhat safer. *The committee concludes that smokers could overestimate any potential benefits of PREPs*, although it is possible that the factors that influence perceptions of traditional tobacco products might not apply to radically different products such as PREPs. In addition, PREP users might be very different from the average smoker.

The committee recommends that smokers be informed at every opportunity that all tobacco products, including modified tobacco PREPs and cigarette-like PREPs, are toxic and poisonous. Health-related statements about PREPs should be accurate and made directly, unaccompanied by terms that imply safety in an oblique manner. It is likely that strategic use of visual images and textual themes of safety would promote harm reduction. Research should be conducted to explore the optimal mix of images and text in communications. In addition, research should be conducted to determine if varying the relative attractiveness of product packaging across product types (i.e., conventional tobacco products, tobacco-related PREPs, and pharmaceutical PREPs) will influence perceptions and behaviors in a health-promoting way.

Future research in this area should be conducted to assess factors that influence perceptions of risk in order to ensure that communications about PREPs—whether made by manufacturers or by health educators—take into account a PREP user's perceptions regarding the risks of conventional tobacco and the potential benefit of using PREPs. The magnitude of misleading optimism bias for each PREP or type of PREP must be known to estimate the use of conventional tobacco and PREPs and, therefore, the possibilities for harm reduction.

The committee recommends strong and decisive efforts to monitor and correct misperceptions that consumers may develop regarding PREPs. These efforts are described in more detail in Chapters 6 and 7.

If PREPs are used by people who otherwise would have smoked conventional products, harm reduction could occur (assuming, of course, that a given PREP actually reduces risk). However, if many people who try these products otherwise would not have experimented with nicotine and then they continue to use a PREP or later switch to cigarettes or other tobacco products, then population harm could increase. Based on data from studies of low-yield conventional cigarettes, *the committee concludes that there is a risk of decreased quitting in current tobacco users if PREPs are available.* Data regarding initiation are less clear. Research is needed to better understand the ways in which various messages might influence children's and adolescents' perceptions and behaviors regarding PREPs and initiation of tobacco use with PREPs. Chapters 6 and 7 include recommendations for research and surveillance (for both public health and regulatory purposes) to better understand and prepare to respond to unintended consequences at the population level of PREP availability.

REFERENCES

030, Minnesota Litigation. Short PL. Smoking and health item 7: The effect on marketing. BAT Co. Ltd. April 14, 1977. 9 pages, (p.3).

081, PSC 60. Research and Development/Marketing Conference. British American Tobacco. 1985. 202 pages, (p.9, 13).

Altman D, Slater M, Albright C, Maccoby M. 1987. How an unhealthy product is sold: cigarette advertising in magazines, 1960-1985. *Journal of Communications* 37:95-106.

Anonymous. 1946. BBDO Newsletter. *Advertising and Selling* 39(12):55.

Anonymous. 1953. Cigarette scare: what'll the industry do? *Business Week*:60.

Arnett JJ. 1999. Winston's "No Additives" campaign: "straight up"? "no bull"? *Public Health Rep* 114(6):522-527.

Ayanian JZ, Cleary PD. 1999. Perceived risks of heart disease and cancer among cigarette smokers. *JAMA* 281(11):1019-1021.

Bolling K, White P, Owen L, McNeil A. 2000. Cigarette labeling: what information do consumers in England want? 11th World Conference on Tobacco OR Health: August 10, 2000; Chicago, Illinois.

Burns D. 2000. Disease risks from low tar and nicotine yield cigarettes. Presentation to the Institute of Medicine: April 25, 2000; Washington, DC.

Caples J. 1947. Best-read ads in the latest newspaper survey. *Advertising and Selling* 40(9):42.

Chickenhead Productions. 2000. [Online]. Available: http://www.chickenhead.com/truth/hool1.html [accessed August 1, 2000].

Cohen JB. 1992. Research and policy issues in Ringold and Calfee's treatment of cigarette health claims. *Journal of Public Policy and Marketing* 11:82-86.

Cohen JB. 1996. Smokers' knowledge and understanding of advertised tar numbers: health policy implications. *Am J Public Health* 86(1):18-24.

Cohn LD, Macfarlane S, Yanez C, Imai WK. 1995. Risk-perception: differences between adolescents and adults. *Health Psychol* 14(3):217-222.

Davis RM. 1987. Current trends in cigarette advertising and marketing. *N Engl J Med* 316(12):725-732.

Fiore MC, Novotny TE, Pierce JP, Giovino GA, Hatziandreu EJ, Newcomb PA, Surawicz TS, Davis RM. 1990. Methods used to quit smoking in the United States. Do cessation programs help? *JAMA* 263(20):358.

FTC (Federal Trade Commission). 1942. In: re: Brown and Williamson Tobacco Corp., DKT 3486. 34 FTC 1689 (Stipulation).

FTC (Federal Trade Commission). 2000. Washington, DC: Federal Trade Commission.

Gallup Organization. 1993. *The Public's Attitudes Toward Cigarette Advertising and Cigarette Tax Increase.* Princeton, New Jersey: The Gallup Organization.

Giovino G. 2000. A review of findings from population-based surveys in the U.S.A. on light cigarettes and smokers of light cigarettes. 11th World Conference on Tobacco OR Health: August 10, 2000; Chicago, Illinois.

Giovino G, Tomar SL, Reddy M, Peddicord JP, Zhu B-P, Escobedo LG, Eriksen MP. 1996. Attitudes, knowledge, and beliefs about low-yield cigarettes among adolescents and adults. National Cancer Institite. *The FTC Cigarette Test Method for Determining Tar, Nicotine, and Carbon Monoxide Yields of U.S. Cigarettes: Report of the NCI Expert Committee. Smoking and Tobacco Control Monograph 7.* Bethesda, MD: National Cancer Institute, U.S Department of Health and Human Services. Pp. 39-57.

Glantz S, Slade J, Bero LA, Hanauer P, Barnes DE. 1996. *The Cigarette Papers.* Berkeley, CA: University of California Press.

Haddock CK, Talcott GW, Klesges RC, Lando H. 1999. An examination of cigarette brand switching to reduce health risks. *Ann Behav Med* 21(2):128-134.

Hahn A, Renner B. 1998. Perception of health risks: how smoker status affects defensive optinism. *Anxiety, Stress, and Coping* 11:93-112.

Hammond EC. 1980. The long-term benefits of reducing tar and nicotine in cigarettes. Gori G, Bock F, eds. *Banbury Report 3: A Safe Cigarette?* Cold Spring Harbor, NY: Cold Spring Harbor. Pp. 13-18.

Harris RW. 1978. *How to Keep on Smoking and Live.* New York, NY: St. Martin's Press.

Herbert J, Kabat G. 1989. Menthol cigarette smoking and oesophageal cancer. *Internal Journal of Epidemiology* 18:37-44.

Hoffman D, Djordjevic M, Brunnemann K. 1996. Changes in cigarette design and composition over time and how they influence the yield of smoke constituents. National Cancer Institute. *The FTC Cigarette Test Method for Determining Tar, Nicotine, and Carbon Monoxide Yields of U.S. Cigarettes: Report of the NCI Expert Committee. Smoking and Tobacco Control Monograph 7.* Bethesda, MD: National Cancer Institute, U.S. Department of Health and Human Services. Pp. 15-37.

Hughes J. 1995. Applying harm reduciton to smoking. *Tobacco Control* 4(suppl):S33-S38.

Hughes JR. 1998. Harm-reduction approaches to smoking. The need for data. *Am J Prev Med* 15(1):78-79.

Hymowitz N, Mouton C, Edkholdt H. 1995. Menthol cigarette smoking in African Americans and Whites (letter). *Tobacco Control* 4:194-195.

Jarvis M, Marsh A, Matheson J. 1989. Factors influencing choice of low-tar cigarettes. Wald N, Froggatt P, eds. *Nicotine, Smoking, and the Low Tar Programme.* Oxford, UK: Oxford University Press.

Kabat GC, Hebert JR. 1991. Use of mentholated cigarettes and lung cancer risk. *Cancer Res* 51(24):6510-6513.

Kabat GC, Hebert JR. 1994. Use of mentholated cigarettes and oropharyngeal cancer. *Epidemiology* 5(2):183-188.

Kluger R. 1996. *Ashes to Ashes: America's Hundred-Year Cigarette War, the Public Health, and the Unabashed Triumph of Philip Morris.* New York, NY: Alfred A. Knopf.

Kozlowski L. 2000. Some lessons from the history of American tobacco advertising and its regulation in the 20th Century. Ferrence R, Slade J, Room R, Pope M, eds. *Nicotine and Public Health*. Washington, DC: American Public Health Association.

Kozlowski LT, Goldberg ME, Sweeney CT, Palmer RF, Pillitteri JL, Yost BA, White EL, Stine MM. 1999. Smoker reactions to a "radio message" that Light cigarettes are as dangerous as Regular cigarettes. *Nicotine Tob Res* 1(1):67-76.

Kozlowski LT, Goldberg ME, Yost BA. 2000a. Measuring smokers' perceptions of the health risks from smoking light cigarettes. *Am J Public Health* 90(8):1318-1319.

Kozlowski LT, Goldberg ME, Yost BA, Ahern FM, Aronson KR, Sweeney CT. 1996. Smokers are unaware of the filter vents now on most cigarettes: results of a national survey. *Tob Control* 5(4):265-270.

Kozlowski LT, Goldberg ME, Yost BA, White EL, Sweeney CT, Pillitteri JL. 1998a. Smokers' misperceptions of light and ultra-light cigarettes may keep them smoking. *Am J Prev Med* 15(1):9-16.

Kozlowski L, Pillitteri J. 1996. Compensation for nicotine by smokers of lower yield cigarettes. National Cancer Institute. *The FTC Cigarette Test Method for Determining Tar, Nicotine, and Carbon Monoxide Yields of U.S. Cigarettes: Report of the NCI Expert Committee. Smoking and Tobacco Control Monograph 7*. Bethesda, MD: National Cancer Institute, U.S. Department of Health and Human Services. Pp. 161-172.

Kozlowski LT, White EL, Sweeney CT, Yost BA, Ahern FM, Goldberg ME. 1998b. Few smokers know their cigarettes have filter vents. *Am J Public Health* 88(4):681-682.

Kozlowski LT, Yost B, Stine MM, Celebucki C. 2000b. Massachusetts' advertising against light cigarettes appears to change beliefs and behavior. *Am J Prev Med* 18(4):339-342.

Levy F, Franceschi S, LaVecchia C, Randimbison L, Te VC. 1997. Lung carcinoma trends by histologic type in Vaud and Neuchatell, Switzerland, 1974-1994. *Cancer* 79:906-914.

Lewine H. 1970. *Good-Bye to All That*. New York, NY: McGraw-Hill.

Martineau P. 1957. *Motivation in advertising: Motives that make people buy*. NY: McGraw-Hill.

McAuliffe R. 1988. The FTC and effectiveness of cigarette advertising regulations. *Journal of Public Policy and Marketing* 7:49-64.

Moore D. 1999. Nine of ten Americans view smoking as harmful. *The Gallup Organization, Gallup News Service*.

Mullen C. 1979. *Cigarette Pack Art*. Toronto, Canada: Totem Books.

OSH (Office on Smoking and Health). 2000. [Online]. Available: http://www.cdc.gov/tobacco/overview/regulate.html [accessed December, 2000].

Pauly JL, Streck RJ, Cummings KM. 1995. US patents shed light on Eclipse and future cigarettes. *Tobacco Control* 1995;4:261-265.

Pierce JP, Lee L, Gilpin EA. 1994. Smoking initiation by adolescent girls, 1944 through 1988. An association with targeted advertising. *JAMA* 271(8):608-611.

Pollay RW. 1989. Filter, flavors . . . flim-flam, too! On "health information" and policy implications in cigarette advertising. *Journal of Public Policy and Marketing* 8:30-39.

Pollay R, Dewhirst T. 2000. *Successful Images and Failed Fact: The Dark Side of Marketing Seemingly Light Cigarettes: History of Advertising Archives, Faculty of Commerce*. Vancouver, CA: University of British Columbia.

R.J. Reynolds Tobacco Company. 2000. Advertisements for Eclipse™. Information provided to IOM committee.

Ringold D, Calfee J. 1989. The informational content of cigarette advertising: 1926-1986. *Journal of Public Policy and Marketing* 8:1-23.

Russell M. 1978. Smoking addiction: some implications for cessation. Progress in smoking cessation. Proceedings of International Conference on Smoking Cessation: New York, NY: American Cancer Society.

Samet J. 1996. The changing cigarette and disease risk: current status of the evidence. National Cancer Institute. *The FTC Cigarette Test Method for Determining Tar, Nicotine, and*

Carbon Monoxide Yields of U.S. Cigarettes: Report of the NCI Expert Committee. Smoking and Tobacco Control Monograph 7. Bethesda, MD: National Cancer Institute, U.S. Department of Health and Human Services.

Shiffman S, Pillitteri JL, Burton SL, Rohay JM, Gitchell JG. (In Press). Smokers' Beliefs About "Light" and "Ultra Light" Cigarettes. *Tobacco Control.*

Sidney S, Tekawa IS, Friedman GD, Sadler MC, Tashkin DP. 1995. Mentholated cigarette use and lung cancer. *Arch Intern Med* 155(7):727-732.

Silverstein B, Feld S, Kozlowski LT. 1980. The availability of low-nicotine cigarettes as a cause of cigarette smoking among teenage females. *J Health Soc Behav* 21(4):383-388.

Slovic P. 1998. Do adolescent smokers know the risks? *Duke Law Journal* 47(6):1133-1141.

Slovic P. 2000. What does it mean to know a cumulative risk? Adolescents' perceptions of short-term and long-term consequences of smoking. *Journal of Behavioral Decision Making* 13: 259-266.

Sobel R. 1978. *They Satisfy: The Cigarette in American Life.* New York, NY: Doubleday.

Swedrock TL, Hyland A, Hastrup JL. 1999. Changes in the focus of cigarette advertisements in the 1950s. *Tob Control* 8(1):111-112.

Thun M, Day-Lally C, Myers DG, Calle EE, Flanders WD, Zhu B-P, Namboodiri MM, Heath CW. 1997a. Trends in tobacco smoking and mortality from cigarette use in Cancer Prevention Studies I (1959 through 1965) and II (1982 through 1988). National Cancer Institute. *Changes in Cigarette-Related Disease Risks and Their Implication for Prevention and Control. Smoking and Tobacco Control Monograph 8.* Bethesda, MD: National Cancer Institute, U.S. Department of Health and Human Services.

Thun MJ, Lally CA, Flannery JT, Calle EE, Flanders WD, Heath CW. 1997b. Cigarette smoking and changes in the histopathology of lung cancer. *J Natl Cancer Inst* 89(21):1580-1586.

Tibor Koeves Associates. 1968. A pilot look at the attitudes of negro smokers towards menthol cigarettes; Twelve page report to Philip Morris. Bates # 1002483819/3830. [Online]. Available: http://www.pmdocs.com/getallimg.asp?DOCID=1002483819/3830.

Townsend DE. 1996. Transcript of discussion. National Cancer Institute. *The FTC Cigarette Test Method for Determining Tar, Nicotine, and Carbon Monoxide Yields of U.S. Cigarettes: Report of the NCI Expert Committee. Smoking and Tobacco Control Monograph 7.* Bethesda, MD: National Cancer Institute, U.S. Department of Health and Human Services.

U.S. DHHS (U.S. Department of Health and Human Services). 1981. *The Health Consequences of Smoking; The Changing Cigarette: A Report of the Surgeon General.* Washington, DC: U.S. DHHS, Centers for Disease Control and Prevention.

U.S. DHHS (U.S. Department of Health and Human Services). 1989. *Reducing the Health Consequences of Smoking; 25 Years of Progress: A Report of the Surgeon General.* Washington, DC: U.S. DHHS, Centers for Disease Control and Prevention.

U.S. DHHS (U.S. Department of Health and Human Services). 1994. *Preventing Tobacco Use Among Young People. A Report of the Surgeon General.* Washington, DC: U.S. DHHS, Centers for Disease Control and Prevention.

U.S. DHHS (U.S. Department of Health and Human Services). 1998. *Tobacco Use Among U.S. Racial/Ethnic Minority Groups-African Americans, American Indians and Alaska Natives, Asian Americans and Pacific Islanders, and Hispanics. A Report of the Surgeon General.* Washington, DC: U.S. DHHS, Centers for Disease Control and Prevention.

Viscusi W. 1992. *Smoking: Making the Risky Decision.* New York, NY: Oxford University.

Viscusi WK. 1998. Constructive cigarette regulation. *Duke Law Journal* 47:1095-1131.

Warner KE. 1985. Tobacco industry response to public health concern: a content analysis of cigarette ads. *Health Education Quarterly* 12:115-127.

Weinstein N. 1999. Accuracy of smokers' risk preceptions. *Nicotine and Tobacco Research* 1:S123-S130.

4

Products for
Tobacco Exposure Reduction

S everal tobacco-related products have been introduced recently with potential exposure reduction properties. Two classes of pharmacutical products approved for smoking cessation are also potential exposure reduction products. The chapter describes these products and concludes with a review of regulatory strategies currently in place.

TOBACCO AND TOBACCO PRODUCTS

At first glance, cigarettes seem very simple in construction and design. Typical cigarettes contain a tobacco blend, flavorings and other additives, filters, and cigarette paper. The impression of simplicity of cigarette construction wanes as each component is taken into consideration. There are various combinations of tobacco blends, filter types, and ventilation methods. Manufacturers have used various means of modifying their products to capture and shape the smokers' preference and acceptability of popular brands. Some modifications have had the potential for harm reduction. This section describes current tobacco products, including the curing and processing of tobacco, design features (both historical and contemporary) with exposure reduction potential, and currently available potential reduced-exposure products (PREPs).

Conventional Tobacco Products

There are a wide variety of tobacco products in the United States including cigarettes, cigars, cigarillos, bidis, kreteks, and different types of smokeless tobacco. The most common form of a tobacco product in the United States is the manufactured cigarette. A cigarette is considered to be any roll of tobacco wrapped in paper or in any substance not containing tobacco. Cigarettes can be either manufactured or individually constructed. Cigarettes are lit, and the burning process allows smoke to be inhaled at the other end. Cigarettes are approximately 8 mm in diameter and 70 mm to 100 mm in length.

A cigar is a roll of tobacco wrapped in leaf tobacco or in any substance containing tobacco. There are four main types of cigars: little cigars, small cigars (sometimes called cigarillos), regular cigars, and premium cigars. Little cigars contain air-cured and fermented tobaccos. Little cigars are wrapped either in reconstituted tobacco or in cigarette paper that contains tobacco and/or tobacco extract. Some little cigars have cellulose acetate filter tips and are shaped like cigarettes. Cigarillos are small, narrow cigars with no cigarette paper or acetate filter. Regular cigars are available in various shapes and sizes and rolled to a tip at one end. The dimensions vary from 110 to 150 mm in length and up to 17 mm in diameter. Regular cigars weigh between 5 and 17 grams. Premium cigars vary in size, ranging from 12 to 23 mm in diameter and 12.7 to 21.4 cm in length.

Bidis are made by rolling dried leaf into a conical shape around approximately 0.2 grams of sun-dried, flaked tobacco and securing the roll with a thread. Bidis are used extensively in India, the rural areas of several countries in Southeast Asia, and some parts of the United States. There have been no scientific studies or assessments of the physical characteristics and pharmacologic properties of Indian bidis. In comparison, American versions of bidis were shown to have higher percentages of tobacco by weight (94% vs. 42.5% respectively) and lower levels of nicotine (16.6 mg/g vs. 21.2 mg/g) than Indian bidis. Kreteks are a type of small cigar containing tobacco, cloves, and cocoa, which gives a characteristic flavor and 'honey' taste to the smoke. Kreteks are indigenous to Indonesia but are also available in the United States.

Smokeless tobacco includes tobacco that is sniffed, dipped, or chewed according to the type and constitution of the tobacco. Smokeless tobacco products are made from dark or burley-leaved tobacco. Smokeless tobacco is often referred to as oral tobacco or spit tobacco.

Snuff is a cured, finely ground, flavored tobacco product that is sold in tins or cans and is available in two main types: dry and moist. Dry snuff is fire-cured powdered tobacco and is sniffed. After initial curing, the tobacco is fermented further and processed into a dry powdered form. It has a moisture content of less than 10%. Moist snuff is a granulated tobacco product that is made from both air-cured and fire-cured tobacco and multiple additives. Moist snuff is used by placing a pinch (or "dip") between the lip or cheek and gum. Moist snuff consists of tobacco stems and leaves that are processed into fine particles or strips. It has a moisture content of up to 50%. It is sold in both loose form and ready-to-use pouches (also called packets or sachets) that contain about 0.5 grams tobacco. Moist snuff is available in two varieties, according to the size and consistency of the tobacco: long cut and fine cut. Moist snuff is by far the more prevalent form of snuff used in the United States.

Chewing tobacco is a coarsely shredded, flavored tobacco that is sold in pouches of tobacco leaves or in "plug" or "twist" form. Chewing tobacco is chewed or held in the cheek or lower lip.

Curing, Blending, and Processing

The genus *Nicotiana* is indigenous to the Americas. A member of the family Solanaceae, *Nicotiana* contains more than 64 species. In the United States, the tobacco used in cigarettes, cigars, and smokeless products comes from the species *N. tabacum* and can be categorized by the three traditional methods used in curing or by the geographic region in which it is grown. These are distinguished by important differences in sugar, nicotine, and nitrogen content (Browne, 1990).

Flue curing uses high heat to speed the curing process and control humidity. The principal chemical change is conversion of starch to sugars. During the aging of cured tobacco, enzymatic oxidation of amino acids and carbohydrates takes place. The water content, acidity, and concentration of malic and citric acids increase. There are other, undefined chemical changes that occur, resulting in increased aroma and a less bitter taste. Flue-cured tobacco is also called Bright (also known as Virginia) tobacco. These plants are grown in sandy soils from Virginia to Florida, but their agriculture is centered in North Carolina. These tobaccos generally have low-nitrogen, high-sugar content. The smoke from Bright is acidic with a light aroma. Bright tobacco has medium nicotine content.

Air curing uses heat only to maintain temperature and humidity, not to speed the curing process. The primary chemical changes that occur during air curing are protein degradation, polyphenol formation, and a change in the composition of organic acids. Air-cured tobacco consists of Burley and Maryland tobacco. These plants are grown in silt loams in Kentucky, Tennessee, and western North Carolina and in sandy loam soils in southern Maryland. These tobaccos have very low-sugar content and are more heavily fertilized with manure and artificial fertilizer than the flue-cured products. Air-cured tobaccos have an alkaline smoke, fuller aroma, and high nicotine content. Maryland tobacco also has the quality of continuing to burn on its own, making it less likely to self-extinguish.

Sun-cured or oriental tobaccos (Oriental) require a Mediterranean climate and come mostly from Turkey, Greece, Yugoslavia, Bulgaria, and Russia. They are cultivated with little fertilizer. Oriental tobaccos have mild, aromatic smoke and low nicotine content.

Cigarettes, cigars, and smokeless products use blended tobaccos. The largest component of most cigarette blends in the United States is Bright tobacco (Browne, 1990). Blends are made to achieve specific pH, taste, burning characteristics, and nicotine content. The type of tobacco blend found in cigarettes significantly affects the pH, nicotine content, and toxicity of the smoke produced. The blend can be manipulated by a choice of 60 species and 100 varieties of tobacco. Almost all commercial tobacco products, however, use *N. tabacum* species and a small amount of *N. rustica* in some specialized tobacco products. Cured tobacco lines can contain between 0.2 and 4.75% nicotine (depending on plant genetics, growing conditions, and place of harvest from the stalk; NIH, 1996).

In addition to tobacco leaf, reconstituted sheet tobacco is also used in most commercial products. Cigarettes primarily made with reconstituted tobacco deliver lower smoke yields of tar, phenols, and benzo [a] pyrene (NIH, 1996). Reconstituted tobacco sheet is also used for economic reasons and for the introduction of additives that change various characteristics of the cigarette. Reconstituted tobacco results from a process that combines stems, leaves, and tobacco scrap into a slurry or from making a tobacco "paper," which is cut (Browne, 1990).

Another alternative to leaf tobacco is puffed, expanded, or freeze-dried tobacco (Hoffman and Hoffman, 1997; NIH, 1996). Less tobacco is therefore needed to fill a cigarette while still providing a sensation of fullness and substance in the smoke. While tobacco is being cured, it loses some of its integrity through water loss. Expanding tobacco increases its filling capacity in the final tobacco column of the cigarette by primarily restoring the original cellular structure. This process is performed by expanding the cells with water, steam, and various organic or inorganic fluids depending on the manufacturers patent (David and Nielsen, 1999; NIH, 1996).

The pH strongly influences the concentration of free nicotine in tobacco smoke. The pH is influenced by the type of tobacco used, as well as by the addition of ammonia to the manufacturing process. Free nicotine has a greater effect on the sensory nerves in the mouth and throat than protonated nicotine, which contributes to the impact or strength of the cigarette. Free nicotine is absorbed more rapidly than protonated nicotine across mucous membranes. The phenomenon of more rapid absorption of nicotine at higher pH has been documented in people using different brands of smokeless tobacco. Free nicotine is absorbed through the mouth from alkaline pipe, cigar, and dark cigarette smoke, but not from acidic smoke of blonde tobacco cigarettes. Free nicotine may be absorbed more quickly from the lungs of cigarette smokers as well, although this has not yet been demonstrated experimentally in smokers.

The nitrate content influences the carcinogenic potential of smoke. Nitrogen oxides formed during pyrolysis are free-radical precursors of the polycyclic aromatic hydrocarbons (PAHs). As nitrate concentration in tobacco increases, the synthesis of benzo[a]pyrene (BaP), a carcinogen, is inhibited. Air-cured tobaccos have higher content of nitrates than sun-cured or flue-cured tobaccos and, therefore, lower BaP content. However, the higher the nitrate concentration, the higher are the levels of tobacco specific-nitrosamines (TSNAs), also a known carcinogen.

Hundreds of additives are used in the manufacturing of cigarettes. Several additives are known to have toxic properties. For example, glycerol is a humectant used in cigarettes. Glycerol may lead to the formation of acrolein, a ciliotoxic agent, and diethylene glycol can be converted to ethylene oxide, a carcinogenic compound (Hoffmann and Hoffmann, 1997). Eclipse, that heats but does not burn tobacco, uses glycerol particles as the carrier for nicotine. The glycerol level in smoke from this product is much higher than in conventional low-yield products. As described above, some additives can influence other tobacco constituents (e.g., the role of ammonia in nicotine protonation and TSNA formation).

Although some additives have toxic potential, the concentration of these compounds is low (other than those listed above) and their relative contribution to overall toxicity compared to compounds such as TSNA, BaP, and carbon monoxide (CO) is not definitively known. The toxicity of individual ingredients is sometimes well described, but little is known about how toxicants affect the body when smoked in combination (U.S. DHHS, 2000). The current emphasis on additive disclosure focuses on the consumer's right to know and on understanding better which additives are used to increase the acceptability (e.g., by improving taste and smoothness) of the product to the consumer.

Menthol is a common additive used for flavor and customer acceptability. Early advertisements for menthol products claimed a "soothing" effect on irritated throats. Menthol can be added to cigarettes in several ways including addition to the tobacco shred through an ethanol spray and addition to the filter or packaging. Approximately 3 mg of menthol is added per cigarette. The rest is lost through the filter, sidestream smoke, and in packaging (Browne, 1990). Because menthol is an anesthetizing agent, it has been hypothesized that it may be easier to inhale deeper when smoking a mentholated cigarette. This might also help explain why there are higher rates of lung cancer among blacks despite their lower daily cigarette consumption than whites, who tend to not smoke mentholated cigarettes. Various studies, however, have produced mixed results on the subject (Carpenter et al., 1999; Clark et al., 1996; Gaworski et al.,

1997; Kabat and Hebert, 1991; McCarthy et al., 1995; Sidney et al., 1995; Women who smoke menthol cigarettes have greater nicotine exposure, 1999).

Cigarette paper is second to tobacco as the most variable component in producing cigarettes. The degree of ventilation allowed by the paper can be manipulated in the production process. More porous cigarette paper has been shown to reduce smoke yields of CO and tar as well as volatile nitrosamines, TSNAs, and BaP through dilution. Increased permeability does not reduce the low-molecular-weight gas-phase components in smoke however (NIH, 1996). For a more detailed discussion of the toxicology of smoke, see Chapter 10.

A new paper has been introduced for Merit cigarettes, which claimed to decrease the smoldering of cigarettes when dropped onto fabric. The new technology consists of a modified wrapping paper that reduces the amount of oxygen entering the cigarette, therefore slowing the rate at which it burns. This could decrease the 25% of fatal residential fires started by smoldering cigarettes (Meier, 2000). New York became the first state to pass legislation imposing fire safety standards on cigarettes (Cigarettes-fire bill, 2000).

Smoke Yields

The procedure for measuring the tar and nicotine yields of mainstream smoke (i.e., the "smoking machine") is standardized for consistency between laboratories and from product to product. There are two methods in widespread use: the Federal Trade Commission (FTC) method and the International Standards Organization (ISO) method. Differences between these two are minor, and puff volume, duration, and interval are common to both standards. Particulate matter is collected on Cambridge filter pads as what is called "wet total particulate matter" (Davis and Nielson, 1999). Tar is a generic term for the total particulate matter minus the nicotine and water. The material that passes through the filter is called the vapor phase. This is described in more detail in Chapters 10 and 11.

Tar yields are influenced primarily through filtration, ventilation (tip ventilation holes and paper porosity), and the choice of tobacco processing and blend. As with any agricultural product, there is natural variation from year to year. In the interest of manufacturing a consistent product, prepared tobacco is blended with stock of crops from previous years to maintain a uniform product line. Finally, the burn rate of cigarettes has been proven to influence smoke yields. The faster the burn rate, the lower the tar yields will be, according to FTC measurements. Shredded tobacco can facilitate a faster burn rate (Davis and Nielsen, 1999), as can the use of accelerants.

The FTC test does not account for the wide range of smoking behaviors and compensation that occurs naturally among smokers, thus, the standardization has come under great criticism. Human smoking behavior can greatly influence the tar, nicotine, and CO levels to which smokers are exposed. The yields of so-called light and ultralight cigarettes have changed over the years but have produced similar nicotine levels across yields. These terms are not official government designates and are often part of the trademarked name of a product. In general, however, ultralights have less than 6 mg of tar, light cigarettes have between 6 and 15 mg of tar, and "regular or full-flavored" cigarettes have more than 15 mg (NIH, 1996; Sweeney et al., 1999). Although there is not a standard nicotine classification, a study by Byrd et al. (1995) comparing measured and FTC-predicted nicotine uptake in smokers described nicotine levels in products they categorized as 1 mg tar, ultralow-tar, full-flavor low-tar, and full-flavor cigarettes. The mean FTC nicotine yields were 0.14, 0.49, 0.67, and 1.13 mg per cigarette, respectively. FTC measures, particularly for low-tar cigarettes or for cigarettes with filter ventilation holes, do not, however, reflect true exposures in humans. Cigarettes are positioned in the smoking machines in a manner that allows air to enter the perforated filters. These holes are often covered by the lips or fingers of smokers (Hoffman and Hoffman, 1997). In addition, smokers of low-yield products compensate by inhaling more deeply, holding a puff in the lungs for longer periods, or puffing more frequently.

Conventional and Historical PREP Technology

Two cigarette design features that reduce toxin yields as measured by the FTC are dilution and filtration. Dilution is achieved primarily with ventilation, although paper porosity can also increase smoke dilution. Cigarette holders popular in the 1930s provided ventilation. Ventilation is achieved today with small holes around the filter. The primary means of toxin reduction however is the addition of a filter component on the mouth end of the cigarette to trap certain components before they are released from the cigarette in the form of smoke.

The majority of cigarettes sold in the United States today have cellulose acetate filters. Most cellulose acetate filters reduce tar and nicotine yields by 40–50% compared to nonfiltered cigarettes (Davis and Neilson, 1999). Because filter materials influence tar and nicotine smoke yields as well as taste to differing degrees, filter preference has become regionalized. Filters that contain charcoal provide selective removal of a range of vapor-phase smoke constituents and are more popular in Japan than in the United States. Although Japanese and American smokers smoke a comparable numbers of cigarettes per day there is a lower incidence of lung cancer in Japan (NIH, 1996). Many factors play a role in the differences in cancer rates between countries. It is speculated that along with diet, genetics, epidemic patterns, and other life-style factors, charcoal filters and tobacco processing may contribute as well (Hoffmann and Hoffmann, 1997).

Traditional cellulose acetate filters treated with certain plasticizers can reduce some additional volatile and semivolatile compounds in smoke. The addition of charcoal particles in the filters reduces volatile smoke constituents such as ciliotoxic hydrogen cyanide, acetaldehyde, and acrolein. It can also reduce some volatile aromatic hydrocarbons, such as benzene and toluene, in the first puffs of a cigarette but becomes less effective in later puffs (NIH, 1996). Segmented filter systems, like charcoal-containing filters, provide a multitude of options. Charcoal, for instance, is available in a variety of activities depending on its surface area and pore volume. The amounts of material used in filters, filter length, and particulate removal efficiency can be adjusted (Davis and Nielsen, 1999).

Since about 1968, many filter tips have been perforated with one or more lines of ventilation holes placed around the middle of the filter. Ventilation holes act slightly differently than the permeability ventilation provided by the cigarette paper. Filter tip ventilation is engineered to dilute the smoke as it travels through the cigarette and the filter. The result is an overall reduction in smoke and tar yields at standard smoking conditions to levels that filters, permeable papers, or processed tobaccos alone could not achieve. Approximately 80% of U.S. cigarettes have tar yields of 15 mg or less (FTC, 2000); most of these cigarettes have filter tip ventilations (Davis and Nielsen, 1999). The ventilation holes are inserted where smokers are likely to place their fingers or lips, which inhibits the intended use of the vents. Machine-generated smoke yield tests, however, position cigarettes so that the ventilation holes are exposed. Jenkins and colleagues conducted a study in 1982 comparing smoke yields between open and blocked tip ventilation. Their abbreviated results can be found in Table 4-1. The results showed that blocking the ventilated filters (VF) increased the tar, nicotine, and carbon monoxide to levels similar to that of regular filter cigarettes (F).

There have been a few notable historic developments regarding novel cigarette filters of unproven efficacy. The Kent cigarette line produced cigarettes with a novel filter of coiled crepe paper with cotton fibers and crocidolite fibers in the 1950s. Crocidolite is a form of asbestos with fibers so thin that they could be arranged and used to trap particles as small as 1 μ (Longo et al., 1995). The so-called Micronite filter eliminated nearly

TABLE 4-1   Effect of Smoking Conditions (Blocked Tip Ventilation) on Smoke Yield (mg per cigarette ± one standard deviation)

Brand	FTC (35 ml, 2 sec, 1 puff per minute)	FTC+ (tip taped 35 ml, 2 sec, 1 puff per minute)
Tar Yield		
VF-A	3.8±0.5	9.4±0.9
VF-C	2.9±0.6	7.6±0.9
VF-D	1.6±0.2	9.7±0.8
F-A	18.5±1.2	ND
F-C	16.4±1.4	13.2±0.6
F-D	9.9±0.8	11.7±1.3
NF-A	22.5±1.0	21.3±1.5
NF-C	19.4±1.1	21.3±1.0
Nicotine Yield		
VF-A	0.40±0.05	0.72±0.05
VF-C	0.25±0.04	0.45±0.03
VF-D	0.19±0.05	0.62±0.07
F-A	1.09±0.07	ND
F-C	0.94±0.02	0.71±0.06
F-D	0.61±0.02	0.68±0.10
NF-A	1.14±0.05	1.37±0.07
NF-C	1.13±0.13	1.6±0.23
Carbon Monoxide Yield		
VF-A	4.1±0.7	12.3±1.5
VF-C	2.1±0.2	8.7±1.2
VF-D	1.0±0.1	10.7±0.4
F-A	15.7±1.8	ND
F-C	13.4±1.2	17.9±1.2
F-D	8.5±0.3	11.5±0.7
NF-A	11.3±1.0	12.8±2.7
NF-C	11.3±1.2	12.9±0.8

NOTE: F=filter; ND=not determined; NF=nonfilter; VF=ventilated filter; A,C,D=different brands.
SOURCE: Modified from Jenkins et al., 1982 in NIH, 1996.

twice as much tar and nicotine delivered to the smoker as any other standard brand of its time. Yet it failed in the marketplace due to the flavorless smoke and difficulty in drawing on it (Kluger, 1997).

Ligget tobacco company experimented with adding palladium and magnesium nitrate to tobacco in efforts to decrease cancer rates in smokers. Preliminary tests on mice resulted in a 95% reduction in tumors compared to other brands. Little else is known about this innovation, because research ended, reportedly due to litigation, in 1988 (Was a safer cigarette research snuffed, 1994).

A new filter treatment called the "Wellstone Filter" has been developed. The cellulose acetate filter has been treated with a nonbiological compound that supposedly removes nearly 90% of tar and carcinogenic compounds while maintaining taste (Fisher, 2000). Patents are still pending.

R.J. Reynolds also experimented with its own variation on filters. Reynolds created an extra corrugated carbon paper filter for the Winston Select EW brand. It was intended to reduce compounds linked to heart problems. The product was test marketed in Oklahoma City in 1996 (Feder, 1996). (This filter was known more commonly as the "carbon scrubber filter" to reduce free radicals linked to cardiovascular disease.)

Other product modifications of interest for historical reasons include the Favor Smokeless Cigarette and Masterpiece Tobacs. The Favor Smokeless Cigarette was introduced in 1985 and was evaluated soon afterwards by the Food and Drug Administration (FDA), which determined the product to actually be a type of drug delivery device also known as a nicotine inhaler. The FDA removed it from the market. Masterpiece Tobacs is a chewing gum containing shreds of tobacco. Pinkerton Tobacco Company introduced it in 1987. This too was withdrawn from the market when the FDA determined the gum to be a food product with an unapproved food additive, tobacco (Fielding et al., 1998).

Philip Morris also experimented with a modified cigarette with denicotinized tobacco called NEXT. NEXT had a tar yield of 9.3 mg but a nicotine yield of only 0.08 mg. Lacking the addictive component, this product was not successful on the market (Ferrence et al., 2000).

Currently Available and Novel PREPs

Modifications of Conventional Tobacco Products (See Table 4-2)

A new curing process is being used for the production of tobacco with substantially reduced tobacco-specific nitrosamines (Star Scientific, 1999). The StarCure technology of Star Scientific (formally called Star Tobacco and Pharmaceutical Co.) has a modified and controlled curing process that has used microwaves to kill the bacteria that convert nitrogen-containing compounds into TSNAs. Star has recently stopped using this method and now uses curing barns that decrease microbial activity (Blackwell, 2000). Star Scientific has been using Virginia flue-cured Kentucky burley tobacco in its process. Star expected to bring to market products with only StarCure tobacco in late-2000. Other tobacco companies are reportedly working on methods for reducing or eliminating TSNAs from tobacco

TABLE 4-2   Tobacco Products

Product	Company	Year	Key Characteristics
Kent	Lorillard Co.	1950s	Micronite filters
Spectra	Kinney	1980s	Contained N bloctin, designed to block nitrosamine absorption in the lungs
Favor Smokeless Cigarette		1985	Nicotine inhaler
Premier Cigarette	R.J. Reynolds	1987	Less combustion than conventional
Masterpiece Tobacs	Pinkerton	1987	Chewing gum with tobacco shreds
NEXT	Philip Morris	1989	Low-nicotine cigarette
Eclipse Cigarette	R.J. Reynolds	1996	Less combustion than conventional
Winston Select	R.J. Reynolds	1996	Extra corrugated carbon paper filter
Accord	Philip Morris	1998	Lower operating temperature than conventional cigarette and heat produced from electrical resistance
Advance	Star Scientific	2000	Low-nitrosamine cigarette

(Fairclough, 2000). Brown and Williamson purchased two million pounds of low-nitrosamine tobacco from Star in 1999 and contracted for millions of additional pounds in 2000 (Fairclough, 2000; Star Scientific, 2000). Nitrosamines are still formed from these tobaccos upon combustion, but the noncombusted product contains very little or no TSNAs.

Star Scientific has used this modified tobacco along with activated charcoal filters in a new line of cigarettes called Advance. Star hopes that by replacing traditional filters with the activated charcoal filter, it will reduce the levels of vapor-phase toxins (Star Scientific, 2000; Fairclough, 2000). Star Scientific gives smoking yields on its Web site and claims to have substantially lower TSNA, CO, and tar levels and similar nicotine levels compared to the average of three leading light brands (Star Scientific, 2000). A package insert for Advance cigarettes lists these findings as reported independently by the FTC and the Massachusetts Department of Public Health.

Smokeless Tobacco Products

Currently, most smokeless tobacco users in the United States use moist snuff. The 1999 National Household Survey on Drug Abuse reported that among the 66.8 million Americans who report current use of tobacco products, 3.4% use smokeless tobacco (SAMHSA, 2000). The highest prevalence of smokeless tobacco use was found in males between 18 and 25 years of age (SAMHSA, 2000). The manufacturing of fine-cut tobacco for the production of moist snuff increased in the 1970s. An active advertising campaign is thought to have led to increased prevalence of smokeless tobacco among 18- to 24-year-old males. From 1970 to 1991, the use of smokeless tobacco in this population increased from 2.2% to 8.9% (U.S. DHHS, 2000). For most of the 1990s, the Centers for Disease Control and Prevention's (CDC's) Youth Risk Behavior Survey has shown a consistent use of smokeless tobacco among male high school students at about 20% (U.S. DHHS, 2000).

Sweden has a high prevalence rate of smokeless tobacco use: 18% of Swedish men and just under 2% of women aged 15-75 dipped snuff in 1995, compared to 3.3% of American adults (less than 1% of American smokers were women) (Ahlborn, 1997; SAMHSA, 1996). In 1996, only 18% of Swedish adults were daily smokers (primary reference: Ramstrom, personal communication in Jimenez-Ruiz, 1998), whereas in the United States in 1998 the smoking prevalence was 24.1% of adults (18 years and older) (CDC, 2000). In fact, Sweden has the largest per capita consumption of moist snuff. In 1990, Swedish citizens consumed 0.68 kg per person per year, or 4,846 tons annually (Ahlborn, 1997). Swedish snuff differs from snuff in the United States in that it has been shown to contain fewer TSNAs. The difference in recent years has become smaller due to a decrease in TSNAs in U.S. products (Ahlborn, 1997). For the health effects and toxicity of smokeless tobacco, see Chapter 10 of this report. There is debate in the tobacco control field whether smokeless tobacco has a role as a PREP and whether controlling factors such as marketing will prevent its use as a "gateway" to cigarette use.

Cigarette-like Products

Aside from modifications of traditional tobacco products, there has been a recent introduction of cigarette-like products. R.J. Reynolds (RJR) was the first of the tobacco companies to develop such a PREP in 1988, called Premier.

An RJR monograph about the development of Premier describes the prototype product as "similar to other cigarettes in that it requires tobacco for taste and enjoyment" (R.J. Reynolds Tobacco Company, 1988). Premier is a cigarette-like product that delivers nicotine with less combustion. This controlled burning reportedly releases volatile, flavorful components, but does not decompose the tobacco. Similar to traditional cigarettes, the smoker inhales an aerosol. The four components of Premier include tobacco and tobacco flavor beads, a volatile liquid to form the aerosol or smoke, a heat source to warm tobacco and vaporize the liquid, and finally a system that condenses the vapor into an aerosol delivered to the user (R.J. Reynolds Tobacco Company, 1988).

The RJR monograph of this cigarette prototype details the "tar," nicotine, and CO levels using a modified FTC rating. The rating is modified to a version that is not based on the butt length. The prototype ranked lower in tar and nicotine levels compared to all the traditional cigarettes to which it was compared: tar=6 mg per cigarette (nicotine-free dry particulate matter); nicotine=0.3 mg per cigarette; CO=12 mg per cigarette.

This prototype has developed into a new product Eclipse, which is being sold over-the-counter in Dallas, TX and other test markets and over the telephone and Internet. Eclipse has recently been marketed as "a better way to smoke. A Cigarette that responds to concerns about certain smoking-related illnesses. Including Cancer" (R.J. Reynolds Tobacco Company, 2000). The advertisements for Eclipse also address social acceptance because it produces no ash and substantially less visible environmental tobacco smoke (ETS).

Eclipse resembles an ordinary cigarette in shape and size just as Premier did. The carbon tip is ignited and heats the mixture of tobacco and glycerin before passing through a charcoal filter. The heating unit consists of a carbon fuel element surrounded by a fiberglass insulator. Glycerin contributes to 50-60% of the composition of the light reconstituted tobacco filling of the product. Research has shown that the glycerin has been treated to prevent it from sweating out of the tobacco blend, yet little else is known regarding what additives have been used or how else the tobacco may have been treated. Eclipse is different from Premier in that it does not have flavor beads or an aluminum cylinder or the alumna beads (Ferrence et al., 2000; Slade, 1996). These changes are likely to enhance the smoke aerosol so that it more closely resembles that of traditional cigarette products. Tests conducted under FTC-like conditions resulted in 3.2 mg tar, 0.18 mg nicotine, and 7.5 mg CO (R.J. Reynolds Eclipse web site, 2000).

In a study conducted by Fagerström et al. (2000) measuring nicotine and CO exposure using Eclipse, nicotine oral inhaler, and traditional cigarettes, it was discovered that there is little difference in nicotine blood concentrations in subjects smoking only Eclipse or their usual brand of cigarettes. Eclipse, however, did produce increased carbon monoxide levels. A recent study commissioned by the Massachusetts Department of Public Health compared Eclipse to two conventional low-yield products. Eclipse produced higher yields of tar and CO than the comparison products. Under more intensive conditions, the Eclipse yields of nicotine were also higher than comparison products, as were specific toxicants, BaP, acroline, and nicotine-derived nitroketone (NNK) (Labstat, 2000). A scientific panel convened by RJR has also studied the toxicity of Eclipse compared to a reference cigarette. Among its many conclusions, this panel reported elevated acrolein, furfural, formaldehyde, and CO in the smoke from Eclipse when compared to a ultralow-tar reference cigarette (Eclipse Expert Panel, 2000). The panel went on to report a significant reduction in evidence of lower respiratory tract inflammation and tumorigenic activity in dermal studies.

Philip Morris Tobacco Company has developed and marketed a PREP with some similarities to Eclipse. It differs significantly in other ways. Preliminary technical information on Accord was publicly presented by Philip Morris scientists in a poster presentation at the Society of Toxicology in 1998. They began consumer testing in the fall of 1997 (Jones, 1998). Like Eclipse, Accord uses the lower temperature and controlled burn of the cigarette to dramatically alter the composition of smoke produced. Accord burns at 950°F, or 700°F lower than the traditional cigarette. The burning device in the Accord cigarette is powered by rechargeable batteries in a beeper-sized unit called a Puff Activated Lighter (Holzman, 1999). This unit fits special Accord cigarettes and powers a microchip that senses when the cigarettes are being drawn on. When signaled, it produces a controlled two-second burn from one of eight heating blades around the cigarette and thus delivers smoke to the user.

The company reports that Accord delivers 3 mg of tar and 0.2 mg of nicotine, similar to other Philip Morris ultralight products (Jones, 1998). Philip Morris scientists report that Accord smoke contains marked reductions in 35 of 53 potentially hazardous compounds and that Accord produces 83-98% less carbon monoxide, benzene, and nitrogen-based compounds than the cigarette smoke of comparison products (Jones, 1998). The data was stated without specifying the nature of the comparison products. Like Eclipse, Accord does not produce ash and produces little ETS (ASH, 1997). A recent study indicates that smokers who switch to Accord under experimental conditions are exposed to minimal CO and less nicotine compared to their usual brand (Buchhalter and Eissenberg, 2000).

PHARMACEUTICAL PRODUCTS

Unlike tobacco products, medications developed to aid smoking cessation have undergone rigorous scientific and regulatory examination. Medications for the treatment of tobacco dependence came into existence in the 1970s in Europe and 1980s in the United States. The first effective medications for the treatment of nicotine dependence were nicotine replacement therapies. Currently, there are four different nicotine replacement products that have been approved by the Food and Drug Administration.

These products include nicotine gum, transdermal nicotine, nicotine inhaler, and nicotine nasal spray. The only nonnicotine medication for smoking cessation that is approved by the FDA is bupropion sustained release (SR), or Zyban. This product, originally marketed as an antidepressant, was initially observed to reduce smoking among depressed patients (Ferry et al., 1992). Subsequent clinical trials among smokers showed that this agent is an effective smoking cessation aid.

In the United States, the FDA has approved these medications only as cigarette smoking cessation aids. Currently, these pharmaceuticals are not recommended solely for the purposes of reducing the number of cigarettes or as a step toward achieving abstinence, to treat withdrawal symptoms or craving in situations when smoking is not allowed, or for quitting tobacco products other than cigarettes. In addition, the safety and efficacy of these medications in pregnant smokers have not been determined, and the use of medications to aid smoking cessation in this population has been delegated to the discretion of the physician.

It is important to note that the pharmaceutical industry was required to provide ancillary behavioral treatments along with the medications to assist the smoker in quitting. These behavioral treatments range from general self-help materials, to tailored self-help materials, to telephone counseling. The FDA imposed this requirement because cigarette smoking not only is considered a physical addiction to nicotine but also is associated with behavioral components. Research results indeed show that behavioral treatment will augment the success rates of medications alone, and the more intensive the treatment is, the greater is the rate of abstinence (Fiore et al., 2000). However, the use of medications with minimal or no behavioral treatments still outperforms placebo treatment, as demonstrated by the efficacy of over-the-counter nicotine medications (Fiore et al., 2000). Therefore, harm reduction approaches with medications may have to be considered in the context of ancillary behavioral treatments.

Nicotine Replacement Products

The concept of the use of nicotine replacements was first fully described by Ferno (1973). The principles behind nicotine replacements are to (1) provide cigarette smokers a sufficient amount of nicotine to allay some of the withdrawal symptoms experienced shortly after tobacco abstinence; (2) to permit progressive reduction of the level of nicotine exposure, leading to eventual ease of totally withdrawing from nicotine products; and (3) to reduce the abuse potential of nicotine due to the slower rate of nicotine absorption. The validity of these principles and mechanisms has been demonstrated by a number of studies (see Henningfield and Keenan, 1993; Hughes et al., 1989). Other potential mechanisms accounting for product efficacy could be the reinforcing effects derived from nicotine or, on the other hand, blocking the reinforcing effects of nicotine from cigarettes. Chronic exposure to nicotine could result in desensitization of nicotinic cholinergic receptors, thereby blocking potentiation of the nerve. In addition to these beneficial effects, the administration of nicotine medications is associated with significantly lower toxicity than cigarettes since the main ingredient in these replacement products is nicotine with no other toxic elements. The toxicity associated with nicotine alone is confined primarily to reproductive disorders and enhancement of cardiovascular risk factors (Benowitz, 1998), although nicotine's contribution to cardiovascular disorders is minimal compared to that of cigarettes.

Nicotine polacrilex, or nicotine gum (Nicorette), manufactured by Pharmacia Upjohn, was introduced first in Europe and approved for marketing in the United States in 1984. Nicotine gum was initially a prescription medication in the United States but, in 1996, became an over-the-counter medication. In 1998, a mint-flavored nicotine gum was approved in the United States and it was marketed in 1999. An orange flavored version has recently been introduced. Two doses of nicotine gum are currently available, 2 and 4 mg. In 1991 and 1992, the United States was introduced to nicotine patches distributed by four different pharmaceutical companies. The first patch introduced in the U.S. market was Habitrol (marketed and manufactured by Ciba-Geigy, and now by Novartis); then Nicoderm (manufactured by Alza Corporation, marketed by Marion Merrell Dow, then Hoescht Marion Rousel, followed by SmithKline Beecham, now GlaxoSmithKline); Nicotrol (manufactured and marketed by Pharmacia Upjohn); and finally Prostep (manufactured by Elan and initially marketed by Lederle as a prescription medication; currently generic patches using a different patch technology are being manufactured by Perrigo). Nicotine patches were initially prescription only, but both Nicoderm CQ and Nicotrol were approved to go over the counter in 1996. Habitrol and a modified Prostep are now sold as generic products through major drugstore chains. Nicotine patches are available in three different doses for Habitrol and Nicoderm CQ (21, 14, and 7 mg). Nicotrol was available in three different doses (15, 10, and 5 mg) by prescription, but only in one dose (15 mg) over-the-counter. Perrigo's patches are available in 22-mg and 11-mg doses. The nicotine inhaler was introduced on the market in 1998 and is regulated as a prescription drug. This product is a tube-shaped device that contains a porous plug saturated with nicotine enclosed in a replaceable plastic cartridge. The nicotine inhaler is puffed upon like a cigarette, but absorption occurs primarily buccally and from the upper airway and not in the lungs because its design does not provide an aerosol that could be substantially inhaled in the lung. Each cartridge contains 10 mg of nicotine, with 13 µg delivered with each puff (compared to 100 µg delivered with cigarettes), which is about one-tenth the nicotine dose from puffing on a cigarette. The nicotine nasal spray was introduced to the U.S. market around 1997 and is also regulated as a prescription drug. The dosing for nicotine nasal spray is one spray in each nostril, delivering 0.5 mg of nicotine per spray for a total dose of 1 mg.

Each product has a different route of administration, instructions for use, amount of nicotine absorbed, and speed of nicotine delivery. Nicotine patches are used once a day, with a patch being placed on the skin in the morning and taken off prior to bedtime or the next morning. For nicotine gum, a fixed schedule of use (e.g., at least one piece every 1-2 hours, with a maximum of 24 pieces per day) is recommended to achieve sufficient levels of nicotine. The nicotine inhaler is puffed on, not inhaled like a cigarette, and used ad libitum. The recommended number of cartridges is 6-16, with each cartridge used in three smoking periods. For the nicotine nasal spray, the recommended initial dosing is 1-2 sprays per hour, with a minimum recommended treatment of 8 doses per day and a maximum of 40 doses per day. The recommended duration of treatment for all these products ranges from 3 to 6 months.

The route of administration will contribute to the side-effect profile and contraindications (see Table 4-3). Contraindications include smokers in the immediate postmyocardial infarction period with serious arrhythmias

TABLE 4-3 Side Effects and Contraindications of Nicotine Replacement Products

Product	Most Frequent Side Effects	Contraindications
Nicotine Gum	Jaw ache, mouth soreness, dyspepsia, hiccups	TMD, Dentures
Nicotine Inhaler	Local irritation of mouth and throat, coughing, rhinitis	Allergy to menthol
Nicotine Patch	Local skin reaction, sleep disruption	Skin disorders
Nicotine Spray	Nasal and airway irritation	Reactive airway disease, sinusitis
Bupropion	Insomnia and dry mouth	Seizures, concurrent use of MAO inhibitors, history of eating disorders

NOTE: MAO=Monoamine Oxidose; TMD=temporomandibulor joint disfunction.
SOURCE: Information gathered from Fiore et al., 2000.

and with serious or worsening angina pectoris. Studies in individuals with stable cardiovascular disease have shown that smokers who used nicotine patches were not at greater risk for cardiovascular events than those assigned a placebo (Joseph et al., 1996), even when used while smoking a cigarette (Working Group for the Study of Transdermal Nicotine in Patients with Coronary Artery Disease, 1994). For pregnant or lactating female smokers, nicotine replacements are recommended only when the use of psychosocial treatment has failed and the benefits of abstinence achieved from using the nicotine replacements outweigh the risk of nicotine replacement and potential concomitant smoking (Fiore et al., 2000). In general, the contraindications and side effects are minimal with these products.

The bioavailability (see Table 4-4) and amount of nicotine absorbed per unit dose (see Figure 4-1) varies across products. The mean nicotine peak concentrations attained from tobacco products are higher than those of all nicotine replacement products. Furthermore, for cigarettes, high arterial plasma nicotine concentrations are achieved that are not reached by nicotine replacement products (Henningfield et al., 1993). The daily dose of nicotine from medications used ad libitum will depend on the frequency of use. Typically, these products are underutilized in terms of both frequency and duration. Rarely does the daily level of nicotine concentration attained by the use of these products exceed three-fourths the level attained with cigarettes and typically ranges from one-fourth to two-thirds of the cigarette level (Benowitz et al., 1988; Hjalmarson et al., 1997; Leischow et al., 1996; Schneider et al., 1995, 1996; Sutherland et al., 1992; Tonnesen et al., 1993).

The time to reach peak blood nicotine concentrations (T_{max}) also varies across products. The nicotine replacement product that has the shortest T_{max} is nicotine nasal spray followed by nicotine inhaler, nicotine gum, and finally the nicotine patch (see Figure 4-1 and Table 4-5). It should be

TABLE 4-4 Bioavailability of Nicotine

Product	Bioavailability per Dose
Cigarette	1-2 mg
Smokeless tobacco	3.6-4.5 mg
Nicotine gum (2 or 4 mg)	1 or 2 mg
Nicotine inhaler	2 mg per cartridge
Nicotine patch	15-22 mg (over 16-24 hr)
Nicotine spray	0.5 mg

SOURCE: Data from Benowitz, 1988; Fagerström, 2000; Fant et al., 1999.

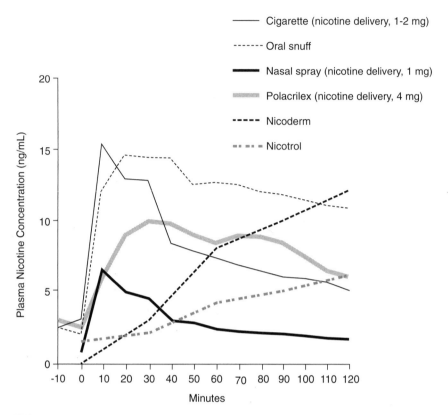

FIGURE 4-1   Venous blood concentrations.

NOTE: Venous blood concentrations in nanograms of nicotine per millimeter (ng/ml) of blood as a function of time for various nicotine delivery systems.
SOURCE: Reprinted with permission from Fant et al., 1999. Copyright 1999 Lippincott, Williams and Wilkins.

noted that a faster T_{max} for nicotine nasal spray was observed with arterial than with venous blood (mean value, 5 vs. 25 minutes, respectively) as well as a higher maximum nicotine concentration (mean value, 10 vs. 5 ng/ml; Gourlay and Benowitz, 1995).

The faster the speed of nicotine delivery, rate of absorption, and attainment of peak level, the greater is the likelihood of continued use or abuse (deWit and Zacny, 1995). The fastest time for maximum nicotine concentration occurs with cigarettes, followed by nicotine nasal spray, smokeless tobacco, then other nicotine replacement agents. Smokeless tobacco appears to take longer or seems equal in time to reach maximum

TABLE 4-5 Pharmacokinetics

Product	Time to Maximum (T_{max}) for Venous Blood
Cigarettes	Within 5 min.
Smokeless tobacco	20-30 min.
Nicotine nasal spray	10 min.
Nicotine gum (2-4 mg)	30 min.
Nicotine inhaler	20-30 min.
Nicotine patch	4-9 hr.

SOURCE: PDR, 2000; Benowitz et al., 1988.

concentrations compared to some nicotine replacement products such as nicotine gum and inhaler. However, the abuse potential is likely to be greater for smokeless tobacco not only because of the amount of nicotine delivered, but also because a nicotine boost of 10 ng/ml has been observed within the first 10 minutes of use (Holm et al., 1992). This venous level is higher than the maximum venous concentration observed with other nicotine products including from nicotine nasal spray.

Although few studies have examined the abuse potential across various nicotine replacement products, it remains significantly lower than that of cigarettes. Of the few nicotine replacements studied, withdrawal symptoms and the rate of use beyond the recommended period for the nicotine patch appear to be minimal and the abuse potential appears to be low (see deWit and Zacny, 1995; Hughes, 1998). In clinical trials, prolonged use of nicotine gum (e.g., 12 months) is around 22% among abstinent smokers, but about 9% of those who have been randomly assigned to the gum (Hughes, 1998). Similar rates of prolonged use are observed for nicotine inhaler (Schneider et al., 1996; Tonnesen et al., 1993). For nicotine nasal spray, about 29-43% of individuals who have quit smoking reported use of the active spray at 12 months compared to 0% in the placebo spray group, even though the recommended period of use was 3 months (Hjalmarson et al., 1994; Sutherland et al., 1992). Of smokers assigned to the active nasal spray, about 10% continued to use it at 12 months (Hjalmarson et al., 1994). This finding would indicate that continued use of this product might be greater than that of other nicotine products. It is important to note that the continued use of nicotine products may be motivated by the desire not to relapse to smoking rather than addiction to the product. Furthermore, the cost of the product plays a role in continued use as well, with higher cost deterring longer use. All of the aforementioned studies provided the nicotine products for free, and therefore the rate of continued use may be highly exaggerated. In a nonresearch setting, the unit dose costs of cigarettes and smokeless tobacco are typically less than or equivalent to those of the medication (see Table 4-6). However, medications must be bought in larger quantities, resulting in yet higher costs per purchase.

Based on a report issued by the Public Health Services (Fiore et al., 2000), the nicotine replacement products have comparable rates of treatment success. The PHS undertook an extensive meta-analysis of studies to determine the efficacy of various medications for smoking cessation. Only published, peer-reviewed and randomized controlled studies with a follow-up time of at least five months post-quit were included in the analyses. The primary outcome variable was point prevalence abstinence (not smoking in the past week), unless only data for continuous abstinence (or of an unknown nature) were available. Data for the efficacy of various medications are presented in Table 4-7. The estimated odds ratios range from 1.5 to 2.7 for active nicotine replacements versus the controls, and the estimated rates of abstinence range from 18 to 31% for active nicotine replacements versus 10 to 17% for the controls. Several trials have been conducted with nicotine patches that simulate over-the-counter use. The estimated rates of abstinence observed in these studies are about 11.8% with active nicotine patch and 6.7% with placebo patch with an estimated odds ratio (OR) of 1.8 (Fiore et al., 2000).

Similar meta-analyses have been conducted by the Cochrane Tobacco Addiction Group (Lancaster et al., 2000). Randomized controlled trials (published and unpublished) with at least six months' follow-up were included in the analyses. Sustained abstinence was the primary outcome examined, rather than point prevalence, although point prevalence rates

TABLE 4-6   Costs for Tobacco Products and FDA-Approved Medications[a]

Product	Cost per Day
Cigarettes (20 per day)	$1.70-$4.85 per pack
Smokeless tobacco[b]	$1.88-$2.00
Nicotine gum	$4.26-$6.87 (10 2–4 mg pieces)
Nicotine inhaler	$10.81-$10.94 for 10 cartridges
Nicotine patch	$2.36-$4.50
Nicotine spray	$4.50-$9.20 for 10-20 doses
Bupropion	$3.33-$3.40

[a]Prices are from U.S. PHS guidelines and based on the retail price of the medication purchased at a national chain pharmacy located in Madison, Wisconsin, and Minneapolis, Minnesota.

[b]Cost per day is based on a mean use of 3.5 tins per week (Hatsukami and Severson, 1999) at $3.75 to $4.00 per tin.

TABLE 4-7   Efficacy of Nicotine Replacement Products

Medication	Abstinence Rate: Active (95% CI)	Abstinence Rate: Placebo	Odds Ratio (95% CI) PHS (Cochrane)
Gum (N=13 studies)	24 (21, 27)	17	1.5 (1.3, 1.8) (1.6 [1.5, 1.8][a])
Inhaler (N=4 studies)	23 (16, 29)	11	2.5 (1.7, 3.6) (2.1 [1.4, 3.0][b])
Nasal spray (N=3 studies)	31 (22, 39)	14	2.7 (1.8, 4.1) (2.3 [1.6, 3.2][c])
Patch (N=27 studies)	18 (16, 20)	10	1.9 (1.7, 2.2) (1.8 [1.6, 1.9][d])

NOTE: CI=Confidence Interval.

[a]N=48 studies
[b]N=4 studies
[c]N=4 studies
[d]N=31 studies

SOURCE: Abstracted from Fiore et al., 2000; Lancaster et al., 2000.

were used when sustained abstinence rates were unavailable. The results are very similar to PHS results. The odds ratio across the nicotine products ranged from 1.7 to 2.3 for active nicotine replacement versus control, and efficacy was similar across products (see Table 4-7).

Other Nicotine Replacement Agents and Nonapproved Methods of Use

Several other nicotine replacement products have been tested, but are not currently on the market. These products include a sublingual nicotine tablet (Molander et al., 2000), the oral nicotine transmucosal (Leischow et al., 1996), and the nicotine lozenge (Foulds et al., 1998). Other nicotine replacement products are likely to be developed in the future. It is likely that nicotine replacements with a faster speed of delivery that mimic the effects of cigarettes will be explored, with the hopes that such delivery devices would be safer than nicotine-containing tobacco products. In addition, fast nicotine delivery devices may lead to greater treatment success or provide a bridge toward using slower nicotine absorption products. Future replacement medications are also likely to become more sophisticated, targeting specific nicotinic receptor subtypes that are associated with specific functions.

Many researchers also believe that combinations of different types of nicotine replacements might be more effective than single agents alone (Fiore et al., 2000). Combination products would result in higher levels of nicotine replacement, which may lead to less desire to smoke and less reinforcement from a cigarette when smoked due to the development of tolerance. Interestingly, doses greater than 21 mg of nicotine generally show minimal increases in long-term abstinence rates (Hughes et al., 1999a,b), although the Cochrane Group found that high-dose nicotine patches were marginally more effective than a standard dose (1.2, 95% CI=1.0 to 1.4; N = 6 trials). However, combinations of nicotine patch with nicotine products that can be used ad libitum have resulted in better treatment success. The use of the nicotine patch would result in a steady-state trough level of nicotine to prevent withdrawal symptoms, whereas the ad libitum product could be used during periods when an urge to smoke is experienced. Treatment studies have been conducted that examine a combination of nicotine gum and patch (Kornitzer et al., 1995); nicotine spray and patch (Blondal et al., 1999); and nicotine inhaler and patch (Westman et al., 2000). Results from the meta-analyses conducted with some of these studies showed that a combination treatment approach was more effective than a single treatment approach (OR=1.9, 28.6% vs. 17.4%, respectively) (Fiore et al., 2000). Furthermore, two studies showed that a combination approach led to greater reductions in withdrawal symptoms compared to a single treatment approach (Fagerström, 1994; Fagerström et al., 1993).

Antidepressants

To date, the antidepressants that have been successful in treating smokers are bupropion SR and nortriptyline. Bupropion SR, or Zyban, which is approved by the FDA, is recommended as a first-line pharmacotherapy similar to other nicotine replacement therapies (Fiore et al., 2000). Nortriptyline, which is not approved for this indication by the FDA, is recommended as a second-line treatment for smokers who were unresponsive to the first-line treatment. The mechanism of action of various antidepressants is unknown. Understanding these mechanisms is important in order to refine and develop drugs that are targeted to specific population of smokers or essential neurotransmitter systems. As one mechanism, it is possible that since a higher incidence of depression is observed among smokers than nonsmokers (Breslau et al., 1991; Glassman et al., 1990; Kendler et al., 1993) and smokers with a history of depression are more likely to relapse (e.g., Anda et al., 1990; Covey et al., 1990; Glassman et al., 1990; Hall et al., 1993) and experience depressive symptoms or mood after cessation (Borrelli et al, 1996; Glassman et al., 1990), antidepressants may be effective in enhancing treatment efficacy among this population. However, clinical studies show that antidepressants are effective for smoking cessation in both nondepressed and depressed populations (Fiore et al., 2000). Another mechanism is the use of anti-depressants to reduce withdrawal symptoms. Tobacco withdrawal symp-toms overlap with symptoms associated with major depression—depressed mood, irritability, low energy, and problems with sleep. A third possible mechanism may be that antidepressants and nicotine release similar neuro-transmitters. For example, since bupropion is a weak dopamine uptake inhibitor (Ascher et al., 1995), the efficacy of this product has been attrib-uted to an increase in dopamine levels. Dopamine levels are increased by nicotine and constituents in tobacco smoke, and this increase is thought to be responsible for some of the positive reinforcing effects of cigarette smoking. Bupropion also weakly blocks the neuronal reuptake of noradrenaline (Ascher et al., 1995; Ferris and Cooper, 1993; Perumal et al., 1986), and increased noradrenaline levels may also serve a reinforcing function. Nortriptyline is a tricyclic antidepressant drug that is also known to enhance levels of noradrenaline and to have some serotonergic activity. Interestingly, both medications also decrease firing of the locus ceruleus. The beneficial effects from inhibiting firing in the locus ceruleus may be derived from blocking the pathways of acute nicotine stimulation or re-sembling the desensitized state seen with continuous nicotine exposure (Benowitz and Peng, 2000). Finally, a preliminary study suggests that bupropion may act as a noncompetitive antagonist of nicotinic acetylcho-line receptors, an action that is independent of its antidepressant mecha-nisms of action (Fryar and Lucas, 1999).

Fluoxetine and sertraline target primarily the serotonin system and are serotonin reuptake inhibitors. Antidepressants that target the seroto-nin system do not show enhanced efficacy over placebo, although some preliminary evidence exists that fluoxetine may be effective as a smoking cessation aid in smokers experiencing symptoms of depression at baseline (Niaura et al., 1995). In a very small trial, doxepin, a tricyclic antidepres-sant, reduced the number of cigarettes smoked and increased short-term success compared to placebo (Edwards et al., 1988), but no larger clinical trials with long-term follow-up have been conducted at this time. Other antidepressants, such as moclobemide, a monoamine oxidase (MAO) in-hibitor, have shown equivocal success (Berlin et al., 1995).

Bupropion was approved as a prescription medication for smoking cessation in 1997. At a constant daily dose, it takes about eight days for bupropion blood levels to reach steady state. There is an active metabolite that takes even longer to reach steady state. Consequently, the dosing procedure involves taking bupropion one to two weeks prior to the quit date. Smokers are instructed to take 150 mg per day during the first three days and 300 mg per day thereafter for seven to twelve weeks. Side effects include primarily dry mouth and insomnia.

In general, dose-related effects are observed with bupropion. At one year, point prevalence smoking cessation rates are significantly different between placebo and 150- or 300-mg bupropion per day, but not between placebo and 100 mg per day (Hurt et al., 1997). For the 150-mg and 300-mg doses, the OR was approximately 2.0, with abstinence rates of around 23% at 12-month follow-up versus 12% with placebo (Hurt et al., 1997). Continuous abstinence rates (biochemically confirmed not smoking at each visit) were highest among the 300-mg (24.4%) and 150-mg (18.3%) groups versus placebo (10.5%). In another study, bupropion and bupropion plus nicotine patch were observed to be more effective than placebo or nicotine patch alone, although in this study the efficacy of the patch was unusually poor (Jorenby et al., 1999). Based on the PHS meta-analyses of these two studies, the OR was 2.1 (95% CI=1.5, 3.0) for bupropion compared to placebo, and the estimated abstinence rate was 30.5% versus 17.3% (Fiore et al., 2000). The Cochrane review (Lancaster et al., 2000) also found a similar OR of 2.7 (95% CI=1.90, 3.9) based on the two published studies and two unpublished smaller studies. The efficacy of bupropion is unrelated to a history of major depression (Hayford et al., 1999).

Nortriptyline also involves a dosing procedure that is initiated 10-28 days prior to quitting to achieve steady-state levels. Dose initiation begins at 25 mg per day and escalates to 75-100 mg per day. The duration of treatment in published trials has been about 12 weeks. Two studies have shown that nortriptyline is more effective than placebo (Hall et al., 1998; Prochazka et al., 1998) with an OR of 3.2 (95% CI=1.8, 5.7) for nortriptyline compared to placebo and estimated abstinence rates of 30.1% and 11.7% respectively, (Fiore et al., 2000). These results were similar to those observed in the Cochrane review (OR 2.8, 95% CI=1.6, 5.0; Lancaster et al., 2000). This rate of success again is independent of a history of depression. Although this medication costs less than bupropion, there is concern about side effects, particularly overdose. Side effects include sedation, dry mouth, blurred vision, urinary retention, lightheadedness, and shaky hands.

Other Medications

Antagonists

Other medications to aid cessation have an antagonist effect, that is, they prevent the drug from acting on the receptor site. This antagonist action would reduce the reinforcing effects from smoking and thereby decrease some of the satisfying aspects of smoking and the desire to smoke. Mecamylamine is a nonspecific nicotinic receptor antagonist that was originally used as an antihypertensive agent. Mecamylamine has been shown to block many of the physiological, behavioral, and reinforcing effects of nicotine (Collins et al., 1986; Corrigall and Coen, 1989; Levin and Rose, 1991; Martin et al., 1989; Stolerman, 1986). Mecamylamine also decreases craving for cigarettes and reduces nicotine preference (Rose et al., 1989). Clinical trials have focused on the use of a combination of mecamylamine and the nicotine patch. The rationale behind this antagonist–agonist combination is that both mecamylamine and nicotine from the patch would block the reinforcing effects of nicotine by occupying the nicotinic receptor sites. In addition, the nicotine patch would reduce the experience of withdrawal and minimize adverse side effects from the peripheral ganglionic blockade produced by mecamylamine. One of the major problems with the use of mecamylamine is constipation as a side effect, although at lower doses this side effect is not as much of a problem. An early smoking cessation trial with a small sample size showed that a combination of mecamylamine and the nicotine patch demonstrated greater success than the patch alone (Rose et al., 1994). In a later trial, precessation treatment with mecamylamine with or without the nicotine patch compared to no-mecamylamine conditions (nicotine patch alone or no drug) led to significantly higher continuous abstinence throughout treatment in smokers, who later all received both the nicotine patch and mecamylamine after the quit date (Rose et al., 1998). The six-month continuous abstinence rates were high only in the nicotine-mecamylamine precessation condition compared to pooled data from the other groups. These trials also showed greater reductions in craving, satisfaction from smoking, and smoking rates during the pre-quit period when mecamylamine and the nicotine patch were administered together compared to any drug alone. A subsequent larger clinical trial showed a higher abstinence rate at seven weeks posttreatment with the combination approach compared to the patch alone, but only in females (Rose et al., 1999).

Another type of antagonist treatment that has been examined only in animal models is immunization to produce nicotine-specific antibodies. Such antibodies would reduce drug entry into the central nervous system by binding to circulating nicotine and thereby decreasing the concentration of unbound nicotine. In animal studies, the nicotine vaccine has been found to reduce nicotine in the brain in a dose-dependent manner (Hieda et al., 1997; Pentel et al., 2000); to block the relief of withdrawal symptoms from nicotine administration in rats undergoing abstinence; and to block behavioral (locomotion), physiological (blood pressure), and neurochemical adrenocorticotropic hormone (ACTH) release effects from nicotine administration (Pentel et al., 2000). Studies are under way to examine the effects of this vaccine on nicotine self-administration. This approach is an attractive intervention because of its specificity and lack of direct impact on the central nervous system, although one concern may be attempts to surmount the effects of the vaccine by intensive compensatory smoking.

Nicotine has been shown to release endogenous opioids, which may be responsible for some of the reinforcing effects from smoking (Pomerleau and Pomerleau, 1984; Taylor and Gold, 1990; Watkins et al., 2000). An opioid antagonist may therefore reduce smoking by blocking the endogenous opioid-induced reinforcing effects. The effect of naloxone and naltrexone on smoking behavior has been variable, with some studies showing that naloxone reduces smoking compared to placebo in a laboratory session (Gorelick et al., 1989; Karras and Kane, 1980), while other studies showed no effect of naloxone (Nemeth-Coslett and Griffiths, 1986) or naltrexone (Sutherland et al., 1995) on smoking rate. Naltrexone has been successfully used for the treatment of opioid and alcohol abuse and dependence. Although, earlier studies showed some promise for naltrexone as a smoking cessation aid, long-treatment outcome success has not been enhanced by naltrexone over placebo, even in combination with the nicotine patch (Wong et al., 1999).

Medications That Target Other Systems

Clonidine is another antihypertensive that has been used to promote smoking cessation. Its mechanism of action is likely through stimulation of the α_2 adrenergic autoreceptors in the brain stem, which results in decreased noradrenergic activity and inhibits firing in the locus ceruleus. In a study conducted about 15 years ago, clonidine was observed to alleviate withdrawal symptoms from opiates, alcohol, and cigarettes (Glassman et al., 1984). Because of the observed reductions of nicotine withdrawal symptoms, several clinical trials have been undertaken to determine the effects of clonidine in the treatment of smokers. The PHS performed a meta-analyses on five randomized, placebo-controlled trials with at least five months follow-up post-quit (Fiore et al., 2000). This analysis found that clonidine, administered either orally or transdermally, is effective as a smoking cessation aid, resulting in a twofold increase in cessation compared to placebo (estimated abstinence rates of 25.6% vs. 13.9%, respectively; OR=2.1 [95% CI=1.4, 3.2]). A Cochrane review (Lancaster et al., 2000) of six trials showed evidence of similar efficacy (OR=1.9, 95% CI=1.3, –2.7). When examining individual studies, some studies showed greater efficacy among women than men (Gourlay and Benowitz, 1995). One study found greater treatment effect in women who are heavily dependent and/or who experience recurrent episodes of depression (Glassman et al., 1993). In another study, more dependent smokers (classified with a Fagerström score of ≥7) experienced greater efficacy with higher compared to lower concentrations of clonidine, whereas efficacy was independent of clonidine concentrations in smokers with low dependence scores (Niaura et al., 1996). Because of the proven efficacy of clonidine, the PHS has recommended using clonidine as a second-line pharmacological treatment. However, the main drawback to using clonidine is the profile of side effects, which include dry mouth, drowsiness, dizziness, sedation, and constipation as well as lowering of blood pressure. In addition, rebound hypertension may occur when the medication is discontinued.

Medications other than antidepressants that target the serotonin system, such as buspirone, a partial serotonin (5-hydroxytryptamine) agonist having anxiolytic effects that may also increase firing of dopaminergic and noradrenergic neurons (Benowitz and Peng, 2000), have produced equivocal results with regards to short-term treatment outcome. One study showed positive results with buspirone (West et al., 1991), and two studies showed negative results (Robinson et al., 1992; Schneider et al., 1996). One other study showed enhanced rates of abstinence at the end of treatment, but not at long-term follow-up, for those individuals with high levels of anxiety but not for those with low anxiety levels (Cincirpini et al., 1995). Effects on withdrawal symptoms have also been equivocal, with some studies showing positive effects of this medication in reducing withdrawal symptoms (Gawin et al., 1989; Hilleman et al., 1992) and other studies showing no effect (Cincirpini et al., 1995; Robinson et al., 1992; Schneider et al., 1996; West et al., 1991).

One innovative proposal for treating cigarette smokers is to change the rate of nicotine metabolism. Nicotine is metabolized primarily by the hepatic CYP2A6 (cytochrome P-450) enzyme. Several studies have examined the effects of having normal homozygous CYP2A6*1 (wild-type) alleles compared to inactive or mutated CYP2A6 alleles on nicotine metabolism and smoking behavior in humans. A prior study with a small sample size showed that smokers homozygous for the CYP2A6 deletion (and therefore having impaired enzyme function) showed lower cumulated urinary cotinine excretion compared to individuals who were homozygous CYP2A6*1 (Kitagawa et al., 1999). In another study, tobacco-dependent smokers who are carriers of null or inactive alleles (CYP2A6*2 or CYP2A6*3) were observed to smoke fewer cigarettes per week than smokers with two active CYP2A6 alleles (Pianezza et al., 1998). Results from both of these studies were duplicated by another study showing that smokers with defective alleles (*4 and *2) smoked fewer cigarettes and demonstrated lower expired CO levels than smokers with homozygous wild-type alleles. Cotinine levels were lower in the group with the defective alleles, and the nicotine-cotinine ratios were higher in this group (Tyndale et al., 2000). One clinical laboratory study examined the effects of methoxsalen, a CYP2A6 inhibitor, on cigarette smoking and found that methoxsalen in combination with oral nicotine decreases carbon monoxide exposure, smoking rate, latency between lighting the first and second cigarettes, and number of puffs taken (Sellars et al., 2000) compared to a placebo-placebo condition. No differences in cigarette consumption or CO exposure were observed with oral nicotine or methoxsalen alone. To date, these results are suggestive of the role of nicotine metabolism in smoking behavior.

Summary and Recommendations:

In summary, many different types of medications have demonstrated efficacy: nicotine replacement therapies, antidepressants, and other medications that target the dopaminergic and noradrenergic systems. These medications have also been shown to be safe and to produce minimal dependence and misuse. New antagonist medications, such as the nicotine vaccine or medications that may alter the metabolism of nicotine, are being evaluated for their effort on smoking cessation. Furthermore, with increasing knowledge of the function of various nicotinic receptor subtypes, medications that target the specific receptor subtypes responsible for the reinforcing effects of nicotine are likely to be developed. Unfortunately, although current smoking cessation aids are effective and relatively safe, the use of these medications by smokers to facilitate cessation attempts is not widespread. New prescription rates represent only a 10% share of the approximately 24 million quit attempts made per year (Shiffman et al., 1998), or 15% of the 16 million who attempt to stop smoking cigarettes for at least 24 hours (CDC, 2000). Obstacles to the use of these medications include cost, consumer concern, and misperception regarding the health effects of nicotine and the limitations presented by prescription status (Shiffman et al., 1998).

Medications that are now available over-the-counter (OTC) have had a significant impact on the number of quit attempts. With the introduction of OTC nicotine patch and nicotine gum, the estimated number of pharmacological quit attempts increased from two million to three million in 1993-1995 to six million in 1996 with numbers increasing in 1997 and remaining stable in 1998 (CDC, 2000). Successful quitting has been estimated to increase by 6%-20% when OTC products are made available compared to when only prescription products are available (Lawrence et al., 1998; Shiffman et al., 1998). If significant reduction in harm is to be achieved, easier access to and reduced costs of these products are necessary.

Another area that requires attention is misinformation regarding the safety of nicotine. In a 1996 survey by Ketchum and Harris, the majority of smokers (86%) perceived nicotine as harmful. Of further concern is the fact that despite the proven efficacy of existing pharmacological agents, relapse rates still remain high at around 60-75%, with even higher rates (exceeding 95%) among those who quit on their own. Furthermore, tobacco cessation treatment targets only a small percentage of smokers who want to quit. Among cigarette smokers, at any one time, only 10% are prepared to take action toward quitting (i.e., they intend to quit in the next 30 days; Prochaska and Goldstein, 1991) and only 20% of tobacco users are willing to quit in the next six months (Etter et al., 1997). For individuals who are unwilling or unable to quit, alternative methods of treatments or tobacco-like nicotine delivery devices must be considered and tested. Although total abstinence should be the ultimate goal in treatment, for those who are unwilling or unable to quit, reduction may be an important alternative to further decrease the mortality and morbidity associated with smoking (Henningfield and Slade, 1998; Hughes, 1995). The importance of this alternative strategy is highlighted by the dose-response relationship that has been observed between the amount of tobacco consumption and morbidity and mortality (Jimenez-Ruiz et al., 1998; Thun et al., 1995). Similarly, models have been developed showing that reduced smoking may lead to reduced of risk premature death (Burns, 1997), although longitudinal studies have not been conducted to date to confirm the results from these models. Therefore, methods to reduce tobacco use or long-term use of products to sustain abstinence require serious consideration.

Harm Reduction Indications for Pharmacological Treatments

The use of medications for harm reduction can be considered in various ways. First, medications can be used to reduce the rate of smoking either as a means toward eventual abstinence or as an end goal. Second, medications can also be used in situations in which the smoker cannot smoke cigarettes and chooses to use a medication, most likely nicotine replacement, to abate craving or withdrawal symptoms. This situational use of nicotine replacement may indirectly lead to reduced smoking, as well. Third, harm reduction may also include long-term maintenance on a medication as a relapse prevention aid and not merely a smoking cessation aid.

Several concerns exist when considering these approaches, particularly when advocating for reduced smoking if smokers are unwilling or unable to quit. One concern that has been raised is the potential for exposure to increased amounts of nicotine beyond those that the smoker normally experiences solely from smoking and the adverse effects that may be experienced when combining smoking with a nicotine replacement. This increased nicotine exposure may have associated toxicity, although the risk depends on the toxicity specific to nicotine, which tends to be confined primarily to reproductive and cardiovascular disorders. Prior studies have examined the use of nicotine replacements with concurrent smoking. These studies have found no major adverse effect even at a very high dose of nicotine patch (Benowitz et al., 1998) or in smokers who have experienced cardiovascular disease (Joseph et al., 1996; Murray et al., 1996; Working Group for the Study of Transdermal Nicotine in Patients with Coronary Artery Disease, 1994).

A second concern is the reduced desire for abstinence as a result of a reduced perception of risk associated with decreased levels of smoking. Individuals who would normally have quit may choose continuing to smoke at lower amounts. Similarly, the desire to quit may also be decreased if the use of nicotine products is encouraged in areas that restrict tobacco use, so that the individual no longer needs to contend with withdrawal symptoms in these situations. A third and related concern is the reduced perception of risk among adolescents and younger adults. If the option of smoking a few cigarettes with reduced health consequences is available, then perhaps a greater number will be more willing to initiate smoking. A fourth concern, which is related to the long-term use of medications, is the potential toxicity that may be associated with chronic use, even though the toxicity is lower than that of tobacco products. An important principle that underlies all of these concerns is that no increase in harm occurs as a result of using a tobacco exposure reduction approach and that a significant and meaningful decrease in actual harm be the outcome.

Several advantages related to the availability of exposure reduction approaches include potentially increasing recruitment into treatment. That is, smokers who are not ready to quit can perhaps be persuaded to begin to reduce their tobacco consumption as a step toward cessation or as a method to reduce harm. This reduction may then reduce mortality and morbidity among individuals who want to continue smoking, and may also reduce environmental tobacco smoke exposure, and eventually facilitate abstinence. Furthermore, the use of medications in situations where smoking may not be allowed (e.g., work environment) may reduce work-related accidents, which have been observed to increase during periods of tobacco withdrawal (Waters et al., 1998).

Use of Pharmacological Agents for Tobacco Exposure Reduction

The use of nicotine replacements for tobacco exposure reduction has been suggested to minimize compensatory smoking behavior when reducing the number of cigarettes smoked (Shiffman et al., 1998). In addition, nicotine replacements are likely to induce minimal harm since it is the cigarette constituents and pyrolysis products other than nicotine that are primarily responsible for the morbidity and mortality associated with smoking (Benowitz, 1998; Henningfield, 1995). Use of nicotine to reduce smoking is not a new concept. One study observed that half of the smokers prescribed gum at a nonresearch routine outpatient setting reported using gum to help them cut down rather than quit smoking (Johnson et al., 1992). Furthermore, nicotine replacement treatment studies have shown that smokers continue to smoke at a reduced rate in conjunction with using nicotine products (e.g., Bjornson-Benson et al., 1993; Transdermal Nicotine Study Group, 1991).

The effects of administering nicotine intravenously or orally on reducing cigarette consumption were observed in the late 1960s and 1970s (e.g., Jarvik, 1970; Lucchesi et al., 1967). Subsequent short-term, laboratory research studies have demonstrated that nicotine administered intravenously (e.g., Benowitz and Jacob, 1990), by nicotine gum (e.g., Herning et al., 1985; Nemeth-Coslett and Henningfield, 1986; Nemeth-Coslett et al., 1987; Russell et al., 1976), and by patch (e.g., Foulds et al., 1992; Pickworth et al., 1994) reduced smoking behavior, although modestly. This modest reduction may be due to the insufficient doses of nicotine that were administered, short-term treatment, and/or enrolling subjects who were not interested in reducing the number of cigarettes smoked. In a recent study, Benowitz et al. (1998) examined the use of high-dose transdermal nicotine (TN) on smoking suppression. In a double-blind, crossover design among smokers with no desire to quit, they observed that TN reduced nicotine intake from cigarette smoking by 3, 10, and 40% on average under the 21, 42, and 63-mg/day conditions, respectively. Peak plasma nicotine concentrations were approximately 40, 60, and 70 ng/ml, respectively, versus 28 ng/ml attained with the placebo patch. They concluded by saying that high-dose nicotine has the potential to substantially reduce smoking and thereby harm.

Current studies, which have examined the use of nicotine replacements for the primary purpose of cigarette use reduction, have shown substantial decreases in the number of cigarettes smoked and levels of carbon monoxide (Fagerström et al., 1997; Rennard et al., 1990). Rennard et al. (1990) studied the effects of treating smokers with at least ten pieces of 2-mg gum per day on lower respiratory tract inflammation in heavy smokers. He observed that cigarette consumption decreased by more than one-half (from 51 to 19 cigarettes per day) and carbon monoxide was similarly reduced (from 49 to 27 parts per million [ppm] after a two-month period). Furthermore, lower respiratory tract inflammation in these heavy smokers was significantly improved.

Fagerström et al. (1997) conducted a crossover study in which dependent smokers had a choice of either 2-mg nicotine gum, a nicotine patch, nasal spray, and 2-mg oral tablet or were assigned to a product for two weeks. This study showed that compared to baseline, there were a 54% reduction in smoking, 35% reduction of CO, and 32% reduction of withdrawal symptoms. About 59% reduced smoking by more than 50%, and 5% smoked no cigarettes at the end of the four-week intervention period. The highest mean cotinine concentration attained during treatment was 373 ng/ml compared to 360 ng/ml obtained during the screening period. Fagerström and associates concluded that a "smoking reduction procedure may help the very recalcitrant smoker gain confidence and increase the control of his/her smoking behavior."

Bolliger et al. (2000) examined the effects of the nicotine inhaler in reducing smoking among smokers who were unwilling or unable to stop smoking immediately (N=400) in a double-blind, placebo-controlled four month trial with a two year follow-up. Participants were allowed access to the inhalers for up to 18 months. At the end of the four-month trial, significantly greater numbers of participants assigned to the active group (26%) were able to sustain reduction (reduced smoking by at least 50% from week six) compared to the placebo condition (9%). This significant difference continued to be observed at 24 months (9.5% vs. 3.0%, active vs. placebo). Significant differences in reduction in carbon monoxide were observed between active and placebo inhaler conditions at six weeks (68.4% vs. 84.1% of baseline, respectively) and four months (58.3% versus 71.1% of baseline, respectively). About 10% of the population was abstinent at the two-year follow-up. No serious adverse events related to treatment occurred during the study. The authors concluded that sustained reduction in smoking can be achieved using the nicotine inhaler. Recent studies have examined the effects of short-term use of the nicotine patch on reducing cigarette consumption among psychiatric patients who are not interested in quitting. These studies have observed reductions in cigarette consumption ranging from 20 to 42% (Dalack and Meador-Woodruff, 1999; Hartman et al., 1989, 1991), with the greatest reductions observed among the heaviest smokers (Dalack and Meador-Woodruff, 1999). Reduction approaches with psychiatric populations may be beneficial.

Other Harm Reduction Strategies Associated with Medication Use

Nicotine replacement agents, used on an ad libitum basis, can be used under circumstances in which smoking is not permitted (e.g., restaurants, airplanes, smoke-free workplaces). Nicotine replacements have been shown to reduce craving as well as withdrawal symptoms (Hughes et al., 1989) and therefore may be beneficial in situations where a smoker cannot smoke. Long-term maintenance of medications may also be considered a harm reduction strategy, similar to methadone use among individuals addicted to opiates. Few studies have examined the long-term use of nicotine replacement products to sustain long-term abstinence.

One study was the Lung Health study in which nicotine gum was available up to six months, but approved for longer use if deemed necessary. Nicotine gum was dispensed in the context of 12 weeks of cognitive behavioral smoking cessation treatment. The results showed a five-year sustained smoking cessation rate of 22% in this intervention group versus 5% in the usual care group (Kanner et al., 1999). An ancillary examination of the data focused on long-term gum use in this population (Bjornson-Benson et al., 1993). Among participants enrolled in the study, 28.9% were using gum at 12 months since study entry. About one-third of the sustained nonsmokers, over one-half of intermittent smokers, and one-fifth of continuing smokers were using gum at 12 months. This rate of long-term gum use among nonsmokers is consistent with other studies that showed rates of use around 22% at one year (Hughes, 1998). As an added note, in the Lung Health Study, continuing smokers using nicotine gum were smoking less than continuing smokers not using nicotine gum (12.4 vs. 23.5 cigarettes per day). No adverse effects were observed, although among sustained nonsmokers, continuous gum users reported more mild side effects than intermittent gum users.

Recently, a trial has been conducted examining the use of bupropion for at least one year as a relapse prevention agent (Hays et al., 2000). All smokers enrolled in the trial were assigned to bupropion SR for a period of seven weeks. Those smokers who were abstinent at the end of seven weeks were randomly assigned to bupropion SR or placebo for 48 weeks. Subjects assigned to the bupropion group had greater success in maintaining abstinence compared to those assigned to the placebo at the end of the 78-week follow-up (47.4% vs. 37.7%, respectively).

Summary and Recommendations

In summary, the results show that nicotine replacements are effective in reducing smoking on a short-term as well as a long-term basis in some smokers. The availability of a reduction or controlled smoking approach does not seem to deter individuals from becoming abstinent. Furthermore, high doses of nicotine do not seem to cause acute adverse events even among smokers who have experienced cardiovascular disease. Long-term use of nicotine replacements may also be effective in sustaining abstinence and is less toxic than a relapse to smoking. There is a great need for large and long-term clinical trials to determine whether different pharmacological agents, including products other than nicotine replacements, can lead to prolonged and significant reductions in smoking and less harm to individuals. Included in potential medications to be examined are some antidepressants, antagonists and medications that alter the metabolism of nicotine. Finally, these treatment methods must be considered carefully for special populations of smokers, including adolescents, individuals with comorbid conditions or medically compromised individuals, and pregnant women.

OTHER POTENTIAL HARM REDUCTION METHODS: BEHAVIORAL STRATEGIES AND TOBACCO CONTROL POLICIES

Altering tobacco products and using pharmacological agents to reduce smoking are not the only methods of harm reduction. Behavioral methods and tobacco control policies have also led to reduced smoking. These approaches have to be considered so that the pharmacological approaches aimed at reducing an individual's smoking behavior can be complemented or augmented by behavioral and public policy approaches. Furthermore, changes in tobacco products aimed at reducing toxicity must be marketed only in the context of a comprehensive tobacco control policy whose primary goals are prevention of smoking initiation and total cessation of smoking. The normative belief that any tobacco use is harmful must be maintained (IOM, 2000).

Harm Reduction Using Behavioral Methods

Even as early as the 1970s, researchers observed that a significant "number of habitual smokers reported that they wanted to give up smoking but found it extremely difficult to reduce their rate of smoking or quit entirely" (Shapiro et al., 1971). This observation led to a number of studies using behavioral interventions aimed at reducing smoking. The behavioral means for achieving a reduction in smoking included smoking at fixed intervals and increasing the intervals between cigarettes; smoking a cigarette only when signaled to smoke (e.g., Levinson et al., 1971; Shapiro et al., 1971); changing smoking behavior, such as taking shorter puffs, reducing the number of puffs, and reducing the percentage of the cigarette smoked (e.g., Frederiksen and Simon, 1978; Glasgow et al., 1983); contingency contracting (Frederiksen and Peterson, 1976); and eliminating smoking in specific situations (e.g., Foxx and Axelroth, 1983; Glasgow, 1978). In addition, although they represent more of a pharmacological than a behavioral approach, gradually lowering the nicotine content in cigarettes (e.g., Foxx and Brown, 1979; Prue et al., 1981) and graduated filters (McGovern and Lando, 1991) have also been used as methods for reducing nicotine. All of these methods have shown some degree of success in reducing the number of cigarettes smoked, with concomitant reductions in extent of nicotine exposure. Two studies have compared the effects of a behavioral intervention of cigarette reduction versus a waiting list control. Each study had a goal of reducing cigarettes by 50% of baseline using a variety of behavioral techniques. In one study, although only 60% of the original sample remained in the program, these smokers achieved a median reduction of 75% compared to a 2% reduction in the waiting list control group at the end of the eight-week treatment period (Shapiro et al., 1971). Furthermore, six weeks after the termination of treatment, the median reduction was 43%. In the other study, a combination of treatment techniques, with and without feedback on carbon monoxide levels, was used sequentially over five weeks: changing brands to machine-measured low-tar and nicotine cigarette, reducing the number of cigarettes, and reducing the percentage of the cigarette smoked (Glasgow et al., 1983). Significant treatment effects were observed at the end of treatment when comparing the waiting list control with controlled smoking treatment groups. In general the results showed a mean reduction of 56% in the nicotine content of cigarettes, 28% in the number of cigarettes, and 19% in the percentage of cigarette smoked in the controlled smoking conditions.

Studies have also been conducted to determine long-term maintenance of reduced smoking and the rate of cessation attempts among these individuals. Hughes et al. (1999a) examined whether cigarette smokers can significantly reduce and sustain reduction by analyzing longitudinal data from subjects involved in the Community Intervention Trial for smoking cessation. He observed that at the two-year follow-up, 17% had decreased smoking by 5-25%, 15% by 24-49%, and 8% by 50% or more. Among smokers who reduced their smoking ($\geq$ 5%) at two years, 52% reported maintaining this reduction at four years. In addition, these investigators found that decreased smoking did not predict an increase or decrease in quit attempts or abstinence, indicating that reduction does not seem to promote or deter cessation.

One study indirectly assessed maintenance of reduction and rate of cessation by conducting a three- to four-year follow-up among smokers who had enrolled in a smoking cessation program (Colletti et al., 1982). About one-third of the smokers who were unable to achieve abstinence, but smoked less than or equal to 50% of baseline smoking at posttreatment, maintained this rate at one-year follow-up. This maintenance rate declined to about 13-18% at three- to four-year follow-up. However, 18-20% of the smokers achieved abstinence. Therefore, among those who had reduced smoking at posttreatment, about 33-36% were able to quit or sustain the reduction for three- to four-years. These results would suggest that a lower smoking rate can be sustained and does not necessarily discourage cessation attempts.

Glasgow et al. (1985) directly examined these issues by conducting a 2½-year follow-up in 48 subjects enrolled in a controlled smoking program rather than a cessation study. The results indicated that 9% became abstinent and 9-36% showed some improvement on various reduced smoking behavior parameters from posttest to follow-up. These results would suggest that further tobacco exposure reduction can occur in about a third or more of the population. In addition, the smoking cessation rate is no lower than that observed among a general population of smokers or general practice intervention with smokers. In a later study, Glasgow et al. (1989) explored how an abstinence-based program, in which smoking was not condoned after the quit date, compared to a program in which participants had the option of complete smoking cessation or controlled smoking. No significant differences in smoking cessation rates were observed between the two conditions at either posttest or six months. This result would indicate that allowing controlled smoking among those who want to quit does not necessarily lead to either less interest in abstinence or a promotion of abstinence. In a review article (Hughes, 2000) also concluded that smokers can sustain reductions in smoking, and reductions in smoking do not undermine cessation.

More recent exploration of reduced smoking has used computerized devices to gradually wean smokers from cigarettes as a means to achieve cessation. One computerized program, Lifesign Computer Assisted Smoking Program, involves a scheduled reduction by increasing the interval between cigarettes and informing individuals when to smoke. The scheduled time-interval approach seems the most promising of the behavioral treatment methods based on studies by Cinciripini and colleagues (Cinciripini et al., 1995 and 1997) compared to abrupt cessation or non-scheduled reduction of cigarettes. This behavioral method systematically reduces the level of nicotine exposure, disrupts habitual smoking patterns, and gives smokers the opportunity to develop new behaviors or skills in response to cues associated with smoking.

Summary and Recommendations

In summary, the results from these studies show that smokers can reduce their smoking rate using behavioral methods, that this rate can be sustained over time, and that reduced smoking does not necessarily compromise cessation efforts. However, more systematic studies focused directly on these issues should be conducted. Furthermore, tobacco addiction involves more than a physical addiction to nicotine, but also behavioral and psychological aspects that also need to be targeted in exposure reduction as well as cessation. Rigorous studies combining behavioral and pharmacological methods for reduced smoking have yet to be conducted. For example, the use of pharmacological agents may have to be embedded in effective behavioral treatment methods to maximize tobacco use reduction and maintain this reduction.

Harm Reduction Using Environmental Methods

Although not directly considered a harm reduction approach, comprehensive tobacco control policies have clearly lead to a reduction in the overall consumption of cigarettes, including smoking intensity (e.g., IOM, 2000; Pierce et al., 1998). Increasing taxes or the price of cigarettes has uniformly reduced their overall consumption (IOM, 2000), with a 10% increase in price resulting in a 4% decrease in total consumption of tobacco. Price elasticity, that is, the degree of responsiveness of demand to price change, may be greater among youth than among older adults. Increased cigarette prices lead to significant reductions in the quantity smoked by youthful smokers as well as a reduction in participation in smoking (Chait, 1994; Chaloupka and Wechsler, 1995; Lewit et al., 1981; Tauras and Chaloupka, 1999). Increased taxes on smokeless tobacco also result in a reduction in the amount used as well as the frequency of use (Chaloupka et al., 1996).

Workplace smoking restrictions have also significantly reduced the consumption of cigarettes (IOM, 2000). This reduction includes the intensity of smoking among employees as well as increased quit rates (e.g., Brownson et al., 1997; Evans et al., 1996; Glantz, 1997; Glasgow et al., 1997; Tauras and Chaloupka, 1999). In one study, the number of cigarettes smoked per day was not significantly different between work sites with restrictive versus unrestrictive smoking policy in a cross-sectional analysis. However, in a longitudinal analysis, work sites that initially had unrestrictive smoking policies but changed to restrictive policies showed reduced smoking compared to those that continued unrestrictive smoking policies (Jeffery et al., 1994). A 10% decrease in smoking was observed in this study, which is similar to if not lower than the reduction in number of cigarettes reported in other studies (Evans et al., 1996; Farrelly et al., 1999).

In another study, smokers who worked in places where the smoking ban was total or partial smoked five fewer cigarettes during the work days than leisure days. No differences were observed in the consumption of cigarettes between work days and leisure days in smokers who were employed in places with no smoking bans (Wakefield et al., 1992). These results were found across all occupational status groups. Reduced work day smoking, 18 months after the initiation of a total ban on smoking in the workplace, was found among 32.3% of smokers, and generalization to nonworkdays occurred in some smokers (Borland et al., 1991).

Another study showed that a greater number of daily smokers were light smokers (<15 cigarettes per day) if they worked in a smoke-free environment and if they lived in a home in which there was a partial or total ban on smoking (Farkas et al., 1999). Among young smokers, limits on smoking in schools and restrictions in public places led to a reduced number of cigarettes consumed (Chaloupka and Grossman, 1996; Chaloupka and Wechsler, 1995). Furthermore, schools that have comprehensive policies, including a high emphasis on prevention education, resulted in lower amounts of smoking by smokers than schools that had less comprehensive policies (Pentz et al., 1989).

Other tobacco control policies include tobacco advertisement bans and limiting access of adolescents to tobacco products. While a limited set of tobacco advertising restrictions have no effect on tobacco consumption, comprehensive tobacco advertising bans have reduced tobacco consumption by more than 6% and counter advertising by 2% (Saffer and Chaloupka, 1999). Whether these bans have any effect on the number of cigarettes smoked is unclear. Limits on the availability of tobacco products to underage youths have no impact on college students (Chaloupka and Wechsler, 1995) and adolescents or youth, which may be a function of poor enforcement of these restrictions (Chaloupka and Grossman, 1996; Rigotti et al., 1997). Other studies show that enforcement of youth access restrictions does reduce tobacco consumption (IOM, 2000). Strong limits on youth access to smokeless tobacco, however, have been observed to reduce the frequency of use of this product (Chaloupka and Grossman, 1996).

Summary and Recommendations

In summary, individual harm reduction strategies must occur in the context of public policy of tobacco control if a significant reduction in death and disease is the primary goal. These policies would set the normative standard that tobacco use is highly discouraged. Other issues that require careful consideration are the cost and availability of products. The costs of pharmacological agents that have less associated toxicity are much higher per unit of purchase than those of the more highly toxic tobacco products. Furthermore, tobacco products are more readily available at a number of outlets than pharmaceutical products. The impact of availability, even within the area of pharmaceuticals, is highlighted by a study showing increased quit attempts among smokers when the nicotine patch was released and when nicotine gum and the patch went over-the-counter (CDC, 2000). The rate of these quit attempts was sustained with the advent of OTC products rather than the introduction of new products. Therefore, if a significant impact is to be made on the negative consequences associated with tobacco use, then the safer nontobacco products must be made more available to consumers who are already addicted to nicotine, whereas the more toxic tobacco products must be made less available.

General Conclusive Statements

Harm reduction is not a new concept, but it is a controversial one, in part because of the previous history with low-tar and nicotine cigarettes. Evidence exists showing that "light" cigarettes may have lead to compensatory smoking behavior and therefore no reduction in harm. Furthermore, many smokers of these light cigarettes believed that they were reducing harm, and this perception may have undermined cessation attempts (Cohen, 1996; Kozlowski et al., 1998). These observations led to recommendations for principles that should be followed to determine the feasibility and effectiveness of tobacco exposure reduction approaches (Henningfield and Slade, 1998). Some of the these principles that should be considered include the following: (1) reduction of exposure to toxins with verification based on biomarkers; (2) no reduction in cessation attempts; and (3) no increased safety risk.

Various methods to reduce harm have been proposed. These include changing tobacco products themselves by adding filters, reducing tar and nicotine, via ventilation, or maintaining nicotine but reducing tar (e.g., reducing tobacco nitrosamines, controlled tobacco burning). Other approaches include reducing tobacco consumption, by use of either pharmacotherapies, behavioral strategies, or polices that restrict access to tobacco products. In addition, long-term use of pharmacotherapies to substitute for tobacco use has also been advocated. Examination of harm reduction products and approaches will present a number of issues such as (1) the extent to which reduction of exposure to tobacco toxins must occur before beneficial effects are observed and treatment can be considered a success, that is, how to define a successful treatment outcome; (2) the length of the follow-up or type of surveillance necessary to monitor for any adverse effects; (3) valid and reliable indices for reduction of tobacco toxins; (4) methods to market and position this approach without compromising the message of abstinence as the primary goal and without increasing the initiation of using tobacco products; (5) the cost-effectiveness of this approach; (6) the willingness of health care providers to advise and assist patients who are reluctant to quit in using this technique if reduction in risk is observed as a result of reduction in toxic tobacco exposure; (7) whether nicotine-containing tobacco products should ever be touted by the tobacco industry as a way to quit smoking or reduce its health consequences; and (8) the industry's positioning of pharmaceutical products and tobacco-containing products with less toxins, especially when the differences between them tend to become blurred. In clinical trials, the issue may arise as to how to identify individuals who are unwilling or unable to quit, that is, the criteria to be used to make this determination as well as how and when to intervene with smokers, who use PREPs, to help them eventually achieve abstinence. Finally, a number of regulatory issues will have to be addressed.

REGULATION OF EXPOSURE REDUCTION PRODUCTS

Drugs

The regulatory system in the United States for therapeutic drugs, administered by the Food and Drug Administration, is the most stringent regulatory system in our society for new products. The scientific, legal, and administrative features of this system have been described in many publications, but a particularly good review for the purposes of this report is that of Page (Page, 1998).

The Food, Drug and Cosmetic Act requires affirmative approval of all new drugs by the FDA before marketing. The scientific information required to support such approval includes proof of the identity and structure of the active ingredient(s); detailed information on the composition of the formulation (e.g., tablet, capsule, solution); reports of toxicology studies in animals, including carcinogenicity and reproductive toxicology when necessary; clinical pharmacology studies in humans to show the pharmacokinetic (blood-level) profile and potential for interactions with other drugs; and most importantly, at least two controlled clinical trials in humans demonstrating the effectiveness of the drug for the claimed indication and an acceptable side-effect profile (21 CFR 314). This information is summarized in the product labeling for the physician in a leaflet popularly known as a package insert. This labeling also serves as the basis for regulating promotion and advertising after marketing.

In addition, all clinical studies in humans sponsored by drug manufacturers are subject to regulatory oversight under the Investigational New Drug (IND) regulations (21 CFR 312). This oversight includes review of each protocol by the FDA and by an institutional review board, submission of periodic reports, and prompt submission of serious adverse events that occur during clinical studies.

After marketing, the manufacturer must continue to submit to the FDA reports on new and unexpected adverse events, changes in manufacturing or formulation, changes in labeling (e.g., new warnings), and new advertising materials. Manufacturing plants are also subject to periodic inspection to ensure compliance with good manufacturing practices.

This regulatory system serves to promote public trust in the quality, effectiveness, and truthful labeling of medicinal products and also makes the pharmaceutical and biotechnology industries among the most heavily regulated businesses in our society.

Essentially all new drugs are first approved as prescription drugs. In time, however, some may be switched to over-the-counter status and be sold directly to consumers. OTC drugs are subject to the same regulatory requirements as prescription drugs except that the regulation of their advertising is under the authority of the FTC rather than the FDA. The tests for determining whether a drug can be sold OTC are (1) whether it can be labeled for use by the consumer without the need for a physician and (2) whether it is safe and effective for OTC use. The FDA has historically limited the use of OTC products to symptomatic conditions such as colds, heartburn, and headaches that can be diagnosed without the need for a physician. Furthermore, to promote safe use, the FDA has typically approved for OTC use only drugs of low inherent risk such as antacids and sunscreens or, in the case of drugs that are potentially toxic such as nonsteroidal anti-inflammatory agents and antihistaminics, lower doses than are available by prescription. Drugs with sufficient abuse potential to be scheduled under the Controlled Substances Act cannot be offered OTC.

The only nicotine-containing products currently approved by the FDA for OTC use are Nicorette gum, Nicoderm CQ patch, and Nicotrol patch. The FDA decision to permit these products to switch from prescription to OTC status required discussion at two meetings of the Non-Prescription Drug Advisory Committee before action was taken. A later decision to permit mint-flavored Nicorette gum also required considerable time and discussion. The concerns raised in these discussions were whether these products would actually be effective in an OTC setting without accompanying professional intervention and whether increased abuse and/or cardiovascular risk might develop. Subsequent experience has been reassuring on all counts (Hughes et al., 1999a).

Nicotine-containing drugs are currently approved by the FDA only to reduce withdrawal symptoms as an aid in smoking cessation. Their labeling clearly states that the goal of treatment is cessation of smoking and subsequent withdrawal from the nicotine-containing tobacco product. The labeling of the prescription products advises against chronic use beyond six months and over-the-counter labeling advises against long-term use while continuing to smoke or use other nicotine-containing products. Although use as part of a comprehensive behavioral smoking cessation program is encouraged, there is no information in the labeling of nicotine-containing products about their effectiveness in combination with other programs or Zyban (bupropion SR). In contrast, the labeling for Zyban the only other drug approved as an aid for smoking cessation, notes an additive effect in combination with the nicotine patch and advises that Zyban may be continued indefinitely in successfully treated patients. Long-term use of these products with cigarettes, which might occur in a harm reduction strategy, is discouraged by the approved labeling.

The other prescription drugs reported to be useful for smoking cessation—clonidine, nortriptyline, and mecamylamine—are not approved by the FDA for this indication. That does not limit the ability of physicians to prescribe them for this use, but it does prohibit manufacturers from promoting them for this use.

Medical Devices

The modern regulatory framework for medical devices derives from the Medical Device Amendments of 1976 to the Food, Drug and Cosmetic Act. This legislation defines a medical device as an "instrument, contrivance or similar article intended to affect the structure or any function of the body" and requires the FDA to classify all medical devices according to their degree of risk (Page, 1998). Class I devices are those of low risk such as crutches and bandages that need meet only general standards. Class II devices are envisioned in the law as devices of intermediate risk that need specific performance standards to ensure their safety and effectiveness. Because performance standards can be established only by regulation and the process is time-consuming and burdensome, this provision of the law has seldom been used. Class III devices are those of highest risk, such as heart valves and pacemakers, and require preclearance by the FDA before marketing.

The device laws have never been applied by the FDA to any therapeutic product intended for smoking cessation. Although the containers for the Nicotrol Inhaler and Nicotrol Nasal Spray may look like devices, they are regulated by FDA as the packaging for a drug and not as medical devices.

In 1996, the FDA claimed regulatory authority over cigarettes and smokeless tobacco products on the grounds that a cigarette is a medical device intended to deliver the drug nicotine. This resulted in extensive litigation between the FDA and the tobacco industry, the details of which are included in the next section.

Tobacco Products

The effort to regulate tobacco has a long history (Jacobson and Wasserman, 1997; Kluger, 1997). By the beginning of the twentieth century, there was an important antitobacco movement in the United States based on the conviction that tobacco use was immoral, uncouth, and corrupting. Many states passed laws prohibiting the production, sale, or use of cigarettes. Smoking, especially by women, was discouraged, and female smoking was equated with low moral character. A few states also banned the sale of tobacco to minors in an effort to combat the "demoralizing" effect of tobacco on children. These state laws, however, were poorly enforced and were eventually overturned as smoking became much more popular during the Great Depression and World War II.

Beginning in the 1930s, the scientific community began linking smoking directly to disease. This evidence mounted in the 1940s and 1950s, culminating in 1964 in the landmark Surgeon General's report (PHS, 1964) outlining the adverse health effects of smoking in terms of cancer, heart disease, and lung disease. Later Surgeon General's reports considered such topics as the adverse health effects of environmental tobacco smoke (U.S. DHHS, 1986), the problem of adolescent smoking (U.S. DHHS, 1994), and tobacco use by minorities (U.S. DHHS, 1998). In response to the 1964 Surgeon General's report, Congress passed the Cigarette Labeling and Advertising Act of 1965, which forced cigarette manufacturers to place the warning, "Caution: Cigarette smoking may be hazardous to your health," on all packaging. This act, however, prevented the states and the FTC from enacting their own rules requiring more explicit warnings on packaging.

In 1969, Congress enacted the Public Health Cigarette Smoking Act, which banned all cigarette advertising on television and radio and modified the warning labels to read, "Warning: The Surgeon General Has Determined That Cigarette Smoking Is Dangerous to Your Health". This law negated a decision by the Federal Communications Commission that would have required, under the "Fairness Doctrine", stations broadcasting tobacco ads to provide equal air time to antitobacco public messages because it also prohibited broadcast advertising of cigarettes after January 1, 1971. In 1972, the FTC began requiring cigarette manufacturers to display a warning label on all cigarette advertising. Further changes in the health warnings came in 1984 and 1986 through the Comprehensive Smoking Health Education Act and the Comprehensive Smokeless Tobacco Health Education Act, respectively, which required the rotation of four specific warnings on cigarette and three rotating warnings on smokeless tobacco packaging and advertising (Jacobson and Wasserman, 1997)

Comprehensive federal tobacco regulation has been limited by the exemption of tobacco from numerous federal acts designed to protect the public from harmful products, including the Controlled Substances Act of 1970, the Consumer Product Safety Act of 1972, and the Toxic Substances Control Act of 1970. The Food, Drug and Cosmetic Act (FDCA), however, is silent with respect to tobacco products and thus stands as a potential regulatory tool for any product, tobacco containing or not, that makes an explicit health claim and also meets the definition of a drug or device. In the 1950s the FDA exerted jurisdiction in two cases in which tobacco companies made explicit claims regarding the health benefits of their products. In 1953, the FDA classified Fairfax cigarettes as drugs when the manufacturer enclosed leaflets with language that implied effectiveness in preventing an array of diseases, and in 1959 (*United States v. 354 Bulk Cartons Trim Reducing-Aid Cigarettes*), the FDA successfully prohibited explicit claims of weight reduction by a cigarette. During the late 1970s, however, when Action on Smoking and Health (ASH) and others petitioned the FDA to assert jurisdiction over cigarettes as drugs and devices, the FDA denied the petition on the grounds that cigarettes did not fall under the statutory definition of a drug. The FDA asserted that evidence of consumer intent to use the product predominantly for the health effects or the effects on the structure or function of the body was not sufficient to infer a similar intent by the manufacturers. Tobacco manufacturers appeared to be free from more comprehensive regulation as long as they did not make explicit claims about health benefits or effects on body structure or function and if they complied with advertising and labeling restrictions enacted by Congress (Slade and Ballin, 1993; U.S. DHHS, 2000).

In 1988, the Coalition on Smoking or Health (CSH) and others petitioned the FDA to classify low-tar and nicotine products as drugs and to classify the new smokeless cigarette product by RJR, "Premier", as an alternative nicotine delivery system and, hence, subject to regulation as a drug. CSH cited indirect claims made through advertising and marketing as evidence of the manufacturer's intent to have the product used for the mitigation or prevention of disease (Slade and Ballin, 1993). Again in 1994, the FDA was petitioned by CSH to classify all cigarette products as drugs as defined in the FDCA. Later that year, the FDA Commissioner announced in a letter to CSH and later in congressional testimony that the FDA, in light of new evidence, would revisit the FDA's authority to regulate tobacco products as drugs and devices as defined in the statute.

Following this investigation, the FDA asserted its jurisdiction and proposed regulation of certain tobacco products in the *Federal Register* in August 1995. The authority for such regulation was based on new evidence showing that cigarettes and smokeless tobacco products are nicotine-containing (i.e., drug-containing) devices as defined by the FDCA of 1938. The FDA determined that nicotine could be classified as drug based on the facts that (1) nicotine causes and sustains addiction, (2) nicotine produces other psychoactive (mood-altering) effects, and (3) nicotine controls weight. The definition of nicotine as a drug as defined by the FDCA includes an intent by the manufacturer for the product to be used as a drug in the bodies of their customers.

The FDA's assertion that cigarettes and smokeless tobacco products may be defined as nicotine delivery devices was based on findings that (1) the effects of nicotine are so widely known that it is foreseeable to a reasonable manufacturer that these products will cause addiction and other pharmacological effects and will be used by the consumers for these effects and to sustain the addiction; (2) consumers use tobacco products mainly to sustain addiction, for the mood-altering effects, and for weight loss; (3) manufacturers of tobacco products know that nicotine has pharmacological effects and that consumers use their products primarily to obtain the pharmacological effects of nicotine; (4) manufacturers design their products to provide consumers with a pharmacologically active dose of nicotine; and (5) as a consequence, consumers keep using cigarettes and smokeless tobacco to sustain their addiction to nicotine (U.S. DHHS, 1995). The agency disclosed new evidence from industry documents of product engineering, nicotine delivery manipulation, and industry research in support of its contention that tobacco products are intended to change the structure or function of the body. This provided the rationale for the FDA's proposed new rules on the advertising, marketing, and sale of tobacco. Many of the proposed actions were directed toward limiting the access of minors to tobacco products and stopping cigarette advertising and promotion targeted at adolescents.

Specifically, the new regulations imposed a ban on the sale of tobacco products to minors; required vendors to check for proof of age; banned cigarette vending machines, banned billboard or other advertisements easily accessible to youth; restricted all advertising to black and white text (except in publications read primarily by adults); banned tobacco manufacturer sponsorship of sporting and entertainment events; banned promotional items displaying a brand name, logo, or free samples; and required tobacco industry financial support for antitobacco education for children (U.S. DHHS, 2000).

On August 28, 1996, after soliciting public comment, the FDA published its final rule, modified only in that adult-only businesses were exempted from certain restrictions. The FDA was met by legal action from the tobacco industry, advertising industry, and tobacco vendors to block implementation of these rules. The case was initially heard by the federal district court in Greensboro, North Carolina (April 1997), which upheld the FDA's regulatory authority over tobacco products and supported the FDA definition of tobacco products as combination drug and drug delivery devices. The court, however, ruled that the FDA had no statutory authority to regulate tobacco advertising or promotion.

The decision was appealed by both sides in August 1997, and in August 1998, the Fourth Circuit overturned the district court decision and revoked FDA's proposed authority to regulate tobacco products. The court found that if cigarettes and smokeless tobacco were under FDA jurisdiction as outlined by the proposed regulations, the agency's only choice would be to ban the products in light of their known dangers to health. Any other consideration would not be within the scope of FDA's regulatory powers. The court concluded that Congress did not intend the FDCA to be used for the regulation of tobacco products and that Congress has never equipped the FDA with the power to regulate tobacco products. This decision was upheld by the Supreme Court on March 21, 2000. The majority opinion stated that the FDA's regulatory actions were incongruous with what was intended by Congress and that Congress has historically denied FDA the authority to regulate traditional tobacco products.

It is important to recognize that this recent Supreme Court decision in no way limited the authority of FDA to regulate any product, tobacco containing or not, that makes an explicit health claim that would bring the product under the definition of a drug or device. For example, an exposure-reducing claim for a smokeless tobacco product, to the effect that it (like nicotine patches) promotes cessation of smoking, could be judged by the FDA as a drug claim requiring approval under a new drug application. Similarly, the newly emerging set of smoked nicotine-containing products is not necessarily excluded from FDA regulatory authority by the recent Supreme Court decision. An interesting consequence of the current regulatory situation is that a tobacco manufacturer may not be able to make a legitimate exposure-reducing or reduced-risk claim for a new product, even if truthful, without bringing the product under the jurisdiction of the FDA.

It is also important to note that as a result of the Supreme Court's decision overturning FDA jurisdiction, all current regulatory provisions over tobacco relate only to labeling, promotion and advertising, and taxation. None relate to the technical or scientific standards required of new products or to product safety. Unlike pharmaceutical or device manufacturers, cigarette manufacturers may introduce new curing, blending, and manufacturing techniques into tobacco products without regulatory oversight of any kind. Similarly, new additives, new filters, new aeration mechanisms, new papers, and new constituents may be introduced without regulatory scrutiny. Neither the extent nor the results of toxicology testing of new ingredients in animals are known outside the manufacturer. The effects of new product design and of changes in constituents on the composition of inhaled smoke are not reported to any health authority. Clinical studies on new products are conducted without regulatory oversight over protocols (except for institutional review board review) or review of the results. Once research and development on a new product has been completed, the product is marketed based on the manufacturer's responsibility, again without regulatory review. Manufacturers are under no regulatory obligation to conduct postmarketing epidemiological studies or to collect and report adverse events.

The contrast between the regulatory systems for drugs or devices and for tobacco has been discussed by a number of authors (Henningfield, and Slade, 1998; O'Reilly, 1989; Page, 1998; Slade and Henningfield, 1998; Warner et al., 1997), who point out the paradox of a stringent regulatory system for exposure reduction products developed by the pharmaceutical industry and a weak regulatory system for exposure reduction products developed by tobacco manufacturers. Table 4-8 illustrates the problem.

TABLE 4-8   Comparison of Two Nicotine Inhalers

Feature	Eclipse (tobacco company)	Nicotrol Inhaler (pharmaceutical company)
Operation	Heat source volatizes nicotine and glycerol, and scorches tobacco	Ambient air passing through nicotine reservoir volatizes nicotine
Dose	Mimics cigarettes (lung delivery of nicotine)	Similar to low Nicorette dose (buccal delivery of nicotine)
Projected abuse liability	High	Low
Contaminants	High CO, acrolein, "soot," and other contaminants	Not allowed
Claims or indications	Reduced delivery (unproven to FDA)	Smoking cessation (FDA-approved studies)
Intent	Cause and sustain dependence	Treat dependence
Cost	More than $3.00 per pack of 20 ($0.15 each)	$55.00 per pack of 42 ($1.30 each)
Modification oversight	Modified to be more palatable (and more toxic) without approval	Any modification requires FDA approval
Premarketing approval data	None submitted to FDA	Conventional new drug application submission and FDA approval
Availability	Over-the-counter	Prescription only

SOURCE: Slade and Henningfield, 1998. Reprinted with permission of the Food and Drug Law Institute.

Chapter 7 of this report discusses the principles of a science-based regulatory system for tobacco products, including those with exposure reduction or risk reduction claims.

REFERENCES

Ahlborn A, Olsson UA, Pershagen G. 1997. *Health Hazards of Moist Snuff.* September, 1996; Sweeden. Sweeden: Socialstyrelsen.

Anda RF, Williamson DF, Escobedo LG, Mast EE, Giovino GA, Remington PL. 1990. Depression and the dynamics of smoking. A national perspective. *JAMA* 264(12):1541-1545.

Ascher JA, Cole JO, Colin JN, et al. 1995. Bupropion: a review of its mechanism of antidepressant activity. *J Clin Psychiatry* 56(9):395-401.

ASH (Action on Smoking and Health). 1997. Everything for everybody concerned about smoking and protecting the rights of nonsmokers. [Online]. Available: http://ash.org/oct 97/10-23-97-3.html [accessed November 24, 1999].

Benowitz N. 1988. Drug therapy: pharmacologic aspects of cigarette smoking and nicotine addiciton. *The New England Journal of Medicine* 319(20):1318-1330.

Benowitz N. 1998. *Nicotine Safety and Toxicity.* New York: Oxford University Press.

Benowitz NL, Jacob P. 1990. Intravenous nicotine replacement suppresses nicotine intake from cigarette smoking. *J Pharmacol Exp Ther* 254(3):1000-1005.

Benowitz NL, Peng M. 2000. Non-nicotine pharmacotherapy for smoking cessation. *CNS Drugs* 13(4):265-285.

Benowitz NL, Porchet H, Sheiner L, Jacob P. 1988. Nicotine absorption and cardiovascular effects with smokeless tobacco use: comparison with cigarettes and nicotine gum. *Clin Pharmacol Ther* 44(1):23-28.

Benowitz NL, Zevin S, Jacob P. 1998. Suppression of nicotine intake during ad libitum cigarette smoking by high-dose transdermal nicotine. *J Pharmacol Exp Ther* 287(3):958-962.

Berlin I, Siad S, Spreux-Varoquaux O, et al. 1995. A reversible monoamine oxidase A inhibitor (moclobemide) facilitates smoking cessation and abstinence in heavy, dependent smokers. *Clinical Pharmacology and Therapeutics* 58:444-452.

Bjornson-Benson W, Nides M, Dolce J, et al. 1993. Nicotine gum use in the first year of the Lung Health Study. *Addict Behav* 18(4):491-502.

Blackwell J. 2000. (September 11). Star Scientific is affecting the way tobacco is grown and sold in region. [Online]. Available: http://www.starscientific.com/main_pages/release1sep11.html [accessed September 25, 2000].

Blondal T, Gudmundsson LJ, Olafsdottir I, Gustavsson G, Westin A. 1999. Nicotine nasal spray with nicotine patch for smoking cessation: randomised trial with six year follow up. *BMJ* 318(7179):285-288.

Bolliger CT, Zellweger JP, Danielsson T, et al. 2000. Smoking reduction with oral nicotine inhalers: double blind, randomised clinical trial of efficacy and safety. *BMJ* 321(7257):329-333.

Borland R, Owen N, Hocking B. 1991. Changes in smoking behaviour after a total workplace smoking ban. *Aust J Public Health* 15(2):130-134.

Borrelli B, Niaura R, Keuthen NJ, et al. 1996. Development of major depressive disorder during smoking-cessation treatment. *J Clin Psychiatry* 57(11):534-538.

Breslau N, Kilbey M, Andreski P. 1991. Nicotine dependence, major depression, and anxiety in young adults. *Arch Gen Psychiatry* 48(12):1069-1074.

Browne CL. 1990. *The Design of Cigarettes.* Charolotte, North Carolina: Hoechst Celanese Corporation.

Brownson RC, Eriksen MP, Davis RM, Warner KE. 1997. Environmental tobacco smoke: health effects and policies to reduce exposure. *Annu Rev Public Health* 18(12):163-185.

Buchhalter AR, Eissenberg T. 2000. Preliminary evaluation of a novel smoking system: Effects on subjective and physiological measures and on smoking behavior. *Nicotine and Tobacco Research* 2:39-43.

Burns D. 1997. Estimating the benefits of a risk reduction strategy. Society for Research on Nicotine and Tobacco 3rd Annual Scientific Conference: Nashville.

Byrd GD, Robinson JH, Caldwell WS, deBethizy JD. 1995. Comparison of measured and FTC-predicted nicotine uptake in smokers. *Psychopharmacology* 122:95-103.

Carpenter CL, Jarvik ME, Morgenstern H, McCarthy WJ, London SJ. 1999. Mentholated cigarette smoking and lung-cancer risk. *Ann Epidemiol* 9(2):114-20.

CDC (Centers for Disease Control). 2000. Use of FDA-approved pharmacologic treatments for tobaco dependence in the United States: 1984-1998. *Morbidity and Mortality Weekly Report* 49(29):665-668.

CFR 312, Code of Federal Regulations, Title 21, Part 213, Investigational New Drug Application.

CFR 314, Code of Federal Regulations, Title 21, Part 314, Applications for FDA Approval to Market a New Drug.

Chait LD. 1994. Reinforcing and subjective effects of methylphenidate in humans. *Behavioral Pharmacology* 5:281-288.

Chaloupka, F and Grossman, M. 1996. Price, tobacco control policies, and youth smoking (NBER Working Paper W5740). [Online]. Available: http://www.papers.nber.org/papers/W5740 [accessed 2001].

Chaloupka, F and Wechsler, H. 1995. Price, tobacco control policies and smoking among young adults (NBER working paper W5012). [Online]. Available: http://www.nbersw.nber.org/papers/W5012 [accessed 2001].

Cigarette-fire bill passes after Pataki agrees to sign. *The New York Times.* June 15, 2000. (New York/Region);B1-24.

Cincirpini PM, Lapitsky L, Seay S, Wallfisch A, Kitchens KVVH. 1995. The effects of smoking schedules on cessation outcome: Can we improve on common methods of gradual and abrupt nicotine withdrawal? *Journal of Counsulting and Clinical Psychology* 63(3):388-399.

Cinciripini PM, Wetter DW, McClure JB. 1997. Scheduled reduced smoking: effects on smoking abstinence and potential mechanisms of action. *Addict Behav* 22(6):759-767.

Clark PI, Gautam S, Gerson LW. 1996. Effect of menthol cigarettes on biochemical markers of smoke exposure among black and white smokers. *Chest* 110(5):1194-1198.

Cohen JB. 1996. Smokers' knowledge and understanding of advertised tar numbers: health policy implications. *Am J Public Health* 86(1):18-24.

Colletti G, Supnick JA, Rizzo AA. 1982. Long-term follow-up (3-4 years) of treatment for smoking reduction. *Addict Behav* 7(4):429-433.

Collins AC, Evans CB, Miner LL, Marks MJ. 1986. Mecamylamine blockade of nicotine responses: evidence for two brain nicotinic receptors. *Pharmacol Biochem Behav* 24(6):1767-1773.

Corrigall WA, Coen KM. 1989. Nicotine maintains robust self-administration in rats on a limited-access schedule. *Psychopharmacology (Berl)* 99(4):473-478.

Covey LS, Glassman AH, Stetner F. 1990. Depression and depressive symptoms in smoking cessation. *Compr Psychiatry* 31(4):350-354.

Dalack GW, Meador-Woodruff JH. 1999. Acute feasibility and safety of a smoking reduction strategy for smokers with schizophrenia. *Nicotine Tob Res* 1(1):53-57.

Davis DL, Nielsen MT. 1999. *Tobacco; Production, Chemisty and Technology.* Osney Mead, Oxford: Blackwell Science.

deWit H, Zacny J. 1995. Abuse potentials of nicotine replacement therapies. *CNS Drugs* 4(6):456-468.

Eclipse Expert Panel. 2000. A safer cigarette? A comparative study. A consensus report. *Inhalation Toxicology* 12(Suplement 5):1-48.

Edwards NB, Simmons RC, Rosenthal TL, Hoon PW, Downs JM. 1988. Doxepin in the treatment of nicotine withdrawal. *Psychosomatics* 29(2):203-206.

Etter JF, Perneger TV, Ronchi A. 1997. Distributions of smokers by stage: international comparison and association with smoking prevalence. *Prev Med* 26(4):580-585.

Evans W, Farrelly M, Montgomery E. 1996. Do workplace smoking bans reduce smoking? (NBER Working Paper W5567). [Online]. Available: http://www.nbersw.nbber.org/papers/W5567 [accessed 2001].

Fagerström K. 1994. Combined use of nicotine replacement products. *Health Values* 18:15-20.

Fagerström K. 2000. Nicotine-replacement therapies. Ferrence R, Slade J, Room R, Pope M, eds. *Nicotine and Public Health.* Washington, DC: American Public Health Association.

Fagerström KO, Hughes JR, Rasmussen T, Callas PW. 2000. Randomised trial investigating effect of a novel nicotine delivery device (Eclipse) and a nicotine oral inhaler on smoking behaviour, nicotine and carbon monoxide exposure, and motivation to quit. *Tob Control* 9(3):327-333.

Fagerström KO, Schneider NG, Lunell E. 1993. Effectiveness of nicotine patch and nicotine gum as individual versus combined treatments for tobacco withdrawal symptoms. *Psychopharmacology (Berl)* 111(3):271-277.

Fagerström KO, Tejding R, Westin A, Lunell E. 1997. Aiding reduction of smoking with nicotine replacement medications: hope for the recalcitrant smoker? *Tob Control* 6(4):311-316.

Fairclough G. 2000. State disputes reynolds claims on new cigarette. *The Wall Street Journal.* October 4, 2000: (B);B9A.

Fant RV, Owen LL, Henningfield JE. 1999. Nicotine replacement therapy. *Prim Care* 26(3):633-652.

Farkas AJ, Gilpin EA, Distefan JM, Pierce JP. 1999. The effects of household and workplace smoking restrictions on quitting behaviours. *Tob Control* 8(3):261-265.

Farrelly MC, Evans WN, Sfekas AE. 1999. The impact of workplace smoking bans: results from a national survey. *Tob Control* 8(3):272-277.

Feder BJ. 1996. A safer smoke or just another smokescreen? *The New York Times.* April 12, 1996: (D);D1.

Ferno O. 1973. A substitute for tobacco smoking. *Psychopharmacologia* 31(3):201-4.

Ferrence R, Slade J, Room R, Pope M. 2000. *Nicotine and Public Health.* Washington, DC: American Public Health Association.

Ferris R, Cooper B. 1993. Mechanism of antidepressant activity of bupropion. *Journal of Clinical Psychiatry Monographs* 11(2-14).

Ferry L, Robbins A, Scariati P, et al. 1992. Enhancement of smoking cessation using the antidepressant bupropion hydrochloride. *Circulation* 86(1):647.

Fielding JE, Husten CG, Eriksen MP. 1998. Tobacco: health effects and control. Wallace RB, Doebbeling BN. *Public Health & Preventive Medicine.* 14th ed. Samford, CT: Appleton & Lange. Pp. 817-846.

Fiore M, Baily W, Cohen S, et al. 2000. *Treating Tobacco Use and Dependence: Clinical Practice Guideline.* Rockville, MD: U.S. Department of Health and Human Services. Public Health Service.

Fisher B. 2000. Reducing risk: a new filter may reduce up to 90 percent of tobacco smoke's tar and carcinogenic compounds. *Tobacco Reporter*:58-61.

Foulds J, Russell MA, Jarvis MJ, Feyerabend C. 1998. Nicotine absorption and dependence in unlicensed lozenges available over the counter. *Addiction* 93(9):1427-1431.

Foulds J, Stapleton J, Feyerabend C, Vesey C, Jarvis M, Russell MA. 1992. Effect of transdermal nicotine patches on cigarette smoking: a double blind crossover study. *Psychopharmacology (Berl)* 106(3):421-427.

Foxx RM, Axelroth E. 1983. Nicotine fading, self-monitoring and cigarette fading to produce cigarette abstinence or controlled smoking. *Behav Res Ther* 21(1):17-27.

Foxx RM, Brown RA. 1979. Nicotine fading and self-monitoring for cigarette abstinence or controlled smoking. *J Appl Behav Anal* 12(1):111-125.

Frederiksen L, Peterson G. 1976. Controlled smoking: development and maintenance. *Addictive Behaviors* 1(193-196).

Fredriksen L, Simon S. 1978. Modification of smoking topography: a preliminary analysis. *Behavior Therapy* 9:946-949.

Fryer JD, Lukas RJ. 1999. Noncompetive functional inhibition of diverse, human nicotinic acetylchole receptor subtypes by bupropion, phencyclidine and ibogaine. *J Pharmacolo Exp Ter* 298:88-92.

FTC (Federal Trade Commission). 2000. Washington, DC: Federal Trade Commission.

Gawin F, Compton M, Byck R. 1989. Buspirone reduces smoking. *Arch Gen Psychiatry* 46(3):288-289.

Gaworski CL, Dozier MM, Gerhart JM, Rajendran N, Brennecke LH, Aranyi C, Heck JD. 1997. 13-week inhalation toxicity study of menthol cigarette smoke. *Food Chem Toxicol* 35(7):683-692.

Glantz SA. 1997. Back to basics: getting smoke-free workplaces back on track. *Tob Control* 6(3):164-166.

Glasgow RE. 1978. Effects of a self-control manual, rapid smoking, and amount of therapist contact on smoking reduction. *J Consult Clin Psychol* 46(6):1439-1447.

Glasgow RE, Cummings KM, Hyland A. 1997. Relationship of worksite smoking policy to changes in employee tobacco use: findings from COMMIT. Community Intervention Trial for Smoking Cessation. *Tob Control* 6 Suppl 2(3):S44-48.

Glasgow R, Klesges R, Godding P, Gagelman R. 1983. Controlled smoking, with or without carbon monoxide feedback, as an alternative for chronic smokers. *Behavior Therapy* 14:386-397.

Glasgow R, Klesges R, Klesges L, Vasey M, Gunnarson D. 1985. Long-term effects of a controlled smoking program: a 2 1/2 year follow-up. *Behavior Therapy* 16:303-307.

Glasgow R, Morray K, Lichtenstein E. 1989. Controlled smoking versus abstinence as a treatment goal: the hopes and fears may be unfounded. *Behavior Therapy* (20):77-91.

Glassman AH, Covey LS, Dalack GW, et al. 1993. Smoking cessation, clonidine, and vulnerability to nicotine among dependent smokers. *Clin Pharmacol Ther* 54(6):670-679.

Glassman AH, Helzer JE, Covey LS, et al. 1990. Smoking, smoking cessation, and major depression. *JAMA* 264(12):1546-1549.

Glassman AH, Jackson WK, Walsh BT, Roose SP, Rosenfeld B. 1984. Cigarette craving, smoking withdrawal, and clonidine. *Science* 226(4676):864-866.

Gorelick DA, Rose J, Jarvik ME. 1989. Effect of naloxone on cigarette smoking. *J Subst Abuse* 1(2):153-159.

Gourlay SG, Benowitz NL. 1995. Is clonidine an effective smoking cessation therapy? *Drugs* 50(2):197-207.

Hall SM, Munoz RF, Reus VI, Sees KL. 1993. Nicotine, negative affect, and depression. *J Consult Clin Psychol* 61(5):761-767.

Hall SM, Reus VI, Munoz RF, et al. 1998. Nortriptyline and cognitive-behavioral therapy in the treatment of cigarette smoking. *Arch Gen Psychiatry* 55(8):683-690.

Hartman N, Jarvik ME, Wilkins JN. 1989. Reduction of cigarette smoking by use of a nicotine patch. *Arch Gen Psychiatry* 46(3):289.

Hartman N, Leong GB, Glynn SM, Wilkins JN, Jarvik ME. 1991. Transdermal nicotine and smoking behavior in psychiatric patients. *Am J Psychiatry* 148(3):374-375.

Hatsukami DK, Severson HH. 1999. Oral spit tobacco: addiction, prevention and treatment. *Nicotine Tob Res* 1(1):21-44.

Hayford KE, Patten CA, Rummans TA, et al. 1999. Efficacy of bupropion for smoking cessation in smokers with a former history of major depression or alcoholism. *Br J Psychiatry* 174(1):173-178.

Hays J, Hurt R, Wolter T, et al. 2000. Bupropion-SR for relapse prevention. Society for Research on Nicotine and Tobacco 6th Annual Conference: Arlington, VA.

Henningfield J. 1995. Introduction to tobacco harm reduction as a complementary strategy to smoking cessation. *Tobacco Control* 4(suppl 2):S25-S28.

Henningfield JE, Keenan RM. 1993. Nicotine delivery kinetics and abuse liability. *J Consult Clin Psychol* 61(5):743-750.

Henningfield JE, Slade J. 1998. Tobacco-dependence medications: public health and regulatory issues. *Food Drug Law J* 53 suppl(1):75-114.

Henningfield JE, Stapleton JM, Benowitz NL, Grayson RF, London ED. 1993. Higher levels of nicotine in arterial than in venous blood after cigarette smoking. *Drug Alcohol Depend* 33(1):23-29.

Hering R, Jones R, Fischmann P. 1985. The titration hypothesis revisited: nicotine gum reduces smoking intensity. Grabowski J, Hall S, eds. *Pharmacological Adjuncts in Smoking Cessation.* Vol. 53 ed. Washington, DC: U.S. GPO. Pp. 27-41.

Hieda Y, Keyler DE, Vandevoort JT, et al. 1997. Active immunization alters the plasma nicotine concentration in rats. *J Pharmacol Exp Ther* 283(3):1076-1081.

Hilleman DE, Mohiuddin SM, Del Core MG, Sketch MH. 1992. Effect of buspirone on withdrawal symptoms associated with smoking cessation. *Arch Intern Med* 152(2):350-352.

Hjalmarson A, Franzon M, Westin A, Wiklund O. 1994. Effect of nicotine nasal spray on smoking cessation. A randomized, placebo-controlled, double-blind study. *Arch Intern Med* 154(22):2567-2572.

Hjalmarson A, Nilsson F, Sjostrom L, Wiklund O. 1997. The nicotine inhaler in smoking cessation. *Arch Intern Med* 157(15):1721-1728.

Hoffmann D, Hoffmann I. 1997. The changing cigarette, 1950-1995. *Journal of Toxicology and Environmental Health* 50:307-364.

Holm H, Jarvis MJ, Russell MA, Feyerabend C. 1992. Nicotine intake and dependence in Swedish snuff takers. *Psychopharmacology (Berl)* 108(4):507-511.

Holzman D. 1999. Safe cigarette alternatives? Industry critics say 'not yet.' *Journal of the National Cancer Institute* 91(6):502-504.

Hughes J. 1995. Applying harm reduction to smoking. *Tobacco Control* 4(suppl 2):533-538.

Hughes J. 1998. Dependence on and abuse of nicotine replacement medications: an update. Benowitz NL, ed. *Nicotine Safety and Toxicity.* New York: Oxford University Press. Pp. 147-157.

Hughes JR. 2000. Reduced smoking: an introduction and review of the evidence. *Addiction* 95 Suppl 1:S3-7.

Hughes JR, Goldstein MG, Hurt RD, Shiffman S. 1999a. Recent advances in the pharmacotherapy of smoking. *JAMA* 281(1):72-76.

Hughes JR, Lesmes GR, Hatsukami DK, et al. 1999b. Are higher doses of nicotine replacement more effective for smoking cessation? *Nicotine Tob Res* 1(2):169-174.

Hughes J, Higgins S, Hatsukami D. 1989. Effects of abstinence from tobacco. Kozlowski L, Annis H, Cappell H, eds. *Research Advances in Alcohol and Drug Problems.* New York: Plenum Press.

Hurt RD, Sachs DP, Glover ED, et al. 1997. A comparison of sustained-release bupropion and placebo for smoking cessation. *N Engl J Med* 337(17):1195-202.

IOM (Institute of Medicine). 2000. *State Programs Can Reduce Tobacco Use*. Washington, DC: National Academy Press.

Jacobson PD, Wasserman J. 1997. *Tobacco Control Laws. Implementation and Enforcement*. Washington, DC: RAND.

Jarvik M. 1970. The role of nicotine in the smoking habit. Hunt W, ed. *Learning Mechanisms in Smoking*. Chicago: Aldine Pub Co.

Jeffery RW, Kelder SH, Forster JL, French SA, Lando HA, Baxter JE. 1994. Restrictive smoking policies in the workplace: effects on smoking prevalence and cigarette consumption. *Prev Med* 23(1):78-82.

Jenkins RA, Pair DD, Guerin MR. 1982. Deliveries of Tar, Nicotine, and Carbon Monoxide of Selected U.S. Commercial Cigarettes Smoked Under 'More Relevant' Smoking Parameters. Oak Ridge National Laboratory (ORNL) Project Topical Report No. 120. Unpublished report, available from the authors. Oak Ridge, TN; 13 pp.

Jimenez-Ruiz C, Kunze M, Fagerström KO. 1998. Nicotine replacement: a new approach to reducing tobacco-related harm. *Eur Respir J* 11(2):473-479.

Johnson R, Stevens V, Hollis J, Woodson G. 1992. Nicotine chewing gum use in the outpatient care setting. *The Journal of Family Practice* 34(1):61-65.

Jones, C. 1998. Low-toxin cigarette created/35 potential hazards reduced, Philip Morris says. [Online]. Available: http://www.gatewayva.com/rtd/special/tobacco/philmo18.shtml [accessed November 24, 1999].

Jorenby DE, Leischow SJ, Nides MA, et al. 1999. A controlled trial of sustained-release bupropion, a nicotine patch, or both for smoking cessation. *N Engl J Med* 340(9):685-691.

Joseph AM, Norman SM, Ferry LH, et al. 1996. The safety of transdermal nicotine as an aid to smoking cessation in patients with cardiac disease. *N Engl J Med* 335(24):1792-1798.

Kabat GC, Hebert JR. 1991.Use of mentholated cigarettes and lung cancer risk. *Cancer Res* 51(24):6510-6513.

Kanner RE, Connett JE, Williams DE, Buist AS. 1999. Effects of randomized assignment to a smoking cessation intervention and changes in smoking habits on respiratory symptoms in smokers with early chronic obstructive pulmonary disease: The Lung Health Study. *Am J Med* 106(4):410-416.

Karras A, Kane JM. 1980. Naloxone reduces cigarette smoking. *Life Sci* 27(17):1541-1545.

Kendler KS, Neale MC, MacLean CJ, Heath AC, Eaves LJ, Kessler RC. 1993. Smoking and major depression. A causal analysis. *Arch Gen Psychiatry* 50(1):36-43.

Kitagawa K, Kunugita N, Katoh T, Yang M, Kawamoto T. 1999. The significance of the homozygous CYP2A6 deletion on nicotine metabolism: a new genotyping method of CYP2A6 using a single PCR-RFLP. *Biochem Biophys Res Commun* 262(1):146-151.

Kluger R. 1997. *Ashes to Ashes*. New York: Vintage Books.

Kornitzer M, Boutsen M, Dramaix M, Thijs J, Gustavsson G. 1995. Combined use of nicotine patch and gum in smoking cessation: a placebo-controlled clinical trial. *Prev Med* 24(1):41-47.

Kozlowski LT, Goldberg ME, Yost BA, White EL, Sweeney CT, Pillitteri JL. 1998. Smokers' misperceptions of light and ultra-light cigarettes may keep them smoking. *Am J Prev Med* 15(1):9-16.

Labstat. 2000. Characterization of three "low/ultra low" tar brands. [Online]. Available: http://tobaccofreekids.org/reports/eclipse/ [accessed October 10, 2000].

Lancaster T, Stead L, Silagy C, Sowden A. 2000. Effectiveness of interventions to help people stop smoking: findings from the Cochrane Library. *BMJ* 321(7257):355-358.

Lawrence WF, Smith SS, Baker TB, Fiore MC. 1998. Does over-the-counter nicotine replacement therapy improve smokers' life expectancy? *Tob Control* 7(4):364-368.

Leischow S, Stine C, Nordbrock J, Marriott J, Zobrist R. 1996. Ninth National Conference on Nicotine Dependence: American Society of Addiction Medicine.

Levin ED, Rose JE. 1991. Nicotinic and muscarinic interactions and choice accuracy in the radial-arm maze. *Brain Res Bull* 27(1):125-128.

Levinson B, Shapiro D, Schwartz G, Tursky B. 1971. Smoking elimination by gradual reduction. *Behavioral Therapy* 2:477-487.

Lewit E, Coate D, Grossman M. 1981. The effects of government regulation on teenage smoking. *Journal of Law and Economics* 24(3):273-298.

Longo WE, Rigler MW, Slade J. 1995. Crocidolite asbestos fibers in smoke from original Kent cigarettes. *Cancer Res* 55(11):2232-2235.

Lucchesi BR, Schuster CR, Emley GS. 1967. The role of nicotine as a determinant of cigarette smoking frequency in man with observations of certain cardiovascular effects associated with the tobacco alkaloid. *Clin Pharmacol Ther* 8(6):789-796.

Martin BR, Onaivi ES, Martin TJ. 1989. What is the nature of mecamylamine's antagonism of the central effects of nicotine? *Biochem Pharmacol* 38(20):3391-3397.

McCarthy WJ, Caskey NH, Jarvik ME, Gross TM, Rosenblatt MR, Carpenter C. 1995. Menthol vs nonmenthol cigarettes: effects on smoking behavior. *Am J Public Health* 85(1):67-72.

McGovern PG, Lando HA. 1991. Reduced nicotine exposure and abstinence outcome in two nicotine fading methods. *Addict Behav* 16(1-2):11-20.

Meier B. 2000. Philip Morris says it has a safer paper. *The New York Times*. January 11, 2000:(A2).

Molander L, Lunell E, Fagerström KO. 2000. Reduction of tobacco withdrawal symptoms with a sublingual nicotine tablet: a placebo controlled study. *Nicotine Tob Res* 2(2):187-191.

Murray RP, Bailey WC, Daniels K, et al. 1996. Safety of nicotine polacrilex gum used by 3,094 participants in the Lung Health Study. Lung Health Study Research Group. *Chest* 109(2):438-445.

Nemeth-Coslett R, Griffiths RR. 1986. Naloxone does not affect cigarette smoking. *Psychopharmacology (Berl)* 89(3):261-264.

Nemeth-Coslett R, Henningfield JE. 1986. Effects of nicotine chewing gum on cigarette smoking and subjective and physiologic effects. *Clin Pharmacol Ther* 39(6):625-630.

Nemeth-Coslett R, Henningfield JE, O'Keeffe MK, Griffiths RR. 1987. Nicotine gum: dose-related effects on cigarette smoking and subjective ratings. *Psychopharmacology (Berl)* 92(4):424-430.

Niaura R, Brown R, Goldstein M, Murphy J, Abrams D. 1996. Transdermal clonidine for smoking cessation: a double-blind ramdomized dose-response study. *Experimental and Clinical Psycopharmacology* 4(3):285-291.

Niaura R, Goldstein M, Depue J, Keuthen N, Kristeller J, Abrams D. 1995. Fluoxetine, symptoms of depression, and smoking cessation. *Annals of Behavioral Medicine* 17(Supplement).

NIH (National Institutes of Health). 1996. *The FTC Cigarette Test Method for Determining Tar, Nicotine, and Carbon Monoxide Yields of U.S. Cigarettes. Smoking and Tobacco Control Monograph 7.* Bethesda, Maryland: National Institutes of Health.

O'Reilly JT. 1989. A consistent ethic of safety regulation: the case for improving regulation to tobacco products. *Administrative Law Journal* 3:235-250.

Page JP. 1998. Federal regulation of tobacco products and products that treat tobacco dependence: are the playing fields level? *Food and Drug Law Journal* 53:11-42.

Pentel PR, Malin DH, Ennifar S, et al. 2000. A nicotine conjugate vaccine reduces nicotine distribution to brain and attenuates its behavioral and cardiovascular effects in rats. *Pharmacol Biochem Behav* 65(1):191-198.

Pentz MA, Brannon BR, Charlin VL, Barrett EJ, MacKinnon DP, Flay BR. 1989. The power of policy: the relationship of smoking policy to adolescent smoking. *Am J Public Health* 79(7):857-862.

Perumal AS, Smith TM, Suckow RF, Cooper TB. 1986. Effect of plasma from patients containing bupropion and its metabolites on the uptake of norepinephrine. *Neuropharmacology* 25(2):199-202.

PHS (U.S. Public Health Service). 1964. *Smoking and Health: Report of the Advisory Committee to the Surgeon General of the Public Health Service.* Atlanta, GA: U.S. Public Health Service.

Physician's Desk Reference (PDR). 2000. Montvale, NJ: Medical Economics Company.

Pianezza ML, Sellers EM, Tyndale RF. 1998. Nicotine metabolism defect reduces smoking. *Nature* 393(6687):750.

Pickworth WB, Bunker EB, Henningfield JE. 1994. Transdermal nicotine: reduction of smoking with minimal abuse liability. *Psychopharmacology (Berl)* 115(1-2):9-14.

Pierce JP, Gilpin EA, Farkas AJ. 1998. Can strategies used by statewide tobacco control programs help smokers make progress in quitting? *Cancer Epidemiol Biomarkers Prev* 7(6):459-464.

Pomerleau OF, Pomerleau CS. 1984. Neuroregulators and the reinforcement of smoking: towards a biobehavioral explanation. *Neurosci Biobehav Rev* 8(4):503-513.

Prochaska JO, Goldstein MG. 1991. Process of smoking cessation. Implications for clinicians. *Clin Chest Med* 12(4):727-735.

Prochazka AV, Weaver MJ, Keller RT, Fryer GE, Licari PA, Lofaso D. 1998. A randomized trial of nortriptyline for smoking cessation. *Arch Intern Med* 158(18):2035-2039.

Prue D, Krapfl J, Martin J. 1981. Brand fading: The effects of gradual changes to low tar and nicotine cigarettes on smoking rate, carbon monoxide and thiocyanate levels. *Behavioral Therapy* 12:400-416.

R.J. Reynolds Eclipse web site. 2000. Determination of Mainstream Smoke Yields by the FTC Method. [Online]. Available: http://www.rjrdirect.com/ECL/sci/default.htm [accessed 2000].

R.J. Reynolds Tobacco Company. 1988. *New Cigarette Prototypes That Heat Instead of Burn Tobacco.* Winston-Salem, NC: R.J. Reynolds Tobacco Company.

R.J. Reynolds Tobacco Company. 2000. Advertisement. Introducing Eclipse, A better way to smoke.

Rennard SI, Daughton D, Fujita J, et al. 1990. Short-term smoking reduction is associated with reduction in measures of lower respiratory tract inflammation in heavy smokers. *Eur Respir J* 3(7):752-759.

Rigotti NA, DiFranza JR, Chang Y, Tisdale T, Kemp B, Singer DE. 1997. The effect of enforcing tobacco-sales laws on adolescents' access to tobacco and smoking behavior. *N Engl J Med* 337(15):1044-1051.

Robinson MD, Pettice YL, Smith WA, Cederstrom EA, Sutherland DE, Davis H. 1992. Buspirone effect on tobacco withdrawal symptoms: a randomized placebo-controlled trial. *J Am Board Fam Pract* 5(1):1-9.

Rose JE, Behm FM, Westman EC, Levin ED, Stein RM, Ripka GV. 1994. Mecamylamine combined with nicotine skin patch facilitates smoking cessation beyond nicotine patch treatment alone. *Clin Pharmacol Ther* 56(1):86-99.

Rose JE, Behm FM, Westman EC. 1998. Nicotine-mecamylamine treatment for smoking cessation: the role of pre-cessation therapy. *Exp Clin Psychopharmacol* 6(3):331-43.

Rose J, Behm F, Westman E. 1999. Brand-switching and gender effects on mecamylamine/ nicotine smoking cessation treatment. *Nicotine and Tobacco Research* 1:286-287.

Rose JE, Sampson A, Levin ED, Henningfield JE. 1989. Mecamylamine increases nicotine preference and attenuates nicotine discrimination. *Pharmacol Biochem Behav* 32(4):933-938.

Russell MA, Feyerabend C, Cole PV. 1976. Plasma nicotine levels after cigarette smoking and chewing nicotine gum. *Br Med J* 1(6017):1043-1046.

Saffer, H and Chaloupka, F. 1999. Tobacco advertising: economic theory and international evidence (NBER Working Paper W 6958. [Online]. Available: http://www.nberws.nber.org/papers/W6958 [accessed 2001].

SAMHSA (Substance Abuse and Mental Health Services Administration). 1996. Preliminary estimates from the 1995 National Household Survey on Drug Abuse. [Online]. Available: http://www.health.org/govstudy/ar018/index.htm [accessed November, 2000].

SAMHSA (Substance Abuse and Mental Health Services Administration). 2000. 1999 National Household Survey on Drug Abuse. [Online]. Available: http://www.samhsa.gov/OAS/NHSDA/1999/html [accessed August, 2000].

Schneider NG, Olmstead R, Mody FV, et al. 1995. Efficacy of a nicotine nasal spray in smoking cessation: a placebo-controlled, double-blind trial. *Addiction* 90(12):1671-1682.

Schneider NG, Olmstead R, Nilsson F, Mody FV, Franzon M, Doan K. 1996. Efficacy of a nicotine inhaler in smoking cessation: a double-blind, placebo-controlled trial. *Addiction* 91(9):1293-1306.

Schneider NG, Olmstead RE, Steinberg C, Sloan K, Daims RM, Brown HV. 1996. Efficacy of buspirone in smoking cessation: a placebo-controlled trial. *Clin Pharmacol Ther* 60(5):568-575.

Sellers EM, Kaplan HL, Tyndale RF. 2000. Inhibition of cytochrome P450 2A6 increases nicotine's oral bioavailability and decreases smoking. *Clin Pharmacol Ther* 68(1):35-43.

Shapiro D, Schwartz GE, Tursky B, Shnidman SR. 1971. Smoking on cue: a behavioral approach to smoking reduction. *J Health Soc Behav* 12(2):108-113.

Shiffman S, Mason KM, Henningfield JE. 1998. Tobacco dependence treatments: review and prospectus. *Annu Rev Public Health* 19(2):335-358.

Sidney S, Tekawa IS, Friedman GD, Sadler MC, Tashkin DP. 1995. Mentholated cigarette use and lung cancer. *Arch Intern Med* 155(7):727-732.

Slade J. 1996. ASAM (American Society of Addiction Medicine) Press Release. [Online]. Available: http://www.asam.org/nic/eclipse.htm [accessed November 19, 1999].

Slade J, Ballin S. 1993. Who's minding the tobacco store? It's time to level the regulatory playing field. Tobacco Use: An American Crisis. Final Report of the Conference: January 9, 1993-January 12, 1993; Washington, DC.

Slade J, Henningfield JE. 1998. Tobacco product regulation: context and issues. *Food and Drug Law Journal* 53:43-74.

Star Scientific, Inc. 1999. What is Star doing to reduce the hazards of tobacco use? [Online]. Available: http://www.starscientific.com/frame_pages/health_frame.htm [accessed 1999].

Star Scientific. 2000. Star Scientific, Inc. Secures Additional Purchase Order for Millions of Pounds of StarCure Processed Tobacco From B&W, Which Will Contain Very Low Levels of Cancer-Causing Tobacco Stecific Nitrosamines (TSNAs). [Online]. Available: http://www.starscientific.com/frame_pages/release_frame.html [accessed May 30, 2000].

Stolerman I. 1986. Could nicotine antagonists be used in smoking cessation? *British Journal of Addiction* 81:47-53.

Sutherland G, Stapleton JA, Russell MA, et al. 1992. Randomised controlled trial of nasal nicotine spray in smoking cessation. *Lancet* 340(8815):324-329.

Sutherland G, Stapleton JA, Russell MA, Feyerabend C. 1995. Naltrexone, smoking behaviour and cigarette withdrawal. *Psychopharmacology (Berl)* 120(4):418-425.

Sweeney CT, Kozlowski LT, Parsa P. 1999. Effect of filter vent blocking on carbon monoxide exposure from selected lower tar cigarette brands. *Pharmacology Biochemistry and Behavior* 63(1):167-173.

Tauras J, Chaloupka F. 1999. Price, clean indoor air, and cigarette smoking: evidence from the longitudinal data for young adults (NBER Working Paper W 6937. [Online]. Available: http://www.nberws.nber.org/papers/W6937 [accessed 2001].

Taylor WA, Gold MS. 1990. Pharmcologic approaches to the treatment of cocaine dependence. *Western Journal of Medicine* 152(Addiction Medicine):573-577.

Thun MJ, Day-Lally CA, Calle EE, Flanders WD, Heath CW. 1995. Excess mortality among cigarette smokers: changes in a 20-year interval. *Am J Public Health* 85(9):1223-1230.

Tonnesen P, Norregaard J, Mikkelsen K, Jorgensen S, Nilsson F. 1993. A double-blind trial of a nicotine inhaler for smoking cessation. *JAMA* 269(10):1268-1271.

Transdermal Nicotine Study Group. 1991. Transdermal nicotine for smoking cessation. Six-month results from two multicenter controlled clinical trials. *JAMA* 266(22):3133-3138.

Tyndale R, Rao Y, Bodin L, et al. 2000. CYP2A6 gene defects and duplications alter objective indices of smoking. *Clincal Pharmacology and Therapeutics* 67(2):113.

U.S. DHHS (U.S. Department of Health and Human Services). 1986. *The Health Consequences of Involuntary Smoking: A Report of the Surgeon General.* Atlanta, GA: U.S. Department of Health and Human Services, Centers for Disease Control and Prevention.

U.S. DHHS (U.S. Department of Health and Human Services). 1994. *Preventing Tobacco Use Among Young People: A Report of the Surgeon General.* Atlanta, GA: U.S. Department of Health and Human Services, Centers for Disease Control and Prevention.

U.S. DHHS (U.S. Department of Health and Human Services). 1995. Analysis regarding the Food and Drug Administration's jurisdiction over nicotine-containing cigarettes and smokeless tobacco products. *Federal Register* 60(155):41453.

U.S. DHHS (U.S. Department of Health and Human Services). 1998. *Tobacco Use Among U.S. Racial/Ethnic Minority Groups.* Atlanta, GA: U.S. Department of Health and Human Services, Centers for Disease Control and Prevention.

U.S. DHHS (U.S. Department of Health and Human Services). 2000. *Reducing Tobacco Use: A Report of the Surgeon General.* Atlanta, GA: U.S. Department of Health and Human Services, Centers for Disease Control and Prevention.

Wakefield MA, Wilson D, Owen N, Esterman A, Roberts L. 1992. Workplace smoking restrictions, occupational status, and reduced cigarette consumption. *J Occup Med* 34(7):693-697.

Warner KE, Slade J, Sweanor DT. 1997. The emerging market for long-term nicotine maintenance. *JAMA* 278(13):1087-1092.

Was Safer Cigarette Research Snuffed? 1994. *Science* 264(5160):766-767.

Waters AJ, Jarvis MJ, Sutton SR. 1998. Nicotine Withdrawal and accident rates. *Nature* 394(137).

Watkins SS, Koob GF, Markou A. 2000. Neural mechanisms underlying nicotine addiction: acute positive reinforcement and withdrawal. *Nicotine Tob Res* 2(1):19-37.

West R, Hajek P, McNeill A. 1991. Effect of buspirone on cigarette withdrawal symptoms and short-term abstinence rates in a smokers clinic. *Psychopharmacology (Berl)* 104(1):91-96.

Westman E, Tomlin K, Rose J. 2000. Combining the nicotine inhaler and nicotine patch for smoking cessation. *American Journal of Health Behavior* 24(2):114-119.

Women who smoke menthol cigarettes have greater nicotine exposure. 1999. *Oncology (Huntingt)* 13(7):915.

Wong GY, Wolter TD, Croghan GA, Croghan IT, Offord KP, Hurt RD. 1999. A randomized trial of naltrexone for smoking cessation. *Addiction* 94(8):1227-1237.

Working Group for the Study of Transdermal Nicotine in Patients with Coronary Artery Disease. 1994. Nicotine replacement therapy for patients with coronary artery disease. *Arch Intern Med* 154(9):989-995.

5

The Scientific Basis for
PREP Assessment

Assessing health risks from conventional tobacco products is similar to that for many environmental and occupational exposures. Tobacco risks, however, are among the more complicated to assess for several reasons. The general components of risk assessment (hazard identification, dose-response assessment, exposure assessment, and risk characterization) described in Chapter 1 are still useful to consider (see Table 5-1).

Hazard identification is challenging because tobacco and the smoke generated upon its combustion are complex mixtures. Some of the hundreds (or thousands) of known or suspected toxicants are fairly well understood; however, the relative contribution to overall toxicity of most of the individual compounds is not. In addition, tobacco products contain added constituents or ingredients, but the identity and concentration of these compounds within a specific tobacco product is unknown, due to proprietary concerns. Animal models of tobacco toxicity are limited, posing additional barriers to complete hazard identification.

Dose-response assessment is complicated. Because the exposure is a complex mixture, the diseases associated with tobacco exposure are many and the dose-response relationships vary significantly. Assessing the dose in epidemiological studies is complicated in part by the factors described for hazard identification. In addition, the dose a tobacco user is exposed to can change often over a long and variable smoking history. Finally, the responses are most often diseases with long periods of disease progression until diagnosis and from time point of dose estimation.

TABLE 5-1 Tobacco and PREP Risk Assessment

	Hazard Identification	Dose Response	Exposure Assessment	Risk Characterization	Risk Management
1983 "Red Book"	Epidemiology Animal bioassay Short-term Studies Comparisons of molecular structure	Epidemiology Low-dose extrapolation Animal to human extrapolation	Dose to which humans are exposed Dose of special populations Estimation of size of population potentially exposed	Estimate of the magnitude of the public health problem	A risk-assessment (qualitative or quantitative) may be one of the bases of risk management
Challenges in risk assessment of tobacco	Complex mixture Animal models are limited Constituents and additives are proprietary information	Dose changes for an individual over time Dose of individual toxicants varies over time Exposure at time of disease progression	Changes in smoking topography Complex mixture	For which disease? At which point in smoking history?	FTC regarding advertising
Additional challenges of PREP risk assessment	Products will change rapidly with time	Assessing effect of moving backwards on a dose-exposure curve, assuming long-time previous higher exposure	Changing exposure after long-term higher dose exposure Some toxicants could increase	Need models to consider effects on initiation, cessation, and relapse	FDA authority currently exerted only over pharmaceutical PREPs

continues

TABLE 5-1 Continued

	Hazard Identification	Dose Response	Exposure Assessment	Risk Characterization	Risk Management
Committee charge	1. Does product decrease exposure to the harmful substances in or produced during use of tobacco?	2. Is decreased exposure associated with decreased harm to health? 3. Are there useful surrogate indicators of disease that could be used?	1. Does product decrease exposure?	4. What are the public health implications?	4. What are the public health implications?
Disease-specific summary data (Chapter 5; Section II)	3. Utility of preclinical research to judge feasibility	1. Dose-response data for conventional tobacco products 2. Validation and development of biomarkers 4. Short-term clinical and epidemiological studies	2. Validation and development of biomarkers 4. Short-term clinical and epidemiological studies	5. Long-term epidemiological studies and surveillance	

NOTE: Numbers correspond to research recommendations listed in the Executive Summary.

Exposure assessment is difficult for some of the same reasons. There is a multiplicity of tobacco products on the market. The specific exposures associated with any one "branded" product could change throughout time because the product can change. Changes in exposure throughout time are not documented. In addition, smokers of "low-yield" products often compensate (change smoking behavior to increase nicotine exposure), so their exposures to nicotine and tobacco/smoke toxicants are often higher than predicted by a common form of exposure assessment, self-report.

The objective of a potential reduced-exposure product (PREP) risk assessment is to determine if the risk of harm from the use of the PREP is less than the risk of harm in the absence of the PREP (see Table 5-1). The risk management objective considered by the committee is not to ban or control the exposure per se, as is the case for environmental and occupational exposures. The risk management objective, as will be made clear in Chapter 7, is primarily to verify whether or not a product is associated with either exposure reduction or harm reduction.

A PREP risk assessment involves lowering the dose of a complex mixture in a person (or population) with varying degrees of pre-existing pathology or cellular damage caused by a complex mixture exposure (that of conventional products) and trying to reverse early damage or to stop disease progression. This is problematic at this point, as there are no adequate human or animal studies that replicate this scenario. While some studies report risks in persons who switch from nonfiltered cigarettes to filtered cigarettes, or from high- to low-tar cigarettes, this "switching" did not reduce exposure (due to compensation) significantly in many people. The reduction in risk, if any, would occur only in persons who do not compensate for lower nicotine levels by smoking more or smoking differently. The basic elements of risk assessment can, however, be still considered. The questions become slightly different, and the data required or the study designs might be different from that required for a tobacco risk assessment.

For *hazard identification*, the questions include:

- Does the PREP contain (or produce during use) toxicants known to cause adverse health effects?
- To what extent are the compounds targeted for reduced exposure causally linked to a tobacco-related disease?
- How does its content compare with the toxicants in the conventional tobacco product to which it is compared?
- Are there unique toxicants in a PREP compared to conventional tobacco products?

For hazard identification, it is essential to know the composition of the material to which people will be exposed from the PREP compared to the standard product. Any new material, such as flavors, added to standard products must be included in the analyses. It is important to analyze the product that actually enters the body (for example, the combustion products that are inhaled) rather than the composition of the product as sold.

The approach to testing the toxicity of the material to which people are exposed in the tobacco-related PREP compared to standard tobacco products is discussed in Chapter 10. The objectives of the toxicity tests are to determine what toxic effects can be induced by the test materials (the tobacco-related PREP compared to the standard product) and how much of the test materials is required to cause the adverse effect, i.e., the dose-response characteristics in animals of the test materials. Data from animal studies can be used to eliminate new products that are much more toxic than existing ones.

A series of comparative potency tests is appropriate. In vitro studies in cultured cells from both animals and humans can be used to determine the ability of the test materials (from the tobacco-related PREP and the standard product) to induce cellular damage, an inflammatory response, or cell death. Assays of the mutagenic or clastogenic activity of the test materials can be done in bacterial or mammalian cell systems.

In animal studies, tests for tobacco-related toxicity should include evaluation of the ability to induce adverse health effects or cancer in the respiratory tract, the nervous system, the cardiovascular system, the reproductive and developmental systems, and other organs. Toxicokinetic studies should be used to determine dosimetry to different organs and to suggest biomarkers of internal dose that can be used in humans. Short-term clinical tests in humans should be done to compare the potencies of the test materials to induce acute adverse health effects (such as reduced pulmonary function) and to determine the toxicokinetics of the tobacco-related PREP compared to the standard product.

For *dose-response assessment*, the questions include:

- What are the dose-response characteristics of the PREP compared to the conventional tobacco product?
- Do smokers use PREPs at a time in their individual smoking history (and therefore of disease progression) that induce different dose-response effects?
- Are the patterns of adverse health effects different from PREPs compared to conventional tobacco products?
- What is the evidence that reduction in exposure to the targeted compounds in the complex mixture or other hazardous material in the PREP will decrease or reverse the development of disease?
- What is the dose-response relationship between the targeted compounds and the disease outcome?

- What quantitatively happens to disease induction if exposures are reduced?
- How much exposure needs to be reduced to result in a measurable benefit?
- Are there individual susceptibilities (age, gender, genetic makeup, and prior use of tobacco products) that change this dose-response relationship?

Some dose-response information can be obtained from standard pre-clinical animal studies. However, there will be uncertainties in extrapolating from animal data to humans. Additionally, it will be difficult, even in animals, to determine the response to a reduction in dose after a period of higher exposure.

For *exposure assessment*, the questions include:

- Do PREP users compensate differently than users of more conventional "low-yield" products?
- Do PREP users exclusively use PREPs or do they switch back and forth between PREPs and conventional products?
- How does the PREP change the exposure pattern for the population?
- Does the introduction of the PREPs increase the number of people initiating use of tobacco products or decrease cessation attempts?
- What is the overall balance in exposures (directly and as environmental tobacco smoke) with or without the PREP?

Past experiences with "low-yield" cigarettes containing less tar and nicotine than other products underscore the need to determine internal dosimetry of toxic material entering the body during use of tobacco-related PREPs. The internal dosimetry of the new products can be compared to that of standard products in short term toxicokinetic studies in humans. Biological markers of internal dosimetry of key ingredients can be used when available.

The risk assessment process will need to rely on the use of animal preclinical and human clinical studies, in which biomarkers of exposure and potential harm can be measured. Biomarkers of exposure to tobacco products have been validated and are in current use. Unfortunately, few specific early indicators of biomarkers have been validated as predictive of later disease development. It is recognized that today, biomarkers of exposure are better validated compared to biomarkers of potential harm, and that it is more feasible to consider exposure reduction in contrast to risk reduction. However, while an assessment of risk reduction through biomarkers will have more uncertainty, these will need to be included in the PREP risk assessment process in order to enhance confidence that there is no worsening risk, in the least.

For *risk characterization*, the questions include:

- Can the exposure levels of the PREP, given its dose-response characteristics, be expected to result in reduced risk of one or more tobacco-related diseases than the standard tobacco product to which it is compared?
- What is the magnitude of any reduction in risk?
- What are the limits in understanding of the risk reduction?
- Do fewer tobacco users quit and use PREPs instead?
- Do former tobacco users relapse to PREPs?
- Do nonusers initiate tobacco use through PREPs?

In order to achieve a level of confidence that a PREP will provide meaningful reductions in risk, especially compared to the real possibility of others using this product to initiate or resume smoking, prospective epidemiological studies are required. This could be done in a timely manner for some disease endpoints, such as birth outcomes or recurrence of myocardial infarction. For many other diseases associated with tobacco use, however, definitive demonstration of reduced harm will require studies of a long duration. Moreover, it is reasonable to anticipate that the design of PREPs would change rapidly in the coming years so that assessing PREP use will be difficult.

The claim that a PREP will result in a reduction of the risk for harm requires scientific evidence for the validity of that claim. A discussion of the information relevant to hazard identification and dose-response information (the first two elements of risk assessment) is given in Chapter 10. Information on the best means of evaluating exposures is given in Chapter 11. In the case of tobacco products, animal models of adverse health effects have been problematic in the past, but new animal models show promise for being useful (see Chapter 10). There currently are no population risk assessment models that mimic the types of predictions hoped for tobacco users who switch to PREPS and how the public health will be effected by the initiation of PREP use by never and former smokers.

The committee has evaluated the science base regarding the toxic effects of tobacco on the major diseases known to be caused by tobacco exposure (i.e., cancer and diseases of the cardiovascular, pulmonary, reproductive systems). The committee has done so to arrive at summary conclusions regarding the evidence base that would directly feed into a risk assessment paradigm, such as that described above (see Table 5-1). Specifically, the committee has elaborated in each of the major disease-oriented chapters of Section II and summarized later in this chapter the evidence regarding:

1) the dose-response relationship between tobacco smoke and/or constituent exposure and health outcomes,
2) identification and development of surrogate markers for disease,
3) the utility of preclinical research in understanding the potential of PREPs to be harm reducing for the disease under review,
4) utility of short-term clinical and epidemiological studies, and
5) the role of long-term epidemiological studies and surveillance.

The review of preclinical research and the material in Chapter 10 (Toxicology) provide information on hazard identification and, in part, dose-response assessment. The material on surrogate markers for disease is informative for both dose-response assessment and exposure assessment. The review of the short-term clinical studies, epidemiology, and surveillance data provide the proof of reduced harm.

In conclusion, the PREP risk assessment process will be challenging because no single definitive study, either human or experimental animal study, would stand up to rigorous scientific scrutiny to be used in such a process. Therefore, today, several types of data will be needed that includes both experimental animal studies and human clinical data, with a definitive plan to conduct epidemiological and surveillance studies. The clinical data is needed because animal studies cannot predict interindividual differences in human behavior that would affect how the PREP is used or cause damage, and because there are too many uncertainties about the use of animal data. During this interim time, and with more research, the regulatory process might be able to identify key types of data needed for the risk assessment process. At this point in time, however, it can only be concluded that both experimental animal and clinical human studies would be needed, and that this would include both the consideration of exposure to individual PREP constituents and as a complex mixture. It is conceivable that the PREP risk assessment process can be simplified, for example by comparing only experimental data for one PREP to another, but this will require substantial experience and characterizing the data for existing PREPs. Sufficient data for streamlining this process in not available today.

The remainder of this chapter provides a summary of the major conclusions and recommendations, arranged by chapter, reached by the committee in Section II of this report.

TOBACCO SMOKE AND TOXICOLOGY
(SEE SECTION II, CHAPTER 10)

Mainstream tobacco smoke and environmental tobacco smoke each is a complex mixture of toxicants composed of carcinogens and other chemicals with health effects that alone or in combination are only partially known (see Table 5-2) (Davis and Nielson, 1999). The evaluation of conventional tobacco products and tobacco-related PREPs is complicated by a lack of adequate in vivo models of tobacco-related morbidities in man.

Toxicology studies, both in vitro and in vivo, provide the opportunity to evaluate the potential harm reduction offered by potential reduced-exposure products. The comparative potency of the PREP can be determined in a series of preclinical studies that include both the PREP and the standard tobacco product that can be replaced by the PREP. The preclinical tests should include in vitro tests in both animal and human cells

TABLE 5-2 List of Selected Tobacco Mutagens and Carcinogens[a]

Constituent Class	Phase	IARC Evaluation[b]	Examples
N-Nitrosamines	Particulate	Sufficient in animals	Tobacco-specific nitrosamines (NNK, NNN), dimethylnitrosamine, diethylnitrosamine
Polycyclic aromatic hydrocarbons	Particulate	Probable in humans	Benzo[a]pyrene, benzo[a]anthracene, benzo[b]fluoranthene, 5-methylchrysene
Aryl aromatic amines	Particulate	Sufficient in humans	4-Aminobiphenyl, 2-toluidine, 2-naphthylamine
Heterocyclic amines	Particulate	Probable in humans	2-Amino-3-methylimidazo[4,5-b]-quinolone (IQ)
Organic solvents	Vapor	Sufficient in humans	Benzene, methanol, toluene, styrene
Aldehydes	Vapor	Limited in humans	Acetaldehyde, formaldehyde
Volatile organic compounds	Vapor	Probable	1,3-Butadiene, isoprene
Inorganic compounds	Particulate	Sufficient in humans	Arsenic, nickel, chromium, polonium-210

NOTE: NNK=nicotine-derived nitroketone; NNN=N-nitrosonornicotine.

[a]This list is intended to provide a conceptual overview of the complexity of tobacco product exposures. It is not all-inclusive but is presented to allow the reader to understand the number of considerations that must be made in assessing harm reduction strategies.

[b]International Agency for Research on Cancer: The classifications here refer to evaluations of the compound from any exposure, not just tobacco. Not all chemicals within the class are considered carcinogenic in humans. There is no consideration in this table to delivered dose or route of exposure (IARC, 1986).

to determine the cytotoxicity and the genotoxicity of the tobacco product to which humans will be exposed. Such tests have recently been reported for a new tobacco-related PREP (Eclipse Expert Panel, 2000). Such a test must include dose-response studies to determine the amount of the exposure material required to cause toxicity. Next, studies should be conducted in vivo in the best animal models available to determine the comparative potency of the PREP versus the standard product in producing: (1) pulmonary inflammation, (2) COPD, (3) cardiovascular disease, (4) reproductive toxicology, and (5) pulmonary neoplasms.

In vitro studies and in vivo animal studies are useful but limited tools in evaluating the toxicity of products that claim to reduce exposure to tobacco toxicants and potentially reduce tobacco-related harm. In vitro studies may allow rapid, low-cost screening for the toxic properties of conventional tobacco products and tobacco-related PREPs, although the relationship between in vitro toxicity and in vivo human response has not been established for most compounds. These assays include cytotoxicity and genotoxicity assays, which are possible screens for the carcinogenic or inflammatory potential of products. In vivo toxicity testing can be developed to supplement in vitro and clinical studies.

Such animal models, if developed, may be useful as a screening assessment of the efficacy of PREPs for reduction of various tobacco-attributable diseases (see Chapters 12-16). The committee concludes that animal models should be used to test for the potential adverse health effects of tobacco smoke or any proposed additives. The A/J mouse model, which is sensitive to induction of lung adenomas, shows promise as an animal model for screening the potential of tobacco products to induce lung tumors (Witschi et al., 2000; Witschi et al., 1999; Witschi et al., 1997a,b). Future studies should validate the model. These studies (Witschi et al., 2000) indicated that removal of single classes of carcinogens, such as nitrosamine or polycyclic aromatic hydrocarbons (PAHs) may not be protective against induction of lung tumors by smoke. Studies also indicate that some animal models show promise for use in studying the development of symptoms similar to those of chronic obstructive pulmonary disease (COPD), the development of cardiovascular disease, adverse effects on the immune system, intrauterine growth retardation, and poor fetal lung maturation from the inhalation of new or existing tobacco products (see Chapter 10).

Testing the general toxicity of smokeless tobacco and evaluating of the potential harm reduction properties of smokeless tobacco (e.g., Swedish snus) use in smokers may greatly benefit from assays for genotoxic and cytotoxic potential and the animal models discussed above.

Details to be considered in determining the specific set of toxicity tests include species and strain of test animal, duration of test, end points of interest, dose-response considerations, biomarkers of dosimetry and response, and standard comparison products to be tested as positive and negative controls.

EXPOSURE AND BIOMARKER ASSESSMENT IN HUMANS (SEE SECTION II, CHAPTER 11)

Accurate measures of exposure and the development of biomarkers of adequate specificity and sensitivity are needed to evaluate the toxicity and harm reduction potential of PREPs. Biomarkers can be defined as measurements of tobacco constituents, tobacco smoke constituents, or changes in body fluids (including exhaled air) and organs. The assessment of a PREP will have to include markers of external exposure and biomarkers indicative of internal exposure, biologically effective dose (Perera, 1987), and potential harm. The definitions of each are provided in Table 5-3. There have been different definitions of types of exposure assessments used previously, but more recent understandings of biomarker uses and limitations, as well as different approaches needed for PREP evaluation lead to a need for clarification and redefinition.

The latter three measurements in Table 5-3 improve upon the first by quantifying exposure at the cellular level to characterize low-dose exposures or low-risk populations, provide a relative contribution of individual

TABLE 5-3 Exposure and Biomarker Assessment Definitions

Exposure or Biomarker Assessment[a]	Definition
External exposure marker	A tobacco constituent or product that may reach or is at the portal of entry to the body
Biomarker of exposure	A tobacco constituent or metabolite that is measured in a biological fluid or tissue that has the potential to interact with a biological macromolecule; sometimes considered a measure of internal dose
Biologically effective dose (BED)	The amount that a tobacco constituent or metabolite binds to or alters a macromolecule; estimates of the BED might be performed in surrogate tissues
Biomarker of potential harm	A measurement of an effect due to exposure; these include early biological effects, alterations in morphology, structure, or function, and clinical symptoms consistent with harm; also includes "preclinical changes"

[a]Categories and definitions reflect concept that the critical exposure is at the level of a biological macromolecule, so that exposure for this discussion is not limited to a measurement at the portal of entry to the body.

chemical carcinogens from complex mixtures (e.g., tobacco-specific N-nitrosamines in cigarette smoke), and estimate total burden of a particular exposure where there are many sources (e.g., benzo[a]pyrene [BaP] from air, tobacco, diet, and occupation) (Vineis and Porta, 1996). In assessing PREPs through biomarkers, understanding the biological effects of a wide range of exposures will be important. Within the context of this discussion, exposure at the level of the cell and critical macromolecules is considered with greater weight, rather than the traditional view of exposure at the portal of entry into a person.

Markers of external exposure attempt to measure the dose of tobacco or tobacco smoke constituents that may enter the body and usually involve machine testing of products and user questionnaires. Internal exposure markers assess the amount of tobacco or tobacco smoke constituents or their metabolites in body fluids or organs. Biomarkers estimating the biologically effective dose measure the internal dose that interacts with cells and macromolecules and may be mechanistically related to disease outcome. Finally, biomarkers of potential harm reflect changes in cells and macromolecules that may lead to disease (see Table 5-4).

Measuring the number of cigarettes per day and smoking duration, estimating lifetime exposure, smoking topography, and so forth can provide an effective indicator of exposure that has been associated with harm. However, these measures may be insensitive to changes in risk, are difficult to assess accurately over time, and have not been tested in the context of harm reduction. Also, because there is interindividual variation in the way the body responds to these exposures, such measures might not be sufficiently accurate for new products intended to decrease exposure. Thus, for new products, the relationship of external exposure markers to disease risk might be less predictable. Currently, there is sufficient evidence to show that biomarkers can provide better estimates of risk in the context of exposure, and therefore they will likely be able to provide improved assessments for harm reduction products. However, no single biomarker has been sufficiently validated and related to disease risk that it can be recommended as an intermediate biomarker of cancer risk. Thus, different types of biomarkers along the pathway from internal exposure, biologically effective dose, and potential harm are needed, and additional research is necessary to identify the best combination of markers to be used. Experimental toxicity testing (in vitro and animal models) are not sufficient to support a PREP claim because only validated biomarkers can show that the PREP reduces exposure adequately enough to imply risk reduction. The use of intermediate biomarkers as surrogate risk factors for disease may possibly overestimate the number of people who actually develop disease, because not all early changes in morphology or function progress to disease. On the other hand, it may underestimate if, as

TABLE 5-4   Biomarkers of Potential Harmful Effects[a,b]

Category	Variables Used in Literature	Dose-Response Data	Associated with Cessation or Half-life	Target Tissue Assay Available	Chemical Specificity
Enzymatic induction	Aryl hydrocarbon hydroxylase	No	>30 d	Yes	Yes
	CYP1A2	No	NDA	Yes	Yes
	DNA repair enzymes	NDA	Yes	Yes	NA
	Microarray assays for mRNA expression and proteomics	NDA	NDA	Yes	NA
Chromosomal alterations	Chromosomal aberrations	Yes	Yes	Yes	No
	Micronuclei	Yes	Yes	Yes	No
	Sister chromatid exchanges	Yes	Yes	No	No
	Loss of heterozygosity	Yes	Yes	Yes	No

Specific to Tobacco	Related to a Disease Risk[c]	Strengths	Limitations
No	Yes	Indicates acquired changes in susceptibility; related to DNA-adduct levels	Technically difficult to assess in large epidemiological studies
No	Yes	Indicates acquired changes in susceptibility; related to DNA-adduct levels	Technically difficult to assess in large epidemiological studies
No	NDA	Indicates acquired changes in susceptibility; provides analysis of what is likely to be critical part of carcinogenesis	Technically difficult
No	NDA	Reflects integrated measure of multiple genotypes, provides complex data potentially usable for rapid identification of important risk factors	Difficult to perform; relationship to disease risk is technically difficult to prove; requires extensive laboratory validation; RNA and protein microarray assays are expensive; large-scale studies are needed; refined bioinformatic analysis required
No	Yes	Can be done in blood as surrogate tissue. Similar lesions observed in cancer. Can be measured in persons without cancer	Very nonspecific; relationship to target organ is not established; significant lack of specificity and wide overlap between smokers and nonsmokers
No	NDA	Facile assay	Lack of specificity
No	No	Easy to do in blood as surrogate tissue. Can be measured in persons without cancer	Very nonspecific; relationship to target organ is not established; predictivity for disease risk not established. Association with cancer in case-control studies may have case bias. Significant lack of specificity and wide overlap between smokers and nonsmokers
No	NDA	Similar lesions observed in cancer	Technically complex; relationship to cancer risk unknown

continues

TABLE 5-4 Continued

Category	Variables Used in Literature	Dose-Response Data	Associated with Cessation or Half-life	Target Tissue Assay Available	Chemical Specificity
	Mutations in reporter genes (*HPRT, GPA*)	Yes	Yes	No	No
	Mutational load in target genes (p53, K-ras)	NA	NDA	Yes	No
Mitochondrial mutations	Deletions, insertions	NDA	NDA	Yes	No
Epigenetic cancer effects	Whole genome methylation	NDA	NDA	Yes	No
	Hypermethylation of promoter regions	NDA	NDA	Yes	No
Lipids	Blood lipids: HDL, LDL, oxidized LDL, triglycerides	Yes	NDA	Yes	Yes
Cardiovascular response	Heart rate, blood pressure	No	Yes	Yes	NA
Thrombosis	Bleeding time	No	NDA	Yes	No
	Fibrinogen	NDA	NDA	Yes	Yes
	Prothrombin time, partial thromboplastin time, plasminogen activator inhibitor, C-reactive protein	Yes	NDA	Yes	Yes

Specific to Tobacco	Related to a Disease Risk[c]	Strengths	Limitations
No	NDA	Facile assay in blood	Relationship to target tissue or blood unknown
No	NDA	Target gene specificity	Very difficult to do in normal tissues
No	NDA	Provides corroborative marker	Relationship to disease not established
No	No	Facile assay	Relationship to disease unknown
No	No	Similar lesions observed in cancers	Technically difficult; relationship to risk unknown
No	Yes	May be directly related to disease risk	Levels among heavy smokers cannot be distinguished. Wide interindividual variation. Many individuals under medication therapy. Significant confounders exist
No	Yes	Easy to measure; intraindividual differences may be important for the individual	Both interindividual and intraindividual differences are significant. Substantial confounders exist, and many persons are on medications
No	No	Minimally invasive	Very nonspecific
No	NDA	Pathogenically related to disease	Does not distinguish levels of smoking. Nicotine might separately affect these parameters so limited use in persons using NRT
No	NDA	Leave a fingerprint at the site of their formation	

continues

TABLE 5-4 Continued

Category	Variables Used in Literature	Dose-Response Data	Associated with Cessation or Half-life	Target Tissue Assay Available	Chemical Specificity
	Urinary thromboxane and prostacyclins	Yes	No	No	Yes
	Platelet activation and survival	Yes	NDA	Yes	No
Blood cell parameters	White blood cell counts (i.e., lymphocytes, neutrophils, total counts)	Yes	Yes	Yes	Yes
	Hematocrit, hemoglobin, red blood cell mass	Yes	Yes	Yes	No
Bronchio-alveolar lavage response	Inflammatory cells, protein, cytokines	Yes	Yes	Yes	No
	Neutrophil elastase a1-antiprotease complex	Yes	Yes	Yes	No
	α1-antitrypsin	No	No	Yes	Yes
Inflammatory mediators of response	Leukotrienes	Yes	NDA	No	Yes
Pulmonary function tests	FEV1, FVC	Yes	Yes	Yes	No
Periodontal disease	Periodontal height	Yes	Yes	Yes	No
	Gum bleeding	Yes	Yes	Yes	No

Specific to Tobacco	Related to a Disease Risk[c]	Strengths	Limitations
No	Yes	May be markers of platelet-vascular interactions; reflect chronic exposure	Technically difficult. Wide overlap of values due to individual differences in response
No	No	Platelet activation in vivo might be pathophysiologically related to cardiac artery thrombosis	Technically difficult to use for large numbers of subjects. Significant number of confounding variables. Smoking increases platelet counts
No	Yes	Can be a surrogate marker for several processes including atherosclerosis and thrombosis	Relationship to disease uncertain, although alterations in levels are linked epidemiologically to disease. Wide interindividual and intraindividual variation and large number of confounders
No	No	Can reflect both cardiac and respiratory disease risk	Insensitive; wide interindividual differences
No	NDA	Provides different types of data with single procedure	Bronchoscopy is too invasive for large epidemiological studies
No	NDA	Provides different types of data with single procedure	Bronchoscopy is too invasive for large epidemiological studies
Yes	NDA	May be specific to tobacco smoke	Requires invasive test; short half-life
No	NDA	May be measured in urine, bronchioalveolar lavage, and serum	Substantial number of confounders
No	Yes	Widely available	Low sensitivity for mild disease. Decrease in function with aging. Large interindividual variation
No	Yes		
No	Yes		

continues

TABLE 5-4 Continued

Category	Variables Used in Literature	Dose-Response Data	Associated with Cessation or Half-life	Target Tissue Assay Available	Chemical Specificity
Osteoporosis	Fractures	Yes	NDA	NA	No
	Bone density	NDA	NDA	Yes	No
Skin	Premature wrinkling	Yes	NDA	NA	No
Fetal and neonatal effects	Birth weight	Yes	Yes	Yes	No
Weight	Weight loss and gain	Yes	Yes	Yes	No

NOTE: NA=not applicable; NDA=no data available; FVC=Forced vital capacity; FEV_1=Forced expiratory volume in 1 sec; HDL=high-density lipoprotein; LDL=low-density lipoprotein.

expected, other mechanisms are involved in the disease process that are not reflected by the biomarkers. Biomarkers may also underestimate the incidence of disease since none are necessarily present in all who develop disease. Therefore, the implication of potential benefit from a harm reduction strategy could be an overestimate or an underestimate, but this limitation in the scientific methodology for identifying sufficiently specific biomarkers of risk requires acceptance at the current time.

Previously, the most common way to infer exposure reduction (e.g., through use of low-tar cigarettes) has been via methods that simulate human smoking behavior, such as the Federal Trade Commission (FTC) method. Although they provide a standardized way to assess cigarettes, it is clear that these methods have limited usefulness because people smoke cigarettes differently from the machine, with resultant qualitative and quantitative differences in exposure.

Specific to Tobacco	Related to a Disease Risk[c]	Strengths	Limitations
No	Yes	Easily measured	Numerous confounders
No	Yes		
No	NA		Lack of specificity; involves subjective evaluation
No	Yes	Data collection is easy	Nonspecific; numerous confounders
No	Yes	Both a biomarker for metabolism and an important outcome for some people.	Some people perceive weight loss as a benefit of smoking, despite significant adverse effects associated with smoking

[a]Selected examples; list is not all-inclusive.

[b]References are not provided in this table but can be found in the text of this and disease-related chapters.

[c]Any report related to a disease outcome associated with tobacco where the report is plausible but has not necessarily been replicated.

Biomarkers may be shown to reveal differences in individual susceptibilities and differences in response depending on dose. Thus, biomarkers that measure both complex exposures and single tobacco product constituents are needed and should be assessed for the range of possible human exposures, and those that assess complex exposures should carry a greater weight. Also, some biomarkers or sets of biomarkers should be developed that reflect exposure to many tobacco constituents in order to monitor for the introduction of new hazards from PREPs.

Today, there remain technical limitations to the use of biomarkers. Depending on the harmful effect, surrogate assays that represent effects in target organs may be easier to perform in humans because the target tissues might not be easily accessible. However, if such is the case, the relevance of the surrogate biomarker to the target organ effect should be demonstrated.

The use of a biomarker for harm reduction assessments should include several qualities including reflection of disease pathogenesis, specificity, and sensitivity. Also, consideration must be given to available harm and harm regression dose-response data, target tissue effects, and validation methods. Each biomarker should be validated for its relationship to exposure and harm, and also as a laboratory assay that provides reliable and reproducible data. Separately, the way interindividual variation in response and smoking behavior affects biomarkers should also be considered.

The assessment of harm and harm reduction should be made through direct human experience, as these products are used by the general population. Most of what is known about harmful tobacco products has resulted from epidemiology and supported by in vitro studies, laboratory animal studies, and human experiments. However, although epidemiological studies can provide the most definitive data about tobacco harm and harm reduction products, the study of diseases with long latency (e.g., cancer, heart disease, COPD) is problematic because such studies often require many years before they provide useful data. Thus, because definitive evidence that a new PREP actually reduces harm will often be unavailable, short-term markers that reflect long-term outcomes are needed. If assessment of the harm reduction potential of a PREP were based only on epidemiological data measuring disease outcome prior to its use by the public, very few if any harm reduction products would be introduced. Importantly, the use of intermediate markers does not replace long-term follow-up and epidemiological surveillance, but it can be a basis for estimating effects before direct evidence from epidemiological studies is available.

Biomarkers of internal exposure, biologically effective dose, or potential harm have been validated to different degrees. It is typically easier to show a relationship between external exposure and biomarkers in the following order: internal exposure, biologically effective dose, and harm. Conversely, it is typically easier to show a relationship between disease outcome and biomarkers in the following order: harm, biologically effective dose, and internal exposure. It might be acceptable to rely on external exposure measurements for considering risk and dose-response, but only with substantial corroborative biomarker data. The best strategy for assessing the claims of risk reduction methods is to have several markers that range from exposure to outcome, one being linked to another, and at least one with which a dose-response risk assessment can be made.

The recommendation that PREPs be assessed by the use of biomarkers should reflect the available data, which show that individual use and response to tobacco products are affected by cultural and heritable traits.

To achieve the greatest confidence that a PREP will reduce risks for persons who cannot stop smoking, well-validated methods for predicting risk, including external exposure indicators, and the best available biomarker assays should be used.

NICOTINE PHARMACOLOGY (SEE SECTION II, CHAPTER 9)

Nicotine is the addictive component of tobacco products, and the strength of this addiction affects the individual's ability to stop smoking (U.S. DHHS, 1988). Nicotine is also a component of most PREPs, and therefore, evaluation of the harm reduction potential of PREPs requires evaluation of nicotine's relative toxicity, especially during long-term use (see Chapter 4).

Structurally, nicotine is very similar to acetylcholine (Ach) and interacts with specific nicotinic receptors (nAchRs) in the central and peripheral nervous systems. The interaction between nicotine and its receptor affects the release of numerous neurotransmitters and results in upregulation of the nicotinic receptors leading to the physiological, cognitive, and sensory effects associated with tobacco use, addiction, and withdrawal. Nicotine also has well-documented effects on metabolism and on the cardiovascular, gastrointestinal, and hormonal systems (see Chapter 9).

Pharmacological nicotine replacement therapy (NRT) has proven to be a remarkably well-tolerated and effective strategy for many, leading to cessation of cigarette smoking at least in the short- to medium-term (Benowitz et al., 1998; Fiore et al., 2000). Although the experience is much more limited, it is also a potential strategy for reducing the number of cigarettes smoked by smokers who cannot or will not quit (Fagerström et al., 1997; Rennard et al., 1990; Shiffman et al., 1998; Transdermal Nicotine Study Group, 1991).

There are important considerations in evaluating nicotine products for possible tobacco harm reduction. First, nicotine is addictive, and although the daily exposure may be reduced by NRT use, continued usage implies psychological dependence, if not physical addiction. It is arguable whether this should be a concern, given the assumption of an undisputed reduction of risk compared to smoking. However, it would seem reasonable to include surveillance of the dependency potential and abuse liability of each NRT product. Furthermore, the effects of long-term nicotine intake on such factors as drug and alcohol consumption, the progression of coincidental diseases, the impact of aging on cognitive and other physiological functions, and susceptibility to other forms of addictive behavior are largely unknown. For example, observations suggesting that nicotine impairs endothelial function, a property it shares with cigarette smoking, raise questions about its effect on atherogenesis during long-term use (Chalon et al., 2000; Gairola and Daugherty, 1999; Sabha et al., 2000). Such an effect may take many years to emerge and highlights the importance of continued postmarketing surveillance of NRT. Although existing data do not suggest that nicotine is carcinogenic in humans, it would be prudent to have continued surveillance of the incidence of cancer among users of NRT.

Studies of long-term nicotine administration on surrogate variables that more closely resemble the mechanism under consideration (e.g., imaging of plaque progression) and attendant studies in animal models seem timely. Increasingly, the application of genomic and proteomic approaches will help clarify the differential effects of smoking and NRT on the expression and translation of genes related to the development of smoking-related diseases. Finally, the understanding of nicotine's effect on inflammation and the immune response is confused and limited (Sopori et al., 1998). More research is needed to clarify its effects on cytokine generation, the formation of nitric oxide (NO) and eicosanoids, and oxidative injury. Research should continue to explore other potential therapeutic efficacies of NRT, including for ulcerative colitis, analgesia, weight reduction, Parkinson's disease, and cognitive disorders associated with aging and schizophrenia.

The continued use of NRT in conjunction with continued, albeit reduced, smoking prompts additional questions. For example, the constituents of cigarette smoke that mediate tissue injury are not all precisely known, and it is also not known if modulating the coincident nicotine level might influence their absorption, metabolic disposition, mechanism of action, or elimination. Design of such studies will rely upon the development of more refined and tractable methodology to investigate the in vivo kinetics and dynamics of other constituents of cigarette smoke and their interactions with nicotine.

Finally, although ethnicity has been shown to be relevant (Sabha et al., 2000), the factors that determine interindividual differences in nicotine efficacy, safety, and addictive potential remain largely unexplored. Particular attention might be paid to genetic variation in proteins relevant to nicotine pharmacokinetics and dynamics and their interaction with environmental variables. As with other drugs, one anticipates increasing individualization of nicotine dosage and/or delivery when given as a therapeutic agent. Insight into the interaction of genetic and environmental factors which influence initiation (Gynther et al., 1999; Heath et al., 1999) of cigarette smoking, latency until the practice becomes habitual (Stallings et al., 1999), and the quantity that is then smoked (Koopmans et al., 1999) has been increasing. Clarification of how these factors interact is also likely to afford insights of value in predicting the individual likelihood of response to the use NRT as a strategy for quitting or reducing tobacco exposure.

CANCER (SEE SECTION II, CHAPTER 12)

Feasibility of Harm Reduction in Therapy

There are sufficient laboratory and human data to suggest that harm reduction for cancer might be an achievable goal for persons who cannot stop smoking. There is evidence of decreased risk of cancer for persons who abstain from smoking, and there is strong evidence of a positive relationship between smoking and risk of developing cancer. The risk varies by the different tobacco products used and how they are used. Clearly, abstinence from smoking is the most effective method for reducing cancer risk, and the cancer risk to former smokers is the lowest-level risk that might occur from the use of any PREP. Importantly, it must be recognized that the use of any harm reduction product will likely increase the risk of cancer at some level as long as there is exposure to tobacco carcinogens, in contrast to abstinence, which stops exposure to all tobacco constituents. Nonetheless, reduction in exposure to tobacco smoke and tobacco products to the lowest possible levels may provide some benefit to individual users and to the general population. However, there are insufficient data from which to conclude how much reduction in exposure would yield a measurable benefit and which individuals would benefit. Currently, it seems likely that methods that reduce exposure to tobacco constituents to the greatest extent would likely provide the greatest benefit, but this remains to be proven.

A systematic and thorough assessment of PREPs and cancer risk will require analysis of data obtained from well-designed laboratory and human studies. In laboratory animals, the shape of the dose-response curves differs for different tobacco constituents, indicating that the dose-response relationship of tobacco smoke is complex.

Dose-Response Relationship

In humans, the carcinogenic response increases most around five cigarettes per day, and there is relatively little increase in carcinogenicity above 20 cigarettes a day. However, while there are sufficient data to conclude that a dose-response relationship exists for the use of tobacco products and cancer risk, the precise dose-response relationship is really not known in part because exposure is not accurately measured without considering actual smoking behavior. There is some evidence to indicate that when internal exposure is considered through biomarkers, the shape of the curve follows a quadratic equation, indicating a greater benefit in exposure reduction for persons who smoke more. Thus, data are insufficient to predict the harm reducing effect of a change from any intensity of smoking to a PREP. There are sufficient data to suggest that dose-response relationships differ as a function of gender, race, age, and ethnicity, although the actual risk levels have not been sufficiently defined to draw definitive conclusions about risks among groups. Based on these types of data and possible modifiers of cancer risk (e.g., genetic susceptibilities, diet, lifestyle, occupation), it is likely that PREPs would affect risk differently in different people and not at all in some.

There is no evidence of a threshold below which tobacco smoking does not increase cancer risk. This conclusion is consistent with the fact that there are many carcinogens in tobacco smoke, and the aggregate might increase risk at any level. Modeling for low-dose exposures indicates that there is an increased risk with less than one cigarette per day. Thus, persons who initiate smoking with harm reduction products that contain tobacco would be likely to have an increased risk for cancer, and there is unlikely to be a "safe" cigarette. Former smokers who resume smoking with such products would increase their risk further.

Regression of risk using PREPs might eventually bring a smoker to a risk equal to some lower level of lifetime exposure to conventional products. However, there are insufficient data to validate this assumption or indicate that a decrease in risk would be measurable for some or all smokers. There are insufficient data to indicate the shape of the curve for regression of risk for any PREP.

The data are sufficient to conclude, with some caveats, that filtered cigarettes compared to nonfiltered cigarettes pose a lower risk of lung cancer and possibly other cancers. The caveats are that this occurs only in persons who do not substantially increase the number of cigarettes they smoke per day or otherwise compensate by their smoking behavior for lower levels of nicotine. Also, these studies may be confounded by diet, lifestyle, or other characteristics of people who use filtered cigarettes, which might be different in smokers of filtered compared to nonfiltered cigarettes. The available data are suggestive, but not sufficient, to conclude that smokers of low-tar cigarettes have a lower cancer risk compared to smokers of higher-tar cigarettes, with the same caveats as for the filter smoking studies. However, there are insufficient data to assess the differences in risk for ultralow-, low-, and high-tar cigarettes that are filtered. These cigarettes only became available more recently, so there has not been a long enough latency period in the general population to assess them until recently. There are insufficient data to adequately consider how risk changes when switching types of cigarettes.

This report has not reviewed potential cancer risks due to fibers released from cigarette filters or tobacco additives, because it is thought that the risk of these exposures are substantially lower than the risk from the constituents of tobacco smoke. However, there are no existing data to prove this assumption. Importantly, as harm reduction products are developed that substantially reduce exposure to tobacco constituents, the relative role of fibers and additives in carcinogenesis might become more important. Thus, fiber and additive exposure should be considered when assessing PREPs.

Utility in a Preclinical Setting

There are some experimental models (e.g., in vitro cell cultures, laboratory animals) that may be useful for the assessment of the carcinogenicity of tobacco-related PREPs. Although there are many reasonable models with which to assess individual tobacco smoke products, better models are needed for assessing exposures to complex mixtures. Such studies are not alone sufficient to support claims of potential harm reduction. No claim of potential harm reduction should be allowed without adequate human clinical and epidemiological studies. In vitro and animal studies, however, are very important for (1) determining those products that are not likely to result in measurable harm reduction (e.g., if the product results in exposures that increase genotoxicity, then there would be less enthusiasm for it and so should not be tested in a human clinical study and should not be introduced into the marketplace); (2) identifying unforeseen reactions (e.g., if a product reduces exposure but does not decrease tumors), then there might be some constituent or combination of constituents that is either new or more important than those changed in the product; (3) providing supportive evidence for the use of a particular bioassay in humans (e.g., if a biomarker predicts cancer risk in experimental animals); and 4) assessing the dose-response and the shape of the regression of risk for the PREP as exposure is reduced, although the data should be considered qualitative or semiquantitative and cannot be extrapolated directly to human smoking risk. Both in vitro cell culture and experimental animal studies should be used in assessing PREPs, where both can assess genotoxic and nongenotoxic end points, and chronic animal bioassays are needed to assess the end point of cancer risk. It is beyond the scope of the committee to recommend the specific panel of assays, but such a panel will need to be developed. Also, these studies should assess changes due to both specific carcinogens and to complex mixtures, where the latter should be mandatory.

Clinical Assessment of Tobacco-Related Disease and Biomarkers of Tobacco-Related Disease

There is sufficient evidence to conclude that human experimental studies and short-term clinical studies provide evidence of the harmful effects of tobacco products. Thus, such studies can be used in assessing harm reduction. These studies, through the use of biomarkers and surrogate indicators of cancer risk, can evaluate the manipulation of carcinogens and nicotine to reduce exposures and how these changes might affect smoking behavior, metabolic activation, enzymatic induction, conjugation, excretion, biologically effective doses (or their validated surrogates), and biomarkers of potential harm. Separately, these studies can assess differences in risk and provide evidence for modifying effects due to genetic susceptibilities, diet, lifestyle, occupation, and so forth. However, at the current time, no single biomarker or panel of biomarkers can be considered sufficient indicators of cancer risk by themselves, in part because most have not been sufficiently validated. New technologies are offering new opportunities for biomarkers. Thus, a panel of experts will be needed to devise a set of biomarkers that reflect different exposures, biologically effective doses, and pathways for potential harm.

It is clearly possible to assess the effects of PREPs on cancer as the ultimate outcome, and only such studies can provide definitive evidence for the success of a product. However, the long latency for cancer makes these studies infeasible for making such claims today or in the near future. This relatively long latency period for cancer and the slower decline, probably years or decades, in risk from exposure reduction compared to cardiovascular disease and other tobacco-related diseases will have an impact not only on the time frame of PREP assessment and but also on the health effects experienced by and apparent to the individual. Preneoplastic lesions or the identification of harmful effects in single cells might be used as indicators for the carcinogenetic pathway, but the technology to identify these in the general population or large epidemiological studies is not yet available. In such studies, the characterization of smoking history and behavior is well validated for recent exposures but less accurate for assessing lifetime exposure. Also, self-reported smoking history is insufficient to adequately assess risk in the context of PREP assessments, so biomarkers also are needed to assess exposure, biologically effective doses, and potential harm.

Currently, the best approach to assessing PREPs and cancer risk is to focus on lung cancer, because this is the most common cancer and so will provide studies with the greatest statistical power. However, data are sufficient to conclude that there is a risk that the widespread use of PREPs will shift the burden of cancer in the population from one type to another or from cancer to a different disease. Thus, a particular cancer type cannot be the sole indicator for the success of a PREP, and other cancers, diseases, and overall mortality must be evaluated as well.

Many studies of nicotine suggest that nicotine is unlikely to be a cancer-causing agent in humans or, at worst, that its carcinogenicity would be trivial compared to that of other components of tobacco. The consideration of nicotine as a carcinogenic agent, if at all, is trivial compared to the risk of other tobacco constituents.

Some smokeless tobacco products increase the risk of oral cavity cancers, and a dose-response relationship exists. However, the overall risk is lower than for cigarette smoking, and some products, such as Swedish snus, may have no increased risk. It may be considered that such products could be used as PREPs for persons addicted to nicotine, but these products should undergo testing as PREPs using the guidelines and research agenda contained herein.

The effects of PREPs on cancer risk from environmental tobacco smoke (ETS) are uncertain because of the difficulties in measuring reductions in exposure. Also, although there is clearly an increased risk of lung cancer from ETS, the determination of changes in risk from the use of PREPs will require studies of large numbers of people, and smoking is currently in this country prohibited in many places where ETS might have occurred.

CARDIOVASCULAR DISEASE (SEE SECTION II, CHAPTER 13)

Dose-Response Relationship

Highly informative information on the existence of a dose-response relationship between cigarette exposure and cardiovascular risk comes from many studies such as the CPS-II (Thun et al., 1997) and Harvard Nurses' Health Studies (Kawachi et al., 1997). In both instances, there is a relationship between the number of cigarettes smoked and the incidence of cardiovascular events. This is illustrated for the incidence of myocardial infarction and stroke. In both cases, the most striking difference is between nonsmokers and individuals who smoke the least number of cigarettes recorded. The relationship becomes somewhat less pronounced as the number of cigarettes smoked per day increases. Interestingly, ex-smokers tend to occupy a space intermediate between nonsmokers and those with the lowest daily smoking frequency.

This dose-response curve prompts several considerations. First, there is no persuasive evidence of a threshold below that a cardiovascular risk does not exist. This observation affirms the primary objective of encouraging smokers to quit completely. Second, the shallow dose-response relationship, with the impression of a plateau, accords with similar observations relating the number of cigarettes smoked and measurements of systemic bioavailability, such as nicotine and cotinine. The same is true of the relationship with CO (Gori and Lynch, 1985). Although this may represent saturation kinetics of nicotine or CO, the most likely explanation is compensation for lower numbers of cigarettes smoked. Thus, the smoker titrates nicotine delivery toward a range of convergence that is reflected by the measurement of nicotine delivery, which in turn, may reflect dose-dependent convergence of the delivery of additional toxic, but unmeasured, constituents of cigarette smoke. The relative contribution of distinct constituents of cigarette smoke to smoking-related cardiovascular morbidity and mortality is unknown. In this regard, the available information is incomplete regarding any differences between the dose-response relationship of filter or low-tar versus high-tar or unfiltered cigarettes.

Feasibility of Harm Reduction in Therapy

There are no data that are directly informative on the issue of harm reduction. Thus, although a dose-response relationship exists for cardiovascular and cerebrovascular events, we do not know if reducing the number of cigarettes smoked results in a quantitative reduction in the risk of these events. However, the intuitively appealing prospect that this is indeed the case is supported by evidence from individuals who quit smoking. Thus, quitting results in a time-dependent reduction in the incidence of myocardial infarction and stroke. The latter is most nicely illustrated by data relating to subarachnoid hemorrhage, which was significantly elevated in women smokers in the Nurses' Health Study. In addition to evidence from such unequivocal clinical events, there is evidence that the increase in biomarkers of oxidant stress, platelet activation, and inflammation (Benowitz et al., 1993), all of potential mechanistic relevance to tobacco-related cardiovascular injury, rapidly falls toward the normal range on quitting cigarettes. The offset kinetics of more functional surrogates, such as endothelial dysfunction, remain to be determined in smokers. In summary, the data from quitters encourage the prospect that a graded reduction in cardiovascular risk and in biomarkers of this risk may accompany a reduction in the number of cigarettes smoked in pursuit of a harm reduction strategy. It is possible, indeed likely, that individuals and perhaps populations, differ in their susceptibility to tobacco-induced cardiovascular risk and, indeed, in their potential benefit from a harm reduction strategy. Data from quitting studies indicate a considerable variance in the rate of offset of risk, which declines with time. No data are available to address such issues across ethnic groups or gender. Acquisition of such information and research on the environmental and genetic factors that condition interindividual variability in exposure-risk relationships are necessary.

Utility in a Preclinical Setting

Studies in cell culture and model systems can afford much needed information on tobacco-related cardiovascular risk. These might include a profiling of gene expression and translation in cardiovascular tissues in response to cigarette smoke, constituents of smoke, and potential harm reduction substituents. These might identify proteins of potential functional relevance to the transduction of cardiovascular risk. Such studies might be coupled with gene inactivation and overexpression studies to address the role of these proteins in vivo. Similarly, studies of exposure to cigarette smoke or to discrete constituents of smoke might be deployed to investigate effects on atherosclerosis progression, susceptibility to vascular injury, thrombotic stimuli, graft rejection, cardiovascular development, or endothelial dysfunction in model systems such as mice. Studies of cardiovascular genomics and ultimately proteomics can also be extended to model systems to investigate gene expression and translation in response to exposure to tobacco-related products in vivo. These observations may, in turn, be related to the pattern of gene expression and translation in cardiovascular tissues obtained from cigarette smokers.

Biomarkers of Tobacco-Related Disease

The predominant mechanisms by which cigarette smoking induces cardiovascular injury is unknown. However, small studies in smokers of potentially relevant biomarkers of platelet and vascular activation, lipid peroxidation, and inflammation afford evidence of a dose-response relationship and a decline on quitting. There is even evidence of a signal in individuals exposed to ETS in the case of some of these markers. More mechanism-based clinical studies are required to confirm and expand these findings. Where possible, these should be related to surrogate measurements of cardiovascular function, such as hemodynamics, flow-mediated endothelial function and estimates of plaque progression by ultrasound or electron-beam computerized tomography (EBCT). Furthermore, biomarker studies can usefully be integrated into many studies in model systems as well as studies of clinical outcome to afford their ultimate validation.

Clinical Assessment of Tobacco-Related Disease

The time course of offset of myocardial infarction and stroke in people who stop smoking suggests that cardiovascular disease represents a tractable scenario in which one might evaluate harm reduction strategies. Clearly, the health effects experienced by the individual and the assessment of the impact of such events can occur in a more reasonable time frame than from cancer in which declines in risk from tobacco exposure reduction may only be apparent after years or decades.

NONNEOPLASTIC RESPIRATORY DISEASE
(SEE SECTION II, CHAPTER 14)

In evaluating harm reduction strategies for tobacco-related lung disease, three major nonneoplastic respiratory diseases linked to cigarette smoking are considered: COPD, asthma, and respiratory infections. Respiratory diseases are major tobacco-related illnesses, and there is a clear need to mitigate the harmful effects of exposure to both mainstream and secondary tobacco smoke. It is generally accepted that cessation of smoking slows or stops the progression of the lung diseases related to smoking and it is plausible that decreasing smoking will reduce the severity of chronic lung diseases and the incidence of respiratory infections. However, there is no adequate scientific evidence to support this because the effects of reduced smoking on harm reduction have not been extensively studied in man.

Dose-Response Relationship

There is a need to determine dose-response relationships more precisely and to develop biomarkers of respiratory disease. Rational design of studies to assess harm reduction requires knowledge of the dose-response relationship. At present, such data for respiratory diseases are limited and of uncertain quality. Study design would also incorporate biomarkers of disease, and the testing of current and new biomarkers might be done concurrently in the models and populations studied for dose-effects. The Cancer Prevention Studies I and II, large-scale prospective studies, however, do suggest a direct dose-response relationship between cigarettes smoked per day and mortality rates from COPD (NIH, 1996, 1997), indicating that decreasing the number of cigarettes smoked may lead to fewer deaths from COPD.

Biomarkers of Tobacco-Related Disease

There are currently no specific molecular biomarkers of the nonneoplastic respiratory diseases due to smoking tobacco products. No unique molecular or genetic defect specific for tobacco-related respiratory disease has been identified. The processes involved, such as inflammation and increased levels of oxidants, are not unique to tobacco-related respiratory diseases. Identifying unique biomarkers is further confounded by the heterogeneous nature of these diseases, the complex mixture of tobacco smoke, and the range of individual susceptibilities to the harmful effects of tobacco smoke. The most widely used markers of tobacco-related respiratory diseases in population studies are symptom questionnaires and pulmonary function testing. These have well-known limitations of specificity and sensitivity, particularly for detecting the early effects of tobacco smoke on lungs (U.S. DHHS, 1989). Subtle effects of tobacco smoke exposure on the lung can be detected by sampling fluid in the lower respiratory tract via a bronchoscope inserted into the airways, but the significance of these changes for clinically important pulmonary disease has not been established. Newer approaches such as sampling the subjects' urine (Pratico et al., 1998) or exhaled gas (Ichinose et al., 2000) for metabolic products due to tissue injury have the advantage of noninvasive sampling but must be validated. Clearly, the greatest obstacle for rational development of a specific biomarker is the lack of fundamental information on mechanisms of how tobacco smoke exposure causes specific respiratory diseases.

The availability of dose-response data and validated biomarkers may improve the design of contemplated intervention studies and allow greater confidence in the results. However, the time frame for generating dose-response data and testing biomarkers is uncertain. The inclusion of dose-response considerations and biomarkers in the design of clinical trials on reduction of harm from respiratory diseases must also be validated.

Clinical Assessment of Tobacco-Related Disease

An alternative is to proceed with interventional trials based on current knowledge if there are uncertainties about the added value of dose-response data or untested biomarkers to study design. As an example, an intervention study of the effect of smoking reduction on COPD could be considered, similar in design to the Lung Health Study (Anthonisen et al., 1994), a large prospective trial of the effects of smoking cessation on rate of decline of FEV_1 (forced expiratory volume at 1 second) in middle-aged smokers with mild COPD. Another approach is to conduct a trial using a low-tar and moderate-nicotine product made available from a noncommercial source to avoid product endorsement issues.

Design of population studies for harm reduction of major respiratory diseases is challenging because of uncertainties about effectiveness and long-term compliance with harm reduction interventions. Reduction in the burden of tobacco-related respiratory diseases through harm reduction strategies should be a major priority for the nation's public health.

REPRODUCTIVE AND DEVELOPMENTAL EFFECTS
(SEE SECTION II, CHAPTER 15)

Feasibility of Harm Reduction in Therapy

Cigarette smoking is a major cause of fetal and infant morbidity and mortality (U.S. DHHS, 1988, 1990; Kleinman et al., 1988). This is particularly true for the associations with low-birthweight and its consequences, as well as preterm delivery and SIDS (CDC, 2000; Leach et al., 1999; Shah and Bracken, 2000; U.S. DHHS, 1983). For several important adverse reproductive effects of maternal smoking, a decrease in smoking has been found to be associated with a decrease in risks to the fetus and infant (Li et al., 1993; Hebel et al., 1988). The greatest benefit, of course, comes from smoking cessation. However, the smoking cessation rate for women smokers who become pregnant is very low and remain comparable to those in the general population, despite knowledge of the harmful effects of smoking and personal experience with adverse fetal and infant conditions. Moreover, as current rates of smoking increase slowly among adolescent women, these adverse effects associated with tobacco smoke exposure while pregnant are likely to worsen.

Dose-Response Relationship

On average, infants exposed to maternal smoking in utero are 200 grams lighter and 1.4 cm shorter than those unexposed (Wang et al., 1997). A strong dose-response relationship has been supported in numerous studies (Li et al., 1993),and a decrease in dose (number of cigarettes) in controlled studies has led to increased birthweights in a predictable pattern (Wang et al., 1997). What is known about the mechanism of effect of cigarette smoke on the fetus suggests that several agents in tobacco smoke contribute to the adverse effects. There is evidence that CO plays a major role in growth retardation through increased tissue hypoxia (Benowitz et al., 2000). Nicotine has also been thought to play a role through increasing vasoconstriction and decreasing perfusion through the placenta.

Although nicotine replacement products and buproprion are currently not approved by the Food and Drug Administration for use by pregnant women, the Agency for Healthcare Research and Quality's (AHCRQ) Clinical Practice Guidelines for Treating Tobacco Use and Dependence (Fiore et al., 2000) recommend that "Pharmacotherapy should be considered when a pregnant woman is otherwise unable to quit, and when the likelihood of quitting, with its potential benefits, outweighs the risks of the pharmacotherapy and potential continued smoking". It is generally thought that NRT can reasonably be used with pregnant patients if prior behavioral modifications have failed and the patient continues to smoke at least 10-15 cigarettes per day (ACOG, 1997). There are no data regarding the efficacy of potential reduced-exposure products (PREPs) during pregnancy, but there is the presumption that the tobacco-related PREPs are likely to have adverse effects at some level and that until further evidence is produced, existing guidelines concerning pharmacologic PREPs still pertain.

Clinical Assessment of Tobacco-Related Disease and Utility in a Preclinical Setting

To practically assess the health effects of PREPs, reliable measures of health outcomes that can be utilized in a relatively short time are desired. Among the reproductive outcomes of maternal smoking, intrauterine growth retardation resulting in low-birthweight babies has been studied extensively, and a large body of evidence has supported a causal link with cigarette smoke exposure. The committee recommends, based on currently available scientific knowledge, that fetal birthweight be used as a reliable outcome measure for evaluating the harm reduction potential of specific PREPs. Study designs should include repeated cohort or case-control studies of pregnant women, with an appropriate distribution of exposures to both PREPs and conventional products, and suitable contrast groups. Concomitant, coordinated toxicological studies should be undertaken to provide biological correlations with clinical outcomes. Such outcomes as fetal birth weight and the incidence of other reproductive and developmental health outcomes (e.g., fertility outcomes, placental complications, gestational age at birth, incidence of sudden infant death syndrome [SIDS], spontaneous abortions) should be considered primary objects of study in order to assess the harm reduction potential of specific PREPs.

Findings in pregnant women exposed to PREPs may have value beyond maternal or fetal outcomes. The nature of adverse effects derived from PREP exposure will likely be determined much sooner in this case than findings on chronic disease outcomes in humans, such as various cancers and cardiovascular disease. Should adverse findings become apparent, there may be substantial implications for chronic illnesses among older adults, and coordinated pathogenic studies might allow conclusions on new tobacco product outcomes in advance of studies exploring longer "incubation periods."

The committee recommends that further basic research be undertaken to elucidate the components of cigarette smoke that are primarily responsible for adverse health outcomes. In order to evaluate the safety of many PREPs, it is important to understand the toxicity of specific smoke components, especially nicotine and CO, on the pathogenesis of intrauterine growth retardation, spontaneous abortion, and other health outcomes. In addition, a better understanding of the risks of bupropion SR use by pregnant women (i.e., seizure risk) and the teratogenic effects of nicotine on the central nervous system (CNS) is needed for adequate risk-benefit analysis of the harm reduction potential of these products.

Surveillance of Tobacco Use Patterns Among Pregnant Women

Central to understanding exposure to tobacco products is continuous population information on usage patterns among pregnant women. This may not be attainable by general population survey methods because of inadequate sample sizes and insufficient representation of various geographic or demographic groups or of the earliest stages of pregnancy. There is a need for surveys devoted specifically to pregnant women in all stages of gestation, irrespective of the receipt of medical care. Survey content should include other known or putative causes of adverse maternal or fetal outcomes, as well as detailed product types and usage patterns. Recommendations for general population surveillance can be found in Chapter 6 of this report.

Biochemical and toxicological exposure measures should be a routine part of surveillance for exposure to conventional products as well as PREPs. These will be necessary to conduct more precise, coordinated toxicological studies and also to assess actual exposure rates more accurately. For example, dose may be measured by maternal serum and urine cotinine levels, which have shown reliable correlations with maternal, and consequently fetal, tobacco smoke exposure. Self-reported data have been found unreliable, since pregnant women tend to underreport tobacco use because of the stigma attached to smoking. Also, self-reports do not adequately account for differences in depth and frequency of puffs among smokers.

OTHER HEALTH EFFECTS (SEE SECTION II, CHAPTER 16)

Feasibility of Harm Reduction in Therapy

Several important diseases and conditions of adults, in addition to cardiovascular diseases, chronic obstructive lung disease and various cancers, have been associated with tobacco use, including—but not limited to—peptic ulcer disease, poor wound healing, inflammatory bowel disease, rheumatoid arthritis, oral disease, dementia, osteoporosis, ocular disease, diabetes, dermatological disease, schizophrenia, and depression (see Chapter 16). Some of these associations are supported by substantial scientific evidence, and a causal linkage is likely. These illnesses must ultimately be subjected to the same evaluation of changing risks and outcomes associated with PREPs, because these are common and clinically important conditions, even if they are not as often fatal as cancer, cardiovascular disease or pulmonary disease. Further, each of the conditions for which the association with tobacco use is substantial also offers the opportunity to address pathogenic mechanisms related to the varying constituents of PREPs, as well as the impact on disease incidence of concomitant behaviors and exposures such as alcohol use, various dietary elements, and certain medications.

Utility of Preclinical Studies and Short-term Indicators of Clinical Harm Reduction

Some of the conditions reviewed in this chapter may be applied as indicators of the general biological effects of new tobacco products. For example, cigarette smoking has been consistently found to be an independent risk factor for an adverse clinical course of both peptic ulcer disease and wound healing. The effects of smoking on ulcer formation and healing have been clearly described clinically and in animal models (Ma et al., 1999). Peptic ulcers have been found to be larger, slower to heal, and more likely to recur among smokers and to exhibit clinically improved healing upon cessation (Tatsuta et al., 1987). Surgical and traumatic wounds heal more slowly among cigarette smokers (Kwaitkowski et al., 1996; Mosely et al., 1978). The committee recommends that rigorous clinical studies be designed and executed to determine whether variations in ulcer and wound-healing rates are related to various categories of tobacco products, including those with claims of harm reduction. This may offer the opportunity to define some clinical outcomes that have clinical relevance in their own right and to identify potential indicators of harm alteration much sooner after the introduction of PREPs than would be possible when evaluating heart disease and cancer.

Other candidate diseases for such evaluation might include periodontal disease (Bergstrom, 2000; Haber, 1994), Crohn's disease (Rhodes and Thomas; 1994), and rheumatoid arthritis (Uhlig et al., 1999). Here the outcomes to assess would be the effect of various conventional tobacco products and PREPs on the natural history of these conditions, including intermittancy, progression or regression, and longitudinally collected biomarkers of disease severity. As noted above, PREPs that alter the history and outcomes of these conditions could be further evaluated for specific constituent exposures associated with this altered history. This may lead to a more refined understanding of pathogenic mechanisms as well.

Clinical and basic research on intermediate clinical outcomes is also needed. For example, as noted in this chapter, the risk of osteoporosis has also been strongly linked to cigarette smoking. In controlled observational studies, bone mineral density has been found to be significantly lower among cigarette smokers, which contributes to a greater risk of osteoporotic fractures among older populations. While the effects of smoking on fracture rates may take a few decades or longer to detect, it is possible that surveillance of bone mineral density among those using PREPs and conventional products may be informative in a shorter time period and, thus, serve to detect important outcomes over an interval in which tobacco policy and clinical preventive interventions may have their greatest effects.

Surveillance

The committee recommends that selected conditions, as reviewed in this chapter, be part of a comprehensive, population-based surveillance program, outlined in Chapter 6. This will allow determination of the relationship between the use of PREPs and of trends in occurrence for these tobacco-related conditions and assessment on a national basis of whether changes in tobacco product use have an effect on these important health problems. Based on these surveillance findings, more specific population, clinical, and basic research studies can be directed to evaluate PREPs to pursue causal mechanisms and to suggest more effective interventions.

REFERENCES

ACOG (American College of Obstetricians and Gynecologists). 1997. Educational Bulletin: Smoking and women's health. *International Journal of Gynecology & Obstetrics* 60:71-82.

Anthonisen NR, Connett JE, Kiley JP, Altose MD, Bailey WC, Buist AS, Conway WA Jr, Enright PL, Kanner RE, O'Hara P, et al. 1994. Effects of smoking intervention and the use of an inhaled anticholinergic bronchodilator on the rate of decline of FEV1. The Lung Health Study. *JAMA* 272(19):1497-1505.

Benowitz NL, Dempsey DA, Goldenberg RL, Hughes JR, Dolan-Mullen P, Ogburn PL, Oncken C, Orleans CT, Slotkin TA, Whiteside HP Jr, Yaffe S. 2000. The use of pharmacotherapies for smoking cessation during pregnancy. *Tob Control* 9 Suppl 3(2):III91-94.

Benowitz NL, Fitzgerald GA, Wilson M, Zhang Q. 1993. Nicotine effects on eicosanoid formation and hemostatic function: comparison of transdermal nicotine and cigarette smoking. *J Am Coll Cardiol* 22(4):1159-1167.

Benowitz NL, Zevin S, Jacob P 3rd. 1998. Suppression of nicotine intake during ad libitum cigarette smoking by high-dose transdermal nicotine. *J Pharmacol Exp Ther* 287(3):958-962.

Bergstrom J, Eliasson S, Dock J. 2000. Exposure to tobacco smoking and periodontal health. *J Clin Periodontol* 27(1):61-68.

CDC (Centers for Disease Control and Prevention). 2000. Tobacco use during pregnancy. *National Vital Statistics Report* 48(3):10-11.

Chalon S, Moreno H Jr, Benowitz NL, Hoffman BB, Blaschke TF. 2000. Nicotine impairs endothelium-dependent dilatation in human veins in vivo. *Clin Pharmacol Ther* 67(4):391-397.

Davis DL, Nielsen MT, eds. 1999. *World Agriculture Series; Tobacco Production, Chemistry and Technology*. London: Blackwell Science.

Eclipse Expert Panel. 2000. A safer cigarette? A comparative study. A consensus report. *Inhalation Toxicology* 12(Supp 5):1-48.

Fagerström, K, Tejding, R, Westin, A, Lunell, E. 1997. Aiding reduction of smoking with nicotine replacement medications: hope for the recalcitrant smoker? *Tobacco Control* 6: 311-316.

Fiore, M, Bailey, W, Cohen, S, Dorfman, S, Goldstein, M, Gritz, E, Heyman, R, Jaen, C, Kottke, T, Lando, H, Mecklenburg, R, Mullen, P, Nett, L, Robinson, L, Stitzer, M, Tommasello, A, Villejo, L, Wewers, M. 2000. *Treating Tobacco Use and Dependence. Clinical Practice Guideline*. Rockville, MD: U.S. Department of Health and Human Services. Public Health Service.

Gairola CG, Daugherty A. 1999. Acceleration of atherosclerotic plaque formation in ApoE- mice by exposure to tobacco smoke. *The Toxicologist* 48:297.

Gori GB, Lynch CJ. 1985. Analytical cigarette yields as predictors of smoke bioavailability. *Regul Toxicol Pharmacol* 5(3):314-326.

Gynther LM, Hewitt JK, Heath AC, Eaves LJ. 1999. Phenotypic and genetic factors in motives for smoking. *Behav Genet* 29(5):291-302.

Haber J. 1994. Smoking is a major risk factor for periodontitis. *Curr Opin Periodontol:* 12-8.

Heath AC, Kirk KM, Meyer JM, Martin NG. 1999. Genetic and social determinants of initiation and age at onset of smoking in Australian twins. *Behav Genet* 29(6):395-407.

Hebel JR, Fox NL, Sexton M. 1988. Dose-response of birth weight to various measures of maternal smoking during pregnancy. *J Clin Epidemiol* 41(5):483-489.

IARC (International Agency for Research on Cancer). 1986. *Tobacco Smoking: Monographs on the Evaluation of Carcinogenic Risk of Chemicals to Humans*. Vol. 38 ed. Lyon, France: IARC.

Ichinose M, Sugiura H, Yamagata S, Koarai A, Shirato K. 2000. Increase in reactive nitrogen species production in chronic obstructive pulmonary disease airways. *Am J Respir Crit Care Med* 162(2 Pt 1):701-706.

Kawachi I, Colditz GA, Stempfer MJ, Willett WC, Manson JE, Rosner B, Hunter DJ, Hennekens CH, Speizer FE. 1997. Smoking cessation and decreased risk of total mortality, stroke, and coronary heart disease incidence in women; a prospective cohort study. National Cancer Institute. *Changes in Cigarette-Related Disease Risks and Their Implication for Prevention and Control. NCI Smoking and Tobacco Control Monograph 8.* Washington, DC: NCI. Pp. 531-565.

Kleinman JC, Pierre MB, Madans JH, Land GH, Schramm WF. 1988. The effects of maternal smoking on fetal and infant mortality. *Am J Epidemiol* 127(2):274-282.

Koopmans JR, Slutske WS, Heath AC, Neale MC, Boomsma DI. 1999. The genetics of smoking initiation and quantity smoked in Dutch adolescent and young adult twins. *Behav Genet* 29(6):383-393.

Kwiatkowski TC, Hanley Jr EN, Ramp WK. 1996. Cigarette smoking and its orthopedic consequences. *Amer J Orthopedics* 25(9):590-597.

Leach CE, Blair PS, Fleming PJ, Smith IJ, Platt MW, Berry PJ, Golding J. 1999. Epidemiology of SIDS and explained sudden infant deaths. CESDI SUDI Research Group. *Pediatrics* 104(4):e43.

Li CQ, Windsor RA, Perkins L, Goldenberg RL, Lowe JB. 1993. The impact on infant birth weight and gestational age of cotinine-validated smoking reduction during pregnancy. *JAMA* 269(12):1519-1524.

Ma L, Chow JY, Cho CH. 1999. Cigarette smoking delays ulcer healing: role of constitutive nitric oxide synthase in rat stomach. *Am J Physiol* 276(1 Pt 1):G238-248.

Mosely LH, Finseth F, Goody M. 1978. Nicotine and its effect on wound healing. *Plastic and Reconstructive Surgery* 61(4):570-574.

NIH (National Institutes of Health), National Cancer Institute. 1996. *The FTC Cigarette Test Method for Determining Tar, Nicotine, and Carbon Monoxide Yields of U.S. Cigarettes. Smoking and Tobacco Control Monograph 7.* Bethesda, MD: National Institutes of Health.

NIH (National Institutes of Health), National Cancer Institute. 1997. *Changes in Cigarette-Related Disease Risks and Their Implication for Prevention and Control. Smoking and Tobacco Control Monograph 8.* Bethesda, MD: National Institutes of Health.

NRC (National Research Council). 1983. *Risk Assessment in the Federal Government: Managing the Process.* Washington, DC: National Academy Press.

Perera FP. 1987. Molecular cancer epidemiology: a new tool in cancer prevention. *J Natl Cancer Inst* 78(5):887-898.

Practico D, Basili S, Vieri M, Cordova C, Violi F, Fitzgerald GA. 1998. Chronic obstructive pulmonary disease is associated with an increase in urinary levels of isoprostane F2alpha-III, an index of oxidant stress. *Am J Respir Crit Care Med* 158(6):1709-1714.

Rennard, S, Daughton, D, Fujita, J, Oehlerking, M, Dobson, J, Stahl, M, Dobson, J, Stahl, M, Robbins, R, Thompson, A. 1990. Short-term smoking reduction is associated with reduction in measures of lower respiratory tract inflammation in heavy smokers. *European Respiratory Journal* 3(7):752-759.

Rhodes J, Thomas GA. 1994. Smoking: good or bad for inflammatory bowel disease? *Gastroenterology* 106(3):807-810.

Sabha M, Tanus-Santos JE, Toledo JC, Cittadino M, Rocha JC, Moreno H Jr. 2000. Transdermal nicotine mimics the smoking-induced endothelial dysfunction. *Clin Pharmacol Ther* 68(2):167-174.

Shah NR, Bracken MB. 2000. A systematic review and meta-analysis of prospective studies on the association between maternal cigarette smoking and preterm delivery. *Am J Obstet and Gynecol* 182(2):465-472.

Shiffman S, Mason KM, Henningfield JE. 1998. Tobacco dependence treatments: review and prospectus. *Annu Rev Public Health* 19:335-358.

Sopori ML, Kozak W, Savage SM, Geng Y, Soszynski D, Kluger MJ, Perryman EK, Snow GE. 1998. Effect of nicotine on the immune system: possible regulation of immune responses by central and peripheral mechanisms. *Psychoneuroendocrinology* 23(2):189-204.

Stallings MC, Hewitt JK, Beresford T, Heath AC, Eaves LJ. 1999. A twin study of drinking and smoking onset and latencies from first use to regular use. *Behav Genet* 29(6):409-421.

Tatsuta M, Iishi H, Okuda S. 1987. Effects of cigarette smoking on the location, healing and recurrence of gastric ulcers. *Hepatogastroenterology* 34(5):223-228.

Thun MJ, Myers DG, Day-Lally C, Namboodiri MM, Calle EE, Flanders WD, Adams SL, Heath CW. 1997. Age and exposure-response relationships between cigarette smoking and premature death in cancer prevention study II. National Cancer Institute. *Changes in Cigarette-Related Disease Risks and Their Implication for Prevention and Control. NCI Smoking and Tobacco Control Monograph 8.* Washington, DC: NCI. Pp. 383-475.

Transdermal Nicotine Study Group. 1991. Transdermal nicotine for smoking cessation. *JAMA* 266(22):3133-3138.

U.S. DHHS (U.S. Department of Health and Human Services). 1983. *The Health Consequences of Smoking; Cardiovascular Disease; A Report of the Surgeon General.* Atlanta, GA: U.S. Department of Health and Human Services, Centers for Disease Control and Prevention.

U.S. DHHS (U.S. Department of Health and Human Services). 1988. *The Health Consequences of Smoking, Nicotine Addiction; A Report of the Surgeon General.* Atlanta, GA: U.S. Department of Health and Human Services, Centers for Disease Control and Prevention.

U.S. DHHS (U.S. Department of Health and Human Services). 1989. *Reducing the Health Consequences of Smoking; A Report of the Surgeon General.* Washington, DC: U.S. Department of Health and Human Services, Centers for Disease Control and Prevention.

U.S. DHHS (U. S. Department of Health and Human Services). 1990. *The Health Benefits of Smoking Cessation; A Report of the Surgeon General.* Rockville, MD: U.S. Department of Health and Human Services, Centers for Disease Control and Prevention.

Uhlig T, Hagen KB, Kvien TK. 1999. Current tobacco smoking, formal education, and the risk of rheumatoid arthritis. *J Rheumatol* 26(1):47-54.

Vineis P, Porta M. 1996. Causal thinking, biomarkers, and mechanisms of carcinogenesis. *J Clin Epidemiol* 49(9):951-956.

Wang X, Tager IB, Van Vunakis H, Speizer FE, Hanrahan JP. 1997. Maternal smoking during pregnancy, urine cotinine concentrations, and birth outcomes. A prospective cohort study. *Int J Epidemiol* 26(5):978-988.

Witschi H, Espiritu I, Maronpot RR, Pinkerton KE, Jones AD. 1997a. The carcinogenic potential of the gas phase of environmental tobacco smoke. *Carcinogenesis* 18(11):2035-2042.

Witschi H, Espiritu I, Peake JL, Wu K, Maronpot RR, Pinkerton KE. 1997b. The carcinogenicity of environmental tobacco smoke. *Carcinogenesis* 18(3):575-586.

Witschi H, Espiritu I, Uyeminami D. 1999. Chemoprevention of tobacco smoke-induced lung tumors in A/J strain mice with dietary myo-inositol and dexamethasone. *Carcinogenesis* 20(7):1375-1378.

Witschi H, Uyeminami D, Moran D, Espiritu I. 2000. Chemoprevention of tobacco-smoke lung carcinogenesis in mice after cessation of smoke exposure. *Carcinogenesis* 21(5):977-982.

6

Surveillance for the Health and Behavioral Consequences of Exposure Reduction

The goal of surveillance systems in epidemiology and public health is to provide timely information from populations on the occurrence of diseases and conditions of interest, the presence of risk factors for those conditions, and the impact of disease control programs. Public health surveillance systems are not the only sources of information on the frequency or causes of various disease nor are they the only indicators of disease control program success or failure, but the population perspective brings focus to the entire community and the totality of the burden of suffering from various conditions.

The Centers for Disease Control and Prevention (CDC) offers the following definition of surveillance (Thacker and Berkelman, 1988):

> Public health surveillance is the ongoing, systematic collection, analysis, and interpretation of health data essential to the planning, implementation, and evaluation of public health practice, closely integrated with the timely dissemination of these data to those who need to know. The final link in the surveillance chain is the application of these data to prevention and control. A surveillance system includes a capacity for data collection, analysis, and dissemination linked to public health programs.

The extent and sophistication of surveillance systems have evolved over the years (Remington and Goodman, 1998). At the turn of the 20th century, they largely involved monitoring of persons with particular infectious diseases and their personal contacts, such as surveillance of persons who came in contact with smallpox or typhus cases. By mid-century, they evolved into monitoring a wide variety of communicable diseases for detection and control purposes. Selected chronic illnesses became the target of surveillance programs beginning in the 1970s. Later, a host of surveillance techniques were used to monitor environmental exposures such as hazardous occupations, personal injuries, and health-related individual behaviors. Tobacco use was first studied in a federal survey in 1955 (Haenszel et al, 1956). In 1996, the Council of State and Territorial Epidemiologists added the state-specific prevalence of cigarette smoking to the list of conditions designated as notifiable by states to the CDC (CDC, 1996).

Among the attributes that are used to evaluate surveillance systems are simplicity, flexibility, acceptability, sensitivity, representativeness, and timeliness (Klaucke et al., 1988). The *simplicity* of a given surveillance system is influenced both by its structure and ease of operation. A given surveillance system will ideally be as simple as possible and still meet all of its objectives. A *flexible* system can economically adapt to changing information needs or operating conditions. *Acceptability* refers to the willingness of organizations and individuals to adopt and/or participate in the surveillance system. In this instance, acceptability will be influenced by whether the system is mandated. *Sensitivity* refers to the ability of a system to detect diseases and conditions, health states, or various health behaviors or attitudes of interest. A *representative* surveillance system will accurately describe the distribution of a health event by person, place, and time. A *timely* system minimizes the delay between occurrence of an event and the initiation and completion of the process of monitoring and reporting of findings.

Another important attribute of surveillance systems is whether the detection targets are collected *actively* or *passively*. Passive surveillance generally involved the collection of spontaneously reported health events from interested health professionals or others. The current system of reporting adverse drug events to the Food and Drug Administration generally falls into this category. On the other hand, active surveillance involves expending the resources to marshal all available data collection modes to assure as complete an ascertainment as possible of the health and behavioral events of interest. Active surveillance would seem to be essential for helping to assess the impact of PREPs in population context.

This chapter reviews existing surveillance systems and activities for monitoring tobacco product exposure and their health consequences, with emphasis on the introduction and use of PREPs and the issue of harm reduction in the United States. Then proposals to enhance existing surveillance programs are offered. While surveillance data provide only one part of the information needed for scientific and regulatory judgments, it is a critical component that complements clinical, basic, and other data collection. In general, a successful surveillance activity would determine amounts and types of tobacco products distributed in the community, population patterns of product use, and rates of smoking-related conditions. Specifically, an ideal surveillance system for evaluation of PREPs and other tobacco products would contain the following elements:

1 *Consumption of tobacco products and PREPs.* A first step to understanding changes in tobacco-attributable diseases and the impact of control programs is to monitor consumption rates for conventional tobacco products and PREPs. The federal government has monitored per capita consumption (in pounds) of tobacco products for over a century (Millmore and Conover, 1956; U.S. Department of Agriculture, 2000). National estimates of consumption use overall include sales data and are adjusted to incorporate estimates of smuggling. Information on the use of pharmaceutical aids for smoking cessation has been published recently (CDC, 2000b).

2 *Specific tobacco constituents of both the products and the smoke they generate.* Central to any surveillance system is accurate characterization of environmental exposures of interest. With respect to conventional tobacco products and PREPs, documenting the physical and chemical content of these products, including additives and structural components, is critical. It is equally important to determine the constituents of the products of tobacco product combustion and other elements otherwise delivered during human consumption.

3 *Tobacco product marketing, including PREPs.* It is similarly extremely important to understand the distribution and availability of PREPs in the community. For example, monitoring of general media advertising, free-sample distribution, and other marketing practices including mass mailings and public relations activities would seem essential to monitor any health claims, implicit or explicit, related to PREPs as well as conventional tobacco products.

4 *Biomarkers of exposure to tobacco products.* Depending fully on personal self-report of tobacco product use is important but not always sufficient. On occasion, individuals may misrepresent their tobacco exposure or may not be fully aware of it. Further, bodily exposure to tobacco constituents may not be fully ascertained from self-report due to variation in smoking behavior and use patterns (i.e., smoking topography). Biomarkers can also provide indication of the degree of exposure to environmental tobacco smoke among nontobacco users. For these and other reasons, population levels of biomarkers of exposure become extremely important.

5 *Personal tobacco product use and related behavioral patterns.* Critical to assessing the health impact of conventional tobacco products and PREPs is the determination of actual products used, including product types and brands. It is also important to understand the impact of PREPs in terms of smoking initiation, quit attempts, maintained abstinence, and personal consumption patterns (Shiffman, 1999). In general, this can only be determined from sample surveys of relevant populations. Attitudes toward tobacco usage and knowledge of actual threats to health would also be important components of such a system.

6 *Disease outcomes.* Current surveillance of tobacco-related illnesses through mechanisms such as vital records and disease registries provide important information. The development of additional types of registries, clinical record monitoring systems, and systems measuring aggregate health outcomes would add further useful information. Supplementary epidemiological studies of PREPs would enhance the ability to determine specific health outcomes. These studies would deal with use of various product lines and with potential confounders and effect modifiers of the associations. Surveillance and other long-term studies are necessary because of the duration of exposure before many chronic diseases appear. These adverse outcomes would include the health consequences that are expected based on the toxicological profile of the PREP, as well as those that are unexpected.

EXISTING TOBACCO SURVEILLANCE SYSTEMS

This section highlights existing systems of surveillance that monitor tobacco product consumption patterns, knowledge, attitudes, behaviors, and health consequences—elements that would inform the evaluation of PREP usage and impact (Giovino, 2000). The section emphasizes national and state level systems. It is possible that local or regional systems may add considerable useful information. Citations or web sites are provided for the reader who desires more detailed information.

Consumption of Tobacco Products and PREPS

The U.S. Department of Agriculture reports consumption data for the various types of tobacco products (U.S. Department of Agriculture, 2000; ERS, 2001). FTC also reports on the characteristics of cigarettes (e.g., length, filtered/non-filtered, mentholated/nonmentholated) sold in the United States (FTC, 2000a). At least one research unit (the Roswell Park Cancer Institute's Department of Cancer Prevention, Epidemiology and Biostatistics) has begun to monitor the introduction of new products.

Specific Tobacco Constituents of Both the Products and the Smoke They Generate

Currently, there is no U.S. nation-wide reporting by tobacco manufacturers of the physicochemical content of tobacco products, nor of additives or structural components. The Federal Trade Commission (FTC) reports on the results of testing of cigarette brands for tar, nicotine, and carbon monoxide (e.g., FTC, 2000a). However, as described elsewhere (Chapter 11), the usefulness of this system has been challenged (NCI, 1996).

The National Center for Environmental Health at CDC is building capacity for monitoring and research on various aspects of product design, including studies of tobacco, tobacco smoke, and biomarkers in human body fluids. Other laboratories (e.g., the American Health Foundation) have the capacity to perform tests of tobacco constituents and combustion product exposure, but they also do not conduct population surveillance.

In the Commonwealth of Massachusetts, cigarette companies (Brown and Williamson Tobacco Company, Lorillard Tobacco Company, Philip Morris USA, and R.J. Reynolds Tobacco Company) provide benchmark indicators on a sample of cigarette brands deemed representative of the U.S. market (Borgerding, 2000). The 1999 Massachusetts Benchmark Study investigated the functional relationships between standard smoke-yield parameters (e.g., "tar," nicotine, and carbon monoxide) and selected smoke constituent (e.g., acetaldehyde, 4-Aminobiphenyl, arsenic, and benzene) yields. Measures were taken on both mainstream and sidestream smoke. However, there are regional variations in tobacco product use and no national system of tobacco product distribution and consumption is in place.

Tobacco Product Marketing

No comprehensive surveillance system exists for monitoring industrial activities. The Federal Trade Commission annually collects brand-specific data but reports only aggregated national data on industry marketing expenditures (FTC, 2000b), in part obtained by subpoena. Several researchers analyze and report industry lobbying, sponsorship, and public relations activities (Glantz and Begay, 1994; Glantz et al., 1996; Siegel, 2000).

Biomarkers of Exposure to Tobacco Products

The National Health and Nutrition Examination Survey (NHANES) assesses self-reported tobacco use and serum cotinine levels annually on nationally representative samples of children, adolescents, and adults (NCHS, 2000). Determination of serum cotinine levels, a nicotine metabolite, permits biochemical validation of active use and assessment of environmental tobacco smoke exposure in persons who don't use tobacco products. However, there is insufficient but growing ascertainment of specific tobacco product brands or detailed smoking behaviors. The National Center for Environmental Health at CDC is building capacity for monitoring and research on tobacco products, including studies of biomarkers in human body fluids.

Personal Tobacco Product Use and Related Behavioral Patterns

Since most tobacco use initiation occurs among adolescents, their knowledge, attitudes and usage patterns become an important part of tobacco and PREP assessment. Three major national surveys of adolescents exist that measure at least some tobacco-related knowledge, attitudes, and behaviors (Table 6-1). These are the National Youth Tobacco Survey (NYTS) (TIPS, 2000), the Monitoring the Future (MTF) surveys of 8th, 10th, and 12th grade students (Monitoring the Future, 2001), and the National Household Survey on Drug Abuse (NHSDA) (SAMHSA, 2000). The NYTS is a categorical survey, dedicated to measuring tobacco-related knowledge, attitudes, and behaviors in middle and high school students. The MTF and the NHSDA are primarily designed to measure illicit drug use, with more limited coverage of tobacco. NHSDA surveys persons aged 12 years old and older. The Youth Risk Behavior Survey (YRBS) (NCCDPHP, 2001b) measures health risk behaviors in high school students. Several states conduct their own versions of the YRBS (Kahn, 1998) and the Youth Tobacco Survey (U.S. DHHS, 2000). MTF includes a longitudinal component, but only for 12th grade students (Johnston et al., 2000).

Three major national surveys of adults (persons aged 18 years and older) ascertain tobacco-related knowledge, attitudes, and behaviors (Table 6-1). The National Health Interview Survey (NHIS) measures several tobacco use indicators on the core instrument every year, and assesses knowledge, attitudes, and additional behavioral measures on periodic supplements (NCHS, 2001). The NHSDA questions for adults are similar to those for adolescents. The National Cancer Institute Tobacco Use Supplement of the Current Population Survey (CPS) provides measures of tobacco-related knowledge and behaviors, as well as opinions about various tobacco control policies for all states and the District of Columbia (Gerlach et al., 1997). In addition, the Behavioral Risk Factor Surveillance System (BRFSS), a set of coordinated statewide health behavior surveys, queries self-reported tobacco use in all states and the District

TABLE 6-1 Inclusion of Key Variables Regarding Tobacco Use on Existing National Surveys

Variable	YRBS	NYTS	MTF	NHSDA	NHIS	BRFSS	CPS
Susceptibility/intentions		X	X	X			
Ever smoke cigarettes (even a puff)	X	X	X	X			
Age/grade of first try/first whole cigarette	X	X	X	X			
Ever smoke regularly/daily	X	X	X	X			X
Age/grade first smoked regularly/daily			X	X	X		X
Smoked 100+ cigarettes		X		X	X	X	X
Detailed # lifetime cigarettes		X					
Current use	X	X	X	X	X	X	X
Patterns of current use	X	X	X	X	X	X	X
Indicators of dependence		X		X			
Duration of abstinence		X		X	X	X	X
Ever tried to quit	X				X		X
# prior attempts (ever)		X			X		
Quit attempt in previous year		X			X		X
# attempts (previous year)							X
Duration of previous quit attempt (most recent)		X					
Stage of change					X		X
Motivation to quit		X					
MD discuss tobacco		X			X		
MD advise quitting							X
Dentist discuss tobacco		X					
Dentist advise quitting							X
Method(s) used to quit		X			X		
Ever use other tobacco products		X		X	X	X	
Current use of other tobacco products	X	X	X	X	X	X	X
Self-esteem			X				
Stress			X				
Depressive symptoms/other mental health indicators				X			
Perception of youth smoking prevalence			X				
Family/peer use of tobacco		X		X			
Parental relationship quality			X				
Parental monitoring			X				
Anti-tobacco socialization by parents		X					
Home bans							X
Home exposure to ETS					X		
Worksite indoor air policy					X		X

continues

TABLE 6-1 Continued

Variable	YRBS	NYTS	MTF	NHSDA	NHIS	BRFSS	CPS
Outcome of last purchase attempt	X			X			
Source(s) of cigarettes	X	X					
Price paid for tobacco	X			X			
Usual brand	X	X		X			
Promotional items (own/would use or wear)	X						
Perceived risks of smoking	X	X		X	X		
Harm reduction mindset	X				X		
Risk orientation		X		X			
Functional utility	X						
Approval/disapproval				X			
Social environment			X	X			
School performance	X	X					
Religiosity	X	X					
Receptivity to marketing	X						

NOTES: YRBS=Youth Risk Behavior Survey—high school students (items listed are on the national YRBS).

NYTS=National Youth Tobacco Survey—middle and high school students.

MTF=Monitoring the Future Surveys—8th, 10th, and 12th grade students (only two tobacco questions are on the core questionnaire: one deals with lifetime use and the other deals with current patterns of use. All others are on subsets of the full sample, meaning that they provide less precise estimates) (Monitoring the Future, 2001).

NHSDA=National Household Survey on Drug Abuse—ages 12 years and older (2000 questionnaire).

NHIS=National Health Interview Survey—ages 18 years and older (NHIS 2000 Cancer Supplements).

BRFSS=Behavioral Risk Factor Surveillance Survey—ages 18 years and older; state-specific estimates.

CPS=Current Population Survey—ages 15 years and older; state-specific estimates (note that CPS uses proxy estimates for some selected sample persons; proxy reports of smoking for teenagers are more likely to lead to under estimates of prevalence than self-reports).

of Columbia (NCCDPHP, 2001a). BRFSS is developing the capacity to provide small area estimates. As noted above, the NHANES assesses adult use and serum cotinine values to biochemically validate active use and assess exposure to environmental tobacco smoke.

Two ongoing surveys provide information on tobacco and reproductive health issues. The National Survey of Family Growth surveys women 15-44 years of age to assess factors affecting pregnancy and women's health (National Vital Statistics System, 2001). The Pregnancy Risk Assessment Monitoring System (PRAMS) provides representative data from

23 states on maternal attitudes, behaviors, and experiences in order to reduce adverse outcomes of pregnancy (NCCDPHP, 1999).

Disease Outcomes

Since tobacco product use has been linked to so many different diseases and conditions, reviewed elsewhere in this report, national determination of tobacco-related morbidity assessment would be a daunting task. For example, not all states have comprehensive cancer surveillance, the most complete of which is sponsored by the registries of the U.S. National Cancer Institute (NCI, 2001) and the CDC cancer surveillance program (CDC, 1999). In addition, birth certificates for such issues as low birthweight (NVSS, 2000) and data from surveys of hospital discharges (Agency for Healthcare Research and Quality, 2000) and medical expenditures (MEPS, 2001) could be used. There is no ongoing national surveillance of incident heart disease and stroke, chronic lung disease, osteoporotic fractures, or most other tobacco-related health outcomes. However, the NHIS and the NHANES do assess self-reported conditions on a regular basis, sometimes supplemented with physiological measurements.

The National Vital Statistics System coordinates data from state operated registration systems (NVSS, 2000). Many states assess tobacco use on the death certificate and other vital records. The universal vital record system in the United States can be extremely useful for tobacco-related outcomes that often lead to death, but leaves the remaining important outcomes unassessed. Further, tobacco usage histories on vital record documents has not been fully validated, and linking mortality to tobacco product use generally requires special studies.

Other Surveillance Activities: The Social and Legislative Environment

Current systems monitor state and local legislation and programmatic activities (CDC, 2001; Robert Wood Johnson Foundation, 2001; Stillman et al. 1999); exposure to pro-health messages (Robert Wood Johnson Foundation, 2001); and tobacco placement in stores, promotions, and prices (Robert Wood Johnson Foundation, 2001). NCI's ASSIST project monitors newspaper stories and editorials, permitting assessment of the print media's coverage of and policy on tobacco control activities (Stillman et al., 1999).

PROPOSED SURVEILLANCE SYSTEM ENHANCEMENTS

The overriding goal of a surveillance system on PREPs should be to maximize the ability to assess the public health impact of the introduction of these products, with the explicit goal of maximizing the health of the public. As derived from the elements of an ideal surveillance system noted in the introduction to this chapter, and existing surveillance activities noted above, the following are suggestions for new or enhanced components to these existing activities.

Consumption of Tobacco Products and PREPs

State and regional information on the consumption of various products would provide useful information, especially if reported on a monthly or quarterly basis. In addition, future reporting systems that include PREPs may also need to be based on milligrams of nicotine consumed per product, as pounds of tobacco may become a less complete marker of consumption.

Specific Tobacco Constituents of Both the Products and the Smoke They Generate

At the time of PREP and other new product release, there should be detailed, manufacturer-derived information on important and major physical and chemical constituents of all tobacco products, including additives and the structural components of the products, such as filters, fibers, and fragments of fibers. Some independent postmarketing monitoring of product constituents may be necessary to ensure that changes are known to the public and the scientific community. For example, a recent letter from the Commissioner of the Massachusetts Department of Public Health to the Chairman of the Federal Trade Commission (Koh, 2000) highlighted the need for such monitoring. Koh points out that R.J. Reynolds' Eclipse product produced higher concentrations of toxic chemicals in 2000 than in 1996, suggesting that consumers would need to be informed of the added dangers from the 2000 version of the product. More details and specific recommendations can be found in Chapter 7, Implementation of a Science-Based Policy of Harm Reduction.

Product constituents can be influenced by agricultural and manufacturing practices. There is currently no systematic surveillance of agricultural practices or curing processes that can influence levels of undesirable constituents (e.g., tobacco-specific nitrosamines), as well as new breeds or hybrids (including genetically-altered) of tobacco that may have implications for human health. Hence, there should be enhanced monitoring of tobacco agricultural practices. General data on the types and amounts of tobacco harvested, as well as curing and processing practices would assist in identifying new and existing potentially undesirable constituents (e.g., tobacco-specific nitrosamines), as well as new breeds or hybrids (including those genetically altered) of tobacco that may have implications for human health. There should be similar information on imported tobacco products. Additionally, surveillance of manufacturing practices, especially those involving ingredients, should be instituted.

Tobacco Product Marketing

The monitoring of tobacco product marketing and public relations strategies will provide policy makers with data upon which to base decisions about the accuracy of information presented to the public and health professionals. The FTC (or another agency) could release brand-specific marketing data, if permitted to do so by legislation. Systematic media and other marketing practice monitoring would allow the assessment of messages conveyed on television, the Internet and in movies, newspapers, magazines, and mass mailings. Some monitoring of the industrially produced technical information may be of value. Another important question is whether industry marketing and public relations strategies undermine explicit public policies, laws, and regulations relevant to tobacco control.

While a research topic for further evaluation, routine message evaluation before release could provide early warnings of future problems. For example, Shiffman (1999) described two methods of testing messages. In the first, people from groups of concern (e.g., adolescents) are exposed to test stimuli and assessed for changes in attitudes, beliefs, and intentions. This system is generally conservative, as laboratory testing situations do not replicate the real world in terms of the number of repetitions of the test message or the number of different messages an individual receives on the same topic from numerous channels. Thus, any indications of future problems should be seriously considered, while false negatives may be common. In the second, expert qualitative analysis is employed to assess likely message impact.

Biomarkers of Exposure to Tobacco Products

Studies of biological fluids should be continued within the context of NHANES, which serves as a robust national sample survey that acquires serum and urine specimens. The specific biomarkers to be determined would evolve over time with scientific advancements and would be aimed at biomarker-based determination of exposure to tobacco products in general, including environmental tobacco smoke, and to specific constituents that might allow determination of specific tobacco product usage or that have predictive value for tobacco-related diseases and conditions. Additional relevant biomarkers are suggested in Section II of this volume. In addition, special studies should be conducted to assess relevant biomarkers in special groups who may not be well-represented in representative national surveys, such as living in a test market area or pregnant women.

Personal Tobacco Product Use and Related Behavioral Patterns

Key predictors of tobacco product usage that are relevant to important changes in population morbidity and mortality, such as changes in prevalence of use, initiation occurrence rates, product quitting behavior rates, and patterns of relapse, should be carefully monitored. Detailed measures of lifetime product use patterns are also needed. Studies of product usage in special populations, such as pregnant women, should be considered as a matter of routine, as well the use of nicotine replacement therapy. Finally, exposure to environmental tobacco smoke should be also be monitored at a level that can estimate the magnitude of population exposure. Whether through basic surveillance or special, it will be important to have estimates, for each product, not only of lifetime exposure, but age-at-initiation, quantitative "person-years" assessments of exposure, including the ages at which these exposures occurred, and age-at-permanent-quitting. Quantitative exposure determinations will be central to understanding whether disease outcomes may have been altered by PREP use.

Tobacco product use, and specifically PREP use, should also be measured in an ongoing and systematic manner. One central question about the net population impact of PREP introduction is whether these products influence patterns of quitting. A comprehensive surveillance system should be able to characterize factors that influence quitting. For example, measuring stages-of-change, motivation to quit, dependence, personal relevance of possible harm from tobacco use, favorable and unfavorable attitudes toward smoking, misperceptions of both tobacco use and PREPs, and reasons for relapse among those who do would be particularly important. Relevant populations of interest include tobacco users who adopt PREPs, tobacco users who don't adopt PREPs, and ex-users at risk for relapse (Shiffman, 1999).

Ancillary prospective studies of representative populations could further inform PREP impact. These studies would ideally be set up prior to the introduction of these products. Baseline data on a number of relevant variables would provide researchers with information that may explain, at least in part, why some tobacco users adopt PREPs, others do not, and others simply quit. This study would need to measure and statistically control for other environmental factors (e.g., prices of tobacco products, policy changes, treatment options, and emerging medical information), thus making it difficult to clearly detect an independent effect for a PREP or set of PREPs. An additional limitation is that these studies would provide information after an undesirable event (e.g., reduced quitting), requiring regulators to attempt to ameliorate the harm already done (Shiffman, 1999). Nevertheless, detection of change in behavioral studies is more rapid than in studies of some of the health outcomes (e.g., lung cancer or emphysema).

Another central question is whether the introduction of PREPs influences the attractiveness of tobacco use among those who have never regularly used PREPs or other tobacco products, particularly adolescents. Only population surveillance of tobacco-naïve populations could address this issue. Again, studies ancillary to the regular surveillance system can provide important and relevant information. For example, Pierce and colleagues (1996) have demonstrated among adolescents the predictive validity of a measure of susceptibility to smoke, which combines the domains of intention to smoke and perceived ability to resist the offer of a cigarette by a best friend. Susceptibility to smoke could be used as an early indicator of future changes in initiation. Another important part of behavioral surveillance is to determine misperceptions about risks from use of tobacco and PREPs as well as attitudes about their use and about persons who use them. Monitoring of these indicators and incorporation of new measures as they develop will optimally assess changes in this construct.

The population-based surveys currently providing data to the public health community are generally released from 7 to 24 months after data collection. In addition, questionnaire content is often inflexible. Prevention programming would be better served if smaller, but more frequent (e.g., monthly) tracking surveys were conducted to assess reactions to new products and campaigns (Giovino, 2000).

Disease Outcomes

As noted above, there is no systematic, ongoing, national morbidity surveillance system for the major illnesses and conditions related to tobacco products, although elements of this information are available from representative federal sample surveys of Americans, regional disease registries, and vital records. National morbidity data could in itself provide important insights into tobacco product and PREP outcomes, but could also be used for other analytical studies. For example, ecological comparisons of lung cancer mortality rates (from the National Vital Statistics System) with historical patterns of cigarette smoking (from the National Health Interview Survey)(e.g., Mannino et al., 2001; Shopland, 1995) are consistent with the interpretation that historical smoking patterns strongly influence rates of lung cancer. Similar analyses to assess the influence of

PREPs, would be more problematic and would thus require a variety of specific epidemiological studies that would not be part of routine surveillance, in part because of the duration between exposure and disease outcomes, and the complexity of multiple product exposure. For lung cancer and chronic obstructive lung disease, mortality data could serve as useful proxies of disease incidence.

As part of a comprehensive scientific program to determine the relation of PREPs to disease outcomes, analytical epidemiological studies would provide most robust and direct findings. As an example, cross-sectional surveys of tobacco product utilization could be turned into population-referent cohorts for determining health outcomes according to types of tobacco or PREP consumed, with over-sampling of persons who use new products. Surveys could also provide the data for case-control studies. A related case-control approach would be to append specific smoking histories to cancer and other disease surveillance systems, with suitable control populations.

For many policy and regulatory purposes, it may be sufficient to know whether PREPs have clinically and epidemiologically important and significant effects on occurrence and mortality for important individual chronic illnesses such as lung cancer, heart attack and chronic obstructive lung disease. However, there are several reasons why addressing these outcomes alone may be an insufficient approach to determining harm reduction potential: this approach does not document symptom patterns, various organ system dysfunctions, and the quality-of-life prior to the occurrence of a major chronic illness. As noted elsewhere in this volume, current tobacco products cause many other important health conditions as well as dysfunction and disability; and the effects of new tobacco products may be in opposite directions, causing lesser incidence of some but greater incidence of other outcomes.

Thus, in addition to specific major disease outcomes, more summary and inclusive measures of health status and outcomes should be used in assessing PREP effects. Those selected should be based on conceptual models of health status (Steinwachs, 1989) as well as the questions to be addressed and methodological considerations and impediments (McHorney, 1999). Some general approaches to these outcomes are suggested:

- Determining the occurrence of other important smoking-related conditions, such as osteoporotic fracture, low birthweight, and cataract can inform the general nature of PREP outcomes.
- Surveying for the occurrence of various symptoms and syndromes related to smoking. Such chronic or persistent conditions such as cough, sputum production, back pain due to osteoporosis, skin

lesions or discoloration, and healing time for surgical wounds and peptic ulcers may be important to individuals and to optimal function. Some of these outcome measures (e.g., cough, sputum production) also have the advantage of requiring a relatively short amount of time to develop.

- Assessing various types of biological function can summarize the net biological and clinical impact of environmental exposures across several organ systems and anatomic sites. For example, common physical functions such as the ability to jog or carry groceries are dependent in part on cardiac, pulmonary, musculo-skeletal and neurological function.

- Various measures of mortality can be of use in addition to cause-specific death rates. The overall mortality rate is increased among cigarette smokers and effect of PREPs should be evaluated in this regard. A mortality assessment approach that combines age-specific mortality rates with general social functioning, such as in the "Years of Potential Life Lost," (Lai and Hardy, 1999) which has been used for several specific causes of death, might be considered. Mortality outcomes may also have an impact on other summary measures of health outcome and the quality-of-life (CDC, 2000a).

- Self-reported health status can be an important summary measure of both general physical health, as well as mental and social functional problems (Cott et al., 1999). A variation that has proven useful occurs when individuals are allowed to assess changes in their health status, such as might occur after a clinical intervention (Fischer et al., 1999).

- There are a number of multivariate approaches to determining general health status, going under the general term "health-related quality-of-life," reflecting symptoms, conditions, dysfunctions, behaviors, and biological markers. These measures have found application in both the clinical and public health settings (Hennessy et al., 1994; Tsevat et al, 1994). Some measures combine a large number of diverse health domains, such as the "SF-36" (Ware and Sherbourne, 1992), and others combine measures of function with mortality (Tsuchiya, 2000).

Other measures exist that can't be summarized here. However, it seems important to define aggregate health measures that are sufficiently comprehensive and sensitive to the changing constituents of new tobacco products, in order to define health problems in global as well as specific terms.

Surveillance systems can also be used to assess the prevalence of non-tobacco risk factors that influence tobacco-induced illnesses (e.g., alcohol use in head and neck cancers). The committee also sees this system as an opportunity to monitor behavioral patterns such as diet and elicit drug use. Although the committee recognizes that available data do not support the hypothesis that illicit drug use increases as tobacco use decreases (Chaloupka et al., 1999; Frosch et al., 2000; Lê et al., 2000; Taylor et al., 2000), the committee notes the ease with which such data could be obtained and recommends surveillance of this undesirable outcome.

ISSUES AND LIMITATIONS REGARDING SURVEILLANCE SYSTEMS FOR ASSESSING TOBACCO-RELATED HEALTH OUTCOMES

One important issue is who would conduct surveillance on conventional tobacco products and PREPs. The types of data recommended above would almost preclude all surveillance being conducted by one organization or agency. It is likely that the elements of surveillance will come from many sources, and a coordinated effort will be needed to plan, assimilate, and interpret information for reasons of efficiency and standardization. As noted elsewhere in this volume, it will be important to include all conventional tobacco products, since they become one critical reference for health outcome studies, and to monitor changes in these products themselves. A part of the surveillance system would be to validate manufacturer claims of product distribution, content and biological and clinical effects.

Another issue is the collection of ancillary information necessary to conduct credible epidemiological studies with disease outcomes, as suggested above. For example, understanding lung cancer causation and changing frequency may require ascertainment of other risk factors such as radon or occupational exposures. Monitoring coronary disease outcomes requires determination of major risk factors other than tobacco exposure, such as those noted in Chapter 13. It may not be the burden of the surveillance system to furnish all relevant risk factors for smoking-related conditions, but where possible, this would be helpful.

There are several limitations and issues with respect to applying surveillance systems to the assessment of tobacco product usage and health. As noted above, there are many tobacco-related health outcomes for which no comprehensive, geographic surveillance system exists, and a great limitation is that such surveillance systems are costly, especially for national ascertainment of tobacco and PREP-related illnesses. However, these are the most common and important preventable conditions and the

investment seems justified. Decreasing the sample sizes in national population surveys or limiting population coverage may cause compromises
in data quality or generalizability. A related issue is that it might take
very large population surveys to adequately cover important demographic subgroups of interest, such as pregnant women or certain minority groups. Thus, it may be more efficient to have separate surveys or
surveillance surveys of special populations than only one large population survey. A comprehensive surveillance system, as described in this
chapter, could also be critical for other disease control activities that are
not tobacco-related, and conceivably the cost of the system could be
shared.

Another important limitation is that many aspects of population surveillance depend largely on self-report, which can be subject to error. In
some instances, tobacco product usage can be validated by external
means, but not in all circumstances. There are also limitations to predicting behaviors based on self-reported personal knowledge and attitudes,
although both are important. Here, too, there are mechanisms to improve
the validity of these reports.

There may not be suitable or logistically feasible biomarkers of exposures for the range of important tobacco products and toxicants to which
users are exposed. Some of the biomarkers of exposure used in the past,
such as cotinine, may still have utility for assessing conventional tobacco
product exposure, but as new PREPs come to the marketplace, these markers may no longer be fully suitable because they won't necessarily serve
as adequate surrogate markers for the range of major tobacco constituents.

Some elements of a comprehensive surveillance system, such as
mandating tobacco manufacturers to report product characteristics, ingredients, additives, and brand-specific sales and distribution data might
require a legislative or regulatory approach to enforce. Without this information, a comprehensive surveillance program would be much weaker.

Finally, it should be noted that it is not the burden of surveillance
systems *per se* to relate PREPs or other tobacco product exposure to specific health outcomes or altered levels of those outcomes. That is usually
the function of targeted epidemiological studies such as cohort studies of
persons using PREPs to monitor for long-term health effects, with suitable contrast groups. Well-designed case-control studies may also be appropriate vehicles for exploring certain tobacco-disease associations, although the retrospective recall of the past product usage may not always
be credible. As always, epidemiological studies should be accompanied
by the best basic science and clinical research to guide the study design,
apply the most modern markers of exposure and disease, and optimally
interpret the findings.

SUMMARY AND RECOMMENDATIONS

The goal of the proposed surveillance system and accompanying epidemiologic studies is to provide much of the data need to determine the ongoing contribution of tobacco products and PREPs to the public's health status and to inform policy initiatives and regulatory judgments. Thus, the system will need to estimate relative changes in the prevalence of tobacco use, as well as changes in the relative harm to users of PREPs. Strong data accumulated over many years are necessary to judge if PREPs (or classes of PREPs) contribute to maximizing the health of the public. Public health officials will need to determine if the prevalence of tobacco use drops to a level at or near what it would have in the absence of PREPs and if the health benefits (if any) caused by switching to PREPs compensate for any decrement in prevalence reduction that they cause. This will be a challenging process, but one that will only be possible if optimal data collection systems are swiftly put in place. Until surveillance mechanisms that would enable prospective assessment of the public health impact are in place, it might be prudent to take an especially risk averse position regarding communications and claims (see Chapter 7). Given this approach, **the committee makes the following recommendations:**

1. There is an urgent need for a national comprehensive surveillance system that collects information on a broad range of elements necessary to understand the population impact of tobacco products and PREPs, including attitudes, beliefs, product characteristics, product distribution and usage patterns, marketing messages such as harm reduction claims and advertising, the incidence of initiation and quitting and nontobacco risk factors for tobacco-related conditions. There should be surveillance of major smoking-related diseases as well as construction of aggregate population health measures of the net impact of conventional product and PREPs.
2. The surveillance system should consist of mandatory, industry-furnished data on tobacco product constituents, additives, and population distribution and sales.
3. Resources should be made available for a program of epidemiological studies that specifically address the health outcomes of PREPs and conventional tobacco products, built on a robust surveillance system and using all available basic and clinical scientific findings.

REFERENCES

Agency for Healthcare Research and Quality. 2000. Healthcare Cost and Utilization Project (HCUP). [Online]. Available: http://www.ahrq.gov/data.hcup [accessed February 11, 2001].

Borgerding M. 2000. State of Massachusetts Regulation of Tobacco Products. Meeting of the Institute of Medicine's Committee to Assess the Science Base for Tobacco Harm Reduction: April 25, 2000; Washington, DC.

CDC (Centers for Disease Control and Prevention). 1996. Addition of prevalence of smoking as a nationally notifiable condition—June 1996. *Morbidity and Mortality Weekly Report* 45:537.

CDC (Centers for Disease Control and Prevention). 1999. Cancer. [Online]. Available: http://www.cdc.gov/nccdphp/survcanc.htm [accessed February 11, 2001].

CDC (Centers for Disease Control). 2000a. Community indicators of health-related quality of life-United States, 1993-1997. *Morbidity and Mortality Weekly Report* 49(13):281-285.

CDC (Centers for Disease Control and Prevention). 2000b. Use of FDA-approved pharmacologic treatments for tobacco dependence—United States, 1984-1998. *Morbidity and Mortality Weekly Report* 49:665-668.

CDC (Centers for Disease Control and Prevention). 2001. STATE (State Tobacco Activities Tracting and Evaluation). [Online]. Available: http://www2.cdc.gov/nccdphp/osh/state [accessed February 11, 2001].

Chaloupka FJ, Pacula RL, Farrelly MC, Johnston LD, O'Malley PM, Bray JW. 1999. Do higher cigarette prices encourage youth to use marijuana? Cambridge, Massachusetts: National Bureau of Economic Research; NBER Working Paper Series 6939.

Cott CA, Gignac MA, Badley EM. 1999. Determinants of self rated health for Canadians with chronic disease and disability. *Journal of Epidemiology and Community Health* 53:731-736.

ERS (Economic Research Service). 2001. Briefing room, tobacco. [Online]. Available: http://www.econ.ag.gov/briefing/tobacco [accessed February 11, 2001].

FTC (Federal Trade Commission). 2000a. "Tar, nicotine, and carbon monoxide of the smoke of 1294 varieties of domestic cigarettes for the year 1998." A Federal Trade Commission Report to Congress (July 2000). Washington, DC: Federal Trade Commission.

FTC (Federal Trade Commission). 2000b. Federal Trade Commission Report to Congress for 1998. Pursuant to the Federal Cigarette Labeling and Advertising Act. Washington, DC: Federal Trade Commission.

Fischer D, Stewart AL, Bloch DA, Lorig K, Laurent D, Holman H. 1999. Capturing the patient's view of change as a clinical outcome measure. *Journal of the American Medical Association* 282:1157-1162.

Frosch D, Shoptaw S, Nahom D, Jarvik JE. 2000. Associations between tobacco smoking and illicit drug use among methadone-maintained opiate-dependent individuals. *Experimental and Clinical Psychopharmacology* 8(1):97-103.

Gerlach KK, Shopland DR, Hartman AM, Gibson JT, Pechacek TF. 1997. Workplace smoking policies in the United States: results from a national survey of more than 100,000 workers. *Tobacco Control* 6:199-206.

Giovino GA. 2000. Surveillance of patterns and consequences of tobacco use—United States. *Tobacco Control* 9:232-233.

Glantz SA, Begay, ME. 1994. Tobacco industry campaign contributions are affecting tobacco control policymaking in California. *JAMA* 272:1176-1882.

Glantz SA, Slade J, Bero LA, Hanauer P, Barnes DE. 1996. *Cigarette Papers.* Berkeley, CA: University of California Press.

Haenszel W, Shimkin MB, Miller HP. 1956. *Tobacco Smoking Patterns in the United States. Public Health Monograph 45.* Washington, DC: U.S. Department of Health, Education, and Welfare.

Hennessy CH, Moriatry DG, Zack MM, Scherr PA, Brackbill R. 1994. Measuring health-related quality-of-life for public health surveillance. *Public Health Reports* 109:66-72.

Johnston LD, O'Malley PM, Bachman JG. 2000. *Monitoring the Future: National Survey Results on Drug Use, 1975-1999.* Volume 1—Secondary School Students. NIH Publication No. 00-4802. Bethesda, MD: USDHHS, NIDA.

Kahn L, Kinchen SA, Williams BI, Ross JG, Lowry R, Hill CV, Grunbaum J, Blumson PS, Collins JL, Kolbe JL, and State and Local YRBSS Coordinators. 1998. Youth risk behavior surveillance—United States, 1997. [Online]. Available: http://www.cdc.gov/nccdphp/dash/MMWRFile/ss4703.htm [accessed February 11, 2001].

Klaucke DN, Buehler JW, Thacker SB, Parrish RG, Trowbridge FL, Berkelman RL, Surveillance Coordinating Group. 1988. Guidelines for evaluating surveillance systems. *Morbidity and Mortality Weekly Report* 37(S-5):1-18.

Koh HK. 2000. Letter to Robert Pitofsky, Chairman of the Federal Trade Commission.

Lai D, Hardy RJ. 1999. Potential gains in life expectancy or years of potential life lost: impact of competing risk of death. *International Journal of Epidemiology* 28:894-898.

Lê AD, Corrigall WA, Watchus J, Harding S, Juzytsch W, Li TK. 2000. Involvement of nicotinic receptors in alcohol self-administration. *Alcohol: Clinical and Experimental Research* 24(2):155-163.

Mannino DM, Ford E, Giovino GA, Thun M. 2001. Lung cancer mortality rates in birth cohorts in the United States from 1960 through 1994. *Lung Cancer* 31:91-99.

McHorney CA. 1999. Health status assessment methods for adults: past accomplishments and future challenges. *Annual Review of Public Health* 20:309-335.

MEPS (Medical Expenditure Panel Survey). 2001. MEPS Homepage. [Online]. Available: http://www.meps.ahrq.gov [accessed 2001].

Millmore BK, Conover AG. 1956. Tobacco consumption in the United States, 1880-1955. Addendum in: US Department of Health, Education, and Welfare. *Tobacco Smoking Patterns in the United States: Public Health Monograph 45.* Washington, DC: Department of Health Education and Welfare, Public Health Service. DHEW publication No. (PHS)463:107-111.

Monitoring The Future. 2001. Monitoring The Future Homepage. [Online]. Available: http://www.monitoringthefuture.org [accessed February 11, 2001].

National Vital Statistics System. 2001. National Survey of Family Growth (NSFG). [Online]. Available: http://www.cdc.gov/nchs/nsfg.htm [accessed February 11, 2001].

NCCDPHP (National Center for Chronic Disease Prevention and Health Promotion). 1999. Maternal and Infant Health Surveillance. [Online]. Available: http://www.cdc.gov/nccdphp/pregnanc.htm [accessed February 11, 2001].

NCCDPHP (National Center for Chronic Disease Prevention and Health Promotion). 2001a. Behavioral Risk Factor Surveillance System. [Online]. Available: http://www.cdc.gov/nccdphp/brfss [accessed February 11, 2001].

NCCDPHP (National Center for Chronic Disease Prevention and Health Promotion). 2001b. Youth Risk Behavior Surveillance System. [Online]. Available: http://www.cdc.gov/nccdphp/youthris.htm [accessed February 11, 2001].

NCHS (National Center for Health Statistics). 2001. National Health Interview Survey (NHIS). [Online]. Available: http://www.cdc.gov/nchs/nhis.htm [accessed February 11, 2001].

NCHS (National Center for Heath Statistics). 2000. National Health and Nutrition Examination Survey. [Online]. Available: http://www.cdc.gov/nchs/nhanes.htm [accessed February 11, 2001].

NCI (National Cancer Institute). 1996. *The FTC Cigarette Test Method for Determining Tar, Nicotine, and Carbon Monoxide Yields of U.S. Cigarettes. Smoking and Tobacco Control Monograph 7.* Bethesda, Maryland: US Department of Health and Human Services, Public Health Service, National Institutes of Health, National Cancer Institute. NIH Publication No. 96-4028.

NCI (National Cancer Institute). 2001. Surveillance, epidemiology and end results. [Online]. Available: http://www-seer.ims.nci.nih.gov [accessed February 11, 2001].

NVSS (National Vital Statistics System). 2000. Surveys and data collection systems. [Online]. Available: http://www.cdc.gov/nchs/nvss.htm [accessed February 11, 2001].

Pierce JP, Choi WS, Gilpin EA, Merritt RK. 1996. Validation of susceptibility as a predictor of which adolescents take up smoking in the United States. *Health Psychology* 15:355-361.

Remington PL, Goodman RA. 1998. Chronic disease surveillance. Brownson RC, Remington PL, Davis JR, eds.: *Chronic Disease Epidemiology and Control.* Washington, D.C.: American Public Health Association. Pp. 55-76.

Robert Wood Johnson Foundation. 2001. ImpacTeen. [Online]. Available: http://www.uic.edu/orgs/impacteen [accessed February 11, 2001].

SAMHSA (Substance Abuse and Mental Health Services Administration). 2000. Preliminary results from the 1996 National Household Survey on Drug Abuse. [Online]. Available: http://www.samhsa.gov/oas/NHSDA [accessed February 11, 2001].

Shiffman S. 1999. *Population Risks of Less-Deadly-Cigarettes: Beyond Toxicology.* Report of Canada's Expert Committee on Cigarette Toxicity Reduction.

Shopland DR. 1995. Tobacco use and its contribution to early cancer mortality with a special emphasis on cigarette smoking. *Environmental Health Perspectives* 103(Suppl 8):131-141.

Siegel M. 2000. *Tobacco Industry Sponsorship in the United States, 1995-1999.* Boston, MA: Boston University School of Public Health. (available at http://dcc2.bumc.bu.edu/tobacco).

Steinwachs DM. 1989. Application of health status assessment measures in policy research. *Medical Care* 27:S12-S16.

Stillman F, Hartman A, Graubard B, Gilpin E, Chavis D, Garcia J, Wun LM, Lynn L, Manley M. 1999. The American Stop Smoking Intervention Study. Conceptual Framework and Evaluation Design. *Eval Rev* 23(3):259-280.

Taylor RC, Harris NA, Singleton EG, Moolchan ET, Heishman SJ. 2000. Tobacco craving: intensity-related effects of imagery scripts in drug abusers. *Experimental and Clinical Psychopharmacology* 8(1):75-87.

Thacker SB, Berkelman RL. 1988. Public health surveillance in the United States. *Epidemiologic Reviews* 10:164-190.

TIPS (Tobacco Information and Prevention Source). 2000. New study provides first comprehensive look at tobacco use among middle school and high school students. [Online]. Available: http://www.cdc.gov/tobacco/nyts2000.htm [accessed February 11, 2001].

Tsevat J, Weeks JC, Guadagnoli E, Tosteson AN, Mangione CM, Pliskin JS, Weinstein MC, Cleary PD. 1994. Using health-related quality-of-life information: clinical encounters, clinical trials and health policy. *Journal of General Internal Medicine* 9:576-582.

Tsuchiya A. 2000. QALYs and ageism: philosophical theories and age weighting. *Health Economics* 9:57-68.

U.S. Department of Agriculture. 2000. *Tobacco Situation and Outlook Report.* Market and Trade Economics Division, Economic Research Service, U.S. Department of Agriculture, TBS-246.

U.S. DHHS (U.S. Department of Health and Human Services). 2000. *Reducing Tobacco Use.* Washington, DC: U.S. DHHS, Centers for Disease Control and Prevention.

Ware JE Jr, Sherbourne CD. 1992. The MOS 360tiem short-form survey (SF-36). I. Conceptual framework and item selection. *Medical Care* 30:473-483.

7

Implementation of a Science-Based Policy of Harm Reduction

Scientific evidence establishes unequivocally that tobacco use causes serious adverse health effects in humans and that the nicotine delivered by tobacco is a highly addictive drug. The ultimate public health goal of tobacco policy is to eliminate the excess morbidity and mortality associated with tobacco use. The only scientifically proven way to accomplish this is to eliminate tobacco use by preventing initiation by those who have never used tobacco, achieving cessation for those who currently use tobacco, and preventing relapse by former users of tobacco. Achieving this goal will protect tobacco users themselves and those affected by environmental tobacco smoke (ETS). But because nicotine is a highly addictive drug, quitting tobacco use is extremely difficult for many people. Despite overwhelming evidence and widespread recognition that tobacco use poses a serious risk to health, some tobacco users cannot or will not quit. For those addicted tobacco users who do not quit, reducing the health risks of tobacco products themselves may be a sensible response. This is why many public health leaders believe that what has come to be called "harm reduction" must be included as a subsidiary component of a comprehensive public health policy toward tobacco.

Some public health officials oppose the adoption of harm reduction strategies because of concerns that promoting this approach will not, over the long term, prove to be beneficial to public health or to the individual tobacco users who might otherwise have quit (Ferrence et al., 2000; Warner et al., 1997). Whatever the merits of this position, marketplace forces already at work have put this issue on the public policy agenda, and new products are being developed and offered as harm-reducing alternatives to conventional tobacco products. The task before this committee is to address the science base for implementing a harm reduction approach, and for assessing the impact of such an approach on public health. The committee's task is not to recommend whether or not tobacco harm reduction should be pursued. Furthermore, the committee's effort to carry out its charge should be understood as only one component of a comprehensive tobacco control policy.

This report shows that the prospect of harm reduction presents both promise and uncertainty. For tobacco and pharmaceutical companies to be investing in products that reduce exposure to tobacco toxicants could be a salutary development. Nonetheless, despite advances in understanding tobacco toxicology and the pathophysiology and epidemiology of tobacco-related diseases, little is known about the health effects of using products that reduce exposure to one or more tobacco toxicants or about the public health consequences of promoting tobacco-related products or pharmaceutical products as potential reduced-exposure products (PREPs). It will take many years of research to develop definitive data. It is also clear, however, that action must be taken now, even as better data are being developed, to respond to the already emerging market for PREPs and to ensure that the necessary data are developed, that consumers are accurately informed, and that the public health is fostered.

Until adequate data are available, individual and regulatory choices will by necessity have to be made on the basis of predictions of risk and harm reduction based on inference from indirect evidence. Action should therefore be taken not only to monitor the market, but also to shape it as scientific knowledge unfolds. Despite continuing ambivalence among some health officials about the wisdom of embracing harm reduction as a public health policy, numerous consumers will be taking steps in this direction, with or without scientific guidance. The aggregate effect of these decisions might make an important contribution to public health— or might further exacerbate the problems posed by tobacco products. Policy makers must use the best that science has to offer to ensure that the harm reduction strategies pursued by tobacco and pharmaceutical manufacturers and by individual consumers will truly reduce harm, and to the greatest extent possible.

The committee was drawn by its charge into considering how best to implement the scientific and policy recommendations in this report and, in so doing, was necessarily required to address some features of a regulatory framework for PREPs. The regulation of tobacco products is most certainly controversial and many approaches have been proposed. Tobacco regulation conceivably includes taxation, access by minors, point of sales, etc. For the purposes of this report the regulatory framework considered is narrow and focused exclusively on evidence and any necessary powers required to assess harm reduction products, be they pharmaceutical or tobacco-related. Other aspects of possible tobacco regulation are outside of the charge of this committee.

The committee did come to conclude that regulation of PREPs is necessary and feasible. First, effective regulation is a necessary precondition for advancing scientific knowledge on the toxicology and clinical effects of these products, for developing the data necessary for systematic risk assessment, and for monitoring the impact of such products on the public health through appropriate postmarketing surveillance and the collection of long-term epidemiological data. Second, regulation is needed to ensure that the product labeling and advertising do not mislead consumers and accurately describe the products' risks, including the uncertainties that can only be resolved after long-term use. Consumers should not use these new products on the basis of explicit or implicit claims that these products carry less risk than traditional tobacco products unless such claims are true. Absent careful regulation of industry claims about these products, informed choices by consumers will not be possible, the potential benefit of a harm reduction strategy is likely to go unrealized, and the long and unsettling saga of light cigarettes may well be repeated. Finally, regulation is also needed to foster integrated, coherent, and equitable policies with respect to the scientific testing, labeling, and advertising of the diverse array of PREPs marketed by tobacco and pharmaceutical companies.

NEXT STEPS: AN OVERVIEW

Although the science base for tobacco harm reduction (summarized in Chapter 5 and described in detail in Chapters 9-16) is extensive, the committee identified many gaps and limitations. To improve the science base for a harm reduction strategy for tobacco products and to protect public health, the committee's scientific findings and conclusions must be translated into a comprehensive policy framework that includes the following elements:

- a substantial and sustained *research* program to address the critical unresolved questions that are susceptible to scientific resolution, as identified in this report;
- a strong *surveillance* program that will serve both as an "early warning system" for identifying problems associated with PREPs and as an epidemiological tool for evaluating long-term health consequences;
- a well-designed program of public health *education* (including media campaigns) to help people understand that preventing the

initiation of smoking and assisting smokers to quit constitute the only proven methods of minimizing tobacco-related harm, that the health benefit of using PREPs remains uncertain, and that this uncertainty will not be resolved for many years; and

• an integrated program of federal *regulation* of both tobacco-related PREPs and pharmaceutical PREPs designed to protect public health and to facilitate the research, surveillance, and public education activities described above.

The remainder of this chapter will set forth the committee's recommendations regarding the regulation of PREPs. As discussed in Chapter 4, the current regulatory situation leaves conventional tobacco products essentially unregulated while imposing stringent regulatory controls on the development and marketing of pharmaceutical PREPs. Modified tobacco products with exposure reduction claims currently fall into a twilight zone of regulatory uncertainty.

Previous reports of the Institute of Medicine (IOM, 1994,1998) have recommended that Congress enact a comprehensive regulatory statute delegating to an appropriate agency, preferably the Food and Drug Administration (FDA), "the necessary authority to regulate tobacco products for the dual purpose of discouraging consumption and reducing the morbidity and mortality associated with use of tobacco products." The need for comprehensive tobacco regulation, as recommended in these previous reports, is reinforced by the Supreme Court's ruling, on March 21, 2000, that the FDA lacks comprehensive authority over tobacco products under the Federal Food, Drug and Cosmetic Act (Food and Drug Administration, et al. v. Brown & Williamson Tobacco Corporation et al., 2000). These reports also recommended that the designated agency be authorized to regulate the design and constituents of tobacco products—for example, by adopting "performance standards" that set limits for the levels of toxicants in all tobacco products and/or by regulating the levels of, or exposure to, nicotine. This committee endorses these recommendations.

This committee's charge focuses specifically on products that purport to reduce exposure to tobacco toxicants. In this context the need for federal regulation is made more urgent by the recent introduction of Advance and the expansion of Eclipse into the marketplace and by the likely introduction of other tobacco-related PREPs, including oral nicotine and Swedish snus, in the future. The FDA already has the authority to regulate certain of these new or modified tobacco products as drug-delivery devices or as "drugs" if the manufacturer claims that the product prevents disease by reducing the health risks of using tobacco. The scope of the FDA's jurisdiction is unclear, though, if the manufacturer goes no further than claiming that the product reduces exposure to known tobacco toxicants (Page, 1998). Whatever the reach of FDA's current jurisdiction over modified tobacco products, the existing regulatory framework for all tobacco products, including tobacco-related PREPs is inadequate to provide a basis for informed consumer choice and to protect the public health. Accordingly, **the committee recommends that Congress enact legislation enabling a suitable agency to regulate tobacco-related products that purport to reduce exposure to one or more tobacco toxicants or to reduce risk of disease, and to implement other policies designed to reduce the harm from tobacco use.**

Regulatory classifications are traditionally based on the use(s) of a product and/or the claims associated with it. The outline of a sensible regulatory classification is therefore as follows:

1. conventional tobacco products and modified products that are marketed without claims of reduced exposure to tobacco toxicants or reduced risk of disease;
2. modified tobacco products or tobacco-like products, whether tobacco containing or not, that are marketed with such claims; and
3. pharmaceutical products and medical devices, whether nicotine containing or not, that are marketed with a claim of effectiveness for cessation of, or significant reduction in, smoking.

The committee's charge is directed to products in categories 2 and 3 (all PREPs). The legal structure for regulating pharmaceuticals in category 3 is already fully in place under the federal Food, Drug and Cosmetic Act (FDCA) and administered by the FDA. The committee has therefore largely focused on products in category 2, a classification that encompasses any tobacco-related product marketed as a PREP—that is, with claims of reduced exposure to tobacco toxicants or reduced risk of disease—presumably involving a novel design or a modification of a conventional tobacco product. In the interest of placing recommendations concerning the regulation of tobacco-related PREPs in the larger regulatory context, the committee has embraced several recommendations in previous reports of the IOM (1994, 1998) with respect to conventional tobacco products (category 1 above). Thus, several of the principles outlined below apply to conventional tobacco products as well as to tobacco-related PREPs.

Taken together with the drug and device laws already in place, the overall regulatory system should be sufficiently broad and flexible to encompass all relevant product innovations and to respond to new scientific knowledge in the years ahead. Manufacturers of tobacco products and pharmaceuticals should be encouraged to develop and introduce new products that will reduce the burden of tobacco-related disease. However, manufacturers of new products should not be allowed to escape regulatory oversight by claiming that their products are dietary supplements, herbals, botanicals, or foods or that they make no explicit claims.

PRINCIPLES FOR REGULATING POTENTIAL REDUCED-EXPOSURE PRODUCTS

In the remainder of this chapter, the committee offers its judgment for consideration of the challenging scientific issues presented by the complex array of new products offered with claims bearing on tobacco harm reduction. Specifically, the committee proposes 11 principles that it believes should govern the regulation of new or modified tobacco products and pharmaceutical products with harm-reducing potential. The overall regulatory structure builds on the foundation of existing regulatory law, with appropriate adaptations to take into account the unique history and toxicity of tobacco products. The committee's approach reflects the following general configuration:

- All tobacco products would be subject to requirements for certain testing and reporting and to the regulation of labeling and advertising.
- A manufacturer who wishes to market a new tobacco product as a PREP—a product with a claim, whether explicit or implicit, of reduced exposure to one or more tobacco toxicants or of reduced risk of adverse health effects (compared with conventional products)—would be required to receive from the regulatory agency prior approval of the claim based on scientific evidence presented by the manufacturer that the claim is not false or misleading; products for which risk reduction claims are made would be subject to postmarketing epidemiological studies.
- New brands and modifications of conventional tobacco products without health claims would be permitted to enter the market without prior regulatory approval if they are certified to present "no greater risk" than products already on the market.
- Pharmaceutical products and medical devices, whether nicotine containing or not, with health claims for reduction of smoking would continue to be subject to the current requirements of the FDCA.

The committee believes that the agency charged with regulating tobacco PREPs should have a public health orientation and should be given authority over all tobacco products, including conventional ones. The scientific expertise needed to regulate both categories of products is similar. This agency will require a competent scientific review staff, including molecular biologists, pharmacologists, toxicologists, physicians, epidemiologists, statisticians, and scientists experienced in the technology of tobacco product design and manufacture. In addition, it will need social scientists and marketing experts experienced in the evaluation of product labeling and the regulation of advertising. The agency must also have an analytical laboratory capable of testing a wide range of conventional and modified tobacco products and a product surveillance program staffed with epidemiologists and data management experts. Finally, it will need an enforcement and legal staff and appropriate administrative and information technology personnel.

Some of these functions already exist for tobacco products at the Food and Drug Administration (disease claims), at the Federal Trade Commission (FTC; regulation of advertising and enforcement staff), and at the Centers for Disease Control and Prevention (CDC; analytical laboratory). But no agency currently has the comprehensive mandate or staff necessary to fulfill the policies recommended by the committee. This committee, like previous IOM committees (IOM, 1994, 1998), believes that the FDA would be an appropriate site for this regulatory function, but other administrative locations are certainly possible. The important point is that the requisite authority be lodged in a suitable federal regulatory agency with sufficient expertise and resources to execute the mission successfully.

The committee notes the efforts of some states, notably Massachusetts in particular, to take on some of the regulatory challenges created by the federal government's failure to establish comprehensive regulatory authority over tobacco products. However, patchwork state legislation is not a satisfactory long-term response to the problem. The only adequate response is for Congress to confer comprehensive authority on a suitable federal agency.

Summary of Regulatory Principles

A science-based regulatory framework for implementing tobacco harm reduction should conform to the following 11 principles:

1. Manufacturers of tobacco products, whether conventional or modified, should be required to obtain quantitative analytical data on the ingredients of each of their products and to disclose such information to the regulatory agency.
2. All tobacco products should be assessed for yields of nicotine and other tobacco toxicants according to a method that reflects actual circumstances of human consumption; when necessary to support claims, human exposure to various constituents of tobacco

smoke should be assessed using appropriate biomarkers. Accurate information regarding yield range and human exposure should be communicated to consumers in terms that are understandable and not misleading.

3. Manufacturers of all PREPs should be required to conduct appropriate toxicological testing in preclinical laboratory and animal models and appropriate clinical testing in humans to support the health-related claims associated with each product and to disclose the results of such testing to the regulatory agency.

4. Manufacturers should be permitted to market tobacco-related products with exposure reduction or risk reduction claims only after agency approval based on scientific evidence (a) that the product substantially reduces exposure to one or more tobacco toxicants and (b) if a risk reduction claim is made, that the product can reasonably be expected to reduce the risk of one or more specific diseases or other adverse health effects, compared with whatever benchmark product the agency requires to be stated in the labeling. The "substantial reduction" in exposure should be sufficiently large that independent scientific experts would anticipate finding a measurable reduction in morbidity and/or mortality in subsequent clinical or epidemiological studies.

5. The labeling, advertising, and promotion of all tobacco-related products with exposure reduction or risk reduction claims must be carefully regulated under a "not false or misleading" standard, with the burden of proof for the claim resting on the manufacturer not the government. The responsible agency should have the authority and resources to conduct surveys of consumer perceptions relating to these claims.

6. The regulatory agency should be empowered to require manufacturers of all products marketed with claims of reduced risk of tobacco-related disease to conduct postmarketing surveillance and epidemiological studies as necessary to determine the short-term behavioral and long-term health consequences of using their products and to permit continuing review of the accuracy of their claims.

7. In the absence of any claim of reduced exposure or reduced risk, manufacturers of tobacco products should be permitted to market new products or modify existing products without prior approval of the regulatory agency after informing the agency of the composition of the product and certifying that the product could not reasonably be expected to increase the risk of cancer, heart disease, pulmonary disease, adverse reproductive effects, or other adverse health effects, compared to similar conventional tobacco

products, as judged on the basis of the most current toxicological and epidemiological information.

8. All added ingredients in tobacco products, including those already on the market, should be reported to the agency and be subject to a comprehensive toxicological review.
9. The regulatory agency should be empowered to set performance standards (e.g., maximum levels of toxicants; definitions of terms such as "low tar") for all tobacco products, whether conventional or modified, or for classes of products.
10. The regulatory agency should have enforcement powers commensurate with its public health mission, including the power to issue subpoenas.
11. Exposure reduction and risk reduction claims for drugs and devices that are supported by appropriate scientific and clinical evidence should be allowed by the FDA.

The following sections elaborate on each of these principles.

Principle 1: Disclosure of Product Ingredients

Manufacturers of tobacco products, whether conventional or modified, should be required to obtain quantitative analytical data on the ingredients of each of their products and to disclose such information to the regulatory agency.

The manufacturers of tobacco products have detailed and extensive knowledge of the composition, curing, and blending of tobacco; the ingredients of their products; and the composition of tobacco smoke. They also have detailed quantitative information on the tobacco in each of their brands of cigarettes and smokeless products, since each is blended to achieve the desired taste and levels of nicotine and other ingredients. Manufacturers also conduct major research and development programs aimed at developing new products. Currently, there is no requirement for disclosure of this information to an appropriate regulatory body. Such disclosure is prerequisite to any meaningful scientific appraisal of the comparative risks of different tobacco-containing products or the potential for risk reduction offered by modified products. For smoked products, analytical information on the smoke and the concentrations of smoke components that are absorbed under actual smoking conditions may be even more important than knowledge of the product ingredients themselves. For smokeless tobacco products, whether conventional or modified, similar information on the concentrations of major ingredients in saliva and blood are equally important. The disclosure of quantitative information on ingredients (with appropriate safeguards to protect trade secrets) is a standard requirement in regulatory laws relating to drugs, biologics, devices, food additives, toxic chemicals, and environmental contaminants and should be required of all tobacco products, whether conventional or modified.

Although FTC regulation requires public reporting of some constituents in cigarette smoke, manufacturers are not required to report brand-specific information about the nicotine content or other properties (e.g., nitrosamine level) of the material that forms the tobacco rod. Under legislation enacted in 1986, manufacturers of smokeless products are required to report total nicotine content to the secretary of the Department of Health and Human Services (HHS), but the secretary may not release the data. Under the same legislation, tobacco manufacturers are required to submit lists of additives to the tobacco (but not to filters or papers) to the secretary of HHS. Information about the quantity of additives and their presence in specific brands is not required, and the secretary is bound to safeguard the information from public disclosure. In 1993, attorneys for six cigarette manufacturers released a combined list of 599 additives. The following year, ten manufacturers of smokeless products released a list of additives in their products. Three states have enacted legislation requiring disclosure of additives in tobacco products (U.S. DHHS, 2000).

Principle 2: Yield Assessment

All tobacco products should be assessed for yields of nicotine and other tobacco toxicants according to a method that reflects actual circumstances of human consumption; when necessary to support claims, human exposure to various constituents of tobacco smoke should be assessed using appropriate biomarkers. Accurate information regarding yield range and human exposure should be communicated to consumers in terms that are understandable and not misleading.

The actual yield—that is, the amount of toxicants inhaled by an individual—from particular cigarettes varies considerably among smokers. As discussed in Chapter 11, this is because the standard yield is measured by a machine that smokes cigarettes in a mechanical and standardized way, whereas smokers can and do smoke their cigarettes with different numbers of puffs and different depths of inhalation. As a result, for any product, the temperature of combustion and the composition of the smoke varies among smokers depending on the pattern of smoking. The information required to address the relative harmfulness of tobacco products will be available only if an improved methodology for ascertaining the range of actual toxicant yields in human consumers is developed and applied; only then will there be a scientific basis for developing some of the features of a harm reduction program, including, for example, accurate labeling, meaningful definition of terms such as "low tar", and reasonable standards for yields of tar, carbon monoxide (CO), or nicotine.

Although the limitations of the FTC's current methodology for assessing cigarette yield is generally acknowledged, no definitive steps have been taken to replace it. In 1994, the President's Cancer Committee and the National Cancer Institute (NCI) issued a report criticizing the current approach and recommending adoption of a new one. In 1997, the FTC issued a proposed revision of the methodology and a proposed format for disclosing test results in advertising and labeling. However, many proponents of a new approach urged the FTC to defer action pending judicial clarification of the FDA's authority and possible congressional enactment of comprehensive regulation for tobacco products. It is time for the FTC—or, ideally, an agency with regulatory authority over tobacco products—to adopt a new standardized methodology to assess the range of toxic exposures to smokers under actual conditions of human smoking.

Although the focus of yield assessment has been smoked tobacco, an analogous methodology should also be developed for smokeless products. Without quantitative information on the absorption of various constituents under normal product use, it is impossible to assess comparative exposures and comparative risks.

The committee embraces the conclusions and recommendation on this issue set forth in the IOM (1994) report *Growing Up Tobacco Free*:

> [T]he regulatory agency, as its first step, should develop a sound methodology for ascertaining the actual yields of nicotine, tar, or any other constituents of tobacco products, based on human consumption. Human exposure to some constituents of tobacco smoke can be assessed by use of biochemical markers of exposure to those constituents, although at present this methodology is technically difficult and still imprecise; however, it is likely that better exposure measures will be developed in the future. In any case, even the currently available measures of human exposure are likely to provide a better indicator of relative risk than do the standard cigarette smoking machine yields. At a minimum, the smoking machine tests can be modified to reflect the range of ways in which people actually smoke, including numbers of puffs and blocking of ventilation holes, to determine likely *ranges* of delivery. In addition, the manufacturers of tobacco products could be directed to submit information to the regulatory agency regarding the actual yields of their products in humans for particular brands of tobacco products, based on use of prescribed protocols.

> A variety of regulatory initiatives relating to tar and nicotine yields can be envisioned. At a minimum, the regulatory agency should take steps to assure that consumers are informed about the meaning of statements regarding tar and nicotine yields, about the behavioral influences on exposure, and about the relative importance of the characteristics of the cigarette and the way it is smoked. The agency might also require that tar and nicotine yields be presented in a standard format, such as

absolute content together with a statement of the range of expected systemic yield determined according to number of puffs or other behavioral factors. Currently, marketed cigarettes typically *contain* 8-9 mg nicotine in the tobacco rod, and have an expected actual yield to the smoker of 0.5-3.0 mg nicotine. Stating the nicotine content of the tobacco contained in the cigarettes is important because the content reflects the maximum possible yield, and reduction of content would be expected to result in a reduction in actual yield. Manufacturers might be required to convey this information on the package or through package inserts. In order to avoid misunderstanding, the agency might require consumers to be told that small differences in nominal yield do not reflect significant differences in health risks. Regulations of this nature will improve risk perception among consumers, and will correct any misleading impression about the relative hazards of cigarettes containing different levels of tar and nicotine. From the same perspective, the agency should be authorized to ban or regulate use of misleading terms (such as "light") in advertising or on packaging, and should be authorized to require the use of standard terms.

Principle 3: Toxicity Testing

Manufacturers of all PREPs should be required to conduct appropriate toxicological testing in preclinical laboratory and animal models and appropriate clinical testing in humans to support the health-related claims associated with each product and to disclose the methods and results of such testing to the regulatory agency.

Under the regulatory arrangements recommended by the committee, tobacco manufacturers would have to obtain appropriate toxicological data from the scientific literature or conduct toxicological testing in connection with any new product or modification of an existing product (1) to support a claim of reduced exposure or reduced risk (see principle 5), (2) to demonstrate that a new ingredient does not increase the product risk (see principle 7), or (3) to satisfy the requirements of the added-ingredients review described in principle 8.

Because clinical trials on the long-term health risks of individual ingredients in tobacco-related PREPs cannot reasonably be conducted in humans, any scientific assessment of the risks, except perhaps low birthweight and sudden cardiac death, of added ingredients must necessarily be based on animal studies, short-term human studies, and epidemiological studies. For ingredients that have not previously been used in conventional or modified tobacco products, animal studies to support their safety should be conducted by the manufacturer prior to human exposure to those ingredients. The committee believes that the practice of introducing new ingredients into smoked products without full and adequate testing of such ingredients, as judged by a competent regulatory agency, poses a substantial risk to the consumer and should cease. A requirement for adequate preclinical testing of potentially toxic ingredients is a basic feature of all regulatory laws relating to drugs, biologics, devices, food additives, toxic chemicals, and environmental contaminants and should apply as well to conventional and modified tobacco products.

Principle 4: Premarket Approval of Claims

Manufacturers should be permitted to market tobacco-related products with exposure reduction or risk reduction claims only after agency approval based on scientific evidence (a) that the product substantially reduces exposure to one or more tobacco toxicants and (b) if a risk reduction claim is made, that the product can reasonably be expected to reduce the risk of one or more specific diseases or other adverse health effects, compared with whatever benchmark product the agency requires to be stated in the labeling. The "substantial reduction" in exposure should be sufficiently large that independent scientific experts would anticipate finding a measurable reduction in morbidity and/or mortality in subsequent clinical or epidemiological studies.

Products accompanied by explicit or implicit claims of reduced exposure to tobacco toxicants or reduced risk of adverse health effects should be subject to premarket approval of the specific claim. Under this approach, manufacturers would be permitted to market products with claims of reduced exposure or reduced risk (compared with whatever benchmark product is stated in the labeling) as long as the claim is approved on the basis of appropriate scientific data by the regulatory agency before marketing and, where judged necessary for risk reduction claims, appropriate plans for postmarketing surveillance are also approved.

When the committee refers to a "reduced risk" claim in this chapter, it is referring to any statement, however qualified, indicating that use of the product presents or may present a reduced risk of disease or other adverse health effect, compared with use of some other tobacco product.

Premarket approval of tobacco-related PREPs with claims of reduced exposure or reduced risk is essential to ensure that full information is presented to potential consumers. Under the current regulatory system, manufacturers may, in the absence of drug claims, introduce new or modified tobacco products into the marketplace without any prior regulatory review, subject only to the FTC's authority to enjoin "deceptive acts and practices" after the deception has already occurred. The result has been the long history, cited elsewhere in this report, of promotional practices by tobacco manufacturers that have, at the least, obscured the adverse health consequences of tobacco use. The committee considers it highly unlikely that tobacco-related PREPs marketed under this system would be labeled and promoted with appropriate caution, especially if the market becomes competitive.

The premarket approval model used for the review of new drugs, while not now applied to new brands and modifications of conventional tobacco products marketed without disease claims, in fact provides the only fail-safe procedure for the review of exposure reduction or risk reduction claims relating to tobacco products, whether made on behalf of novel products like Eclipse or in connection with products already on the market (e.g., smokeless products). The committee thus recommends that any new legislation must provide for premarket approval of claims of exposure reduction or risk reduction. The medical device provisions of the FDCA provide a model for this policy in that high-risk products are subject to premarket approval, while products of lesser risk are subject only to premarket notification. The marketing of a new tobacco-related product with an unsubstantiated claim of reduced disease risk (including a claim of reduced exposure that implies a reduced disease risk) is probably a greater threat to public and personal health than the marketing of a product already known to be dangerous, unless the claim is proven to be accurate. The scientific evidence behind such a claim thus deserves full and independent scrutiny.

The committee judges that exposure reduction and risk reduction claims for tobacco-related PREPs should be treated as analogous to "effectiveness" claims for drugs and devices. Just like drug claims, they are intended to assure consumers of a potential health benefit if the product is used as directed, and just like drug claims, they should be presented accurately and supported by appropriate scientific data. As discussed in Section II of this report, the committee is not aware of a sound scientific basis at the present time for an unqualified risk reduction claim for any tobacco-related PREP. Although the potential for reduced-risk products exists, data supporting such claims have never been subject to comprehensive, independent scientific scrutiny or regulatory review. In no case has the degree of potential risk reduction been estimated from animal or human studies, nor have surveillance programs been implemented to monitor the long-term outcome of the use of any of these products. Uncertainty about the health effects of all PREPs will therefore continue for years even if an effective regulatory framework is created in the near term. If no such framework is created, consumers will continue to act on deficient information.

With respect to the tobacco-related PREPs known to the committee, the most that can reasonably be said today on the basis of the publicly available data supplied by manufacturers is that these products reduce exposure to some but not all of the toxic ingredients in cigarettes, assuming no increase in the total number of cigarettes smoked or in other forms of compensatory smoking. Direct evidence will not be available for many years, if ever, to prove that the degree of exposure reduction achieved, if maintained over years, is sufficient to result in a significant reduction in morbidity and mortality. The use of reduced-nitrosamine tobacco, for example, seems likely to reduce the exposure of smokers to nitrosamines, but whether this will reduce lung cancer rates, given the other carcinogens in tobacco smoke, is unknown (see Chapter 12). Similarly, the use of the cigarette-like product Eclipse may reduce exposure to carcinogens but increase exposure to CO. Whether these and similar products, if smoked chronically like cigarettes, will provide a net benefit in terms of all-cause mortality is unknown and can only be determined by long-term epidemiological studies.

Products that increase exposure to one or more tobacco toxicants while reducing exposure to others will present a complex challenge to the regulatory agency. At a minimum, approved claims of reduced exposure should be accompanied by warning statements regarding any increased exposure. Moreover, in the committee's view, reduced-risk claims should be permitted for such a product only if the agency finds, based on scientific evidence, that the use of the product can reasonably be expected to reduce the overall risks of tobacco-related disease or death. While recognizing that the proposed premarket approval requirement raises the possibility of regulatory delays and may inhibit the marketing of new tobacco-related PREPs, the committee believes that undue delay can be minimized by an efficiently administered agency for the following reasons:

- The body of scientific information submitted for review of each new product is likely to be substantially less than that in a typical new drug application to the FDA. For most such products the scientific evidence supporting exposure reduction claims will come from in vitro studies, animal studies, and pharmacokinetic studies in humans. Depending upon the claim, evidence may also come from short-term clinical studies of pulmonary function and/or cardiovascular function. The huge clinical trial programs required to show the effectiveness and safety of drugs in humans would not be necessary to support exposure reduction claims for tobacco-related PREPs.
- Only a handful of tobacco-related PREPs are currently known to be under development, suggesting a relatively low volume of new applications compared with the hundreds of applications for drugs and devices that are handled annually by FDA.
- The success at FDA of the user fee mechanism, in combination with a budgetary appropriation, in providing resources to support an adequate staff suggests a model of public-private funding for the review of claims for tobacco-related PREPs.

Principle 5: Labeling, Advertising, and Promotion

The labeling, advertising, and promotion of all tobacco-related products with exposure reduction or risk reduction claims must be carefully regulated under a "not false or misleading" standard, with the burden for proof of the claim resting on the manufacturer, not the government. The responsible agency should have the authority and resources to conduct surveys of consumer perceptions relating to these claims.

Labeling requirements and restrictions on advertising and other promotional activity are core elements of product safety regulation. In a previous report focusing on preventing initiation of tobacco use (IOM, 1994), another IOM committee recommended that warnings concerning the risks of using tobacco products be strengthened and made more salient, that advertising be restricted to a text-only format, and that many types of promotional activity be severely curtailed. These measures are still needed to reduce youth uptake, to promote cessation, and—most pertinent to this committee's charge—to ensure that public understanding of the dangers of tobacco use is not obscured or compromised by the marketing and promotion of a new generation of modified tobacco products purporting to be "safer" than conventional products. From this perspective, adequate regulatory authority to promote accurate public understanding of the health effects of tobacco-related PREPs is perhaps the single most important feature of effective regulation.

The burden of proving any claim of reduced exposure or reduced risk should rest on the manufacturer. The evidentiary base for such claims must meet contemporary scientific standards and be fully available for independent review by a responsible regulatory body. Independent review and replication of key findings—the "gold standard" for validating scientific assertions—is a compelling necessity for assessing health-related claims associated with tobacco products. Unfortunately, in today's regulatory environment for tobacco products, this proper burden of proof is reversed. Manufacturers are free to market new tobacco products with explicit or implied health claims until challenged by the FTC, which must then show in court, through expert testimony and consumer surveys, that a particular advertising claim is deceptive and not adequately substantiated. Regulation of claims through this process may take years and offers no chance for prompt correction of misleading advertising. The committee anticipates that under the current regulatory conditions, tobacco manufacturers are no more likely to present—accurately and with appropriate caveats, warnings, and uncertainties—what is known, and unknown, about the potential health effects of PREPs than they have been to describe accurately the risks of conventional cigarettes.

The committee therefore strongly recommends that new legislation be enacted to ensure that the labeling, advertising, and promotion of all

tobacco-related products are carefully regulated so that exposure reduction and risk reduction claims are supported by adequate scientific evidence and are not false or misleading. Claims made in the product labeling should be subject to premarket approval by the regulatory agency, as
explained in connection with principle 4. Although a routine requirement
for prior approval of advertising might not be constitutionally permissible, the agency should have the authority to take prompt action against
specific advertising or promotional campaigns or practices that are false
or misleading. This authority should include the power to require corrective advertising.

The committee considered the alternative of recommending that the
advertising and promotion of tobacco-related PREPs be included under
the present system of regulation for tobacco products in general. For the
reasons noted previously, however, the committee considers this system
to be insufficient to the task. In the judgment of the committee, a harm
reduction strategy using tobacco-related PREPs cannot be implemented
successfully unless consumers are fully informed through accurate labeling and advertising about the health consequences of using all types of
tobacco products and about the substantial gaps in scientific knowledge.
The committee finds no justification for continuing the current situation
in which pharmaceutical products intended to aid smoking cessation—
the desired public health goal—must be labeled and promoted truthfully
under a "not false or misleading" standard, while tobacco products claiming to reduce exposure or risk—at best a partial and uncertain step toward disease reduction—can be promoted under generic consumer deception laws, with the burden of proof on the FTC to police violations
after the fact. Unless this imbalance in the regulation of labeling and
advertising is corrected, the committee finds it likely that aggressive promotion of tobacco-related PREPs, accompanied by incomplete and possibly inaccurate or misleading health-related claims, may well undermine
other public health efforts to promote cessation of smoking and other
forms of tobacco use. In short, unless the regulation of labeling and advertising of all tobacco-related products with health claims can be brought to
parity, the impending availability of tobacco-related PREPs could cause
more harm than good.

As a corollary, the committee concludes that the evidentiary base for
regulating labeling and advertising must include not only scientific studies related to toxicology and human biology but also scientific analyses of
the impact of particular claims, as well as messages and images in advertising and promotional activity, on risk perception and risk communication. Such analyses should take into account the overall public health goal
of reducing tobacco-related morbidity and mortality both in individuals
and in the population as a whole as well as the goal of enhancing the

capacity of each adult to make an informed decision concerning the use of tobacco. The regulatory agency should have the resources and funding to conduct relevant consumer surveys and analyses as well as the authority to require such studies from manufacturers.

The committee has aimed to balance two possible types of errors: (1) allowing reduced-risk claims that turn out to be erroneous, and (2) deterring the development of modified products that would reduce exposure to known toxicants (and that would eventually be shown to reduce disease risk) if they were marketed. The committee acknowledges that manufacturers might forego the necessary investment in product development because of doubts that the accuracy of their claims can be defended successfully to unduly skeptical regulators.

However, the most important considerations are unproven or premature reduced-risk claims and reduced-exposure claims that are mistakenly interpreted as reduced-risk claims. The regulatory process should not discourage or impede scientifically grounded claims of reduced exposure, as along as steps are taken to ensure that consumers are not misled into believing (in the absence of sound evidence) that smoking the modified product is (or is likely to be) less hazardous than smoking the conventional product. How the complex of claims and caveats associated with PREPs can best be articulated in labeling is one of the major challenges facing the regulatory agency. On the one hand, the public health is not well served by the continued use of poorly defined terms such as "light," "low tar," or other phrases that imply a benefit when none has been proven to exist. On the other hand, neither is the public health served if smokers are discouraged by unduly cautionary language from using a new product with the potential for real risk reduction. The problem of conveying balance in communicating health benefits and risks is not unique to tobacco-related PREPs, and the large body of experience in other areas of health and safety regulation may be applicable to these products as well.

The agency will have to direct its attention to the language used as well as the labeling format. Some illustrations, based on existing formats, follow:

- Current cigarette labeling contains warnings that smoking causes lung cancer, heart disease, and emphysema and may cause birth defects. If warranted by scientific evidence, such warnings could be accompanied by a statement that the modified product might carry a reduced risk of one or more of these conditions.
- Current food labeling has on each package a table displaying a quantitative analysis of nutritional content. This approach could be applied to selected ingredients and constituents of tobacco products or tobacco smoke, such as known carcinogens and CO.

- Food-labeling tables express nutritional content not only as grams per serving but also as a percentage of average daily intake. One could envision tobacco product labeling with a similar table that shows exposure or yield ranges for particular toxicants or perhaps ranks the levels of exposure or yield as high, average, or low. These terms could be specifically defined and would perhaps be less misleading than such terms as "light."
- One could also envision the use of words such as "high," "average," or "low," again carefully defined, in a "risk" column in such a table. Pictograms, such as those that appear in poison control warnings, or icons might be used instead of words. It is essential that such labeling in the end be perceived as denoting degrees of risk, not as signifying or implying safety. The message that cessation is the only safe choice must not be obscured or lost.
- The agency should also consider requiring that labels for PREPs that make exposure reduction claims disclose that the reduction in exposure depends upon the user not compensating for the reduction by increasing use or by inhaling more deeply. Consideration should also be given to a disclosure that the health benefits of the product's exposure reduction have not yet been established in scientifically recognized tests or ongoing studies. Such a disclosure would guard against consumer confusion that risk reduction benefits have been proven. Furthermore, such a disclosure would provide an incentive to manufacturers to do more research on the health effects of exposure reduction. The FTC should consider requiring similar disclosures under its existing authority if new legislation is not adopted.

The committee does not have recommendations on the specific form of labeling that is optimal for tobacco-related PREPs. (The formats cited in the foregoing examples are meant only to be illustrative.) This topic deserves the sustained attention of experts in communication and marketing, with the intent of identifying the optimal approach for conveying in a simple format specific information on exposure reduction, the potential for risk reduction, and the associated uncertainty. This information must be balanced, accurate, informative, and not misleading.

Growing Up Tobacco Free (IOM, 1994) reviewed the history of mandated tobacco health warnings; called attention to the inadequacy of the existing warnings (both in content and format); reviewed recent initiatives in other countries, especially Canada; and recommended a number of legislative measures to increase the salience and effectiveness of health warnings on packaging and in advertising. Specifically, the IOM (1994) report made the following recommendations:

Based on current knowledge, Congress should enact specific warning and format requirements now, and should delegate regulatory responsibility for future modifications to the secretary of health and human services. The secretary should ensure that ongoing research is conducted on the effectiveness of prescribed warnings. Congress lacks the institutional expertise and flexibility to monitor the efficacy of regulatory innovations and to respond to new information. Authority to amend the warnings should be delegated to an appropriate regulatory agency. The secretary should be empowered to modify the format or content of existing warnings and to prescribe additional warnings, whenever such action is reasonably necessary to achieve effective communication of significant information regarding the health consequences of tobacco use or to discourage consumption of tobacco products.

The committee endorses this recommendation. The need to implement this approach is all the more evident now in light of the emerging market for products that purport to reduce the risk of tobacco use.

One of the recurrent questions in the regulation of health claims is when to permit a claim that a product *may* reduce the risk of disease. Such a statement may be misleading even though it is not false. Assuming substantial proof of exposure reduction, what type of evidence must be produced in support of a claim that the product *may*, as a result, reduce the risk of disease? Is inference based solely on the existence of a dose-response relationship sufficient? How strong must the evidence be? What is the significance of counterevidence? How much weight should be given to favorable changes in biomarkers that are biologically rational but not yet proven to be true surrogate end points? Although these are questions that will have to be addressed by the regulatory agency in due course, the committee itself is disinclined to permit claims that the product "may" or "might" reduce the risk of disease (or that "growing evidence suggests that the risk is reduced") unless, as judged by the regulatory agency after review of data by an independent scientific advisory committee, the degree of exposure reduction is likely to result in a meaningful decrease in tobacco-related morbidity or mortality that can be measured in subsequent epidemiological studies.

Another question must then be faced. Assume that the agency finds insufficient evidence under such a demanding standard to support a claim that the exposure-reducing product "may reduce the risk" of a particular disease. Should the manufacturer be permitted to make the statement (which is not false) as along as it is accompanied by a disclaimer to the effect that the regulatory agency has not approved the statement? This approach is receiving increasing attention in other regulatory contexts in light of a few judicial decisions holding that regulatory agencies must allow manufacturers to make the unproven statement, together with a disclaimer, under the First Amendment (Gilhooley, 2000). The committee assumes that the First Amendment does not require such a permissive approach and is inclined to be much more cautious in light of the history of public health mistakes in tobacco product innovation. In the committee's judgment, in the absence of sufficient scientific evidence to support a claim that a product may reduce the risk of disease, any claim of reduced exposure to tobacco toxicants should be accompanied by a statement that the health consequences of the change are unproven (or unknown).

Given that considerable time will be needed to enact appropriate legislation, to staff a new regulatory agency, and to bring whole classes of products into a new regulatory system, a number of modified tobacco products with risk-reduction claims may enter the marketplace in coming years without regulatory review, assuming they are positioned not to fall under the drug or device laws. The committee views this situation with alarm but cannot offer a ready solution. Perhaps tobacco manufacturers would be willing to demonstrate their good faith by agreeing to voluntarily submit claims of reduced exposure or reduced harm to FDA, FTC, CDC, or some other appropriate agency for its review and comment, and to conform to agency suggestions pending legislative action. The committee emphasizes, however, that any purported harm-reducing products that have been put on the market before enactment of regulatory legislation should not be grandfathered or exempted from rigorous scientific and regulatory scrutiny.

Principle 6: Postmarketing Surveillance

The regulatory agency should be empowered to require manufacturers of all products marketed with claims of reduced risk of tobacco-related disease to conduct postmarketing surveillance and epidemiological studies as necessary to determine the short-term behavioral and long-term health consequences of using their products and to permit continuing review of the accuracy of their claims.

The committee has repeatedly emphasized that since definitive information on the ultimate value of a PREP cannot be determined at the time of its initial marketing, the true extent of its long-term benefit (or possible harm) will remain uncertain unless specific surveillance and epidemiological studies are conducted after the product is marketed. The FDA already has the authority to require pharmaceutical manufacturers to conduct such studies for the purpose of monitoring drug-related health risks. The designated regulatory agency for tobacco-related PREPs should have equivalent authority to require manufacturers who seek to make reduced-risk claims to sponsor surveillance and epidemiological studies in appropriate cases. These studies should be designed to assess not only the benefit of the product in reducing the risk of smoking but also the degree to which the product has served to maintain as smokers those who might otherwise have quit and/or to recruit new smokers. The quality of these studies should be high so that the results can permit modification of claims as knowledge evolves.

The committee envisions that a plan or protocol for such a study (or studies) would commonly be required at the time of approval of the claim. Whether a postmarketing study is necessary should be judged by the regulatory agency on a case-by-case basis. The agency can also be expected to take steps to ensure the scientific integrity of such studies by requiring appropriate monitoring and, where necessary, an independent safety monitoring board. In addition to the authority to require manufacturers to sponsor postmarketing studies, the regulatory agency should also have in-house expertise in epidemiology. Such expertise is needed to deal competently with scientists in industry and peers in the field, to operate an adverse event reporting system, and to sponsor selected studies when necessary. As discussed in Chapter 6, the regulatory agency, in collaboration with other public health agencies, should conduct an active program of surveillance for all tobacco products, including conventional ones, modified products that are not accompanied by health claims, and PREPs. It is particularly important for public health authorities to monitor patterns of initiation and progression of use, including the interactions of product use (substitution and supplementation), for all PREPs.

Principle 7: "No Increased Risk" Threshold for All Tobacco Products

In the absence of any claim of reduced exposure or reduced risk, manufacturers of tobacco products should be permitted to market new products or modify existing products without prior approval of the regulatory agency after informing the agency of the composition of the product and upon certifying that the product could not reasonably be expected to increase the risk of cancer, heart disease, pulmonary disease, adverse reproductive effects, or other adverse health effects, compared to similar conventional tobacco products, as judged on the basis of the most current toxicological and epidemiological information.

Conventional tobacco products are well established in the marketplace, and the committee recognizes that, regardless of the adverse health consequences of using these products, this situation is likely to continue for the foreseeable future. These products are already subject to a limited regulatory system that includes certain health warnings and other labeling requirements and some restrictions on advertising and promotion. The committee considered whether modified versions of established tobacco products, or new brands, that are marketed without health claims should be required, on public health grounds, to satisfy any new regulatory requirements and, if so, whether manufacturers should continue to be permitted to put such conventional products on the market without prior agency approval.

The committee concludes that any new regulatory system should ensure that all newly marketed tobacco products pose no increase in health risks, and ideally a new system should move products toward reduced potential for harm. To this end, the committee recommends that manufacturers be permitted to introduce a new product or make significant changes to an already marketed conventional tobacco product only if, on the basis of currently known toxicological and epidemiological information, the product could not reasonably be expected to increase the risk of disease. Implementation of this regulatory principle requires the definition of a standard for comparison. Although the responsibility for developing such a standard will rest with the regulatory agency, the committee can envision several possibilities: the most popular brand, the most (or least) harmful brand on specified dimensions, a toxicity profile that is representative of a sales-weighted average brand, or a standardized product of known composition. Because toxicant yields do not necessarily reflect health effects and because different standards may be appropriate for different products, the establishment of standards or reference products may require the type of information that would evolve from the regulatory agency's experience with the disclosure of ingredients in individual products (principle 1) and the added-ingredients review (principle 8). Such standards might be promulgated through guidelines and performance standards (principle 9).

To implement this approach, manufacturers should be required, before marketing a product, to submit ingredient information to the agency and to certify that the criterion of no increased risk is met (premarket notification). Whatever comparative standard is adopted for the product, the committee believes that a judgment of no increased risk should be based on the most current toxicological and epidemiological data. Manufacturers should not use the most dangerous products of the past as an appropriate standard for any new product. Consumers should rightfully expect that all new tobacco products, even those marketed without health claims, are at least no more hazardous than similar contemporaneously marketed products.

The committee considered whether premarket approval by the regulatory agency should be required for modifications and new brands of conventional tobacco products without claims of reduced exposure or reduced risk, and rejected this approach in favor of premarket notification and certification by the manufacturer. Although some product modifications may be undertaken to improve taste or other features linked to consumer satisfaction, future innovation in the tobacco market in this country is likely to focus on changes that purport to reduce risk. Because the main regulatory purpose of a requirement for no increased risk is to help establish the baseline for future comparisons and to stimulate development and use of protocols for tobacco product risk assessment, premarket notification is sufficient. The committee also did not want to see the regulatory agency burdened with a high volume of premarket approval decisions for products that have no claims of and no potential for improving the public health. Nevertheless, the regulatory agency should have the authority to seek the removal of a product from the market in the event that the "no increased risk" standard is not met.

Principle 8: Added-Ingredient Review

All added ingredients in tobacco products, including those already on the market, should be reported to the agency and be subject to a comprehensive toxicological review.

The ingredients added to tobacco products should be subject to a review that is similar to the FDA review of food additives conducted more than two decades ago to determine if those additives were generally recognized as safe and not subject to additional regulation. Although many of the ingredients added to tobacco products are generally recognized as safe when used as food additives, such an understanding may not apply when these substances are combusted and/or inhaled in smoke. A review of added ingredients would establish a baseline of knowledge about the ingredients, including chemicals, papers, and filters, as well as the toxicological testing that has been conducted on them; it would thereby facilitate an informed appraisal of the potential effects of product modifications. This review should be conducted by independent panels of experts reporting to the regulatory agency, with the objective of identifying those ingredients that add no significant toxicity to tobacco products and therefore can be considered safe in the context of this use. Sufficient knowledge of the toxicity of many of the constituents and ingredients of tobacco products may already be available (see Chapter 10) to begin this review. If adequate information is not yet available, the regulatory agency should have the authority to require the in vitro and animal testing that is judged necessary by the expert panels conducting the review. Such a review would begin to rationalize the risk assessment of ingredients in tobacco products. It would also open this field to the same scientific discourse that is now applied to food additives, environmental contaminants, and toxic chemicals, substances that are no longer introduced into use without independent scientific scrutiny or regulatory review. Tobacco products are the last remaining legally marketed toxic products in our marketplace that are tolerated without such review. A review of added ingredients would provide a scientific basis for guidelines or performance standards to limit possible toxicants like pesticides or filter fibers, if such limits are judged necessary for health reasons. It would also facilitate the review of new products by establishing conditions under which added ingredients, such as flavoring agents and papers, would be considered safe for use without additional data or review.

An added-ingredients review would be a large administrative undertaking requiring several years of effort. It would also require the cooperation of the tobacco industry in submitting data. The committee believes that such a review would be in the interests of both manufacturers and consumers. It would permit credible scientific judgment regarding the extent to which nontobacco ingredients in tobacco products contribute to overall toxicity, would build scientific bridges between the tobacco industry and the toxicology community at large, and as noted above, would facilitate the regulatory review of new products. The committee notes that the review of food additives has been conducted without public disclosure of trade secrets, such as the composition of spices and flavoring agents, and believes that this principle should be respected in any review of the ingredients added to tobacco products.

Principle 9: Performance Standards

The regulatory agency should be empowered to set performance standards (e.g., maximum levels of toxicants; definitions of terms such as "low tar") for all tobacco products, whether conventional or modified, or for classes of products.

Performance standards, including maximum permissible levels, are promulgated under many regulatory laws, including those related to drugs, medical devices, and environmental pollutants. As noted above, successful implementation of a harm reduction strategy as an element of the nation's tobacco policy will require proactive government efforts in research, education, and surveillance as well as regulation of specific products. Accordingly, as scientific knowledge evolves regarding product risks and consumer preferences, the regulatory authority should be empowered to require product modifications to eliminate unreasonable risks. (A similar recommendation appears in the 1994 IOM report.) The committee has not attempted to draft precise statutory criteria to guide the agency's discretion in adopting performance standards. However, the committee does assume that the agency would not be empowered to ban nicotine from tobacco products.

Performance standards might take the form of limits on the concentrations of toxic ingredients in the product or the smoke; sets of limits that taken together would qualify a smoked product to be labeled as high, average, or low on a risk scale; or a list of reviewed ingredients that could be used without challenging the no increased-risk standard. A performance standard cannot be adopted without good scientific data, deliberate planning, and careful monitoring to ensure that it is achieving the desired goal. As the FTC test for tar and nicotine illustrates, even well intended performance standards can sometimes be subverted, with perverse and unintended health consequences. Performance standards aimed at setting definitions for terms must therefore be thought through with great care and be subject to change as experience is gained. Such standards will undoubtedly require a public rule-making process, meaning considerable time for their adoption.

Principle 10: Enforcement Powers

The regulatory agency should have enforcement powers commensurate with its public health mission, including the power to issue subpoenas.

The committee envisions that any regulatory agency for tobacco-related products would need the usual enforcement authorities conferred on public health regulatory agencies, such as the FDA, FTC, CPSC, and Environmental Protection Agency (EPA). The committee anticipates that this agency would also have an appropriate technical and legal staff concerned with issues of enforcement.

The committee specifically recommends that the agency have among its powers the authority to require the monitoring of scientific studies sponsored by manufacturers and to inspect manufacturers, investigators, and contractors to verify the data submitted in these studies. Such powers are necessary to ensure the quality and integrity of these studies. The agency should also have the authority to remove from the market ingredients or products that do not meet the test of no increased risk; to prohibit claims that are not supported by adequate scientific data; to seize products that are improperly labeled; to act promptly against advertising campaigns and promotional materials that are false or misleading; and to require corrective action.

Principle 11: Regulation of Drugs and Devices

Exposure reduction and risk reduction claims for drugs that are supported by appropriate scientific and clinical evidence should be allowed by the FDA.

The committee recommends no new legislation regarding the regulation of pharmaceutical products and medical devices intended to assist people to stop smoking. What will be needed is effective policy coordination between FDA regulators responsible for drugs and devices and whatever agency is charged with administering any new legislation related to tobacco products.

The committee also recommends that the FDA be prepared attitudinally and technically to approve both exposure reduction and risk reduction claims for drugs, when such claims are supported by appropriate scientific data. It is clear that the current FDA standard for all approved smoking cessation products is "quit rates" and that long-term use of pharmaceutical PREPs with cigarettes, which might be employed in a harm reduction strategy, is discouraged by the labeling approved for these products (Hughes et al., 1999). While the committee again emphasizes that cessation of smoking is the desired goal for all smokers, it also concludes that drugs with exposure reduction and risk reduction claims, if properly regulated and proved efficacious, have a place among the treatment modalities that should be available to current smokers. The committee judges, for example, that statements such as "helps recruitment into treatment clinics" or "reduces the use of cigarettes without compensation" are appropriate indications if supported by adequate data from clinical trials. The committee also concludes that for persons addicted to nicotine, a nicotine-containing drug product is preferable to a cigarette or other tobacco-containing product as a chronic source of nicotine. The FDA should therefore be prepared to consider the chronic administration of nicotine products as a reasonable exposure reduction strategy, again if supported by valid clinical data.

Transitional Regulation of Tobacco-Related PREPs as Drugs

Although the committee believes that adequate regulation of tobacco-related PREPs requires new legislation conforming to the preceding principles, the potential utility of the existing authority of the FDA and FTC should not be overlooked, and some of the committee's recommendations could be implemented by these agencies in the absence of new authority. The FDA presently has the authority to regulate any product, including new or existing tobacco products, as drug-delivery devices or as drugs, on the basis of an expressed or implied claim by the manufacturer that the product prevents disease by reducing the health risks of tobacco (Page, 1998). FDA can find that a manufacturer intends to make such a claim by looking at any relevant source, including advertising. See United States v. "Sudden Change" (United States v. "Sudden Change," 1969) and National Nutritional Foods Assn. v. FDA (National Nutritional Foods Assn. v. FDA) (objective evidence). Claims that PREPs respond to concerns about smoking-related illnesses or cancer can be regarded as disease prevention claims.

In its recommendations for regulating the risk reduction claims for PREPs, this report identifies the type of testing and post-approval studies that should be conducted. The committee recommends that FDA use these testing standards under its existing authority in regulating PREPs that make claims for reduced risk of disease. FDA can regulate products that are both drugs and devices under its authority over medical devices. In the case of medical devices that require premarket approval, the agency can determine what testing needs to be done to provide a "reasonable assurance" of safety and effectiveness and can set performance standards for class II devices where appropriate (21 U.S.C.; Page, 1998). The products can be considered safe on a relative basis in light of their capacity for risk reduction, even though tobacco itself was not considered safe in FDA v. Brown & Williamson (Food and Drug Administration, et al. v. Brown & Williamson Tobacco Corporation, et al., 2000). Post-approval testing has been allowed and required for new drugs (Page, 1998).

Exposure reduction claims, such as low tar, have not been regulated by FDA as disease claims, and its authority to do so is uncertain (Page, 1998). A new law designed specifically to regulate tobacco-related PREPs can give a regulatory agency clear authority with respect to exposure-related and smokeless claims. Even for disease prevention claims, new legislation can confirm the agency's authority and the type of tests and manner in which the agency should regulate claims. Moreover, the legislation recommended by the committee would provide the agency new authority to obtain data on ingredients of all tobacco products, permit additional disclosures to consumers about toxicants in tobacco, give enhanced authority (and review when needed) over advertising and labeling claims for PREPs, guard against increased risks from new or modified tobacco products, and authorize a toxicology review and performance standards for tobacco products.

FDA's authority with respect to disease claims would continue if the new authority recommended under this report were given to FDA, with whatever modifications were made by statute. If the authority were given to a new agency, FDA's authority over disease claims for PREPs should be revoked only if the new agency has authority that is at least as extensive as FDA's present authority.

SUMMARY

In this chapter, the committee has addressed key elements of a regulatory framework for implementing the scientific and policy recommendations made in the body of the report. In the committee's judgment, harm reduction is a feasible and justifiable public health policy—but only if it is implemented carefully to achieve the following objectives:

- manufacturers have the necessary incentive to develop and market products that reduce exposure to tobacco toxicants and that have a reasonable prospect of reducing the risk of tobacco-related disease;

- consumers are fully and accurately informed of all of the known, likely, and potential consequences of using these products;
- promotion, advertising and labeling of these products are firmly regulated to prevent false or misleading health claims, whether these claims are explicit or implicit;
- health and behavioral effects of all PREP use are monitored on a continuing basis;
- basic, clinical, and epidemiological research is conducted to establish with reasonable scientific certainty the actual health benefits of PREPs to individuals and the population as a whole; and
- harm reduction is implemented as a component of a comprehensive national tobacco control program that emphasizes abstinence-oriented prevention and treatment.

The committee nevertheless acknowledges that a regulatory system of the type outlined in this report will require sustained congressional support and substantial public funding. It will also impose new direct and indirect costs on the tobacco industry. The committee emphasizes, however, that the regulatory system proposed in this report is not to be viewed in isolation. It is proposed as an essential component of a package of public policy initiatives (including research, education, and surveillance) that this committee believes is necessary to realize whatever benefit tobacco or pharmaceutical product innovation can offer in reducing the nation's burden of tobacco-related illness and death. The committee notes again that public health benefits may not emerge, even if the public and private investment in these initiatives is made. In the absence of this investment, however, the hoped-for benefits are highly unlikely to materialize.

REFERENCES

21 U.S.C. 360e(d)(2)

Ferrence R, Slade J, Room R, Pope M. 2000. *Nicotine and Public Health.* Washington, DC: American Public Health Association.

Food and Drug Administration et al. v. Brown & Williamson Tobacco Corporation et al., 2000. 529 U.S. 120.

Gilhooley, 2000. Constitutionalizing food and drug law. *Tulane Law Review* 74:815-882

Hughes JR, Cummings KM, Hyland A. 1999. Ability of smokers to reduce their smoking and its association with future smoking cessation. *Addiction* 94(1):109-114.

IOM (Institute of Medicine). 1994. *Growing Up Tobacco Free.* Washington, DC: National Academy Press.

IOM (Institute of Medicine). 1998. *Taking Action to Reduce Tobacco Use.* Washington, DC: National Academy Press.

National Nutritional Foods Assn. v. FDA, 504 F.2d 761, 789 (2d Cir. 1974)

Page, JA. 1998. Federal regulation of tobacco products and products that treat tobacco dependence: are the playing fields level? *Food and Drug Law Journal* 53: Supp 11-42.

U.S. DHHS (U.S. Department of Health and Human Services). 2000. *Reducing Tobacco Use; A Report of the Surgeon General.* Atlanta, GA: USDHHS, Centers for Disease Control and Prevention.

United States v. "Sudden Change," 409F.2d 734, 739 (2d Cir. 1969).

Warner KE, Slade J, Sweanor DT. 1997. The emerging market for long-term nicotine maintenance. *JAMA* 278(13):1087-1092.

8

Principal Conclusions

The science base for assessing tobacco harm reduction is incomplete. Nonetheless, the presence of potential reduced-exposure products (PREPs) on the market suggests an urgent need for proactive plans to evaluate the potential risks and benefits. The potential for reduction in morbidity and mortality that could result from the use of less toxic products by those who do not stop using tobacco justifies inclusion of harm reduction as a component in a broad program of tobacco control. To date there are two general types of PREPs: pharmaceuticals and modified tobacco products. The pharmaceuticals include, for example, nicotine replacement therapy (NRT) and bupropion, while modified tobacco products include products with modified tobacco and those with modified delivery systems.

Having identified conceptual and operating precepts as stated at the end of Chapter 1, the committee concludes that there can be a successful, scientifically-based harm reduction program that is justifiable and feasible—but only if implemented carefully and effectively and only if:

- manufacturers have the necessary incentive to develop and market products that reduce exposure to tobacco toxicants and that have a reasonable prospect of reducing the risk of tobacco-related disease;
- consumers are fully and accurately *informed* of all of the known, unknown, likely, and potential consequences of using these products;

- promotion, advertising and labeling of these products are firmly *regulated* to prevent false or misleading claims, explicit or implicit;
- health effects of using PREPs are *monitored* on a continuing basis;
- basic, clinical, and epidemiological *research* is conducted to establish the potential use of PREPs for reducing risks for disease in individuals and for reducing harm to the population as a whole; and
- harm reduction is implemented as a component of a comprehensive national tobacco control program that emphasizes abstinence-oriented prevention and treatment.

The 7 chapters of the committee's report that precede this and the extensive reviews found in Section II provide the documentation for the following principal conclusions regarding the four questions posed within the charge, as outlined in Chapter 1. Specific recommendations can be found within the body of the report.

Conclusion 1. *For many diseases attributable to tobacco use, reducing risk of disease by reducing exposure to tobacco toxicants is feasible.* This conclusion is based on studies demonstrating that for many diseases, reducing tobacco smoke exposure can result in decreased disease incidence with complete abstinence providing the greatest benefit. Key to this conclusion is the assumption that compensatory increase in exposure does NOT occur with the use of these products.

Conclusion 2. *PREPs have not yet been evaluated comprehensively enough (including for a sufficient time) to provide a scientific basis for concluding that they are associated with a reduced risk of disease compared to conventional tobacco use.* (One exception is the use of nicotine replacement therapy for maintenance of cessation in the Lung Health Study. See Chapters 13 and 14.) Carefully and appropriately conducted clinical and epidemiological studies could demonstrate an effect on health. However, the impact of PREPs on the incidence of most tobacco-related diseases will not be directly or conclusively demonstrated for many years. Tobacco use causes very serious morbidity and mortality due to several different diseases. Cancer (e.g., of the lung, oral cavity, esophagus, and bladder), cardiovascular disease, chronic obstructive pulmonary disease, and low birthweight are all well-established effects of tobacco use. The conditions can be diagnosed, the natural history of the diseases is reasonably well understood, and scientifically appropriate studies of tobacco users who switch to PREPs could be designed. See Chapters 4 and 11-16 for supporting material.

However, such research will be difficult. For example, tobacco users may not use a particular PREP for long enough to see health impact; tobacco PREPs will undoubtedly change substantially over the next decade; many subjects would be required for adequate statistical power. For all these and other reasons, conclusive proof of the health effects of PREPs will not be available in the near future, as new PREPs are entering the marketplace. Thus, for purposes of educating the public about PREPs and for purposes of regulating health claims, surrogate measures of health effects must be considered.

Conclusion 3. *Surrogate biological markers that are associated with tobacco-related diseases could be used to offer guidance as to whether or not PREPs are likely to be risk-reducing.* However, these markers must be validated as robust predictors of disease occurrence, and should be able to predict the range of important and common conditions associated with conventional tobacco products in order to be useful for PREP evaluation and regulation. PREPs may differentially affect risk of tobacco-related diseases. Furthermore, the efficacy of PREPs will likely depend on user population characteristics, e.g., those defined by gender, genetic susceptibility, ethnicity, tobacco history, and medical history. Chapters 12-16 describe clinical studies using surrogate indicators that could be conducted to better understand whether or not PREPs would decrease specific adverse health outcomes. The potential studies vary in terms of the length of time that would be required to document the effect, the number of patients, and the power of the study to predict disease outcome. There is no one panel or group of tests that the committee could recommend at this time that would, as a whole, serve to assure that morbidity and mortality would decrease with use of PREPs.

Conclusion 4. *Currently available PREPs have been or could be demonstrated to reduce exposure to some of the toxicants in most conventional tobacco products.* There are many techniques to assess exposure reduction, but the report contains many caveats about the use of all of them, including usually an unknown predictive power for harm. Long-term use of pharmaceutical preparations for maintenance of tobacco cessation will clearly achieve exposure reduction. The safety of these products for long-term use, however, is not well established. For example, it is well known that nicotine affects the autonomic nervous system, with uncertain long-term consequences. However, even if NRT use for maintenance of cessation results in nicotine exposure equivalent to that achieved with conventional tobacco products, exposure to the most harmful tobacco toxicants is avoided. See Chapters 4, 9, and 11 for supporting material.

There is insufficient evidence to decide whether concomitant use of NRT or bupropion with decreased tobacco use will lead to significantly decreased exposure to tobacco toxicants such as tar and carbon monoxide. Nor is there sufficient evidence to determine how much this PREP strategy will decrease conventional tobacco use or how much compensation will occur. However, there are exposure assessment tools to assess this issue, as described in Chapter 7.

Tobacco-related PREPs pose different exposure assessment problems. PREPs characterized by the reduction (or, conceivably, elimination) of one class of toxicants, such as the reduced-nitrosamine preparations in varying stages of development and marketing, do result in decreased exposure per cigarette to specific toxicants. Analytic techniques exist to demonstrate this. However, the smoking behavior of people who use these PREPs has not been researched well enough to know whether or not compensation occurs (thus increasing net exposure to other toxicants and possibly maintaining exposure to the potentially reduced chemical). Furthermore, there are insufficient data to allow scientific judgement or prediction of the health effects of removal of one class of chemicals from tobacco products.

The cigarette-like PREPs that use heat or reduced burn temperature of tobacco and deliver aerosolized nicotine pose other exposure assessment problems. The prototypes available now have only just begun to be studied by researchers other than the manufacturers. It is clear that the yield of some of these products is different from that of conventional cigarettes. The pattern of yield changes suggests differential reduction in exposure to toxicants. Some preliminary data suggest increased yield of specific toxicants concomitant with no change or decreases in others.

There does not exist a standard reference product for comparison with tobacco-related PREPs. Assessment of the risk from use of a PREP requires comparison to the risk of the product avoided AND to the risk of the product (including no product, or abstinence) the PREP user would switch to if the PREP were NOT available.

Conclusion 5. *Regulation of all tobacco products, including conventional ones as recommended in IOM, 1994, as well as all other PREPs is a necessary precondition for assuring a scientific basis for judging the effects of using PREPs and for assuring that the health of the public is protected.* Regulation is needed to assure that adequate research (on everything from smoke chemistry and toxicology to long-term epidemiology) is conducted and to assure that the public has current, reliable information as to the risks and benefits of PREPs. Careful regulation of claims is needed to reduce misperception and misuse of the products. If a PREP is marketed with a claim that it reduces (or could reduce) the risk of a specific disease(s) compared to the risk of the product for which it substitutes, regulation is needed to assure that the claim is supported by scientifically sound evidence and that pertinent epidemiological data is collected to verify that claim. The regulation proposed by this committee is narrowly focused on assuring that the products reduce risk of disease to the user and accumulating data that would indicate whether or not the products are harm-reducing for the population in the aggregate. Other potential regulatory approaches to tobacco control are not addressed within this report. See Chapter 7 for supporting and explanatory material.

Conclusion 6. *The public health impact of PREPs is unknown. They are potentially beneficial, but the net impact on population health could, in fact, be negative.* The effect on public health will depend upon the biological harm caused by these products and the individual and community behaviors with respect to their use. Assessing the public health impact will be difficult and will require classic public health tools of surveillance, research, education, and regulation to assure that the impact is positive. The major concern for public health is that tobacco users who might otherwise quit will use PREPs instead, or others may initiate smoking, feeling that PREPs are safe. That will lead to less harm reduction for a population (as well as less risk reduction for that individual) than would occur without the PREP, and possibly to an adverse effect on the population. PREPs should be a last resort only for people who absolutely can not or will not quit. Population-based research and surveillance can determine whether the intended impact is achieved. However, measurements of health impact at the population level can take years to document, as described in previous sections of this chapter and in the report as a whole.

Regulation of PREPs can only assure that a specific PREP could be risk-reducing for a person who uses it compared to the conventional product it replaces. Regulation cannot assure that the availability of risk-reducing PREPs will lead to reduced tobacco-related disease in the population as a whole. However, a regulatory agency can assure that data are gathered that would permit the population effects to be monitored. If population tobacco product use increases or tobacco-related disease increases, these data would serve as a basis for developing and implementing appropriate public health interventions. See Chapters 3, 6, and 7 for supporting material.

Studies using surrogate indictors of population impact could be designed. For example, monitoring the perception that the public, particularly tobacco users and adolescents, has of the risks and benefits of PREPs is possible. Research indicating that people perceive PREPs to be more beneficial than scientific judgment indicates would provide early evidence of the risk for an adverse public health impact. Action of various sorts (e.g., regulatory review of claims, public health education campaigns) could then be taken.

Chapters 1 and 5 include discussions of the utility of a risk assessment framework for organizing the scientific basis for evaluation of PREPs. It is useful to return to that framework to put the committee's conclusions and recommendations into the proper light. Although the committee did not perform a risk assessment for any existing PREP, the committee's conclusions and recommendations provide a means to assure that a risk assessment can be done in the future. As Table 8-1 illustrates, the committee's principal conclusions (discussed in a preceding section of this chapter) assume use of the conventional risk assessment framework, and the conclusions and recommendations for surveillance and regulation point a way to develop the necessary data for such an evaluation.

Hazard identification is inherent in the first question of the committee's charge, Does the product decrease exposure to the *harmful substances* in tobacco? The principal conclusions that harm reduction is feasible and that exposure reduction can be demonstrated require identification of the toxicants within or produced by use of the PREP. The element of the proposed surveillance system related to specific tobacco constituents and several of the regulatory principles (#1, #3, #7, #8, #9) will assure that the necessary toxicology data are gathered, validated, and made available to scientists, public health officials, and regulators.

Dose-response assessment is inherent in the second question of the charge, Is decreased exposure associated with decreased harm to health? An important issue when considering this question in the context of PREP assessment is that while some data are available when assessing a dose-response relationship, there are virtually no data describing the change in response due to dose reduction after a period of higher exposure. This data would reflect the extent of disease reversibility or halting of disease progression possible from exposure reduction. Dose-response assessment is also inherent in the third question of the charge, Are there useful surrogate indicators of disease that could be used? The principal conclusion that surrogate measures could be used to predict harm reduction requires the development of surrogate disease indicators (response) so that a dose-response assessment (and therefore a risk characterization) could be made in some reasonable timeframe, without waiting decades to assess cancer morbidity and mortality. The surveillance system component addressing disease outcomes will help provide some of these necessary data and the regulatory principle #6 requiring postmarketing surveillance and epidemiologic studies for PREPs with claims will assure that the data are collected.

TABLE 8-1 Relationship of Conclusions and Recommendations for PREP Risk Assessment

	Hazard Identification	Dose Response	Exposure Assessment	Risk Characterization	Risk Management
Committee charge	1. Does product decrease exposure to the **harmful substances** in or produced during use of tobacco?	2. Is decreased exposure associated with decreased harm to health? 3. Are there useful surrogate indicators of disease that could be used?	1. Does product **decrease exposure?**	4. What are the public health implications?	4. What are the public health implications?
Principal conclusions	1. Risk reduction is feasible 4. Exposure reduction can be demonstrated	3. Surrogate measures could be used to predict risk reduction	4. Exposure reduction can be demonstrated	1. Risk reduction is feasible 2. Risk reduction not yet demonstrated 6. Public health impact is unknown	5. Regulation is a necessary precondition for assuring a science base and for assuring protection of the health of the public
Elements of surveillance system	Specific tobacco constituents of both the products and the smoke they generate	Disease outcomes	Consumption of tobacco products and of PREPs Biomarkers of exposure to tobacco products Personal tobacco product use and related behavioral patterns	Disease outcomes	Tobacco product marketing, including PREPs

TABLE 8-1 Continued

	Hazard Identification	Dose Response	Exposure Assessment	Risk Characterization	Risk Management
Regulatory principles (all refer to tobacco-related PREPS, except for 11)	1. Ingredient disclosure 3. Preclinical testing required to support health-related claims 7. Evidence for no increased risk 8. Added ingredient review 9. Performance Standards	6. Products with claims would require post-marketing surveillance and epidemiological studies	2. Yield assessment 4. With specific claims, no increased exposure to unclaimed compounds 9. Performance standards 11. Exposure reduction claims for pharmaceutical PREPs	5. Labeling for products with claims cannot be false or misleading	10. Enforcement power
Research and other recommendations	3. Develop appropriate animal models and in vitro assays of pathogenesis	1. Sufficient data to allow estimation of dose-response 2. Develop validated biomarkers of disease	4. Clinical and epidemiological studies in human are required	Comprehensive surveillance is recommended	Regulation is recommended

Exposure assessment is inherent also in the first question of the committee's charge. The principal conclusion that exposure reduction can be demonstrated is fairly straightforward. Several components of the proposed surveillance system will provide important exposure information and at least four regulatory principles (#2, #4, #9, #11) would assure that relevant data are collected.

Risk characterization is the central question of the report and, indeed, of harm reduction writ large. The fourth question of the committee's charge regarding the public health impact of the products is perhaps the most important asked of the committee. The principal conclusions that harm reduction is feasible but not yet convincingly demonstrated and that a beneficial public health impact is not assured are two that are most easily misunderstood as contradictory if not carefully considered. They drive important considerations of the report—harm reduction should be pursued and encouraged but every aspect of it should be watched and studied vigilantly. Appropriate tools of public health must be available and must be powerful. Surveillance of personal tobacco product use and related behavioral patterns and of disease outcomes will provide some of the data necessary to assure a positive population impact. The regulatory principle that labeling for PREPs with claims cannot be false or misleading is another necessary safeguard against a negative public health impact.

Risk management, the culmination of the risk assessment process, is directly related to the committee's principal conclusion that regulation is a necessary precondition for advancing knowledge and for ensuring a public health benefit. Two of the most important tools for a risk manager, are knowledge, which will be developed if the research and surveillance recommendations are followed and if the regulatory principles #1-9 are adhered to, and enforcement power, which is called for in regulatory principle #10. A properly conducted risk assessment outlines gaps in the knowledge required by the risk manager and the assumptions used for the risk characterization in the absence of complete data. Explicit description of these assumptions can help identify the research that will most significantly improve understanding of risk and, thereby, affect public policy.

Questions asked by a risk manager help to integrate the scientific data and assumptions provided by the formal risk assessment into the desired public policy. The questions also assure that regulation, a risk-management tool for tobacco harm reduction proposed by this committee, is based on and informed by the risk assessment process. Questions might include:

- Which of the thousand known tobacco-related toxicants are most important to consider in the assessment of risk? Is the scientific data available for adequate hazard identification?

- Are the data presented by the manufacturer based on assays reflecting the manner in which the product is actually used by the consumer?
- Are the claims by the manufacturer adequately supported by the scientific data? Is the risk characterization accurately conveyed in a manner understandable to the consumer?
- What constitutes a substantial degree of overall risk reduction?
- Who has the burden of proof for each type of claim? Is the burden of proof sufficient to assure the products will provide a benefit to the user? Is the burden of proof so high that innovation will be stifled and the possible benefit never realized?
- What can be done immediately to manage the possible risks of these products, given that the science base is currently inadequate?
- Are there parties responsible for assessing and assuring harm reduction outside this regulatory agency? And if so, are the boundaries of risk-management responsibility and authority clear to all parties?

The data presented and scientific limitations identified in Chapter 5, the surveillance system outlined in Chapter 6, and the regulatory framework described in Chapter 7 provide a sound basis for the risk management for tobacco harm reduction.

In summary, tobacco harm reduction could lead to reduced risk of disease for those who cannot give up tobacco. Unfortunately, without the appropriate public health tools of research, surveillance, education, and regulation, tobacco harm reduction could result in a personal and public health disappointment.

REFERENCES

IOM (Institute of Medicine). 1994. *Growing Up Tobacco Free.* Washington, DC: National Academy Press.
NRC (National Research Council). 1983. *Risk Assessment in the Federal Government. Managing the Process.* Washington, DC: National Academy Press.

Section II

Evidence for
the Science Base

9

Nicotine Pharmacology

BASIC AND HUMAN PHARMACOLOGY

It is some 550 years since the eponymous Jean Nicot sent tobacco and seeds from Portugal to Paris, passing *Nicotiana tabacum* from the Americas to Northern Europe by way of the Iberian peninsula. Nicotine itself was subsequently extracted and synthesized, culminating in the identification of the spatial orientation of the natural *(S)* isomer in the late 1970s (Domino, 1999). Up to 10% of the nicotine in tobacco smoke is the *(R)* isomer, probably arising from racemization during combustion (Benowitz, 1986). Nicotine has gained particular prominence as the addictive constituent of most tobacco based products and, to a lesser extent, as an effective insecticide.

Nicotine, 3-(1-methyl-2-pyrrolidinyl) pyridine, has a molecular weight of 162.23 and is a volatile, colorless base (pK_a=8.5) that turns brown and acquires the typical odor of tobacco on exposure to light. Roughly 69% of its pyrrolidine nitrogen is ionized (positively charged) at pH 7.4 and 37°C, whereas its pyridine nitrogen is un-ionized. This feature of nicotine renders its absorption and renal excretion highly pH dependent, because uncharged lipophilic bases pass easily over lipoprotein membranes and charged organic bases do not. For example, nicotine is primarily ionized at the pH (5.5) of smoke from the flue-cured tobaccos in most American cigarettes, and buccal absorption is minimal (Gorrod and Wahren, 1993). By contrast, smoke from air-cured tobaccos in pipes, cigars, and many European cigarettes is less acidic and is well absorbed through the mouth (Armitage et al., 1978; Gori et al., 1986). Nicotine constitutes about 95% of the total alkaloid content of commercial cigarette tobacco (Gorrod and Jenner, 1975).

The mechanisms by which nicotine exerts its actions at a molecular level are complex. Dale (Dale, 1914) noticed the structural similarity between nicotine and acetylcholine (Ach) and the resemblance of the effects of nicotine in vivo to those of Ach after pretreatment with the muscarinic antagonist atropine. The muscarinic effects of Ach are now recognized to be mediated via one of the five heptahelical muscarinic receptors (M1-M5). Ligation of these receptors may activate downstream signaling pathways via their interaction with diverse G proteins. Nicotinic receptors (nAchRs), by contrast, are ligand gated ion channels (Domino, 1999; Lena and Changeux, 1998). These pentamers are comprised of various combinations of α, β, γ, and δ subunits. Recent studies have demonstrated that specific configurations of these subunits mediate the diverse effects of nicotine. Although this area of research is evolving, the neuronal subunits that appear to be primarily responsible for the effects of nicotine contain $\alpha_{3,4,7}$ and $\beta_{2\,and\,4}$ subunits. The $\alpha_4\beta_2$ subtype is particularly prevalent in the brain and may be responsible for the self-administration of nicotine. Mice deficient in the β_2 subunit do not self-administer nicotine (Cordero-Erausquin et al., 2000), suggesting that this subunit in particular may be important in reinforcing the effects of nicotine. In addition, some preliminary evidence suggests that the α_7 subunit may play a significant role in withdrawal and sensory gating functions of schizophrenics (Adler et al., 1998; Nomikos et al., 2000; Panagis et al., 2000). Localization studies have identified nAchRs in the brain, neuromuscular junctions, autonomic ganglia, and adrenal medulla (Gundisch, 2000). Ligation of nAchRs by nicotine opens the channel, and the ionic influx activates signal transduction pathways, culminating in release of a number of different neurotransmitters, which have been related to nicotine's pharmacodynamic effects. These include dopamine (pleasure and appetite suppression), serotonin (appetite suppression and mood modulation), epinephrine and norepinephrine (arousal and appetite suppression), Ach (arousal and cognitive enhancement), vasopressin (memory improvement), glutamate (improvement in learning), β-endorphin (mood modulation and analgesia), and δ-aminobutyric acid. Nicotine also increases nAchR expression. For example, prenatal nicotine exposure upregulates the pulmonary expression of the α_7 receptor subunit and consequently affects fetal lung development in monkeys (Sekhon et al., 1999). Nicotine caused lung hypoplasia and reduced surface complexity of developing alveoli in this model. Collagen surrounding the large airways and vessels was increased, as was the number of type II cells and neuroendocrine cells in neuropepithelial bodies. Many animal studies have also demonstrated that nicotine administration upregulates expression of nAchRs in the brain. Similarly, ligand-binding studies have demonstrated an increase in binding sites for nicotine analogues in the cerebral cortex and hippocampus of smokers compared to nonsmokers (Perry et al., 1999), although the extent to which this may contribute to the differential central effects of nicotine observed in smokers is unknown.

Dopamine is believed to be the dominant neurotransmitter in the maintenance of drug-taking behavior (DiChiara, 1999; Koob, 1992). The area of the brain that is responsible for the reinforcing effects of all drugs of abuse is the mesolimbic pathway, which contains the ventral tegmental area (VTA), nucleus accumbens, amygdala, cingulate gyrus, and frontal lobe and is rich in dopamine. The VTA and nucleus accumbens seem particularly important in nicotine's reinforcing effects. Activation of nAchRs in the VTA and other parts of the midbrain, modulates the ascending mesolimbic dopamine system, including the nucleus accumbens (George et al., 2000; Yu et al., 2000). Nicotine self-administration behavior is diminished by either surgical or chemical ablation of dopaminergic pathways or by treatment with dopamine antagonists (Kameda et al., 2000). Nicotine evokes an increase in dopamine levels in brain microdialysis studies (Fu et al., 2000). In addition, monoamine oxidase A and B, responsible for the metabolism of dopamine, are reduced by a compound in tobacco smoke that also results in higher levels of neurotransmitters (Quattrocki et al., 2000).

The release or inhibition of other transmitters may also play a role in nicotine addiction. They may be responsible for mood modulation, the modest enhancement of performance, and the weight-reducing effects of nicotine (Benowitz, 1999; Chiodera et al., 1990; Chowdhury et al., 1989; U.S. DHHS, 1988). Mood modulation by nicotine has been a controversial topic, since laboratory studies do not validate the smoking-induced enhancement of mood self-reported by smokers. Furthermore, individuals experience greater positive affect when smoking after a period of abstinence. The relief of negative affect by tobacco use may be more a function of abating withdrawal symptoms (Cinciripini et al., 1997; RCP, 2000). Finally, in addition to its traditional pre- and postsynaptic actions at synapses and at chemoreceptors in the carotid and aortic bodies, nicotine also evokes the release of epinephrine from the adrenal medulla and may act directly to activate ion channels distinct from nAchRs. For example, nicotine has been shown to block directly inward rectifier potassium channels, an effect of potential relevance to cardiac arrhythmogenesis (Wang et al., 2000b).

PHARMACOKINETICS

Absorption

Although buccal absorption is influenced by the pH of tobacco or tobacco smoke, tobacco smoke from all sources is rapidly absorbed from the large surface area of the small airways and the alveoli, following dissolution in the pulmonary fluid at pH 7.4 (Zevin et al., 1998). Nicotine is also readily absorbed from the skin; this property has been exploited in the use of patch delivery in nicotine replacement therapy (NRT) for cigarette smokers (Figure 9-1). Nicotine is well tolerated as a dermal application, even in individuals who suffer from irritant skin disorders (Benowitz, 1995).

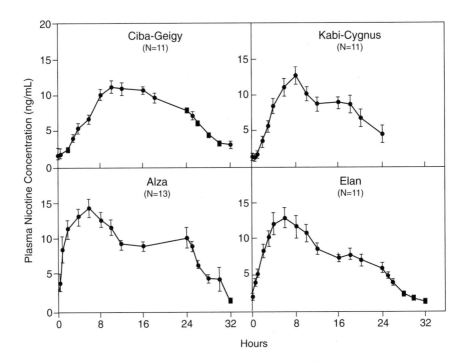

FIGURE 9-1 Time curves of plasma nicotine concentrations.

NOTE: Time curves of plasma nicotine concentration after application of four different transdermal nicotine delivery systems. The Ciba-Geigy Habitrol, the Alza (SmithKline Beecham) Nicoderm, and the Elan Prostep patches were worn for 24 hours, while the Kabi-Cygnus (Pharmacia) Nicotrol patch was worn for 16 hours.
SOURCE: Benowitz, 1993. Reprinted from *Drugs* 1993; 45(2):157-170 with permission. © Adis International, Inc.

Following ingestion, as with chewing tobacco, the pH dependence of nicotine ionization favors its absorption by the small intestine, rather than the stomach, although a timed-release preparation, which permits colonic absorption (Green et al., 1999), has also been developed for use in ulcerative colitis as discussed below. Although the peak levels of nicotine attained after chewing tobacco may approximate those after smoking, the shape of the curve of plasma concentration versus time is quite different (Figure 9-2). Thus, after smoking a cigarette, plasma levels of nicotine rise rapidly to a peak, which is maintained transiently after the cigarette is inhaled, rather than the more gradual and sustained elevation after oral

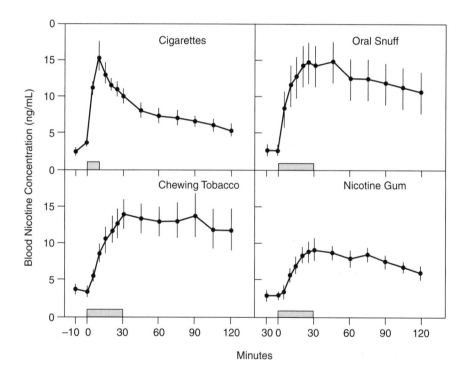

FIGURE 9-2 Blood nicotine concentrations.

NOTE: Blood nicotine concentrations during and after cigarette smoking for nine minutes, oral snuff (2.5 grams), chewing tobacco (average 7.9 grams), and nicotine gum (two 2 mg pieces). Data represent average values for 10 subjects (±SEM). Horizontal bars above time axis indicate period of tobacco or nicotine gum exposure.

SOURCE: Benowitz et al., 1988. Reprinted with permission from Mosby, Inc.

ingestion. The evoked liking, systemic response, and addiction potential of the former pattern of nicotine delivery exceed those of the latter (Benowitz et al., 1988). It takes roughly 10-15 seconds for nicotine, inhaled by puffing a cigarette, to reach the brain, and puffing is associated with a marked arterial-venous gradient of nicotine (Benowitz, 1995; Benowitz et al., 1988; Guthrie et al., 1999). This rapid central nervous system (CNS) delivery permits the smoker to adjust the nicotine dosage to a desired effect, reinforcing self-administration and facilitating the development of addiction (Benowitz, 1995). This contrasts with the slower increase and lesser increment in brain nicotine attained after transdermal delivery, which facilitates the development of tolerance (see below). The average cigarette contains 10-15 mg of nicotine and delivers, on average, roughly 1-2 mg of nicotine systemically to the smoker. However, smoking habit— puff intensity, duration, and so forth—can markedly alter nicotine bio-availability. By comparison, the systemic doses of nicotine from other delivery systems are roughly 1 mg from a 2 mg gum; 5-22 mg per day from transdermal patches; 0.5 mg per dose from one spray per nostril; 3.6 mg from 2.5 grams of snuff, held in the mouth for 30 minutes and 4.5 mg from 7.9 grams of chewing tobacco chewed for 30 minutes (Benowitz and Jacob, 1999).

Following its absorption, nicotine circulates with roughly 60% in the ionized form. It is poorly (around 5%) protein bound (Benowitz et al., 1982) and widely distributed, at least in rats and rabbits, particularly in liver, lungs, and brain (Benowitz et al., 1990).

Distribution and Metabolism

The presence of both aromatic and aliphatic carbon and nitrogen atoms in nicotine affords multiple sites for metabolic oxidation and subsequent conjugation reactions (Figure 9-3). The disposition of nicotine has been reviewed in depth elsewhere (Benowitz and Jacob, 1999; Gorrod and Schepers, 1999). Briefly, roughly 80% of the metabolic inactivation of nicotine involves oxygenation of the 5'-carbon to yield cotinine. This appears to involve an intermediate cytochrome P-450 (CYP)- derived 1',5'-imminium ion, which is further metabolized by aldehyde oxidase to yield cotinine (Brandange and Lindblom, 1979; Gorrod and Hibberd, 1982; Murphy, 1973). This iminium ion is an alkylating agent and has been speculated to have relevance to the carcinogenicity of tobacco, although this is not established (Hibberd and Gorrod, 1983). Oxidation of this radical may also yield nornicotine or 4-(3-pyridyl)-4-oxo-N-methylbutylamine. CYP2A6 and, to a lesser extent, 2B6 appear to play the predominant roles in nicotine carbon oxidation in humans (Benowitz and Jacob, 1999; Nakajima et al., 2001; Nakajima et al., 2000; Yamazaki et al., 1999). Although roughly

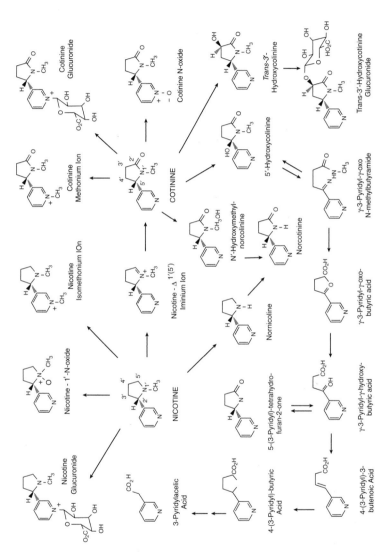

FIGURE 9-3 Nicotine metabolic pathways.

SOURCE: Benowitz and Jacob, 1999. Copyright (1999) Wiley-Liss, Inc. Reprinted with permission of Wiley-Liss, Inc., a subsidiary of John Wiley & Sons. Translated by permission of John Wiley & Sons, Inc. All rights reserved.

249

15% of a nicotine dose is excreted in human urine unchanged, all of the primary metabolites, including cotinine, are subject to further oxidation reactions. Oxidation appears to involve only the alicyclic pyrrolidine nitrogen in biological systems (Gorrod and Schepers, 1999). Nicotine-1'-*N*-oxide may be reduced to nicotine in man by gut bacteria (Dajani et al., 1975).

Phase two metabolites can be formed by methylation, glucuronidation, sulfation, or glutathione conjugation reactions with primary oxidation metabolites. Formation of such polar, water soluble molecules facilitates excretion. Although great interindividual variation is noticeable, glucuronides may account for roughly 40% of the urinary nicotine metabolites in humans. This variability may also be apparent among ethnic groups. Thus, the most abundant phase 2 metabolite in the urine of North Americans is the *N*-glucuronide of cotinine, whereas in Europeans the *O*-glucuronide of *trans*-3'-hydroxycotinine predominates (Gorrod and Schepers, 1999). Similarly, the metabolism of nicotine is slower in African Americans than in Caucasians, due to both slower oxidative metabolism of nicotine to cotinine and slower *N*-glucuronidation (Benowitz et al., 1999; Caraballo et al., 1998; Perez-Stable et al., 1998). Asian Americans also metabolize both nicotine and cotinine more slowly than do Caucasians. Nicotine clearance declines with age (Molander et al., 2001). Although there is some evidence for differences between men and women in the pharmacodynamic response to nicotine (Pomerleau et al., 1991), this does not appear to reflect systematic differences in nicotine pharmacokinetics. Nicotine readily crosses the placental barrier, although there is no apparent conversion of nicotine to cotinine by placental tissues or microsomal fractions (Pastrakuljic et al., 1998). Although the potential for fetal toxicity must be considered in women undergoing NRT (as discussed below), this consideration usually occurs in the context of relativity. Thus, the hazard to the fetus of maternal cigarette smoking is well established (Oncken et al., 1998; Robinson et al., 2000), whereas the theoretically much smaller risk of NRT remains entirely notional.

Clearance of nicotine falls with hepatic blood flow during sleep (Gries et al, 1996) and a circadian pattern in both circulating nicotine and cotinine is evident (Figure 9-4). Although the half-life of nicotine is about 2-3 hours when based on plasma levels, it approximates 11 hours when based on urinary excretion (Benowitz and Jacob, 1994, 1999), so circulating levels tend to accumulate during the day. The afternoon levels of nicotine in the plasma of smokers generally range from 10 to 50 ng/ml, whereas steady-state levels with patches range from 10 to 20 ng/ml and with the nasal spray from 5 to 15 ng/ml (Benowitz and Jacob, 1999).

Sophisticated approaches to analysis not just of nicotine and cotinine, but of minor oxidative metabolites and many phase 2 metabolites, have

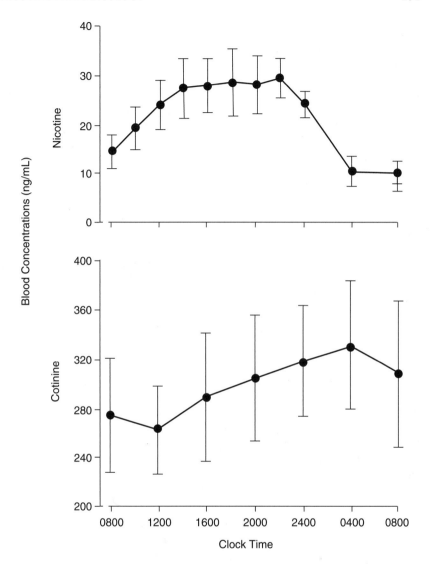

FIGURE 9-4 Circadian blood concentrations of nicotine and cotinine during unrestricted smoking.

NOTE: Data are mean ± SE for eight subjects.
SOURCE: Benowitz et al., 1983. Reprinted with permission from Mosby, Inc.

now been established, albeit in few laboratories. These encompass sensitive and specific methodologies, such as tandem mass spectrometry (Byrd et al., 1994). These tools will afford a more comprehensive approach to investigating genetic (e.g., nAchR or CYP2A6 polymorphisms; McKinney et al., 2000; Nakajima et al., 1996) and environmental factors (e.g., CYP2A6 induction or repression by alcohol or other drugs; Niemela et al., 2000), which might contribute to interindividual variability in nicotine pharmacokinetics. For example, they will include the integration of assessment of nitrosamines formed from nicotine into long-term studies of the safety of NRT.

Excretion

Nicotine is excreted by both glomerular filtration and tubular secretion. Acidification of urine greatly increases its renal clearance, which impedes tubular reabsorption by ionizing the nicotine (Benowitz and Jacob, 1999). Urinary excretion of cotinine is less influenced by pH since it is more basic. However, renal clearance of both compounds is influenced highly by urinary flow rates (Benowitz and Jacob, 1999).

PHARMACODYNAMICS

Cardiovascular Effects

The factors that mediate the effects of nicotine are complex, confounded as they are in the cardiovascular system by direct and reflex effects, acute effects and long-term desensitization, and secondary effects due to sympathoadrenal activation. Acute delivery of nicotine in a cigarette results in a transient tachycardia, cutaneous vasoconstriction, and a rise in blood pressure (Cryer et al, 1976). By contrast, desensitization of vascular or central receptors by nicotine may contribute to the lower blood pressure observed in chronic smokers (Charlton and While, 1995).

The mechanisms involved in mediating the adverse effects of cigarette smoking and of smokeless tobacco on the cardiovascular system are poorly understood, but are thought to include induction of an adverse lipoprotein profile (Allen et al., 1994), induction of a chronic inflammatory response (Strandberg and Tilvis, 2000) including oxidative tissue injury (Morrow et al., 1995; Patrignani et al., 2000; Reilly et al., 1996; Traber et al., 2000), activation of platelets and other hemostatic variables (Benowitz et al., 1993; Ludviksdottir et al., 1999; Whiss et al., 2000), and impairment of endothelial function (Raitakari et al., 2000). Following the introduction of NRT, there was considerable concern about the cardiovascular safety of this intervention, especially in patients with preexisting cardiovascular disease. However, NRT has been shown to be effective, without an apparent cardiovascular hazard, not only in the general population (Benowitz and Gourlay, 1997), but also in patients with established coronary vascular disease (Greenland et al., 1998; Joseph et al., 1996; Nitenberg and Antony, 1999). Controlled studies have demonstrated that switching from cigarette smoking to NRT is associated with amelioration of the lipoprotein and hemostatic profiles (Allen et al., 1994; Ludviksdottir et al., 1999; Winther and Fornitz, 1999) and a reduction in platelet activation (Nowak et al., 1987).

Although evidence of a clinical cardiovascular hazard of NRT has yet to emerge, several observations suggest that aspects of the cardiovascular effects of nicotine merit further research. For example, cigarette smoking impairs endothelial function (Celermajer et al., 1993; Raitakari et al., 1999), which appears to be a surrogate marker of future clinical vascular disease (Dugi and Rader, 2000). Interestingly, nicotine has been reported to impair flow mediated brachial arterial endothelial function (Chalon et al., 2000) and bradykinin stimulated endothelial function in dorsal hand veins of volunteers (Sabha et al., 2000). On the other hand, short term administration of nicotine gum did not alter the coronary constrictor response to the cold pressor test—reflective of the effects of sympathadrenal activation—in patients with established coronary artery disease (Nitenberg and Antony, 1999). Although exposure of mice deficient in apoenzyme E (Apol E) to cigarette smoke accelerates atherogenesis (Gairola and Daugherty, 1999), there have been no analogous studies of nicotine and no studies of atherosclerotic plaque progression in individuals receiving long-term NRT.

Switching to NRT in the short term does not appear to correct the systemic markers of inflammation in cigarette smokers (Nilssen et al., 1996), and there are conflicting reports of the direct effects of nicotine on free radical generation in vitro (Gouaze et al., 1998; Guatura et al., 2000; Wetscher et al, 1995). Interestingly, many cigarette smokers also drink alcohol (Swahn and Hammig, 2000), and alcohol is a potent prooxidant in humans (Meagher and FitzGerald, 2000). While ethanol increases the clearance of nicotine, the acute hemodynamic effects of ethanol and nicotine are additive (Soderpalm et al., 2000), perhaps reflecting a common mechanism. However, there are no reports of the effects of NRT on contemporary markers of oxidative stress or a comprehensive assessment of its effects on markers of inflammation.

Central Effects

General Effects

Nicotine is a CNS stimulant. Paradoxically, it is perceived to be relaxing in stressful situations and to enhance gating of relevant stimuli. The smoker does not react as much as the nonsmoker to external distractions—hence its use by those trying to relax or concentrate. Stress increases the smoker's nicotine consumption. An increase in respiratory rate and heart rate has been observed with NRT. Sleeplessness has also been reported in patients using NRT (Gourlay et al., 1999). Nicotine overdose is remarkably difficult to achieve with NRT (Labelle and Boulay, 1999), however, it occasionally complicates the use of nicotine-containing insecticides. In these cases, central symptoms—initially tremors, nausea, vomiting, and possibly convulsions—give way to signs of central depression and neuromuscular blockade (Saxena and Scheman, 1985).

Memory and Cognition

Although there is considerable interest in the potential effects of nicotine on cognition (Emilien et al., 2000; Waters and Sutton, 2000), this has not been formally evaluated in individuals receiving NRT. Activation of nAchRs containing the α_7 subunit results in Ach release and calcium activation, and both effects have been implicated in memory formation and cognition (Kem, 2000). Recent interest has been focused by co-immunoprecipitation of the amyloid A β(1-42)-fragment and the α_7nAchR from the dendtritic plaques of Alzheimer's disease (AD) lesions (Wang et al., 2000a). The β-fragment of amyloid A binds to the receptor and prevents its activation by nicotine, potentially implicating defective nAchR activation in the pathogenesis of AD. Again, although there is some evidence for a slowing of deterioration of AD in individuals who smoke (Debanne et al., 2000; Doll et al., 2000; Jarvik, 1991; Lopez-Arrieta et al., 2000; Merchant et al., 1999), along with a considerable literature relating to the use of cholinesterase inhibitors for this condition, NRT has not been formally evaluated in AD.

Addiction

The Biological Basis of Addiction. Although tobacco products contain several thousand chemicals, nicotine is considered to be the principal constituent in tobacco that leads to the persistent use of tobacco products (U.S. DHHS, 1988). However, other yet unknown constituents in tobacco may also have a role in the maintaining the use of tobacco. For example, smokers experience a reduction of monoamine oxidase (MAO) activity in the brain (Berlin et al., 1995) as a result of some constituent in smoke (Fowler et al., 1996); inhibition of MAO may result in antidepressant activity (Oxenkrug, 1999) and contribute to the high prevalence in smoking among individuals with depressive disorders. Physical addiction to nicotine is associated with euphoriant and other psychoactive effects, the development of tolerance, and the experience of withdrawal symptoms when the tobacco product is no longer used (U.S. DHHS, 1988). In addition, the rate of absorption and therefore the speed of delivery of nicotine to the brain also play a significant role in the addictive potential of nicotine (Henningfield and Keenan, 1993). These factors contribute to the reinforcing effects or persistent use of nicotine and also may be responsible for day-to-day regulation of nicotine levels in tobacco users.

Psychoactive and reinforcing effects from nicotine are the result of the release of a number of neurotransmitters and hormones (Benowitz, 1999; U.S. DHHS, 1988; Watkins et al., 2000). This cascade of events is associated with mood modulation, cognitive and motor performance enhancements, and weight reduction. These effects may contribute to the initiation and maintenance of tobacco use. Chronic administration of nicotine can lead to neuroadaptation. One of the effects of neuroadaptation is the development of tolerance. Adaptation occurs so that the brain can maintain a state of homeostasis despite an increased release of neurotransmitters. This process includes receptor inactivation and desensitization and an increase or upregulation in receptor number (Benowitz, 1999). The extent of these changes could vary depending on the receptor subtype and site (Watkins et al., 2000). Tolerance may lead to individuals' using more of the tobacco product or switching to higher nicotine-containing products.

Neuroadaptation may subsequently lead to withdrawal symptoms when the tobacco user is no longer exposed to the product. Withdrawal symptoms include negative affect (e.g., irritability, frustration or anger, anxiety, dysphoric or depressed mood), restlessness, difficulty in concentrating, insomnia, decreased heart rate, and increased appetite or weight (APA, 1994). These symptoms occur among regular users of cigarettes and smokeless tobacco (Hughes and Hatsukami, 1992). They are less pronounced with nicotine gum use, but this distinction blurs with prolonged use of the gum (Hughes et al., 1986b; West and Russell, 1985). Approximately 49% of self-quitting smokers and 87% of tobacco cessation program attendees meet Diagnostic and Statistical Manual of Mental Disorders (DSM) IIIR (APA, 1987, 1994) criteria for nicotine withdrawal syndrome (Hughes and Hatsukami, 1992). These withdrawal symptoms peak during the first week of abstinence and return to baseline levels by four weeks (Hughes et al., 1990a). The intensity of these symptoms is further reduced over the course of time. The only exception to this pattern is increased weight. Weight may continue to increase over six months, and a reduction may not be seen at all or only after several months of abstinence (Hughes et al., 1990a).

A major determinant of whether nicotine is likely to be addictive is the amount and speed of nicotine delivery. The route of delivery also determines the pattern of nicotine delivery (as discussed earlier). For example, each puff of a cigarette delivers a bolus dose of nicotine, resulting in a rapid peak, which then falls to a trough level. The time between these bolus doses allows for resensitization of brain nAchRs, so that each delivery can remain reinforcing (Benowitz, 1999). In addition, this route of administration allows the delivery of a greater number or frequency of reinforcements. Other delivery routes result in a slow and persistent absorption of nicotine. Subjective effects, the desire to use more of a drug, and the actual self-administration of a drug are functions of absorption rate (Henningfield and Keenan, 1993). Therefore, whereas cigarettes have high abuse potential, nicotine patches have lower abuse potential.

It is also important to note that addiction to nicotine is not just a biological phenomenon, but also one in which learning or conditioning has taken place. Nicotine self-administration comes under the control of stimuli that have been associated with smoking or tobacco use. These stimuli can precipitate a strong desire for nicotine, withdrawal symptoms, or drug effects. Exposure to these stimuli may lead to the same biological effect on neural substrates as observed from the direct actions of the drug (Childress et al., 1999). Furthermore, stimuli associated with tobacco use, such as the sensory aspects of smoking, can become reinforcing as well; that is, they become secondary reinforcers. In addition a tobacco user develops expectancies regarding the use and effects of the substance, leading to a psychological reliance on the drug.

The susceptibility to nicotine addiction is thus a result of both the biological effects of the drug and learning history. In addition, environmental factors (e.g., access to tobacco, restrictions on tobacco, social modeling) and genetic or organismic factors (e.g., rate of nicotine metabolism, psychiatric disorders, personality factors) may play a significant role. Specific populations might be more vulnerable to nicotine addiction. Genetic twin studies have shown heritability estimates that range from 28 to 84%, with a mean estimate of 53% (Hughes, 1986). Genetic heritability has been associated with the onset as well as the persistence of smoking (Heath et al., 1998, 1999). Examples of what is inherited may be differences in sensitivity to nicotine (Pomerleau, 1995), the rate of nicotine metabolism (Tyndale et al., 1999), or other mechanisms such as genetic polymorphisms in the dopamine transporter and subtypes of dopamine receptors (Lerman et al., 1999; Shields et al., 1998). In addition, individuals with comorbid disorders tend to have a high prevalence of smoking. For example, high prevalence of smoking is found in individuals with depressive disorders, schizophrenia, and alcohol or drug abuse disorders (Breslau, 1995; Hughes et al., 1986a). The mechanisms responsible for susceptibility to smoking may differ across disorders. The nicotine-associated release of neurotransmitters is similar to those found with antidepressants and may be responsible for the association between smoking and depression and for the recurrence of depressive disorders after smoking cessation. Furthermore, studies have shown a genetic linkage between smoking and depression (Kendler et al., 1993), and observations have been made that depression can predate smoking or smoking predate depression (Breslau et al., 1993, 1998). For individuals with schizophrenia, the sensory gating effects of nicotine via the α_7 nicotinic receptor may provide some symptomatic relief (Dalack et al., 1998; Freedman et al., 1997). A genetic link also seems to exist for alcohol and nicotine addiction (Hughes, 1986), along with commonality in the release of dopamine across all drugs leading perhaps to increased sensitivity to the reinforcing effects of drugs or the potential for substitution. Furthermore, nicotine can be used to offset the aversive effects of drug use (Benowitz, 1999).

   Assessment of Addiction. Various measures and methods have been developed to measure dependence on a drug and its abuse or addiction potential (see Table 9-1). These measures and methods are important in examining harm reduction products since addiction to a drug is one of the determining factors associated with its harmful consequences. The addictive potential of a drug can be determined by examining the number of individuals, within the general population and among those exposed to the drug, who are regular users of the drug or are considered dependent on a drug, using specific criteria. Determination of the abuse potential of nicotine replacement agents has also relied on examining whether users of the product escalate their use over time or continue its use beyond a recommended period. However, deciphering whether this persistent

TABLE 9-1   Measures of Dependence or Addiction and Abuse Liability

Measures for Dependence or Severity of Dependence
   Daily or regular smoking (cotinine level)
   DSM criteria
   International Diagnostic Code
   Surgeon General's report, 1988
   Fagerström Tolerance Questionnaire
   Fagerström Nicotine Dependence Test
Methods to Assess for Addiction or Abuse Liability Surveys
   Daily use or dependence among the general population
   Daily use or dependence among those exposed to the drug
   Escalation of drug use
   Relapse rates
Laboratory models
   Psychoactive or subjective effects
   Drug discrimination
   Conditioned place preference
   Drug self-administration
   Withdrawal

medication use is a result of the desire to prevent relapse to cigarettes or an addiction to the product can be difficult. The "addictiveness" of a drug can also be determined by the extent to which relapse occurs among those individuals who have tried to stop using it. In addition, various animal and human laboratory methods have been developed to assess the abuse liability of a drug, including measurement of psychoactive or stimulus effects and determination of whether a drug is a reinforcer (positive or negative) leading to preference for a drug or drug self-administration (Bozarth, 1987; Balster, 1991; U.S. DHHS, 1988).

According to Food and Drug Administration (FDA) guidelines, abuse liability is determined by two primary factors (see deWit and Zacny, 1995). One is the likelihood of repeated use, which is determined by the drug's psychoactive, positive reinforcing effects and the extent to which it can relieve withdrawal symptoms as a result of chronic use. Repeated use may also be determined by the degree of unpleasant effects associated with drug use. The second factor is the incidence of adverse short- and long-term consequences as a result of use. Drugs with a greater number of adverse consequences are thought to be more likely to have abuse liability than those with fewer adverse effects.

Measures and Surveys of Dependence. Surveys and instruments have been used to assess the amount and frequency of use (e.g., daily use, regular use) and whether an individual is dependent on a drug based on specific diagnostic criteria. These measurement tools have been used to determine the extent to which dependence occurs within a general population and among those who have been exposed to or have experimented with the drug. In addition, these diagnostic tools for dependence have been used to determine whether dependence on nicotine is a dose-related phenomenon. Both DSM-IIIR and DSMIV (APA, 1987,1994) and the World Health Organization (WHO) International Diagnostic Code-10 (IDC-10) (WHO, 1991) are the commonly used criteria to assess for nicotine dependence. According to the DSM and the IDC-10, substance dependence, including nicotine, results in several behavioral and cognitive characteristics and physiological manifestations (see Table 9-2). The primary criteria for dependence based on these definitions include a strong desire to take the drug for periods longer than intended, problems controlling its use, use despite negative consequences or having a higher priority than other activities or obligations, tolerance, and physical withdrawal (APA, 1994; WHO, 1991). Not all criteria have to be met, nor is any one criteria critical to satisfy a diagnosis of dependence. In the 1988 Surgeon General's report *The Health Consequences of Smoking: Nicotine Addiction,* the primary criteria for drug dependence included (1) highly controlled or compulsive use of a drug, (2) psychoactive effects from the drug, and (3) drug-reinforced behavior. Additional criteria, similar to those listed in DSM-IV and IDC,

TABLE 9-2 Criteria for Substance Dependence from DSM IV

DSM-IV	IDC-10
A maladaptive pattern of substance use, leading to clinically significant impairment or distress, as manifest by three (or more) of the following, occurring at any time in the same 12-month period	
Tolerance—need increased amounts of substance to achieve desired effect, or diminished effect with continued use of same amount	Increased tolerance
Withdrawal	Sometimes, physical withdrawal
Substance often taken in larger amounts or over a longer period than intended	A strong desire to take the drug
Persistent or unsuccessful efforts to cut down or control substance use	Difficulty controlling use
Great deal of time spent in activities necessary to obtain the substance or recover from its effects	
Important social, occupational, or recreational activities given up or reduced because of substance use	Higher priority given to drug use than to other activities and obligations
Substance use continued despite knowledge of having a persistent or recurrent physical or psychological problem likely to have been caused or exacerbated by the substance	Persisting use despite harmful consequences

SOURCE: Adapted from RCP, 2000.

were also included. The number or type of symptoms experienced varies across different drugs of abuse. The major difference between nicotine and some other drugs of abuse is the lack of intoxication in regular tobacco users that results in behavioral and cognitive disruption (U.S. DHHS, 1988). However, this makes nicotine no less an agent of addiction or dependence than other drugs (Stolerman and Jarvis, 1995). In fact, many cigarette smokers exhibit at least as many indicators of dependence as other drug users and abusers (CDC, 1995b; U.S. DHHS, 1988). Assessment of nicotine dependence using these criteria can be made by a number of diagnostic structured instruments including the Composite International Diagnostic Interview-Substance Abuse Module, the National Institute of

Mental Health-Diagnostic Interview Schedule (NIMH-DIS), and the NIMH-DIS for children (see Colby et al., 2000, for review).

Other methods have been used to assess addiction or dependence on nicotine or tobacco products. For example, population surveys such as the National Household Survey on Drug Abuse (NHSDA) assess for symptoms of tobacco dependence and include such items as how many current tobacco users (1) reported daily use of the product, (2) have tried to cut down, (3) were unable to cut down or quit or experienced difficulty quitting, (4) felt a need for more tobacco for the same effect, (5) felt dependent, or (6) felt sick or experienced withdrawal symptoms when stopping smoking and met at least one or more of these indicators (CDC, 1995a, b; CDC, 1994). Researchers have used meeting a specified number of these symptoms as proxy measures for the DSM-IV criteria for substance dependence. In some assessments, individual items, such as experiencing withdrawal symptoms or difficulty quitting have been of particular focus as indicators of dependence (CDC, 1994, 1995a, b).

Other reports assessing nicotine dependence determine the number of smokers who meet criteria for high level nicotine dependence according to the Fagerström Tolerance Questionnaire (FTQ; Fagerström, 1978; Fagerström and Schneider, 1989) or the revised version, the Fagerström Test for Nicotine Dependence (FTND) (Heatherton et al., 1991). Several adolescent versions have also been developed (Prokhorov et al., 1996, 1998; Rojas et al., 1998). Although this scale is continuous, a cut-off score of 6-7 or higher has been used to separate low and high level of dependence.

Based on the measures of dependence described above, the percentage of cigarette users that report dependence on their tobacco product varies according to the population examined (e.g., total populations, daily smokers, ever smokers, and so forth) and the definition of dependence used. According to the NHSDA, a population survey of noninstitutionalized civilians 12 years and older, the proportion of respondents who reported experiencing at least one indicator of dependence was 75.2% among those individuals who used cigarettes one or more times during the 30 days preceding the survey and 90.9% among daily users (reporting daily use for ≥ 2 consecutive weeks during the 12 months preceding the survey) (CDC, 1995b). In another study, the estimated prevalence of dependence according to the DSM-IIIR criteria (APA, 1987) among Americans 15-54 years old sampled for the National Comorbidity Survey was about 24.1% (Anthony et al., 1994). The lifetime prevalence of dependence among middle-aged male ever smokers in Japan was 42, 26, and 32% according to IDC-10, DSM-IIR, and DSM-IV criteria, respectively (Kawakami et al., 1998). In another study, very high rates were observed with 90% of a general sample of middle-aged male smokers meeting

DSM-III criteria for dependence (Hughes et al., 1987). Kandel and associates (1997) used the indicators listed in the NHSDA (see above) including items assessing for frequency and quantity of use and problems related to use in order to diagnose nicotine dependence. The criteria for diagnosis were based on the DSM-IV method in which smokers must experience three or more of seven indicators of dependence. The findings showed that while 8.6% of the general population 12 years and older met criteria for nicotine dependence, 28% of those who had used tobacco products in the past year experienced nicotine dependence. A few studies have also been conducted with adolescents. The study conducted by Kandel and associates (1997) using the NHSDA examined the prevalence of nicotine dependence by age. They observed that about 19.9% of adolescents who smoked any cigarette met criteria for nicotine dependence, compared to rates ranging from 26.4 to 32.7% among smokers between the ages of 18 and 49 years and 23.7% among those 50 and older. In a study conducted in New Zealand, about 20% of a general sample of 18-year-olds were dependent on tobacco and more than half (56.4%) of the sample who smoked daily met DSM-IIIR criteria for nicotine dependence.

In another survey that used the FTQ with a score of 7 or more (indicative of a high level of dependence, not dependence per se), only 19% of Japanese male ever-smokers age 35 and older met this criteria (Kawakami et al., 1998), but 36% of U.S. males did (Hughes et al., 1987). Among adolescent smokers, the prevalence of high level of dependence according to the FTND or FTQ has also been wide-ranging. Many of the studies assessed prevalence of high level of dependence in special populations of adolescents. The highest percentage of adolescents with a score of 7 or more on the FTQ was observed among a heavy-smoking group who participated in a nicotine patch trial, with an observed rate of 68% (Smith et al., 1996). The lowest rate was 20% using a modified FTQ with a cutoff score of 7 or higher, which was observed in vocational technical high school student smokers (Prokhorov et al., 1996). This proportion was lower than the 50% rate that the investigators observed among adult smokers.

An indicator of the addiction potential of a drug is the development of daily or regular use or dependence among those who have been exposed to it. There is strong evidence to show that a significant number who are exposed to cigarettes may become daily smokers or dependent on them. Among high school students participating in the 1997 Youth Risk Behavior Survey (YRBS), of the 70.2% who tried cigarette smoking, 35.8% went on to smoke daily. This rate of escalation from trying cigarette smoking to regular use of tobacco is similar to the 33-50% observed in other studies (U.S. DHHS, 1994). The development of dependence among those who tried tobacco products is similarly high. In one population-based study of adult smokers, about 31.9% of those who tried tobacco became dependent on it based on the DSM-III criteria (Anthony et al., 1994). In another study of young adults aged 21-30, of the 74% who had smoked tobacco at least once, 27% developed DSM-IIIR criteria for tobacco dependence (Breslau et al., 1993). Similar data are not available for smokeless tobacco users. Existing data are limited to the number of individuals who report having used smokeless tobacco in the past month versus the number who report lifetime use of smokeless tobacco; this method of calculation represents about 18% for smokeless tobacco users. This figure is compared to 37% for cigarette smokers using a similar method of calculation (U.S. DHHS, 2000).

Relapse rates among those who tried to quit have been considered another indicator of dependence on or addiction to a drug. Relapse is high among a general population of smokers who have tried to quit smoking, with only 2.5% being able to sustain abstinence for a year (CDC, 1994). One study showed that among self-quitters, about two-thirds reported smoking within two days postquit (Hughes et al., 1992). The rate of relapse among a population of smokers who have undergone clinical treatment tends to be about 75%, with a significant number of these relapses occurring within the first few weeks. These rates and patterns of relapse are similar to those observed with smokeless tobacco (Hatsukami and Severson, 1999) and other drugs of abuse (Hunt and Matarazzo, 1973; Maddux and Desmond, 1986; Wallace, 1989). High rates of relapse are also observed among youth that smoke. Based on results from the YRBS, among high school students who smoked daily, 72.9% had ever tried to quit smoking and only 13.5% were former smokers (CDC, 1998).

Most research on the dependence on nicotine replacement products has examined the persistence of use or escalation of use over time. No data are available on the prevalence of daily use in the general population or on dependence on these products according to diagnostic criteria for dependence or FTND scores. The rate of persistent use of nicotine replacement products among smokers enrolled in clinical trials who were assigned these products is much lower than the rate of persistent use of cigarettes, ranging from 9% for nicotine gum to 18% for nicotine nasal spray (Hughes, 1998). With nicotine nasal spray the rates of persistent use are higher, and there is evidence to show that a small number escalate the amount of use over time (deWit and Zacny, 1995). In general, addiction to these products is significantly less than addiction to cigarettes due to the relatively slow absorption of nicotine, the side effects that sometimes results from use, and the cost per unit of purchase.

In summary, research shows that nicotine delivered via cigarettes and smokeless tobacco is likely to lead to a high prevalence of use and dependence. One third to one-half of individuals who experiment with cigarette products are likely to become regular users and dependent on them. No data are available on the initiation of nicotine replacement product use among tobacco-naïve individuals or rates of diagnosable dependence, although these rates are likely to be low (Shiffman et al., 1998). The number of new NRT users among those attempting to quit was approximately 10% per year prior to over-the-counter (OTC) nicotine replacement products and 26% per year after OTC availability (Shiffman et al., 1998). Therefore, increased availability has led to increased use of these products among smokers, however, the rate of use still remains quite low. Furthermore, among smokers who use nicotine replacement products, persistent use tends to be low. Future research endeavors should concentrate on developing uniform methods and measures for assessing nicotine dependence so comparisons can be made across products and studies. The present measures are limited to assessing the extent of dependence and limited by being designed to diagnose other drugs of abuse and not specifically to diagnose nicotine dependence. In addition, as new products evolve, rates of initiation, regular use or persistent use and dependence, or progression to dependence as a result of experimentation should be assessed.

Models of Addiction. Several methods have been developed using clinical and animal models to determine the addiction potential or abuse liability of a drug. These include models of self-administration, drug discrimination, and conditioned drug placement. Models to examine withdrawal have also been developed. For humans, subjective responses to drugs can also be determined, although these responses may not necessarily be associated with actual drug-taking behavior.

When a drug is reinforcing, it is more likely to be self-administered or preferred compared to a control drug that has no abuse potential. The subject is exposed to a drug, typically, at varying doses and then required to choose between this particular drug and a control drug or an alternative reinforcer (e.g., sucrose for animals, money for humans), or between different doses of the drug.

- In *self-administration models*, the animal is required to perform a particular maneuver, such as lever pressing, to obtain the drug, which is typically administered intravenously. This lever pressing could be based on a fixed ratio (a specific number of responses are required prior to drug delivery), a progressive ratio (more responses are required after each drug delivery), or an interval schedule (a certain time interval is necessary before drug delivery), or a combination of these. Scheduled reinforcement in response to environmental stimuli associated with drug administration are called second-order schedules (Goldberg et al., 1981). Drugs can be made available for a fixed amount of time or throughout the day. Drugs that are reinforcing prompt the subject to work more or pay a higher cost for them than for the control; reinforcing drugs also lead to a greater persistent responding for them even when they are no longer available (Henningfield et al., 1991). Typically the dose-response curve is U-shaped (Risner and Goldberg, 1983; Rose and Corrigall, 1997) with low and high doses resulting in reduced drug self-administration. Low doses may produce limited or undetectable effects and high doses may produce adverse effects.

- *Drug discrimination models* involve training the subject to discriminate the stimulus properties of drug A from drug B. A third drug may be introduced, and the animal or human subject is asked to decide whether the drug is more like drug A or drug B (Bigelow and Preston, 1989; Preston, 1991). Subjects can also be trained to discriminate among several sets of drugs or different doses of a drug. This model allows determination of the mechanism of action of a drug. For example, if one wanted to determine whether an opioid has μ agonist or κ agonist activity, an experiment can be developed in which the subject is trained to discriminate between drugs that are known to have each of these activities. After this period of training, the drug in question is introduced and the subject has to indicate whether the drug is more like drug A (e.g., a μ agonist) or drug B (a κ agonist). This model can be also used to determine whether a drug has the stimulus properties of a particular pharmacological class of drugs that are abused. A similar method is used in a *drug preference procedure*, in which subjects are exposed to drug A and drug B, and are required to self-administer each of these drugs during separate experimental sessions. After the drug exposure or sampling period, subjects are then asked to choose between drugs A and B, to determine their preference for one drug or another (de Wit, 1991). Drug A or B can be two different doses of a drug, different types of drugs, or an active and placebo drug.

- The *conditioned place preference model* also is used to determine the abuse liability of a drug. Animals are trained that the drug is available only in a particular place (e.g., a specific chamber). Then a determination is made of how frequently the animal is willing to go to this place. If it is chosen significantly more frequently than the other place which is associated with a control drug or no drug administration, the experimental drug may have abuse potential (Bozarth, 1987).

- *Drug withdrawal models* have typically involved observing signs and symptoms during a period of abstinence after repeated administration of a drug (U.S. DHHS, 1988; Hughes et al., 1990; Malin et al., 1992). These withdrawal symptoms can be precipitated by antagonist drugs or allowed to occur naturally. Although the occurrence of withdrawal signs and symptoms does not necessarily indicate that that the drug will be abused, it may be one indicator of the potential for abuse.

• Finally, among humans, *subjective responses* to drugs can be determined (Jasinski and Henningfield, 1989; Fischman and Foltin, 1991; Jaffe and Jaffe, 1989). Subjects can be asked to indicate the intensity of experiencing different subjective effects, such as the degree of euphoria, liking of a drug, "high," desire for a drug, or "head rush." Comparisons can be made across different drugs and across doses within a particular drug. Subjects can also be asked to rate the effects of a drug using various standardized measures that have been developed to assess a drug profile (e.g., stimulant-like effects, depressant effects) such as the Addiction Research Inventory (Haertzen et al., 1963).

Self-administration paradigms have been used to demonstrate that a wide range of species (monkeys, mice, dogs, and rats) exhibit preference for administering nicotine over a control vehicle (Henningfield and Goldberg, 1983; RCP, 2000; Rose and Corrigall, 1997; Swedberg et al., 1990; U.S. DHHS, 1988). Studies have shown that these animals are willing to lever-press several hundred times in order to receive an injection of nicotine (Goldberg et al., 1981; Risner and Goldberg, 1983). However, unlike other drugs such as cocaine, the range of environmental conditions under which nicotine serves as a reinforcer is more restricted (Henningfield and Goldberg, 1983). In laboratory studies, human smokers have also been found to lever-press for intravenous doses of nicotine (Henningfield and Goldberg, 1983) as well as to self-administer greater number of doses of nicotine nasal spray (Perkins et al., 1997) and nicotine gum (Hughes et al., 1990b) compared to the respective placebo conditions. Clinical trials for the nicotine spray (Sutherland et al., 1992) and gum (Hughes et al., 1991) have also observed greater self-administration of active compared to placebo doses. Most human studies, however, have focused on assessing smoking behavior, looking at various indices of exposure, including number of cigarettes, number of puffs, puff volume, puff duration, inhalation duration, and intercigarette interval as well as biochemical indices of exposure such as cotinine or nicotine concentrations. Smoking behavior has been examined in response to changes in dose of cigarettes, preloading with nicotine or administering nicotinic antagonists and other drugs that may affect the reinforcing effects of nicotine (U.S. DHHS, 1988). Self-administration of nicotine is dose related in both humans and animals, although there is lesser dose-dependency than other drugs in animals, and the curve is somewhat flat for humans (Corrigall, 1999). Nonetheless, reduced nicotine self-administration in humans is observed with nicotine preloading and compensation with changing nicotine doses in cigarettes. Speed of nicotine delivery also plays a role in the extent to which nicotine is self-administered. Rapid bolus injections of nicotine result in greater self-administration than a slow infusion (Wakasa et al., 1995). Self-administration can be blocked by mecamylamine, a nonspecific nAchR antagonist or by dopamine receptor antagonists (see earlier discussion of the biological basis of addiction). Self-administration can be facilitated not only by the dosing characteristics of cigarettes or nicotine but also by the sensory characteristics of cigarettes (Henningfield and Goldberg, 1983; Rose and Corrigall, 1997).

Smokers tend to report dose-related subjective effects such as drug liking, drug strength, head rush, and feeling dizzy or aroused as a result of inhaled, buccal (smokeless tobacco), intravenous, or nasal spray nicotine administration (Fant et al., 1999; Henningfield et al., 1985; Jones et al., 1999; Perkins et al., 1994a, 1994b). Smokers who have a history of drug dependence exhibit a similar dose-related increase in "liking" and other subjective responses for intravenously administered nicotine as observed for cocaine, amphetamine, morphine, pentobarbitol, and heroin (Jasinski et al., 1984; Jones et al., 1999; Keenan et al., 1994). Findings from another study also revealed that intravenous nicotine was misidentified as cocaine or amphetamine by the study participants who had histories of drug use (Henningfield et al., 1985; Jones et al., 1999). Subjective responses to nicotine gum, patch, spray and inhaler have been less pronounced than responses to cigarettes or intravenous nicotine (deWit and Zacny, 1995; Henningfield and Keenan, 1993; Schuh et al., 1997).

The occurrence of withdrawal symptoms after cessation of continuous nicotine infusion in rodents has been demonstrated (Malin et al., 1992). In humans, withdrawal symptoms upon cigarette smoking cessation has also been well established (Hughes et al., 1990a). However, fewer studies have been conducted with other tobacco products or nicotine replacement agents. Cessation of smokeless tobacco use generally produces less intense withdrawal symptoms than cessation of cigarette smoking (Hatsukami et al., 1987; Keenan et al., 1989). However, in a population of smokeless tobacco users enrolled in clinical trials, the severity and number of withdrawal symptoms from smokeless tobacco were comparable to those experienced by cigarette smokers who were trying to quit (Hatsukami et al., 2000). Nicotine gum withdrawal symptoms also tend to be significantly less intense in number and severity than cigarette withdrawal symptoms (Hatsukami et al., 1991, 1993, 1995), and higher doses of gum produce greater withdrawal than lower doses of gum (Hatsukami et al., 1991). On the other hand, among those who have used the product for a prolonged period, nicotine gum may be comparable to cigarettes in the number of withdrawal symptoms experienced (Hughes et al., 1986b; West and Russell, 1985).

In summary, various laboratory studies have observed that nicotine is self-administered, produces psychoactive effects, and produces withdrawal symptoms. The route of delivery can determine the extent to which nicotine-containing products can produce these effects and lead to addiction, with cigarettes showing the highest potential for addiction.

Future studies on new products should routinely measure the abuse potential of a drug by using the various methods that have been described. Furthermore, these paradigms could be considered to test medications focused at reducing frequency of tobaccco use.

Gastrointestinal Tract

Nicotine exerts its effects on the gastrointestinal (GI) tract mainly via the activation of parasympathetic ganglia. Generally, it increases tone and contractility, and nausea, vomiting, and diarrhea can result from an overdose. However, GI irritation, other than mild nausea, rarely complicates NRT (Wong et al., 1999). Salivation evoked by cigarette smoking also rarely accompanies the doses used in NRT. Nicotine slows gastric emptying and reduces gastric and pancreatic secretions.

Given the association of smokeless tobacco with oral cancer (Schildt et al., 1998; Winn, 1997), there was initial concern that this might pose a risk with NRT. Follow-up studies of intermediate duration do not substantiate this concern (Wallstrom et al, 1999). In recent years, the observation that ulcerative colitis appears to be ameliorated in smokers has prompted the evaluation of NRT for this condition and controlled studies support its efficacy (Guslandi, 1999; Sandborn, 1999) and a delivery system permitting controlled release of nicotine in the colon has been developed.

Other Effects of Nicotine

There is much speculation about the existence of gender-specific effects of nicotine and their implications for NRT strategies. There is some evidence of differences in the pharmacodynamic effects of nicotine between genders and of an influence of timing in the menstrual cycle on the response to NRT and the success of attempts to quit (Pomerlau et al., 1991; Gritz et al., 1996). Women appear to have more pronounced withdrawal symptoms during the late luteal phase of the menstrual cycle, and it has been suggested that fear of weight gain, confidence in the ability to quit, and readiness to quit smoking might be differentially related to gender. Maternal smoking has adverse effects on the fetus, including the risk of spontaneous abortion, abruptio placentae, reduced weight at birth, and deformities (Haustein, 1999). In animal models, maternal consumption of nicotine results in hyperactivity in the neonate (Tizabi et al., 2000). Maternal smoking has been associated with sudden infant death syndrome and appears to result in an intellectual deficit, apparent in children at least as old as 6-7 years of age (Frydman, 1996).

Chronic smokers tend to have lower blood pressure than nonsmokers (Charlton and While, 1995). Maternal smoking has been associated with a reduced incidence of preeclampsia, but the mechanism is unclear (Lain et al., 1999). Nicotine does cross the placental barrier unchanged and maternal passive smoking raises nicotine levels in breast milk and in suckling infants. No linkage of nicotine consumption to birth deformities has been established; however, its contribution to the other effects of smoking during pregnancy is less clear (Haustein, 1999). For example, nicotine inhibits placental aromatase, reduces uteroplacental blood flow, and may adversely affect endothelial function in animal models (Torok et al., 2000). Presently, the experience with short-term NRT has not been associated with reports of adverse effects on fetal outcome, however; the number of individuals evaluated in this setting has been small.

Cigarette smoking is associated with lower body weight and quitting is associated with weight gain. Involvement in a weight control program amplifies the efficacy of NRT (Danielsson et al., 1999). Although the mechanisms are likely to be complex, nicotine is of some substantial relevance to this effect of smoking. Aside from its stimulatory effect on basal metabolic rate, nicotine reduces the synthesis of neuropeptide Y (NPY) in the arcuate neurons which project into the paraventricular nucleus (PVN). Injection of NPY into the PVN results in hyperphagia and obesity in rats (Frankish et al., 1995). Smoking is associated with insulin resistance, which improves after cessation (Kong et al., 2000). Elevated leptin levels have been related to weight loss in smokers, and levels appear to correlate with the degree of insulin resistance (Assali et al., 1999). Crossover studies in volunteers suggest that plasma leptin levels correlate with changes in insulin sensitivity and that intermediate levels are found in subjects on NRT (Oeser et al., 1999).

Nicotine has diverse effects on other hormonal systems in the brain that are presently poorly understood. For example, chronic nicotine administration stimulates mediobasohypothalamic tyrosine hydroxylase and suppresses pro-opiomelanocortin mRNAs. Suppression of forebrain β-endorphins may be relevant to maintaining nicotine self-administration (Rasmussen, 1998). Similarly, smoking is extremely prevalent among schizophrenics and may modulate the response to certain antipsychotics, such as clozapine (McEvoy et al., 1999). It has been speculated that this behavioral response may represent an attempt at self-medication, and some evidence for a disease-related abnormality in central nAchR sensory gating in schizophrenia has begun to emerge (Breese et al., 2000; Dalack et al., 1998).

Much less information is available concerning the effects of nicotine on other systems. Examples of potentially important observations include impairment of the immune response (Sopori et al., 1998), adverse effects on bone formation (Fung et al., 1999), and testicular hypogonadism (Kavitharaj and Vijayammal, 1999; Reddy et al., 1998), all observed with nicotine in model systems. The relevance of these observations, if any, to the doses of nicotine achieved in humans during NRT is unknown and should be evaluated.

Finally, cigarette smoking may result in drug interactions. While polycyclic hydrocarbons in cigarette smoke induce CYP isozymes of potential relevance to carcinogenesis, nicotine itself can induce CYP2E1, CYP2A1/2A2, and CYP2B1/2B2 in animal studies. Cutaneous vasoconstriction due to nicotine can delay the absorption of transdermal and subcutaneously administered medication, including insulin and heparin, and the stimulant effects of nicotine can diminish the analgesic effects of some opioids and the sedative effects of benzodiazapines (Zevin and Benowitz, 1999). Cigarette smoking reduces the hypotensive response to β-blockers (Fox et al., 1984), but the contribution of nicotine to this effect is unknown. Smoking reduces portal blood flow velocity and volume in humans and may modulate the disposition of drugs subject to hepatic metabolism (Rapaccini et al., 1996).

RESEARCH AGENDA

NRT has proven an effective strategy in the cessation of cigarette smoking that is remarkably well tolerated at least in the short to medium term. Although the experience is much more limited, NRT also holds promise as a strategy for reducing the number of cigarettes smoked by those who cannot or will not quit.

Both of these observations prompt considerations for future research. Thus, for those who quit smoking but continue to take NRT indefinitely, are there reasons to be concerned? First, nicotine can be addictive and although the daily exposure may be lower on NRT than when the individual was smoking, continued use implies psychological dependence, if not physical addiction. It is arguable whether this should be a concern, given the marked reduction of risk compared to smoking. However, it would seem reasonable to include surveillance of the dependence potential and various methods to determine abuse liability of various nicotine products. Furthermore, the implications of long-term nicotine intake for such factors as the safety of drug and alcohol consumption, progression of incidental diseases, impact of aging on cognitive and other physiological functions, and susceptibility to other forms of addictive behavior are largely unknown. For example, observations suggesting that nicotine impairs endothelial function, a property it shares with cigarette smoking, raise concerns about its effect on atherogenesis during long-term usage. Such an effect may take many years to emerge and highlights the importance of continued postmarketing surveillance of NRT. This is also true of carcinogenesis. For example, nicotine may be metabolized to nitrosamines (e.g. nicotine-derived nitroketone) with carcinogenic potential (Hecht, 2001; Hoffman et al., 1991). However, the methodology to assess their formation is just emerging, and the concentration-effect relationships and individual patterns of susceptibility are far from established. Studies of long term nicotine administration on surrogate variables that more closely resemble the mechanism under consideration (e.g., imaging of plaque progression) and attendant studies in animal models seems timely. Increasingly, the application of genomic and proteomic approaches is likely to clarify the differential effects of smoking and NRT on the expression and translation of genes related to the development of smoking-related diseases. Finally, the picture of nicotine's effect on inflammation and the immune response is confused and limited. More research is needed to clarify its effects on cytokine generation, the formation of nitric oxide and eicosanoids and oxidative injury. Research should continue to explore other potential therapeutic efficacies of NRT, including its use in ulcerative colitis, analgesia, weight reduction, Parkinson's disease, and cognitive disorders associated with aging and schizophrenia. Broadly speaking, the experience with long-term experience with nicotine via Swedish snus is reassuring with respect to safety, but formal evaluations of such risk from long-term use under controlled conditions have been scant (Idris et al., 1998; Raw and Macneil, 1990).

Continued use of NRT in conjunction with ongoing, albeit reduced, smoking prompts additional questions. For example, the constituents of cigarette smoke that mediate tissue injury are largely unknown, and it is also unknown if modulating the coincident nicotine level might influence their absorption, metabolic disposition, mechanism of action, or elimination. Design of such studies will rely on the development of more refined and tractable methodology to investigate the in vivo kinetics and dynamics of other constituents of cigarette smoke and their interactions with nicotine.

Finally, although ethnicity already seems relevant, other factors that determine interindividual differences in nicotine efficacy, safety, and addictive potential remain largely unexplored. Particular attention might be paid to genetic variation in proteins relevant to nicotine pharmacokinetics and dynamics and their interaction with environmental variables. As with other drugs, one anticipates increasing individualization of nicotine dosage and/or delivery when given as a therapeutic agent. Insight into the interaction of genetic and environmental factors that influence initiation (Gynther et al., 1999; Heath et al., 1999) of cigarette smoking, latency until the practice becomes habitual (Stallings et al., 1999), and the quantity then smoked (Koopmans, 1999) has been increasing. Clarification of how these factors interact is also likely to afford insights of value in predicting the individual likelihood of response to the use of NRT as a strategy for quitting or reducing tobacco exposure.

REFERENCES

Adler LE, Olincy A, Waldo M, et al. 1998. Schizophrenia, sensory gating, and nicotinic receptors. *Schizophr Bull* 24(2):189-202.

Allen SS, Hatsukami D, Gorsline J. 1994. Cholesterol changes in smoking cessation using the transdermal nicotine system. Transdermal Nicotine Study Group. *Prev Med* 23(2):190-196.

Anthony J, Warner L, Kessler R. 1994. Comparative epidemiology of dependence on tobacco, alcohol, controlled substances, and inhalants: basic findings from the national comobidity survey. *Experimental and Clinical Psychopharmacology* 2(3):244-268.

APA (American Psychiatric Association). 1987. *Diagnostic and Statistical Manual of Mental Disorders, Third Edition-Revised*. Washington, DC: APA.

APA (American Psychiatric Association). 1994. *Diagnostic and Statistical Manual of Mental Disorders, Fourth Edition*. Washington, DC: APA.

Armitage A, Dollery C, Houseman T, Kohner E, Lewis PJ, Turner D. 1978. Absorption of nicotine from small cigars. *Clin Pharmacol Ther* 23(2):143-151.

Assali AR, Beigel Y, Schreibman R, Shafer Z, Fainaru M. 1999. Weight gain and insulin resistance during nicotine replacement therapy. *Clin Cardiol* 22(5):357-360.

Balster RL. 1991. Drug abuse potential evaluation in animals. *Br J Addict* 86(12):1549-1558.

Benowitz NL. 1986. Clinical pharmacology of nicotine. *Annu Rev Med* 37:21-32.

Benowitz NL. 1993. Nicotine replacement therapy. What has been accomplished—can we do better? *Drugs* 45(2):157-170.

Benowitz NL. 1995. Clinical pharmacology of transdermal nicotine. *Eur J Pharm Biopharm* 41:168-174.

Benowitz NL. 1999. Nicotine addiction. *Prim Care* 26(3):611-631.

Benowitz NL, Fitzgerald GA, Wilson M, Zhang Q. 1993. Nicotine effects on eicosanoid formation and hemostatic function: comparison of transdermal nicotine and cigarette smoking. *J Am Coll Cardiol* 22(4):1159-1167.

Benowitz NL, Kuyt F, Jacob P, Jones RT, Osman AL. 1983. Cotinine disposition and effects. *Clin Pharmacol Ther* 34(5):604-611.

Benowitz NL, Gourlay SG. 1997. Cardiovascular toxicity of nicotine: implications for nicotine replacement therapy. *J Am Coll Cardiol* 29(7):1422-1431.

Benowitz NL, Jacob P 3rd. 1994. Metabolism of nicotine to cotinine studied by a dual stable isotope method. *Clin Pharmacol Ther* 56(5):483-493.

272

CLEARING THE SMOKE

Benowitz NL, Jacob P 3rd. 1999. Pharmacokinetics and metabolism of nicotine and related alkaloids. Arneric SP, Brioni JD, eds. *Neuronal Nicotinic Preceptors Pharmacology and Therapeutic Opportunities.* Boston: Wiley-Liss, Inc. Pp. 213-234.

Benowitz NL, Jacob P 3d, Jones RT, Rosenberg J. 1982. Interindividual variability in the metabolism and cardiovascular effects of nicotine in man. *J Pharmacol Exp Ther* 221(2):368-372.

Benowitz NL, Jacob P 3d, Savanapridi C. 1987. Determinants of nicotine intake while chewing nicotine polacrilex gum. *Clin Pharmacol Ther* 41(4):467-473.

Benowitz NL, Perez-Stable EJ, Fong I, Modin G, Herrera B, Jacob P 3rd. 1999. Ethnic differences in N-glucuronidation of nicotine and cotinine. *J Pharmacol Exp Ther* 291(3):1196-1203.

Benowitz NL, Porchet H, Jacob PI. 1990. Pharmacokinetics, metabolism and pharmacodynamics of nicotine. Wonnacott S, Russell MAH, Stolerman IP, eds. *Nicotine Psychopharmacology: Molecular, Cellular and Behavioural Aspects.* New York: Oxford University Press. Pp. 112-157.

Benowitz NL, Porchet H, Sheiner L, Jacob P 3d. 1988. Nicotine absorption and cardiovascular effects with smokeless tobacco use: comparison with cigarettes and nicotine gum. *Clin Pharmacol Ther* 44(1):23-28.

Berlin I, Said S, Spreux-Varoquaux O, Olivares R, Launay JM, Puech AJ. 1995. Monoamine oxidase A and B activities in heavy smokers. *Biol Psychiatry* 38(11):756-761.

Bigelow GE, Preston KL. 1989. Drug discrimination: methods for drug characterization and classification. *NIDA Res Monogr* 92(11):101-122.

Bozarth MA. 1987. *Methods of Assessing the Reinforcing Properties of Abused Drugs.* New York: Springer-Verlag.

Brandange S, Lindbolm L. 1979. The enzyme 'aldehyde oxidase' is an iminium oxidase. Reaction with nicotine-$\Delta^{1'(5')}$-iminium ion. *Biochem Biophys Res Comm* 91:991-996.

Breese CR, Lee MJ, Adams CE, et al. 2000. Abnormal regulation of high affinity nicotinic receptors in subjects with schizophrenia. *Neuropsychopharmacology* 23(4):351-364.

Breslau N. 1995. Psychiatric comorbidity of smoking and nicotine dependence. *Behav Genet* 25(2):95-101.

Breslau N, Fenn N, Peterson EL. 1993. Early smoking initiation and nicotine dependence in a cohort of young adults. *Drug Alcohol Depend* 33(2):129-137.

Breslau N, Peterson EL, Schultz LR, Chilcoat HD, Andreski P. 1998. Major depression and stages of smoking. A longitudinal investigation. *Arch Gen Psychiatry* 55(2):161-166.

Byrd GD, Caldwell WS, Beck DJ, Kaminski DL, Li AP. 1994 (Oct 23-27). Profile of nicotine metabolites from human hepatocytes. In: *Abstract 173; ISSX Proceedings. Volume 6.* Raleigh, NC: Sixth North American ISSX Meeting.

Caraballo RS, Giovino GA, Pechacek TF, et al. 1998. Racial and ethnic differences in serum cotinine levels of cigarette smokers: Third National Health and Nutrition Examination Survey, 1988-1991. *JAMA* 280(2):135-139.

CDC (Centers for Disease Control and Prevention). 1994. Reasons for tobacco use and symptoms of nicotine withdrawal among adolescent and young adult tobacco users-United States, 1993. *Morbidity and Mortality Weekly Report* 43(41):745-750.

CDC (Centers for Disease Control and Prevention). 1995a. Indicators of nicotine addiction among women-United States, 1991-1992. *Morbidity and Mortality Weekly Report* 44(6):102-105.

CDC (Centers for Disease Control and Prevention). 1995b. Symptoms of substance dependence associated with use of cigarettes, alcohol, and illicit drugs—United States, 1991-1992. *Morbidity and Mortality Weekly Reports* 44(44):830-839.

CDC (Centers for Disease Control). 1998. Selected cigarette smoking initiation and quitting behaviors among high school students—United States, 1997. *Morbidity and Morality Weekly Reports* 47(19):386-389.

Celermajer DS, Sorensen KE, Georgakopoulos D, et al. 1993. Cigarette smoking is associated with dose-related and potentially reversible impairment of endothelium-dependent dilation in healthy young adults. *Circulation* 88(5 Pt 1):2149-2155.

Chalon S, Moreno H Jr, Benowitz NL, Hoffman BB, Blaschke TF. 2000. Nicotine impairs endothelium-dependent dilatation in human veins in vivo. *Clin Pharmacol Ther* 67(4):391-397.

Charlton A, While D. 1995. Blood pressure and smoking: observations on a national cohort. *Arch Dis Child* 73(4):294-297.

Childress AR, Mozley PD, McElgin W, Fitzgerald J, Reivich M, O'Brien CP. 1999. Limbic activation during cue-induced cocaine craving. *Am J Psychiatry* 156(1):11-18.

Chiodera P, Capretti L, Davoli C, Caiazza A, Bianconi L, Coiro V. 1990. Effect of obesity and weight loss on arginine vasopressin response to metoclopramide and nicotine from cigarette smoking. *Metabolism* 39(8):783-786.

Chowdhury P, Hosotani R, Rayford PL. 1989. Weight loss and altered circulating GI peptide levels of rats exposed chronically to nicotine. *Pharmacol Biochem Behav* 33(3):591-594.

Cinciripini PM, Wetter DW, McClure JB. 1997. Scheduled reduced smoking: effects on smoking abstinence and potential mechanisms of action. *Addict Behav* 22(6):759-767.

Colby SM, Tiffany ST, Shiffman S, Niaura RS. 2000. Measuring nicotine dependence among youth: a review of available approaches and instruments. *Drug Alcohol Depend* 59 Suppl 1:S23-39.

Cordero-Erausquin M, Marubio LM, Klink R, Changeux JP. 2000. Nicotinic receptor function: new perspectives from knockout mice. *Trends Pharmacol Sci* 21(6):211-217.

Corrigall WA. 1999. Nicotine self-administration in animals as a dependence model. *Nicotine Tob Res* 1(1):11-20.

Cryer PE, Haymond MW, Santiago JV, Shah SD. 1976. Norepinephrine and epinephrine release and adrenergic mediation of smoking-associated hemodynamic and metabolic events. *N Engl J Med* 295(11):573-577.

Dajani RM, Gorrod JW, Beckett AH. 1975. Reduction in vivo of (minus)-nicotine-1'-N-oxide by germ-free and conventional rats. *Biochem Pharmacol* 24(5):648-650.

Dalack GW, Healy DJ, Meador-Woodruff JH. 1998. Nicotine dependence in schizophrenia: clinical phenomena and laboratory findings. *Am J Psychiatry* 155(11):1490-1501.

Dale HH. 1914. The action of certain esters and ethers of choline and their relation to muscarine. *J Pharmacol Exp Ther* 6:147-190.

Danielsson T, Rossner S, Westin A. 1999. Open randomised trial of intermittent very low energy diet together with nicotine gum for stopping smoking in women who gained weight in previous attempts to quit. *BMJ* 319(7208):490-493; discussion 494.

de Wit H. 1991. Preference procedures for testing the abuse liability of drugs in humans. *Br J Addict* 86(12):1579-1586.

de Wit H, Zacny J. 1995. Abuse potentials of nicotine replacement therapies. *CNS Drugs* 4(6):456-468.

Debanne SM, Rowland DY, Riedel TM, Cleves MA. 2000. Association of Alzheimer's disease and smoking: the case for sibling controls. *J Am Geriatr Soc* 48(7):800-806.

DiChiara G. 1999. Drug addiction as dopamine-dependent associative learning disorder. *Eur J Pharmacol* 375(1-3):13-30.

Doll R, Peto R, Boreham J, Sutherland I. 2000. Smoking and dementia in male British doctors: prospective study. *BMJ* 320(7242):1097-1102.

Domino EF. 1999. Pharmacological significance of nicotine. Gorrod JW, Jacob III P, eds. *Analytical Determination of Nicotine and Related Compounds and Their Metabolites.* Amsterdam: Elsevier. Pp. 1-11.

Dugi KA, Rader DJ. 2000. Lipoproteins and the endothelium: insights from clinical research. *Semin Thromb Hemost* 26(5):513-519.

Emilien G, Beyreuther K, Masters CL, Maloteaux JM. 2000. Prospects for pharmacological intervention in Alzheimer disease. *Arch Neurol* 57(4):454-459.

Fagerström KO. 1978. Measuring degree of physical dependence to tobacco smoking with reference to individualization of treatment. *Addict Behav* 3(3-4):235-241.

Fagerström KO, Schneider NG. 1989. Measuring nicotine dependence: a review of the Fagerström Tolerance Questionnaire. *J Behav Med* 12(2):159-182.

Fant RV, Henningfield JE, Nelson RA, Pickworth WB. 1999. Pharmacokinetics and pharmacodynamics of moist snuff in humans. *Tob Control* 8(4):387-392.

Fischman MW, Foltin RW. 1991. Utility of subjective-effects measurements in assessing abuse liability of drugs in humans. *Br J Addict* 86(12):1563-1570.

Fowler JS, Volkow ND, Wang GJ, et al. 1996. Inhibition of monoamine oxidase B in the brains of smokers. *Nature* 379(6567):733-736.

Fox K, Deanfield J, Krikler S, Ribeiro P, Wright C. 1984. The interaction of cigarette smoking and beta-adrenoceptor blockade. *Br J Clin Pharmacol* 17(Suppl 1):92S-93S.

Frankish HM, Dryden S, Wang Q, Bing C, MacFarlane IA, Williams G. 1995. Nicotine administration reduces neuropeptide Y and neuropeptide Y mRNA concentrations in the rat hypothalamus: NPY may mediate nicotine's effects on energy balance. *Brain Res* 694(1-2):139-146.

Freedman R, Coon H, Myles-Worsley M, et al. 1997. Linkage of a neurophysiological deficit in schizophrenia to a chromosome 15 locus. *Proc Natl Acad Sci U S A* 94(2):587-592.

Frydman M. 1996. The smoking addiction of pregnant women and the consequences on their offspring's intellectual development. *J Environ Pathol Toxicol Oncol* 15(2-4):169-172.

Fu Y, Matta SG, Gao W, Brower VG, Sharp BM. 2000. Systemic nicotine stimulates dopamine release in nucleus accumbens: re-evaluation of the role of N-methyl-D-aspartate receptors in the ventral tegmental area. *J Pharmacol Exp Ther* 294(2):458-465.

Fung YK, Iwaniec U, Cullen DM, Akhter MP, Haven MC, Timmins P. 1999. Long-term effects of nicotine on bone and calciotropic hormones in adult female rats. *Pharmacol Toxicol* 85(4):181-187.

Gairola CG, Daugherty A. 1999. Acceleration of atherosclerotic plaque formation in ApoE-mice by exposure to tobacco smoke. *The Toxicologist* 48:297.

George TP, Verrico CD, Picciotto MR, Roth RH. 2000. Nicotinic modulation of mesoprefrontal dopamine neurons: pharmacologic and neuroanatomic characterization. *J Pharmacol Exp Ther* 295(1):58-66.

Goldberg SR, Spealman RD, Goldberg DM. 1981. Persistent behavior at high rates maintained by intravenous self-administration of nicotine. *Science* 214(4520):573-575.

Gori GB, Benowitz NL, Lynch CJ. 1986. Mouth versus deep airways absorption of nicotine in cigarette smokers. *Pharmacol Biochem Behav* 25(6):1181-1184.

Gorrod JW, Hibberd AR. 1982. The metabolism of nicotine-$\Delta^{1'(5')}$-iminium ion, in vivo and in vitro. *Eur J Drug Metab Pharmacokinet* 7:293-298.

Gorrod JW, Jenner P. 1975. The metabolism of tobacco alkaloids. Hayes WJ Jr, ed. *Essays in Toxicology.* New York: Academic Press. Pp. 35-78.

Gorrod JW, Schepers G. 1999. Biotransformation of nicotine in mammalian systems. Gorrod JW, Jacob III P, Eds. *Analytical Determination of Nicotine and Related Compounds and Their Metabolites.* Amsterdam: Elsevier.

Gorrod JW, Wahren J. 1993. *Nicotine and Related Alkaloids: Absorption, Distribution, Metabolism and Excretion.* London: Chapman and Hall.

Gouaze V, Dousset N, Dousset JC, Valdiguie P. 1998. Effect of nicotine and cotinine on the susceptibility to in vitro oxidation of LDL in healthy non smokers and smokers. *Clin Chim Acta* 277(1):25-37.

Gourlay SG, Forbes A, Marriner T, McNeil JJ. 1999. Predictors and timing of adverse experiences during trandsdermal nicotine therapy. *Drug Saf* 20(6):545-555.

Green JT, Evans BK, Rhodes J, et al. 1999. An oral formulation of nicotine for release and absorption in the colon: its development and pharmacokinetics. *Br J Clin Pharmacol* 48(4):485-493.

Greenland S, Satterfield MH, Lanes SF. 1998. A meta-analysis to assess the incidence of adverse effects associated with the transdermal nicotine patch. *Drug Saf* 18(4):297-308.

Gries JM, Benowitz N, Verotta D. 1996. Chronopharmacokinetics of nicotine. *Clin Pharmacol Ther* 60(4):385-395.

Gritz ER, Nielsen IR, Brooks LA. 1996. Smoking cessation and gender: the influence of physiological, psychological, and behavioral factors. *J Am Med Womens Assoc* 51(1-2):35-42.

Guatura SB, Martinez JA, Santos Bueno PC, Santos ML. 2000. Increased exhalation of hydrogen peroxide in healthy subjects following cigarette consumption. *Sao Paulo Med J* 118(4):93-98.

Gundisch D. 2000. Nicotinic acetylcholine receptors and imaging. *Curr Pharm Des* 6(11):1143-1157.

Guslandi M. 1999. Long-term effects of a single course of nicotine treatment in acute ulcerative colitis: remission maintenance in a 12-month follow-up study. *Int J Colorectal Dis* 14(4-5):261-262.

Guthrie SK, Zubieta JK, Ohl L, et al. 1999. Arterial/venous plasma nicotine concentrations following nicotine nasal spray. *Eur J Clin Pharmacol* 55(9):639-643.

Gynther LM, Hewitt JK, Heath AC, Eaves LJ. 1999. Phenotypic and genetic factors in motives for smoking. *Behav Genet* 29(5):291-302.

Haertzen CA, Hill HE, Belleville, RE. 1963. Development of the Addiction Research Center Inventory (ARCI): selection of items that are sensitive to the effects of various drugs. *Psychopharmacologia.* 4:155-166.

Hatsukami DK, Grillo M, Boyle R, et al. 2000. Treatment of spit tobacco users with transdermal nicotine system and mint snuff. *J Consult Clin Psychol* 68(2):241-249.

Hatsukami DK, Gust SW, Keenan RM. 1987. Physiologic and subjective changes from smokeless tobacco withdrawal. *Clin Pharmacol Ther* 41(1):103-107.

Hatsukami DK, Severson HH. 1999. Oral spit tobacco: addiction, prevention and treatment. *Nicotine Tob Res* 1(1):21-44.

Hatsukami DK, Skoog K, Huber M, Hughes J. 1991. Signs and symptoms from nicotine gum abstinence. *Psychopharmacology (Berl)* 104(4):496-504.

Hatsukami D, Huber M, Callies A, Skoog K. 1993. Physical dependence on nicotine gum: effect of duration of use. *Psychopharmacology (Berl)* 111(4):449-456.

Hatsukami D, Skoog K, Allen S, Bliss R. 1995. Gender and the effects of different doses of nicotine gum on tobacco withdrawal symptoms. *Experimental and Clinical Psychopharmacology* 3(2):163-173.

Haustein KO. 1999. Cigarette smoking, nicotine and pregnancy. *Int J Clin Pharmacol Ther* 37(9):417-427.

Heath AC, Kirk KM, Meyer JM, Martin NG. 1999. Genetic and social determinants of initiation and age at onset of smoking in Australian twins. *Behav Genet* 29(6):395-407.

Heath AC, Madden PA, Martin NG. 1998. Statistical methods in genetic research on smoking. *Stat Methods Med Res* 7(2):165-186.

Heatherton TF, Kozlowski LT, Frecker RC, Fagerström KO. 1991. The Fagerström Test for Nicotine Dependence: a revision of the Fagerström Tolerance Questionnaire. *Br J Addict* 86(9):1119-1127.

Hecht SS. 2001. Carcinogen biomarkers for lung or oral cancer chemoprevention trials. *IARC Sci Publ.* 154:245-255.

Henningfield JE, Cohen C, Heishman SJ. 1991. Drug self-administration methods in abuse liability evaluation. *Br J Addict* 86(12):1571-1577.

Henningfield JE, Goldberg SR. 1983. Nicotine as a reinforcer in human subjects and laboratory animals. *Pharmacol Biochem Behav* 19(6):989-992.

Henningfield JE, Heishman SJ. 1995. The addictive role of nicotine in tobacco use. *Psychopharmacology* 117:11-13.

Henningfield JE, Keenan RM. 1993. Nicotine delivery kinetics and abuse liability. *J Consult Clin Psychol* 61(5):743-750.

Henningfield JE, Miyasato K, Jasinski DR. 1985. Abuse liability and pharmacodynamic characteristics of intravenous and inhaled nicotine. *J Pharmacol Exp Ther* 234(1):1-12.

Hibberd AR, Gorrod, JW. 1983. Enzymology of the metabolic pathway from nicotine to cotinine, in vitro. *Eur J Drug Metab Pharmacokinet.* 8:151-162.

Hoffmann D, Rivenson A, Chung FL, Hecht SS. 1991. Nicotine-derived N-nitrosamines (TSNA) and their relevance in tobacco carcinogenesis. *Crit Rev Toxicol* 21(4):305-311.

Hughes JR. 1986. Genetics of smoking. A brief review. *Behavioral Therapy* 7:335-345.

Hughes, J. 1998. Dependence on and abuse of nicotine replacement medications: an update. In Benowitz NL, ed. *Nicotine Safety and Toxicity.* New York: Oxford University Press. Pp. 147-157.

Hughes JR, Gulliver SB, Fenwick JW, et al. 1992. Smoking cessation among self-quitters. *Health Psychol* 11(5):331-334.

Hughes JR, Gust SW, Keenan RM, Fenwick JW. 1990b. Effect of dose on nicotine's reinforcing, withdrawal-suppression and self-reported effects. *J Pharmacol Exp Ther* 252(3):1175-1183.

Hughes JR, Gust SW, Keenan R, Fenwick JW, Skoog K, Higgins ST. 1991. Long-term use of nicotine vs placebo gum. *Arch Intern Med* 151(10):1993-1998.

Hughes JR, Gust SW, Pechacek TF. 1987. Prevalence of tobacco dependence and withdrawal. *Am J Psychiatry* 144(2):205-208.

Hughes JR, Hatsukami DK. 1992. The nicotine withdrawal syndrome: a brief review and update. *International Journal of Smoking Cessation* 1(2):21-26.

Hughes JR, Hatsukami DK, Mitchell JE, Dahlgren LA. 1986a. Prevalence of smoking among psychiatric outpatients. *Am J Psychiatry* 143(8):993-997.

Hughes JR, Hatsukami DK, Skoog KP. 1986b. Physical dependence on nicotine in gum. A placebo substitution trial. *JAMA* 255(23):3277-3279.

Hughes JR, Higgins ST, Hatsukami D. 1990a. Effects of tobacco abstinence. Kowzlowski LT, Annis H, Cappell, et al., eds. *Research Advances in Alcohol and Drug Problems.* New York: Plenum Press.

Hunt WA, Matarazzo JD. 1973. Three years later: recent developments in the experimental modification of smoking behavior. *J Abnorm Psychol* 81(2):107-114.

Idris AM, Ibrahim SO, Vasstrand EN, et al. 1998. The Swedish snus and the Sudanese toombak: are they different? *Oral Oncol* 34(6):558-566.

Jaffe JH, Jaffe FK. 1989. Historical perspectives on the use of subjective effects measures in assessing the abuse potential of drugs. *NIDA Res Monogr* 92(10):43-72.

Jarvik ME. 1991. Beneficial effects of nicotine. *Br J Addict* 86(5):571-575.

Jasinski DR, Henningfield JE. 1989. Human abuse liability assessment by measurement of subjective and physiological effects. *NIDA Res Monogr* 92(10):73-100.

Jasinski DR, Johnson RE, Henningfield JE. 1984. Abuse liability assessment in human subjects. *Trends in Pharm Science* 5:196-200.

Jones HE, Garrett BE, Griffiths RR. 1999. Subjective and physiological effects of intravenous nicotine and cocaine in cigarette smoking cocaine abusers. *J Pharmacol Exp Ther* 288(1):188-197.

Joseph AM, Norman SM, Ferry LH, et al. 1996. The safety of transdermal nicotine as an aid to smoking cessation in patients with cardiac disease. *N Engl J Med* 335(24):1792-1798.

Kameda G, Dadmarz M, Vogel WH. 2000. Influence of various drugs on the voluntary intake of nicotine by rats. *Neuropsychobiology* 41(4):205-209.

Kandel D, Chen K, Warner LA, Kessler RC, Grant B. 1997. Prevalence and demographic correlates of symptoms of last year dependence on alcohol, nicotine, marijuana and cocaine in the U.S. population. *Drug Alcohol Depend* 44(1):11-29.

Kavitharaj NK, Vijayammal PL. 1999. Nicotine administration induced changes in the gonadal functions in male rats. *Pharmacology* 58(1):2-7.

Kawakami N, Takatsuka N, Shimizu H, Takai A. 1998. Life-time prevalence and risk factors of tobacco/nicotine dependence in male ever-smokers in Japan. *Addiction* 93(7):1023-1032.

Keenan RM, Hatsukami DK, Anton DJ. 1989. The effects of short-term smokeless tobacco deprivation on performance. *Psychopharmacology (Berl)* 98(1):126-130.

Keenan RM, Jenkins AJ, Cone EJ, Henningfield JE. 1994 Nov. Smoked and intravenous nicotine, cocaine and heroin have similar abuse liability. Submitted for presentation at American Society of Addiction Medicine.

Kem WR. 2000. The brain alpha7 nicotinic receptor may be an important therapeutic target for the treatment of Alzheimer's disease: studies with DMXBA (GTS-21). *Behav Brain Res* 113(1-2):169-181.

Kendler KS, Neale MC, MacLean CJ, Heath AC, Eaves LJ, Kessler RC. 1993. Smoking and major depression. A causal analysis. *Arch Gen Psychiatry* 50(1):36-43.

Kong C, Elatrozy T, Anyaoku V, Robinson S, Richmond W, Elkeles RS. 2000. Insulin resistance, cardiovascular risk factors and ultrasonically measured early arterial disease in normotensive Type 2 diabetic subjects. *Diabetes Metab Res Rev* 16(6):448-453.

Koob GF. 1992. Drugs of abuse: anatomy, pharmacology and function of reward pathways. *Trends Pharmacol Sci* 13(5):177-184.

Koopmans JR, Slutske WS, Heath AC, Neale MC, Boomsma DI. 1999. The genetics of smoking initiation and quantity smoked in Dutch adolescent and young adult twins. *Behav Genet* 29(6):383-393.

Labelle A, Boulay LJ. 1999. An attempted suicide using transdermal nicotine patches. *Can J Psychiatry* 44(2):190.

Lain KY, Powers RW, Krohn MA, Ness RB, Crombleholme WR, Roberts JM. 1999. Urinary cotinine concentration confirms the reduced risk of preeclampsia with tobacco exposure. *Am J Obstet Gynecol* 181(5 Pt 1):1192-1196.

Lena C, Changeux JP. 1998. Allosteric nicotinic receptors, human pathologies. *J Physiol Paris* 92(2):63-74.

Lerman C, Caporaso NE, Audrain J, et al. 1999. Evidence suggesting the role of specific genetic factors in cigarette smoking. *Health Psychol* 18(1):14-20.

Lopez-Arrieta JM, Rodriguez JL, Sanz F. 2000. Nicotine for Alzheimer's disease. *Cochrane Database Syst Rev* (2):CD001749.

Ludviksdottir D, Blondal T, Franzon M, Gudmundsson TV, Sawe U. 1999. Effects of nicotine nasal spray on atherogenic and thrombogenic factors during smoking cessation. *J Intern Med* 246(1):61-66.

Maddux J, Desmond D. 1986. Relapse and recovery in substance abuse careers. Tims F, Leukefeld C, eds. *Relapse and Recovery in Drug Abuse*. Rockville, MD: U.S. Department of Health and Human Services, Public Health Service, Alcohol, Drug Abuse and Mental Health Administration.

Malin DH, Lake JR, Newlin-Maultsby P, et al. 1992. Rodent model of nicotine abstinence syndrome. *Pharmacol Biochem Behav* 43(3):779-784.

McEvoy JP, Freudenreich O, Wilson WH. 1999. Smoking and therapeutic response to clozapine in patients with schizophrenia. *Biol Psychiatry* 46(1):125-129.

McKinney EF, Walton RT, Yudkin P, et al. 2000. Association between polymorphisms in dopamine metabolic enzymes and tobacco consumption in smokers. *Pharmacogenetics* 10(6):483-491.

Meagher EA, FitzGerald GA. 2000. Indices of lipid peroxidation in vivo: strengths and limitations. *Free Radic Biol Med* 28(12):1745-1750.

Merchant C, Tang MX, Albert S, Manly J, Stern Y, Mayeux R. 1999. The influence of smoking on the risk of Alzheimer's disease [see comments]. *Neurology* 52(7):1408-1412.

Molander L, Hansson A, Lunell E. 2001. Pharmacokinetics of nicotine in healthy elderly people. *Clin Pharmacol Ther* 69(1):57-65.

Morrow JD, Frei B, Longmire AW, et al. 1995. Increase in circulating products of lipid peroxidation (F2-isoprostanes) in smokers. Smoking as a cause of oxidative damage. *N Engl J Med* 332(18):1198-1203.

Murphy PJ. 1973. Enzymatic oxidation of nicotine to nicotine-$\Delta^{1'(5')}$-iminium ion. A newly dicovered intermediate in the metabolism of nicotine. *J Biol Chem* 248: 2796-2800.

Nakajima M, Kwon JT, Tanaka N, et al. 2001. Relationship between interindividual differences in nicotine metabolism and CYP2A6 genetic polymorphism in humans. *Clin Pharmacol Ther* 69(1):72-78.

Nakajima M, Yamagishi S, Yamamoto H, Yamamoto T, Kuroiwa Y, Yokoi T. 2000. Deficient cotinine formation from nicotine is attributed to the whole deletion of the CYP2A6 gene in humans. *Clin Pharmacol Ther* 67(1):57-69.

Nakajima M, Yamamoto T, Nunoya K, et al. 1996. Role of human cytochrome P4502A6 in C-oxidation of nicotine. *Drug Metab Dispos* 24(11):1212-1217.

Niemela O, Parkkila S, Juvonen RO, Viitala K, Gelboin HV, Pasanen M. 2000. Cytochromes P450 2A6, 2E1, and 3A and production of protein-aldehyde adducts in the liver of patients with alcoholic and non-alcoholic liver diseases. *J Hepatol* 33(6):893-901.

Nilsson P, Lundgren H, Soderstrom M, Fagerström KO, Nilsson-Ehle P. 1996. Effects of smoking cessation on insulin and cardiovascular risk factors—a controlled study of 4 months' duration. *J Intern Med* 240(4):189-194.

Nitenberg A, Antony I. 1999. Effects of nicotine gum on coronary vasomotor responses during sympathetic stimulation in patients with coronary artery stenosis. *J Cardiovasc Pharmacol* 34(5):694-699.

Nomikos GG, Schilstrom B, Hildebrand BE, Panagis G, Grenhoff J, Svensson TH. 2000. Role of alpha7 nicotinic receptors in nicotine dependence and implications for psychiatric illness. *Behav Brain Res* 113(1-2):97-103.

Nowak J, Murray JJ, Oates JA, FitzGerald GA. 1987. Biochemical evidence of a chronic abnormality in platelet and vascular function in healthy individuals who smoke cigarettes. *Circulation* 76(1):6-14.

Oeser A, Goffaux J, Snead W, Carlson MG. 1999. Plasma leptin concentrations and lipid profiles during nicotine abstinence. *Am J Med Sci* 318(3):152-157.

Oncken CA, Hardardottir H, Smeltzer JS. 1998. Human studies of nicotine replacement during pregnancy. Benowitz NL, ed. *Nicotine Safety and Toxicity*. New York: Oxford University Press.

Oxenkrug GF. 1999. Antidepressive and antihypertensive effects of MAO-A inhibition: role of N-acetylserotonin. A review. *Neurobiology (Bp)* 7(2):213-224.

Panagis G, Kastellakis A, Spyraki C, Nomikos G. 2000. Effects of methyllycaconitine (MLA), an alpha 7 nicotinic receptor antagonist, on nicotine- and cocaine-induced potentiation of brain stimulation reward. *Psychopharmacology (Berl)* 149(4):388-396.

Pastrakuljic A, Schwartz R, Simone C, Derewlany LO, Knie B, Koren G. 1998. Transplacental transfer and biotransformation studies of nicotine in the human placental cotyledon perfused in vitro. *Life Sci* 63(26):2333-2342.

Patrignani P, Panara MR, Tacconelli S, et al. 2000. Effects of vitamin E supplementation on F(2)-isoprostane and thromboxane biosynthesis in healthy cigarette smokers. *Circulation* 102(5):539-545.

Perez-Stable EJ, Herrera B, Jacob P 3rd, Benowitz NL. 1998. Nicotine metabolism and intake in black and white smokers [see comments]. *JAMA* 280(2):152-156.

Perkins KA, Grobe JE, Fonte C. 1994a. Chronic and acute tolerance to subjective, behavioral and cardiovascular effects of nicotine in humans. *Journal of Pharmacology and Experimental Therapeutics* 270:628-638.

Perkins KA, Sanders M, D'Amico D, Wilson A. 1997. Nicotine discrimination and self-administration in humans as a function of smoking status. *Psychopharmacology (Berl)* 131(4):361-370.

Perkins KA, Sexton JE, Reynolds WA, Grobe JE, Fonte C, Stiller RL. 1994b. Comparison of acute subjective and heart rate effects of nicotine intake via tobacco smoking versus nasal spray. *Pharmacol Biochem Behav* 47(2):295-299.

Perry DC, Davila-Garcia MI, Stockmeier CA, Kellar KJ. 1999. Increased nicotinic receptors in brains from smokers: membrane binding and autoradiography studies. *J Pharmacol Exp Ther* 289(3):1545-1552.

Pomerleau CS, Pomerleau OF, Garcia AW. 1991. Biobehavioral research on nicotine use in women. *Br J Addict* 86(5):527-531.

Pomerleau OF. 1995. Individual differences in sensitivity to nicotine: implications for genetic research on nicotine dependence. *Behavior Genetics.* 25(2):161-177.

Preston KL. 1991. Drug discrimination methods in human drug abuse liability evaluation. *Br J Addict* 86(12):1587-1594.

Prokhorov AV, Koehly LM, Pallonen UE, Hudmon, KS. 1998. Adolescent nicotine dependence measured by the modified Fagerström Tolerance Questionnaire at two time points. *J Child Adol Subst Abuse* 7(4):35-47.

Prokhorov AV, Pallonen UE, Fava JL, Ding L, Niaura R. 1996. Measuring nicotine dependence among high-risk adolescent smokers. *Addict Behav* 21(1):117-127.

Quattrocki E, Baird A, Yurgelun-Todd D. 2000. Biological aspects of the link between smoking and depression. *Harv Rev Psychiatry* 8(3):99-110.

Raitakari OT, Adams MR, McCredie RJ, Griffiths KA, Celermajer DS. 1999. Arterial endothelial dysfunction related to passive smoking is potentially reversible in healthy young adults. *Ann Intern Med* 130(7):578-581.

Raitakari OT, Adams MR, McCredie RJ, Griffiths KA, Stocker R, Celermajer DS. 2000. Oral vitamin C and endothelial function in smokers: short-term improvement, but no sustained beneficial effect. *J Am Coll Cardiol* 35(6):1616-1621.

Rapaccini GL, Pompili M, Marzano MA, et al. 1996. Doppler ultrasound evaluation of acute effects of cigarette smoking on portal blood flow in man [see comments]. *J Gastroenterol Hepatol* 11(11):997-1000.

Rasmussen DD. 1998. Effects of chronic nicotine treatment and withdrawal on hypothalamic proopiomelanocortin gene expression and neuroendocrine regulation. *Psychoneuroendocrinology* 23(3):245-259.

Raw M, McNeill A. 1990. Britain bans oral snuff. *BMJ* 300(6717):65-66.

Reddy S, Londonkar R, Ravindra, Reddy S, Patil SB. 1998. Testicular changes due to graded doses of nicotine in albino mice. *Indian J Physiol Pharmacol* 42(2):276-280.

Reilly M, Delanty N, Lawson JA, FitzGerald GA. 1996. Modulation of oxidant stress in vivo in chronic cigarette smokers. *Circulation* 94(1):19-25.

Risner ME, Goldberg SR. 1983. A comparison of nicotine and cocaine self-administration in the dog: fixed-ratio and progressive-ratio schedules of intravenous drug infusion. *J Pharmacol Exp Ther* 224(2):319-326.

Robinson JS, Moore VM, Owens JA, McMillen IC. 2000. Origins of fetal growth restriction. *Eur J Obstet Gynecol Reprod Biol* 92(1):13-19.

Rojas NL, Killen JD, Haydel KF, Robinson TN. 1998. Nicotine dependence among adolescent smokers. *Arch Pediatr Adolesc Med* 152(2):151-156.

Rose JE, Corrigall WA. 1997. Nicotine self-administration in animals and humans: similarities and differences. *Psychopharmacology (Berl)* 130(1):28-40.

RCP (Royal College of Physicians). 2000. *Nicotine Addiction in Britain. A Report of the Tobacco Advisory Group of the Royal College of Physicians.* Sudbury, Suffolk: Lavenham Press, Ltd.

Sabha M, Tanus-Santos JE, Toledo JC, Cittadino M, Rocha JC, Moreno H Jr. 2000. Transdermal nicotine mimics the smoking-induced endothelial dysfunction. *Clin Pharmacol Ther* 68(2):167-174.

Sandborn WJ. 1999. Nicotine therapy for ulcerative colitis: a review of rationale, mechanisms, pharmacology, and clinical results. *Am J Gastroenterol* 94(5):1161-1171.

Saxena K, Scheman A. 1985. Suicide plan by nicotine poisoning: a review of nicotine toxicity. *Vet Hum Toxicol* 27(6):495-497.

Schildt EB, Eriksson M, Hardell L, Magnuson A. 1998. Oral snuff, smoking habits and alcohol consumption in relation to oral cancer in a Swedish case-control study. *Int J Cancer* 77(3):341-346.

Schuh KJ, Schuh LM, Henningfield JE, Stitzer ML. 1997. Nicotine nasal spray and vapor inhaler: abuse liability assessment. *Psychopharmacology (Berl)* 130(4):352-361.

Sekhon HS, Jia Y, Raab R, et al. 1999. Prenatal nicotine increases pulmonary alpha7 nicotinic receptor expression and alters fetal lung development in monkeys. *J Clin Invest* 103(5):637-647.

Shields PG, Lerman C, Audrain J, et al. 1998. Dopamine D4 receptors and the risk of cigarette smoking in African-Americans and Caucasians. *Cancer Epidemiol Biomarkers Prev* 7(6):453-458.

Shiffman S, Mason KM, Henningfield JE. 1998. Tobacco dependence treatments: review and prospectus. *Annu Rev Public Health* 19:335-358.

Smith TA, House RF Jr, Croghan IT, et al. 1996. Nicotine patch therapy in adolescent smokers. *Pediatrics* 98(4 Pt 1):659-667.

Soderpalm B, Ericson M, Olausson P, Blomqvist O, Engel JA. 2000. Nicotinic mechanisms involved in the dopamine activating and reinforcing properties of ethanol. *Behav Brain Res* 113(1-2):85-96.

Sopori ML, Kozak W, Savage SM, et al. 1998. Effect of nicotine on the immune system: possible regulation of immune responses by central and peripheral mechanisms. *Psychoneuroendocrinology* 23(2):189-204.

Stallings MC, Hewitt JK, Beresford T, Heath AC, Eaves LJ. 1999. A twin study of drinking and smoking onset and latencies from first use to regular use. *Behav Genet* 29(6):409-421.

Stolerman IP, Jarvis MJ. 1995. The scientific case that nicotine is addictive. *Psychopharmacology (Berl)* 117(1):2-10; discussion 14-20.

Strandberg TE, Tilvis RS. 2000. C-reactive protein, cardiovascular risk factors, and mortality in a prospective study in the elderly. *Arterioscler Thromb Vasc Biol* 20(4):1057-1060.

Sutherland G, Russell MA, Stapleton J, Feyerabend C, Ferno O. 1992. Nasal nicotine spray: a rapid nicotine delivery system. *Psychopharmacology (Berl)* 108(4):512-518.

Swahn M, Hammig B. 2000. Prevalence of youth access to alcohol, guns, illegal drugs, or cigarettes in the home and association with health-risk behaviors. *Ann Epidemiol* 10(7):452.

Swedberg MDB, Henningfield JE, Goldberg SR. 1990. Nicotine dependency: animal studies. Wonnacott S, Russell MAH, Stolerman IP., eds. *Nicotine Psychopharmacology: Molecular, Cellular and Behavioural Aspects.* Oxford: Oxford Science Publications. Pp. 38-76.

Tizabi Y, Russell LT, Nespor SM, Perry DC, Grunberg NE. 2000. Prenatal nicotine exposure: effects on locomotor activity and central [125I]alpha-BT binding in rats. *Pharmacol Biochem Behav* 66(3):495-500.

Torok J, Gvozdjakova A, Kucharska J, et al. 2000. Passive smoking impairs endothelium-dependent relaxation of isolated rabbit arteries. *Physiol Res* 49(1):135-141.

Traber MG, van der Vliet A, Reznick AZ, Cross CE. 2000. Tobacco-related diseases. Is there a role for antioxidant micronutrient supplementation? *Clin Chest Med* 21(1):173-187, x.

Tyndale RF, Pianezza ML, Seelars EM. 1999. A common genetic defect in nicotine metabolism decrease smoking. *Nicotine and Tobacco Research* 1(S2):S61-S67.

U.S. DHHS (United States Department of Health and Human Services). 1988. *The Health Consequences of Smoking: Nicotine Addiction; A Report of the Surgeon General.* Rockville, MD: U.S. Department of Health and Human Services, Centers for Disease Control and Prevention.

U.S. DHHS (U.S. Department of Health and Human Services). 1994. *Preventing Tobacco Use Among Young People; A Report of the Surgeon General.* Washington, DC: U.S. Department of Health and Human Services, Centers for Disease Control and Prevention.

U.S. DHHS (U.S. Department of Health and Human Services). 2000. *Summary of Findings From the 1999 National Household Survey on Drug Abuse.* Rockville, MD: Substance Abuse and Mental Health Services Administration.

Wakasa Y, Takada K, Yanagita T. 1995. Reinforcing effect as a function of infusion speed in intravenous self-administration of nicotine in rhesus monkeys. *Nihon Shinkei Seishin Yakurigaku Zasshi* 15(1):53-59.

Wallace BC. 1989. Psychological and environmental determinants of relapse in crack cocaine smokers. *J Subst Abuse Treat* 6(2):95-106.

Wallstrom M, Sand L, Nilsson F, Hirsch JM. 1999. The long-term effect of nicotine on the oral mucosa. *Addiction* 94(3):417-423.

Wang HY, Lee DH, D'Andrea MR, Peterson PA, Shank RP, Reitz AB. 2000a. beta-Amyloid(1-42) binds to alpha7 nicotinic acetylcholine receptor with high affinity. Implications for Alzheimer's disease pathology. *J Biol Chem* 275(8):5626-5632.

Wang H, Yang B, Zhang L, Xu D, Wang Z. 2000b. Direct block of inward rectifier potassium channels by nicotine. *Toxicol Appl Pharmacol* 164(1):97-101.

Waters AJ, Sutton SR. 2000. Direct and indirect effects of nicotine/smoking on cognition in humans. *Addict Behav* 25(1):29-43.

Watkins SS, Koob GF, Markou A. 2000. Neural mechanisms underlying nicotine addiction: acute positive reinforcement and withdrawal. *Nicotine Tob Res* 2(1):19-37.

West RJ, Russell MA. 1985. Effects of withdrawal from long-term nicotine gum use. *Psychol Med* 15(4):891-893.

Wetscher GJ, Bagchi M, Bagchi D, et al. 1995. Free radical production in nicotine treated pancreatic tissue. *Free Radic Biol Med* 18(5):877-882.

Whiss PA, Lundahl TH, Bengtsson T, Lindahl TL, Lunell E, Larsson R. 2000. Acute effects of nicotine infusion on platelets in nicotine users with normal and impaired renal function. *Toxicol Appl Pharmacol* 163(2):95-104.

WHO (World Health Organization). 1991. International classification of Disease (ICD-10). Geneva: World Helath Organization.

Winn DM. 1997. Epidemiology of cancer and other systemic effects associated with the use of smokeless tobacco. *Adv Dent Res* 11(3):313-321.

Winther K, Fornitz GG. 1999. The effect of cigarette smoking and nicotine chewing gum on platelet function and fibrinolytic activity. *J Cardiovasc Risk* 6(5):303-306.

Wong PW, Kadakia SC, McBiles M. 1999. Acute effect of nicotine patch on gastric emptying of liquid and solid contents in healthy subjects. *Dig Dis Sci* 44(11):2165-2171.

Yamazaki H, Inoue K, Hashimoto M, Shimada T. 1999. Roles of CYP2A6 and CYP2B6 in nicotine C-oxidation by human liver microsomes. *Arch Toxicol* 73(2):65-70.

Yu H, Matsubayashi H, Amano T, Cai J, Sasa M. 2000. Activation by nicotine of striatal neurons receiving excitatory input from the substantia nigra via dopamine release. *Brain Res* 872(1-2):223-226.

Zevin S, Benowitz NL. 1999. Drug interactions with tobacco smoking. An update. *Clin Pharmacokinet* 36(6):425-438.

Zevin S, Gourlay SG, Benowitz NL. 1998. Clinical pharmacology of nicotine. *Clin Dermatol* 16(5):557-564.

10

Tobacco Smoke and Toxicology

PHYSICAL AND CHEMICAL CHARACTERISTICS
OF TOBACCO SMOKE

Tobacco smoke is a complex mixture of toxicants and the chemical properties change—rapidly in some cases—as smoke ages. Toxicants measured at one point in time may not be what the smoker actually experiences. It is estimated that there are more than 2,000 chemical constituents of tobacco. Almost twice that number results when tobacco is burned incompletely during smoking. Three kinds of smoke can be described, each differing in terms of toxicant concentration, size of particles, effects of temperature, and a host of other characteristics. Mainstream smoke (MS) is what emerges from the "mouth" or butt end of a puffed cigarette. Sidestream smoke (SS) is what arises from the lit end of a cigarette, mostly between puffs. Environmental tobacco smoke (ETS), smoke present in air, consists of exhaled mainstream smoke and sidestream smoke.

Smoking machines are used to analyze mainstream smoke. A set of parameters has been agreed on by various international organizations. These parameters are 35-cm^3 puff volume, 2-second puff duration, once per minute puff interval, and smoking to a butt length of 23 mm for nonfiltered or 3 mm above the filter overlap for filter-tipped cigarettes. The "yields" of toxicants in the MS are frequently reported by the standard-setting organization. The two most well known organizations are the U.S. Federal Trade Commission (FTC) and the International Organization of Standardization (ISO). The controversy regarding the standard parameters and reporting of values is covered in other chapters of this report.

Mainstream smoke is pulled through the mouth end of the cigarette and then through a "Cambridge filter pad." Aerosol particles in the smoke larger than 1 μm in diameter are trapped with 99% efficiency. The material is referred to as cigarette smoke condensate or total particulate matter (TPM). "Tar" is the weight of TPM minus nicotine and water. The material that passes through the filter pad is the gas or vapor phase of cigarette smoke. In general, the vapor phase consists predominantly of compounds with a molecular weight <60 and the particulate phase consists of compounds with a molecular weight >200.

The yields of MS increase with successive puffs as the cigarette is machine-smoked due to the decrease in filtration provided by the cigarette rod itself. However, smoking behavior studies coupled with yield measurements suggest that yields remain consistent from first puff to last when assessed under real-life smoking conditions.

When tobacco is heated, moisture and volatile material are distilled, and combustion leads to the generation of volatile gases and the residual, carbonized char. Char reacts with oxygen in the air during puffing and smoldering, producing volatile gases (carbon dioxide, carbon monoxide, and water) and the inorganic material known as ash.

The highest temperature reached during the burning of tobacco is approximately 800°C in the center of the burning zone during smolder. During puff, a solid-phase temperature of approximately 910°C is reached at the burning zone periphery, while the gas temperatures are lower. They vary between 600 and 700°C as the puff progresses. After the puff ends, solid-phase temperatures rapidly cool to approximately 600°C. This greatly influences particle formation, particle size, and toxicant formation. These temperatures contrast with that achieved with a newly marketed cigarette-like device, Eclipse, that combusts differently than conventional cigarettes and aerosolizes nicotine and glycerin.

The chemical nature of MS changes as smoke ages. The burning zone generates a highly concentrated vapor that is drawn down the cigarette to form mainstream smoke. The vapor cools quickly (in milliseconds) due to diluting air. Less volatile compounds quickly condense, mostly in airborne state. A combination of physical size and concentration affects both thermal and mechanical properties, which influence the number of particles in smoke. Droplets of less than about 0.1 μm will attach to the tobacco through which they pass or to other particles, which continue on into MS. Particles with sizes around 1 μm are "filtered" out by depositing onto the tobacco surface.

MS is a highly concentrated aerosol mixture. Smoke particles are liquid, consisting of approximately 20% water by volume. The particles vary from less than 0.1 to 1.0-μm diameter. The small size and high concentration promote rapid coagulation, leading to decreased concentration and increased size of the resulting particles within less than a second. The size of particles also increases due to absorption of water, which is relevant for human smoking because of the high relative humidity of the human respiratory tract.

Sidestream smoke particles are smaller than MS particles initially. However, the aging of SS over a few minutes leads to an increase in particle size of ETS due to coagulation of particles and removal of smaller particles that attach to surfaces in the environment. Particle size in smoke is important, because it influences where within the respiratory tract a toxicant is deposited. Smaller particles, in general, deposit further down into the lungs.

Inhaled particles of the size found in tobacco smoke would be predicted to deposit mainly in the alveolar region of the lung. However, cigarette smoke-induced tumors are more prevalent in the bronchial region, suggesting that smoke particles deposit higher up in the respiratory tract than would be predicted from the initial particle size. (Recent increases of adenocarcinomas in lower airways of smokers are hypothesized to be due to so-called smoking compensation of low-yield products. Smokers of these products inhale more deeply to increase their nicotine dose.) Mucociliary clearance of inhaled particles up the respiratory tract may also increase the dose of particles to the upper airways. More importantly, the cloud-like nature of MS (see below) and the increased size of smoke particles on aging are responsible for this finding. Specific factors influencing the site of deposition in airways include coagulation of fresh smoke particles, absorption of water in the humid respiratory tract, human breathing patterns, aerodynamic interactions between nearby particles, electrostatic charge, and vapor deposition on airway walls (Dendo et al., 1998). Theoretical models of particle deposition predict that MS particles would have approximately 20% deposition in the respiratory tract. Measurements in humans suggest that deposition is actually much higher, from 50 to 95% (Phalen, 2000).

The explanation for this high deposition rate is that cigarette smoke is so dense, that it acts as a cloud. Clouds are high concentrations of aerosol particles surrounded by relatively clean air. They behave as entities that are much larger than the individual components. In depositing, the cloud behaves as if it were a much larger particle, with an aerodynamic diameter of approximately 6 or 7 μm. This particle size (see Figure 10-1) results in high total deposition in the respiratory tract, with especially high deposition in the tracheobronchial region.

Deposition in the respiratory tract is also influenced by the size of the person. Smaller individuals have greater tracheobronchial and less

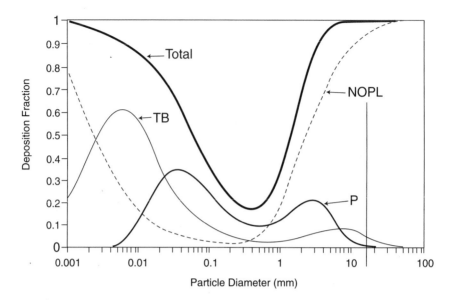

FIGURE 10-1   Aerosol deposition curves.

NOTE: applies to the Reference Man; NOPL=nasal oral pharyngeal laryngeal regions; TB=tracheobronchial region.

SOURCE: Adapted from National Council on Radiation Protection and Measurements report #125, 1997, with permission from the National Council on Radiation Protection and Measurements.

pulmonary deposition. Smaller individuals also have greater minute ventilation normalized for body weight, which is especially important for SS. This is hypothesized to be one reason infants might be more vulnerable to ETS. Importantly the small airway diameter of infants leads to high disposition in the upper airways, where it can be an irritant (Phalen, 2000).

Whole MS consists (by weight) of mostly air (nitrogen N_2, oxygen O_2, argon). The vapor phase constitutes approximately 20% by weight of the smoke, with the particulate phase accounting for approximately 5%.

The majority of cigarettes sold today have filters to remove portions of the smoke. Cellulose acetate filters are used almost exclusively in the United States, whereas charcoal filters are popular in Japan (Norman, 1999). Cellulose acetate filters remove some of the particulate phase of the smoke but have little influence on the vapor phase. The efficiency of particle removal depends on particle size and is minimal at the number-average particle size found in cigarette smoke (about 0.3 μm diameter). In general, cellulose acetate or paper filters remove tar and nicotine particles in this size range with an efficiency of 40–50%. Charcoal filters influence the retention of vapor-phase components and are made by adding up to 60 mg of activated charcoal in a segment of the cellulose acetate filter. Factors that influence the retention of vapor-phase compounds include the amount of charcoal used, the activity of the charcoal (based on surface area and pore volume), and the smoke velocity through the segment. Low-molecular-weight compounds with low boiling points (less than 150°C), which are not retained by the cellulose acetate filters, are partially removed by the charcoal filters. Removal efficiencies vary with the compound but are reported to be from 30 to 90%. As an example, benzene may be removed at an efficiency of 67%. Vented filters are designed to improve filter efficiency by decreased smoke flow through and increased residence time in the filter. However, vent holes can readily be covered by the fingers of the smoker, who may be inclined to do this in order to get the maximum amount of nicotine during smoking.

Under similar smoking conditions, filtered cigarettes will have lower MS yield relative to nonfiltered analogues. SS yields will not vary much, since they are reflective of tobacco weight burned during smolder. In general, more tobacco is consumed during smolder than during puffing. However, SS generally contains more alkaline and neutral compounds. SS smoke contains less or equal amounts of acids, phenols, and phytosterols than MS. Differences are due to temperature and mechanisms of chemical transfer (release) from the unburned tobacco. The approximate chemical composition of MS is given in Table 10-1. The relative concentration of specific constituents in MS versus SS tobacco smoke is shown in Table 10-2.

The pH of cigarette smoke influences the degree of protonation of the active addictive chemical, nicotine. The free-base form of nicotine is favored at a higher pH (more basic) and is more rapidly absorbed into the bloodstream than the mono- or diprotonated salt forms of nicotine that exist at lower pH (more acidic). Tobacco blends with a high sugar content produce a more acidic smoke; a basic cigarette smoke can be achieved by addition of ammonia.

TOXICITY OF TOBACCO SMOKE

The health effects of tobacco smoke in humans are well known from both clinical and epidemiological studies; such information is summarized in later chapters of this report. Animal studies of the toxicity of tobacco smoke are reviewed in the present section. The purpose of this review of animal models of tobacco smoke toxicity is to determine the potential usefulness of such models for assessing the toxicity of new and

TABLE 10-1 Approximate Chemical Composition of Whole Mainstream Smoke

Constituent	% by Weight	
Air		
N_2	62	
O_2	13	75.9
Ar	0.9	
Vapor Phase		
Water	1.3	
CO_2	12.5	
CO	4	
H_2	0.1	
CH_4	0.3	
Hydrocarbons	0.6	
Aldehydes	0.3	
Ketones	0.2	19.6
Nitriles	0.1	
Heterocyclics	0.03	
Methanol	0.03	
Organic acids	0.02	
Esters	0.01	
Other compounds	0.1	
Particulate Phase		
Water	0.8	
Alkanes	0.2	
Terpenoids	0.2	
Phenols	0.2	
Esters	0.2	
Nicotine	0.3	4.5
Other alkaloids	0.1	
Alcohols	0.3	
Carbonyls	0.5	
Organic acids	0.6	
Leaf pigments	0.2	
Other compounds	0.9	

NOTE: Ar=Argon; CH_4=methane; CO=carbon monoxide; CO_2=carbon dioxide; H_2=hydrogen; N_2=nitrogen; O_2=oxygen.
SOURCE: Dube and Green, 1982. Reprinted with permission from the authors.

TABLE 10-2   Some Typical SS/MS Yield Ratios for Plain Cigarettes

Substance	MS Yield	SS/MS
Small Molecules		
Carbonyl sulfide	18-42 µg	0.03-0.1
HCN	160-500 µg	0.06-0.5
CO	10-23 mg	2.5-4.7
Hydrazine	20-43 µg	3
Methane	600-1000 µg	3.1-4.8
Acetylene	20-40 µg	0.8-2.5
Nitrogen oxides	100-600 µg	4-13
CO_2	20-50 mg	8-11
H_2O (gas phase)	3-14 mg	24-30
NH_3	50-130 µg	40-170
N_2 (generated)	<10 µg	>270
Neutral Heteroatom Organics		
Acetonitrile	50-130 µg	40-170
Benzonitrile	5-6 µg	7-10
Acetamide	70-100 µg	0.8-1.7
Methyl chloride	150-600 µg	1.7-3.3
Aldehydes, Ketones, Alcohols		
Acetaldehyde	0.5-1.2 mg	1.4
Propionaldehyde	175-250 µg	2.4-2.8
Acetone	100-250 µg	2-5
Acrolein	60-100 µg	8-15
2-Butanone	~30 µg	2.9-4.3
2-Furaldehyde	15-43 µg	4.9-7.4
Furfuryl alcohol	18-65 µg	3.0-4.8
Cyclotene[a]	3-5 µg	6-10
Pyranone[b]	13-150 µg	0.1-1.2
Phytosterols		
β-Sitosterol	59 µg	0.5
Campesterol	43 µg	0.6
Cholesterol	22 µg	0.9
Phenols		
Phenol	60-140 µg	1.6-3.0
Cresols (o-,m-,p-)	11-37 µg	1.0-1.4
Catechol	100-360 µg	0.6-0.9
Hydroquinone	110-300 µg	0.7-1.0
Acids		
Formic acid	210-490 µg	1.4-1.6
Acetic acid	270-810 µg	1.9-3.9
3-Methylvaleric acid	20-60 µg	0.8-1.5

continues

TABLE 10-2 Continued

Substance	MS Yield	SS/MS
Lactic acid	60-170 µg	0.5-0.7
Benzoic acid	14-28 µg	0.7-1.0
Phenylacetic acid	11-38 µg	0.6-0.8
Succinic acid	70-140 µg	0.4-0.6
Glycolic acid	40-130 µg	0.6-1.0
Amines, Pyridines, Alkaloids		
Methylamine	12-29 µg	4.2-6.4
n-Propylamine	1.6-3.4 µg	2.8-3.8
n-Butylamine	0.5-1.5 µg	2.2-4.0
Aniline	360 ng	30
Pyridine	16-46 µg	6.5-20
3-Ethenylpyridine	11-30 µg	20-40
Methylpyrazine	2-5 µg	3-4
Pyrrole	16-23 µg	9-14
Nicotine	0.8-2.3 mg	2.6-3.3
Myomine	13-33 µg	4.0-7.5
Nicotyrine	4-40 µg	5-14
Anatabine	2-20 µg	0.1-0.5
2,3'-Bipyridyl	16-22 µg	2-3
Aza-arenes		
Quinoline	0.5-2.0 µg	8-11
Isoquinoline	1.6-2.0 µg	2.5-5
Benzo[h]quinoline	10 ng	10
Indole	16-38 µg	2.1-3.4
Hydrocarbons		
Isoprene	330-1100 µg	13-19
Benzene	36-68 µg	5-10
Toluene	100-200 µg	6-8
Limonene	15-50 µg	4-12
Neophytadiene	66-230 µg	1-2
Polynuclear Aromatic Hydrocarbons		
Naphthalene	2.6 µg	17
Pyrene	45-140 µg	2-11
Benzo[a]pyrene	9-40 µg	2-20
Anthracene	24 ng	30
Phenanthrene	77 ng	2-30
Fluoranthene	60-150 ng	11
Nitrosamines[c]		
N-Nitrosodimethylamine	10-40 ng	10-50
N-Nitrosodiethylamine	nd-25 ng	3-35
N-Nitrosopyrrolidine	6-30 ng	6-30

continues

TABLE 10-2 Continued

Substance	MS Yield	SS/MS
N-Nitrosodiethanolamine	0-70 ng	1.2
N'-Nitrosonomicotine	0.2-3 µg	0.5-3
NNK[d]	0.1-1 µg	1-4
N'-Nitrosoanatabine	0.3-5 µg	0.3-1
Inorganic Constituents		
Cadmium	100 ng	4-7
Nickel	20-80 ng	0.2-30
Zinc	60 ng	0.2-7

NOTE: CO=carbon monoxide; CO_2=carbon dioxide; HCN=hydrogen cyanide; H_2O=water; N_2=nitrogen; nd=not detected; NH_3=ammonia; NNK=nitrosonornicotine ketone.

[a]Hydroxy-3-methyl-2-cyclopentanone.
[b]5m6-Dihydro-3,5-dihydroxy-2-methyl-4H-pyran-4-one.
[c]Much of the data in the literature on the smoke levels of volatile and tobacco-specific nitrosamines may be in error, due to artifact formation on the Cambridge pad part of the smoke collection procedure (Caldwell and Conner, 1990).
[d]Nitrosonornicotine ketone or 4-(methylnitrosamino)-1-(3-pyridyl)-1-butanone.

SOURCE: Reprinted, with permission from Davis, DL and Nielsen, MT eds. *Tobacco: Production, Chemistry and Technology.* Pp. 418. Copyright 1999 by Blackwell Science.

existing products that claim to reduce harm from tobacco use. The possibilities and the limits of using animals to test for toxic effects related to tobacco use are discussed. The major adverse tobacco smoke-induced health effects that require evaluation because of their prominence in humans are pulmonary inflammation, induction of lung cancer, chronic obstructive pulmonary disease (COPD), cardiovascular disease, reproductive and developmental effects, and the suppression of the immune system.

In Vitro Toxicity Tests

Toxicity tests that can be performed on cell systems in vitro have the advantage of being done rapidly and with relatively low cost. Such tests can be used to screen for general toxic properties of a chemical or a chemical mixture, such as the cytotoxicity of the material or its ability to alter the genetic material, DNA. The cytotoxicity of a compound can be predictive of its ability to induce inflammation; the genotoxicity of a compound suggests its potential to induce cancer. The limitation of such tests is that the results are based on the response of single cell types and do not include the influence of the whole-body system on the response. Nevertheless, such tests can be valuable in excluding products from further development if they either are extremely cytotoxic or have a high potential for producing mutations in DNA. Cellular screening assays should include benchmark materials of known cytotoxic or genotoxic potential (based on both in vitro and in vivo studies) for comparison to the test material and to allow better interpretation of the results.

Cytotoxicity Tests

Cytotoxicity tests (Balls and Clothier, 1991) are based on either primary cultures or established cell lines from the target organ of interest. Dye exclusion or the release of cytoplasmic enzymes is used to measure damage to cell membranes. In dye exclusion tests, the ability of cells to exclude extracellular dyes such as trypan blue or neutral red is measured. Release of the cytoplasmic enzyme lactate dehydrogenase (LDH) is commonly measured as an indicator of cell membrane damage. For specific purposes, the release of other enzymes, such as the hydrolytic enzymes of pulmonary macrophages, can be useful. The exclusion of the dye, neutral red, and the release of LDH have been recently used to compare the cytotoxicity of smoke condensates from standard cigarettes and a new tobacco-related PREP for which harm reduction was claimed (Eclipse Expert Panel, 2000)

Genotoxicity Tests

Several in vitro tests designed to assay for the mutagenic potential of a chemical or mixture were recently reviewed in an International Workshop on Genotoxicity Test Procedures (Lovell et al., 2000). Perhaps the most commonly used screening tool is the *Salmonella typhimurium* bacterial mutagenicity assay or Ames test (Ames et al., 1987). In addition to assays for point mutations, there are assays of clastogenic DNA damage as indicated by chromosomal aberrations, sister chromatid exchanges (SCEs), micronuclei formation (Hayashi et al., 2000), or single strand breaks via the Comet assay (Tice et al., 2000). Assays for specific chemical adducts to DNA can be used as a measure of dosimetry and, in a few cases, as predictors of adverse effects (Phillips et al., 2000). Measures of oxidized bases in DNA can be used to assay for oxidative stress (Cadet et al., 1998). In recent comparative potency studies comparing the genotoxicity of smoke condensates from standard and new tobacco-related PREP, a battery of in vitro assays included the Ames test, SCEs, and chromosomal aberrations (Eclipse Expert Panel, 2000).

In Vivo Toxicity Tests

Animals do not smoke cigarettes in the same manner as humans, and much effort has been expended in the past in animal studies to mimic human exposures to intermittent puffs of smoke. It is impractical to replicate all of the parameters of human smoking in animals. One problem is that rodents tend to hold their breath during puffs of irritating tobacco smoke and thus avoid receiving a high dose of smoke (Kendrick et al., 1976). Studies were conducted comparing three modes of exposure of rats to cigarette smoke: nose-only intermittent, nose-only continuous, and whole-body continuous (Chen et al., 1995; Mauderly et al., 1989). Plasma nicotine was higher by a factor of 3 in whole-body exposed rats compared to nose-only exposed rats. This suggests that dermal absorption and grooming as well as inhalation contributed to the dose of nicotine received in rats exposed in the whole-body mode. Urinary cotinine was not higher in the whole-body exposed group compared to the nose-only intermittent exposure group but was higher by a factor of 1.5 compared to the nose-only continuous exposure group. This study demonstrated few significant differences in either smoke characteristics or biological effects among the three exposure modes. The biological effects thought to be related to chemical carcinogenesis (cell transformation, chromosomal damage, DNA adducts) and chronic lung disease (cell proliferation, inflammation, respiratory function) were similar for all groups. Whole-body exposures were less labor intensive and less stressful to the rats (based on body weights) and avoided the reduction in breathing known to occur during puff-by-puff exposures. Thus, whole-body exposures may be useful as a method to achieve dosing of tobacco smoke in small laboratory test animals.

ASSESSMENT OF POTENTIAL EXPOSURE REDUCTION PRODUCTS

Lung Cancer

Animals have not proven to be good models for the type of lung tumors induced by cigarette smoke in humans. Rodents tend to develop peripherally arising lung adenomas rather than centrally arising bronchial tumors when exposed to chemicals. Exposure of animals to tobacco smoke has not often produced an excess of lung tumors of any type. In 1986, the International Agency for Research on Cancer (IARC, 1986) critically reviewed animal studies on tobacco smoke; out of four rat studies judged to be adequate for analysis, only one yielded unequivocal evidence for tobacco smoke as a respiratory tract carcinogen. One problem may be that rats build up carboxyhemoglobin faster than humans when exposed to the same level of carbon monoxide (CO) (Guerin et al., 1974; Silbaugh and Horvath, 1982). This results in their not being able to tolerate the level of exposure to cigarette smoke that humans can. Other factors undoubtedly contribute to this species difference in response to cigarette smoke. However, rodent models can be used to test the ability of tobacco products to cause alterations in DNA, and recent studies (discussed below) indicate that the A/J strain of mouse shows promise as a model for in vivo carcinogenesis induced by tobacco smoke.

Short-term exposure of rodents followed by analysis of DNA isolated from the lungs for DNA modifications, such as oxidative damage or methylation, can determine the ability of the product to damage DNA in vivo. Aberrant methylation of DNA can be used as a marker for early stages of oncogenesis in both rats and humans (Belinsky et al., 1998; Nuovo et al., 1999; Swafford et al., 1997). Oxidative damage to DNA is used to monitor oxidative stress (Halliwell, 1998; Loft et al., 1998).

The A/J mouse strain, which is sensitive to induction of lung adenomas, has been used in a series of studies by Witschi to test for the carcinogenic potential of tobacco smoke and the effectiveness of chemopreventive measures (Witschi et al., 1997a, 1999, 2000). The A/J mice exposed to 87 mg/m^3 of environmental tobacco smoke for five months and allowed to recover for four months had a statistically significant elevation in number of lung tumors (Witschi et al., 1997a). The same strain of mice exposed similarly to filtered and unfiltered tobacco smoke suggested that the particulate phase was not required for carcinogenicity (Witschi et al., 1997b). In a chemoprevention study, acetylsalicylic acid, an agent known to protect against nicotine-derived N-nitrosaminoketone (NNK)-induced tumors in the same strain of mice, had no protective effect against tobacco smoke (Witschi et al., 1999). A second chemoprevention study indicated that a diet containing myoinisitol-dexamethasone was effective in preventing tobacco smoke-induced lung tumors but that agents known to protect against NNK-or polycyclic aromatic hydrocarbon (PAH)-induced tumors did not protect against tobacco smoke (Witschi et al., 2000).

The committee concludes that these studies indicate that removal from tobacco smoke of single classes of carcinogens, such as nitrosamines or PAHs, may not be protective against the induction of lung tumors by smoke. These studies also suggest that the A/J mouse, used in "stop-start" studies, shows promise as an animal model of value in screening for the potential of tobacco products to induce lung tumors. Future studies should determine if the model is robust enough to be repeated in other laboratories. In recent comparative potency studies for a newly developed tobacco-related PREP, the potency for smoke condensates to induce cancer was evaluated in 30-week dermal tumor-promotion studies in mice (Eclipse Expert Panel, 2000). Such skin painting studies provide information for hazard identification.

Chronic Obstructive Pulmonary Disease

An early response to any inhaled toxicant is pulmonary inflammation, which if persistent may lead to more severe alterations in the structure and function of the lung. Animal models can readily be used to detect and quantitate the pulmonary inflammatory response to inhaled compounds or mixtures. Analysis of bronchoalveolar lavage (BAL) fluid for cellular and biochemical indicators of inflammation has become a common tool for quantitation of the pulmonary inflammatory response of rodents to inhaled toxicants (Henderson, 1989), including tobacco smoke (Mauderly et al., 1989; Sjostrand and Rylander, 1997; Subramaniam et al., 1996). The differential cell count and the functioning of cells obtained by the BAL technique can be used to classify the type of inflammatory response. The biochemical content of BAL fluid can be used to detect the release of various cytokines and alterations in the pulmonary surfactant.

Chronic obstructive pulmonary disease in the form of either emphysema, bronchitis, or both is a well-recognized sequela of cigarette smoking (Vial, 1986). Harm reduction strategies must take into account the degree to which this type of chronic lung disease is reduced in new products. Animals models to study the degree of COPD induced by the use of new products have been suggested by the work of March et al. (1999a). F344 rats exposed to cigarette smoke over a two-week period showed enhanced pulmonary epithelial cell replication and alterations in axial airway mucosubstances—changes consistent with the development of chronic bronchitis (March et al., 1999a). Both B6C3F1 mice and F344 rats exposed to cigarette smoke over a longer period (7–13 months) were found to develop morphological evidence of emphysema. Mice developed more pronounced signs of emphysema than rats, and the condition progressed with time in mice (March et al., 1999b). In earlier studies, rats exposed for three months to sidestream smoke were reported to have emphysematous changes in the lung (Escolar et al., 1995). Comparative potency studies for a newly developed tobacco-related PREP made use of 90-day inhalation studies in hamsters and rats to test for inflammation as well as epithelial hyperplasia and metaplasia (Eclipse Expert Panel, 2000).

The committee concludes that these studies indicate there are animal models that show promise for use in screening for development of COPD-like symptoms from the inhalation of new or existing tobacco products.

Cardiovascular Disease

Cockerels and rabbits are two animal models that have been used to test for the cardiovascular effects of tobacco smoke. Penn and coworkers (1993, 1994) found that 16-week exposure of cockerels to tobacco smoke (2-3 mg/m^3) increased the size of arteriosclerotic plaques in the aorta. Rabbits fed a cholesterol-rich diet and exposed for 10 weeks to 4 and 33 mg/m^3 showed a dose-dependent increase in the size of arteriosclerotic lesions in the aorta and pulmonary artery, as well as increased stickiness of platelets (Zhu et al., 1994). C57BL/6 mice have also been used to test for the effect of inhaled pollutants on the induction of atherosclerosis and enhancement of arterial fatty deposits in animals fed a high-fat diet (Lewis et al., 1999).

The committee concludes that such studies suggest animal models can be used to detect the potential for tobacco products to enhance the development of atherosclerosis.

Immune System Dysfunction

Smoking-related changes in the peripheral immune system in humans include elevated white blood cell counts, increased cytotoxic or suppressor and decreased inducer or helper T-cell numbers, slightly suppressed T-lymphocyte activity, significantly decreased natural killer (NK) cell activity, lowered circulating immunoglobulin titers (except for IgE, which is elevated), and increased susceptibility to infection. Similar effects have been observed in animals (Johnson et al., 1990; McAllister-Sistilli et al., 1998; Sopori et al., 1994), suggesting that animal models can be used to test for harm reduction to the immune system from use of new tobacco products or nicotine delivery devices. The major areas of interest are reduced host resistance to infections and tumors, suppression of the cellular and humoral immune system, and interference with macrophage cell function.

The effect of tobacco smoke on the immune system of humans and rodents depends on the duration and level of exposure. In general, short-term, low-level exposures do not affect the immune system or may be stimulatory, whereas longer-term exposures (six months or more) or high levels of exposure are immunosuppressive.

The committee concludes that this finding indicates that long-term animal studies are required to evaluate adverse effects of tobacco products on the immune system.

Animals exposed to cigarette smoke for extended periods are more susceptible than naïve animals to tumor and infectious agent challenge. Mice exposed to cigarette smoke for six months or longer were more susceptible to intratracheally instilled Lewis lung or TKL5 tumor cells in terms of increased tumor growth, metastases, and early death than unexposed mice (Chalmer et al., 1975; Thomas et al., 1974b). Such changes are not observed in mice exposed to cigarette smoke for short periods of time (days). Chronic exposure of mice to cigarette smoke results in increased susceptibility to infectious agents such as murine sarcoma virus (Thomas et al., 1974a) and influenza virus (Mackenzie et al., 1976; Mackenzie and Flower, 1979).

Cellular immunity, as evaluated by phytohemagglutinin (PHA)-induced lymphoproliferative response or development of tumor-specific cytotoxic T cells, was initially increased but, on continued exposure, greatly decreased in mice exposed to cigarette smoke (Chalmer et al., 1975; Holt et al., 1975; Thomas et al., 1973). Lymphocytes from mice exposed chronically to tobacco smoke have a decreased response to the mitogen PHA and release factors that inhibit the cytotoxic activity of NK cells against tumor cells. T-cell suppression may be due to defective antigen processing or antibody production.

The humoral immune response is also suppressed by chronic exposure of mice to tobacco smoke, while acute exposures may stimulate the humoral response. The primary and secondary antibody production by lymphocytes in the lung, lymph nodes, and spleen of mice exposed to tobacco smoke for longer than 26 weeks and challenged by inoculation with sheep erythrocytes was decreased (Thomas et al., 1975).

Laboratory test animals can be used to demonstrate the ability of cigarette smoke to slow the mucociliary clearance of particles from the lung and to alter the function of pulmonary macrophages. This effect has been observed in humans (Bohning et al., 1982; Cohen et al., 1979) and animals (Mauderly et al., 1989). In the latter study, rats exposed for eight weeks to cigarette smoke were exposed one time to cerium[144] dioxide particles. Smoking increased the half-time of the short-term clearance of these particles by 63% and long-term clearance twofold. The slowing of clearance of inhaled particles is an adverse health effect that should be considered in studies of tobacco product toxicity.

Alveolar macrophages from rats exposed to tobacco smoke for six months have a decreased ability to phagocytize *S. tuphylocoesus aureus* (Drath et al., 1979; Huber et al., 1980). Alveolar macrophages from rats exposed to tobacco smoke for 36 days or longer had an increased ability to release reactive oxygen species, a property dependent on the particulate faction of the smoke and not the gases. Clearance of *Pseudomonas aeruginosa* from rodents exposed to cigarette smoke for 36 weeks was slower than in controls (Holt and Keast, 1977). The decreased ability to clear particles, including pathogens, and the increased release of reactive oxygen species contribute to enhancement of inflammatory processes in the lung.

Reproductive and Developmental Effects

There are several reports that exposure of pregnant rats to either sidestream or mainstream tobacco smoke results in decreased birthweight of the pups (Leichter, 1989; Rajini et al., 1994; Reznik and Marquard, 1980; Witschi et al., 1994). Thus, animal models are capable of detecting tobacco smoke-induced growth retardation in utero. Rat models have also been used to demonstrate the adverse effects of maternal tobacco smoke exposure on lung maturation in utero (Lichtenbeld and Vidic, 1989), leading to an increase interstitial volume in the lung parenchyma. Another study demonstrated that postnatal rats exposed to tobacco smoke had reduced proliferation of their bronchiolar epithelial cells accompanied by increased levels of cytochrome P-450 enzymes (Ji et al., 1994). Studies in rodents have shown that rat pups exposed in utero to tobacco smoke have altered composition of pulmonary surfactant (Subramaniam et al., 1999).

The committee concludes that such studies indicate the potential usefulness of animal models to detect the interference of tobacco smoke products on airway epithelial cell development.

SYNERGISTIC EFFECTS WITH OTHER POLLUTANTS

Occupational exposures to materials such as asbestos or radon daughters have proven to have a synergistic interaction with tobacco smoke leading to greatly increased production of lung tumors in exposed workers who also smoke. Although rats are not good models for detecting the induction of lung tumors from cigarette smoke alone, rats exposed to both cigarette smoke and plutonium oxide particles clearly revealed the synergistic effects of cigarette smoke on the induction of lung tumors in combined exposures (Finch et al., 1998). In the past, it has not been customary for regulatory agencies to require testing for synergistic effects of a new product with other substances.

The committee concludes that in the case of tobacco smoke, for which several synergisms are known, it would be wise to consider adding such a test to the standard regimen.

MOLECULAR BIOLOGY TESTING TOOLS

Recent advances in the area of molecular biology hold promise as future tools for toxicity screening. The technology for producing transgenic mice allows one to gain gene functions, while the development of knockout mice allows one to delete gene functions (Arbeit and Hirose, 1999). At present these tools are better suited for mechanistic studies than for screening purposes, but in the future, genetically altered animals may become the standards for testing for specific types of toxicity, just as genetically altered *Salmonella* bacteria have become standards for testing the mutagenic potential of xenobiotics.

DNA microarray chips, which consist of an array of thousands of specific cDNA sequences or genes on a chip, allow one to detect and quantitate messenger RNAs that are the transcription products of the specific cDNA samples on the surface of the chips. Thus, if one knows the specific genes that are upregulated in association with the onset of a disease process, one could theoretically use the microarray technique to detect some of the earliest indicators that a disease process has begun (Nuwaysir et al., 1999). This type of tool should be invaluable in developing rapid screens for early indicators of developing disease in exposed animals (or for clinical purposes in humans) in contrast to the long time frame required to detect indicators of established disease in laboratory animals. The field is developing rapidly, and some microarrays designed to detect squamous cell carcinomas of the lung have already been reported (Wang et al., 2000). Future research will be required to determine which genes are upregulated at different times during the progression of specific diseases so that microchip arrays can be designed as accurate and specific preclinical indicators of developing disease.

SMOKELESS TOBACCO TOXICITY

Smokeless tobacco products, traditionally, are differentiated into snuff and chewing tobacco; are not combusted but exert their effects by direct mucosal contact and consequent entry of toxicants into the bloodstream. Snuff is typically a finely ground tobacco product that is used orally or nasally. Snuff is manufactured in a variety of forms including moist, dry, and fine cut (Connolly et al., 1986). The oral tobacco form that is chewed or simply kept in the mouth is generally known as chewing tobacco. Chewing tobacco is also produced in different forms including plug, loose-leaf, and twist varieties (Connolly et al., 1986). (See Chapter 4 for a more in-depth description of smokeless tobacco products and use statistics.)

Smokeless tobacco products are composed primarily of fire or air-cured dark tobacco (Wahlberg and Ringberger, 1999). The tobacco then undergoes an extended aging process that involves heating or fermentation depending on the product. During production, various additives are used for the desired flavor and aroma. The chemical composition of smokeless tobacco products varies due to differences in tobacco composition and cut, additives, and curing or processing conditions. The differences are also found among countries for similar reasons. In Sweden, for example, moist snuff (snus) has a lower level of tobacco-specific nitrosamines (TSNAs) due to processing differences and lack of fermentation compared to snuff in other countries. Generally, the TSNA levels in both U.S. and Swedish products have decreased over the last decade secondary to changes in processing methods, and TSNA levels in certain U.S. brands of snuff have approached the Swedish variety (Ahlborn et al., 1997; Wahlberg and Ringberger, 1999).

The exact chemical composition of smokeless tobacco, as in tobacco smoke, is difficult to assess. The main target of exposure in the smokeless tobacco user is the oral cavity and the upper aerodigestive tract. The lower digestive tract, however, is exposed at a certain level because of the swallowing of snuff particles within saliva. Common carcinogens found in smokeless tobacco include TSNAs, PAHs (especially benzo[a]pyrene [BaP]), and polonium -210. The concentration of TSNAs in snuff ranges from 5,280 to 141,000 parts per billion (ppb), which is hundreds to thousands times higher than that allowed in other consumer and food products (Connolly et al., 1986). TSNAs are thought to be important carcinogens in humans and have been proven to be potent carcinogens in animal studies. Among the TSNAs, NNK and N-nitrosonornicotine (NNN) have proven to be the most important carcinogens in smokeless tobacco in Europe and North America (Hoffmann et al., 1987; Nilsson, 1998). Hecht et al (1986) showed that oral exposure to NNK and NNN in rats caused lung tumors as well as oral tumors at the site of exposure. A snuff user (10 grams of snuff per day) is exposed to 24-46 μg of TSNAs per day compared to a pack per day smoker who is exposed to, on average, 16.2 μg of nitrosamines (Hoffmann et al., 1995). Snuff use also exposes the user to trace amounts of lead, cadmium, and selenium (Hoffmann et al., 1987).

Dark tobacco has a high level of nicotine, with 3.5-4.0% reported in certain brands (Wahlberg and Ringberger, 1999). In general the nicotine content per dose of smokeless tobacco product is higher than that of cigarettes, but the maximum serum nicotine levels are similar among all tobacco users (Benowitz, 1997). While there are interindividual differences in nicotine absorption and metabolism, nicotine is absorbed more gradually from smokeless tobacco than from smoking, and blood concentration persists over a longer period of time and even overnight (NIH Consensus Report, 1986). Overall, smokeless tobacco users are exposed to a greater amount of nicotine because of continued slow absorption of nicotine up to an hour after the tobacco is taken out of the mouth as well as the more alkaline pH, causing nicotine to be present in its unprotonated form contributing to better absorption (Benowitz et al., 1988; Hoffmann and Djordjevic, 1997).

Long-term smokeless tobacco use has been linked to oropharyngeal cancer (IARC, 1985; Mattson and Winn, 1989). The evidence has been more convincing for snuff than for chewing tobacco. As outlined in the 1986, National Institute of Health (NIH) Consensus Development Conference on the Health Implications of Smokeless Tobacco, case reports and controlled studies have consistently reported tumor growth in the location of smokeless tobacco contact with mucosa or skin, resulting in a risk of oral cancer up to 4.2 times that of nontobacco users as reported in one influential study (Winn et al., 1981). The most common type of cancer attributable to smokeless tobacco is oral squamous cell carcinoma, but verrucous carcinoma has also been reported (Connolly et al., 1986). In contrast, recent epidemiological studies from Sweden have failed to confirm a link between Swedish snus use and cancer (Lewin et al., 1998; Schildt et al., 1998). In a large population-based study looking at risk factors for squamous cancer of the head and neck, Lewin et al. (1998) found no increased risk with the use of Swedish snuff.

Results of animal studies were initially mixed regarding the effects of oral, subcutaneous, or topical administration of smokeless tobacco in rodents (Main and Lecavalier, 1988; Pershagen, 1996). More recently, however, as noted in a review of the health hazards of moist snuff by Ahlborn et al., (1997), there as been increased experimental support for the carcinogenicity of snuff. Surgically formed canals in the lips of rodents into which snuff and snuff extracts are placed have been used to more closely model the human practice of snuff dipping (Hoffmann and Djordjevic, 1997). Studies in rats and hamsters have shown a higher incidence of malignancy when there was exposure to both tobacco and herpes simplex type 1 virus or a cancer initiator (Ahlborn et al., 1997; Connolly et al., 1986). There has been inconclusive evidence linking snuff use to a variety of other cancers including prostate, pancreas, bladder, stomach, and kidney (IARC, 1985; Nilsson, 1998; Winn, 1997).

Smokeless Tobacco Research Recommendations

In terms of smokeless tobacco use as a strategy for harm reduction, more research is needed to investigate further some of the contradictory findings regarding the risk of oral cancer and cardiovascular disease. Swedish snus (lower TSNA and nicotine levels than American brands) should be evaluated as a possible harm reduction product since two recent epidemiological studies have suggested that it does not increase the risk of oral cancer and has favorable cardiovascular risk outcomes. More investigations into the cellular toxicity and genotoxic potential of smokeless tobacco extracts are needed. Smokeless tobacco may be a valid substitute for cigarette smoking but would pose specific risks in certain groups including pregnant women, those with inflammatory bowel disease, and those with established cardiovascular disease. Also, the population risks include concomitant smoking and adolescent use of smokeless tobacco as a gateway to cigarette smoking.

The same types of animal studies used to evaluate the toxicity of inhaled tobacco smoke could be done to evaluate the toxicity of smokeless tobacco products, such as Swedish snus or snuff, with a change in emphasis to the oral route of delivery. Based on known adverse health effects in humans, animal tests would be needed to evaluate the potential for smokeless tobacco to cause chronic inflammation or cancer in tissues of the oral cavity. Toxicokinetic studies would be required to determine other potential target organs for extracts of smokeless tobacco. Examples of in vitro studies include reports showing that smokeless tobacco causes pro-inflammatory changes in cultured endothelial cells (Furie et al., 2000) and activates the complement cascade, suggesting an inflammatory potential (Chang et al., 1998). Animal studies have been used to evaluate the potential of smokeless tobacco to induce oral cancer, as reviewed by Grasso and Mann (1998).

Thus, the committee concludes that preclinical toxicity testing should be of value for assessing the potential adverse health effects from use of smokeless tobacco.

GENERAL RESEARCH AGENDA AND RECOMMENDATIONS

Toxicology studies, both in vitro and in vivo, provide the opportunity to evaluate the potential harm reduction offered by potential reduced-exposure products (PREPs). The comparative potency of the PREP can be determined in a series of preclinical studies that include both the PREP and the standard tobacco product that can be replaced by the PREP, particularly tobacco-related PREPs (Figure 10-2). Such tests have recently been reported for a new cigarette-like product (Eclipse Expert Panel, 2000). The preclinical tests should include in vitro tests in both animal and human cells to determine the cytotoxicity and the genotoxicity of the tobacco product to which humans will be exposed. Such a test must include dose-response studies to determine the amount of the exposure material required to cause toxicity. Next, studies should be conducted in vivo in the best animal models available to determine the comparative potency of the PREP versus the standard product in producing: (1) pulmonary inflammation, (2) COPD, (3) cardiovascular disease, (4) reproductive toxicology, and (5) pulmonary neoplasms. If these preclinical studies indicate that the PREP is less potent than the standard tobacco product, clinical studies should be conducted to determine acute toxic effects, the toxicokinetic properties, or the adverse effects of the PREP in humans. The determination

	In Vitro Studies		In Vivo Studies	Clinical Studies In Humans	Epidemio- logical Studies
	Animal Cells	Human Cells	Animals	Acute Effects	Chronic Effects
Standard Tobacco Product	✓	✓	✓	✓	✓
Modified Tobacco Product	✓	✓	✓	✓	?

FIGURE 10-2 Preclinical studies for standard and new tobacco products.

of human health effects from chronic use of the new product can only be inferred from comparisons of the results of comparative tests in the old and the new product, but cannot be determined directly. Thus the testing approach will allow the rejection of risk reduction claims for products that are as toxic or more toxic in preclinical tests compared to products already on the market; however, only after long-term use of the product by many people could it be determined if the chronic toxicity of the new product is less than that of the standard product.

Based on the above information, it is clear that preclinical toxicity testing in vitro and in vivo can be done to assess the potential health effects of new products before they are released for human use. It is beyond the scope of the committee's task to recommend the specific set of toxicity tests that should be done on new or existing tobacco products. The committee recommends that a panel of experts be convened to determine the specific set of toxicity tests and details of the testing regimens. Details to be considered include species and strains of test animals, duration of tests, end points of interest, dose-response considerations, biomarkers of dosimetry and response, and standard comparison products to be tested as positive and negative controls.

REFERENCES

Ahlborn A, Olsson UA, Pershagen G. 1997. *Health Hazards of Moist Snuff.* September, 1996: National Board on Health and Welfare. Socialstyrelsen.

Ames BN, Magaw R, Gold LS. 1987. Ranking possible carcinogenic hazards. *Science* 236(4799):271-280.

Arbeit JM, Hirose R. 1999. Murine mentors: transgenic and knockout models of surgical disease. *Ann Surg* 229(1):21-40.

Balls M, Clothier R. 1992. Cytotoxicity assays for intrinsic toxicity and irritancy. Watson RR, ed. *In Vitro Methods of Toxicology.* Boca Raton: CRC Press. Pp. 37.

Baker RR. 1999. Smoke chemistry. Davis DL, Nielsen MT, eds. *World Agriculture Series; Tobacco Production, Chemistry and Technology.* London: Blackwell Science. Pp 418-419.

Belinsky SA, Nikula KJ, Palmisano WA, et al. 1998. Aberrant methylation of p16(INK4a) is an early event in lung cancer and a potential biomarker for early diagnosis. *Proc Natl Acad Sci U S A* 95(20):11891-11896.

Benowitz NL. 1997. Systemic absorption and effects of nicotine from smokeless tobacco. *Adv Dent Res* 11(3):336-341.

Benowitz NL, Porchet H, Sheiner L, Jacob P. 1988. Nicotine absorption and cardiovascular effects with smokeless tobacco use: comparison with cigarettes and nicotine gum. *Clin Pharmacol Ther* 44(1):23-28.

Bohning DE, Atkins HL, Cohn SH. 1982. Long-term particle clearance in man: normal and impaired. *Ann Occup Hyg* 26(1-4):259-271.

Cadet J, D'Ham C, Douki T, Pouget JP, Ravanat JL, Sauvaigo S. 1998. Facts and artifacts in the measurement of oxidative base damage to DNA. *Free Radic Res* 29(6):541-550.

Caldwell WS, Conner, JM. 1990. Artifact formation during smoke trapping: an improved method for the determination of N-nitrosamines in cigarette smoke. *J Assoc Off Anal Chem* 73:783-789.

Chalmer J, Holt PG, Keast D. 1975. Cell-mediated immune responses to transplanted tumors in mice chronically exposed to cigarette smoke. *J Natl Cancer Inst* 55(5):1129-1134.

Chang T, Chowdhry S, Budhu P, Kew RR. 1998. Smokeless tobacco extracts activate complement in vitro: a potential pathogenic mechanism for initiating inflammation of the oral mucosa. *Clin Immunol Immunopathol* 87(3):223-229.

Chen BT, Benz MV, Finch FL, et al. 1995. Effect of exposure mode on amounts of radiolabeled cigarette particles in lungs and astrointestinal tracts of F344 rats. *Inhalat Toxicol* 7:1095-1108.

Cohen D, Arai SF, Brain JD. 1979. Smoking impairs long-term dust clearance from the lung. *Science* 204(4392):514-517.

Connolly GN, Winn DM, Hecht SS, Henningfield JE, Walker B, Hoffmann D. 1986. The reemergence of smokeless tobacco. *N Engl J Med* 314(16):1020-1027.

Davis DL, Nielsen MT. 1999. *World Agriculture Series; Tobacco Production, Chemistry and Technology.* London: Blackwell Science.

Dendo RI, Phalen RF, Mannix RC, Oldham MJ. 1998. Effects of breathing parameters on sidestream cigarette smoke deposition in a hollow tracheobronchial model. *Am Ind Hyg Assoc J* 59(6):381-387.

Drath DB, Karnovsky ML, Huber GL. 1979. Tobacco smoke. Effects on pulmonary host defense. *Inflammation* 3(3):281-288.

Dube MF, Green CR. 1982. Methods of collection of smoke for analytical puproses. *Rec Adv Tob Sci* 8:42-102.

Eclipse Expert Panel. 2000. A safer cigarette? A comparative study. A consensus report. *Inhalation Toxicology* 12 (Supplement 5):1-48.

Escolar JD, Martinez MN, Rodriguez FJ, Gonzalo C, Escolar MA, Roche PA. 1995. Emphysema as a result of involuntary exposure to tobacco smoke: morphometrical study of the rat. *Exp Lung Res* 21(2):255-273.

Finch GL, Lundgren DL, Barr EB, Chen BT, Griffith WC, Hobbs, CH, Hoover, MD, Nikula, KJ, Mauderly JC. 1998. Chronic cigarette smoke exposures increases the pulmonary retention and radiation dose of 239Pu inhaled as 239PuO2 by F344 rats. *Health Phys* 75(6):597-609.

Furie MB, Raffanello JA, Gergel EI, Lisinski TJ, Horb LD. 2000. Extracts of smokeless tobacco induce pro-inflammatory changes in cultured human vascular endothelial cells. *Immunopharmacology* 47(1):13-23.

Grasso P, Mann AH. 1998. Smokeless tobacco and oral cancer: an assessment of evidence derived from laboratory animals. *Food Chem Toxicol* 36(11):1015-1029.

Guerin M, Maddox WL, Stokely J. 1974. Tobacco Smoke Inhalation Exposure: Concepts and Devices. Proceedings of the Tobacco Smoke Inhalation Workshop on Experimental Methods in Smoke and Health Research. DHEW Publication No. (NIH) 75-906. Pp. 31-48.

Halliwell B. 1998. Can oxidative DNA damage be used as a biomarker of cancer risk in humans? Problems, resolutions and preliminary results from nutritional supplementation studies. *Free Radic Res* 29(6):469-486.

Hayashi M, MacGregor JT, Gatehouse DG, et al. 2000. In vivo rodent erythrocyte micronucleus assay. II. Some aspects of protocol design including repeated treatments, integration with toxicity testing, and automated scoring. *Environ Mol Mutagen* 35(3):234-252.

Hecht SS, Rivenson A, Braley J, DiBello J, Adams JD, Hoffmann D. 1986. Induction of oral cavity tumors in F344 rats by tobacco-specific nitrosamines and snuff. *Cancer Res* 46(8):4162-4166.

Henderson RF. 1989. Bronchoalveolar Lavage: A tool for assessing the health status of the lung. McClellan RO, Henderson RF, eds. *Concepts in Inhalation Toxicology*. New York: Hemisphere Publishing Corporation. Pp. 415-444.

Hoffmann D, Adams JD, Lisk D, Fisenne I, Brunnemann KD. 1987. Toxic and carcinogenic agents in dry and moist snuff. *J Natl Cancer Inst* 79(6):1281-1286.

Hoffmann D, Djordjevic MV. 1997. Chemical composition and carcinogenicity of smokeless tobacco. *Adv Dent Res* 11(3):322-329.

Hoffmann D, Djordjevic MV, Fan J, Zang E, Glynn T, Connolly GN. 1995. Five leading U.S. commercial brands of moist snuff in 1994: assessment of carcinogenic N-nitrosamines. *J Natl Cancer Inst* 87(24):1862-1869.

Holt PG, Chalmer J, Keast D. 1975. Development of two manifestations of T-lymphocyte reactivity during tumor growth: altered kinetics associated with elevated growth rates. *J Natl Cancer Inst* 55(5):1135-1142.

Holt PG, Keast D. 1977. Environmentally induced changes in immunological function: acute and chronic effects of inhalation of tobacco smoke and other atmospheric contaminants in man and experimental animals. *Bacteriol Rev* 41(1):205-216.

Huber GL, Drath D, Davies P, Hayashi M, Shea J. 1980. The alveolar macrophage as a mediator of tobacco-induced lung injury. *Chest* 77(2 Suppl):272.

IARC (International Agency for Research on Cancer). 1986. *Tobacco Smoking: Monographs on the Evaluation of Carcinogenic Risk of Chemicals to Humans*. Vol. 38 ed. Lyon, France: IARC.

IARC. 1985. Tobacco habits other than smoking; betel-quid and areca-nut chewing; and some related nitrosamines. IARC Working Group. 23-30 October 1988. *IARC Monogr Eval Carcinog Risk Chem Hum* 37:1-268.

Ji CM, Plopper CG, Witschi HP, Pinkerton KE. 1994. Exposure to sidestream cigarette smoke alters bronchiolar epithelial cell differentiation in the postnatal rat lung. *Am J Respir Cell Mol Biol* 11(3):312-320.

Johnson JD, Houchens DP, Kluwe WM, Craig DK, Fisher GL. 1990. Effects of mainstream and environmental tobacco smoke on the immune system in animals and humans: a review. *Crit Rev Toxicol* 20(5):369-395.

Kendrick J, Nettesheim P, Guerin M, et al. 1976. Tobacco smoke inhalation studies in rats. *Toxicol Appl Pharmacol* 37(3):557-569.

Leichter J. 1989. Growth of fetuses of rats exposed to ethanol and cigarette smoke during gestation. *Growth Dev Aging* 53(3):129-134.

Lewin F, Norell SE, Johansson H, Gustavsson P, Wennerberg J, Biorklund A, Rutqvist LE. 1998. Smoking tobacco, oral snuff, and alcohol in the etiology of squamous cell carcinoma of the head and neck: a population-based case-referent study in Sweden. *Cancer* 82(7):1367-1375.

Lewis JG, Graham DG, Valentine WM, Morris RW, Morgan DL, Sills RC. 1999. Exposure of C57BL/6 mice to carbon disulfide induces early lesions of atherosclerosis and enhances arterial fatty deposits induced by a high fat diet. *Toxicol Sci* 49(1):124-132.

Lichtenbeld H, Vidic B. 1989. Effect of maternal exposure to smoke on gas diffusion capacity in neonatal rat. *Respir Physiol* 75(2):129-140.

Loft S, Deng XS, Tuo J, Wellejus A, Sorensen M, Poulsen HE. 1998. Experimental study of oxidative DNA damage. *Free Radic Res* 29(6):525-539.

Lovell DP, Yoshimura I, Hothorn LA, Margolin BH, Soper K. 2000. Report and summary of the major conclusions from statistics in genotoxicity testing working group from the International Workshop on Genotoxicity Test Procedures (IWGTP), March 1999. *Environ Mol Mutagen* 35(3):260-263.

Mackenzie JS, Flower RL. 1979. The effect of long-term exposure to cigarette smoke on the height and specificity of the secondary immune response to influenza virus in a murine model system. *J Hyg (Lond)* 83(1):135-141.

MacKenzie JS, MacKenzie IH, Holt PG. 1976. The effect of cigarette smoking on susceptibility to epidemic influenza and on serological responses to live attenuated and killed subunit influenza vaccines. *J Hyg (Lond)* 77(3):409-417.

Main JH, Lecavalier DR. 1988. Smokeless tobacco and oral disease. A review. *J Can Dent Assoc* 54(8):586-591.

March TH, Barr EB, Finch GL, et al. 1999a. Cigarette smoke exposure produces more evidence of emphysema in B6C3F1 mice than in F344 rats. *Toxicol Sci* 51(2):289-299.

March TH, Kolar LM, Barr EB, Finch GL, Menache MG, Nikula KJ. 1999b. Enhanced pulmonary epithelial replication and axial airway mucosubstance changes in F344 rats exposed short-term to mainstream cigarette smoke. *Toxicol Appl Pharmacol* 161(2):171-179.

Mattson ME, Winn DM. 1989. Smokeless tobacco: association with increased cancer risk. *NCI Monogr* (8):13-16.

Mauderly JL, Chen BT, Hahn FF, et al. 1989. The effect of chronic cigarette smoke inhalation on the long-term pulmonary clearance of inhaled particles in the rat. Wehner AP, ed. *Biological Interaction of Inhaled Mineral Fibers and Cigarette Smoke*. Richland, WA: Battelle Press. Pp. 223-239.

McAllister-Sistilli CG, Caggiula AR, Knopf S, Rose CA, Miller AL, Donny EC. 1998. The effects of nicotine on the immune system. *Psychoneuroendocrinology* 23(2):175-187.

National Council on Radiation Protection and Measurements. 1997. Deposition, Retention, and Dosimetry of Inhaled Radioactive Substances. Report #125. Bethesda, MD.

Nilsson R. 1998. A qualitative and quantitative risk assessment of snuff dipping. *Regul Toxicol Pharmacol* 28(1):1-16.

NIH (National Institutes of Health). 1986. Health implications of smokeless tobacco use. *NIH Consensus Statement* 6(1):1-17.

Norman A. 1999. Cigarette design and materials. Davis DL, Nielson MT, eds. *Tobacco: Production, Chemistry, and Technology*. Oxford: Blackwell Science Ltd. Pp. 353-387.

Nuovo GJ, Plaia TW, Belinsky SA, Baylin SB, Herman JG. 1999. In situ detection of the hypermethylation-induced inactivation of the p16 gene as an early event in oncogenesis. *Proc Natl Acad Sci U S A* 96(22):12754-12759.

Nuwaysir EF, Bittner M, Trent J, Barrett JC, Afshari CA. 1999. Microarrays and toxicology: the advent of toxicogenomics. *Mol Carcinog* 24(3):153-159.

Penn A, Chen LC, Snyder CA. 1994. Inhalation of steady-state sidestream smoke from one cigarette promotes arteriosclerotic plaque development. *Circulation* 90(3):1363-1367.

Penn A, Snyder CA. 1993. Inhalation of sidestream cigarette smoke accelerates development of arteriosclerotic plaques. *Circulation* 88(4 Pt 1):1820-1825.

Pershagen G. 1996. Smokeless tobacco. *Br Med Bull* 52(1):50-57.

Phalen R. 2000. Physical-chemical characteristics of tobacco smoke. Open Meeting of the IOM Committee to Assess the Science Base for Tobacco Harm Reduction: Washington, DC.

Phillips DH, Farmer PB, Beland FA, et al. 2000. Methods of DNA adduct determination and their application to testing compounds for genotoxicity. *Environ Mol Mutagen* 35(3):222-233.

Rajini P, Last JA, Pinkerton KE, Hendricks AG, Witschi H. 1994. Decreased fetal weights in rats exposed to sidestream cigarette smoke. *Fundam Appl Toxicol* 22(3):400-404.

Reznik G, Marquard G. 1980. Effect of cigarette smoke inhalation during pregnancy in Sprague-Dawley rats. *J Environ Pathol Toxicol* 4(5-6):141-152.

Schildt EB, Eriksson M, Hardell L, Magnuson A. 1998. Oral snuff, smoking habits and alcohol consumption in relation to oral cancer in a Swedish case-control study. *Int J Cancer* 77(3):341-346.

Silbaugh SA, Horvath SM. 1982. Effect of acute carbon monoxide exposure on cardiopulmonary function of the awake rat. *Toxicol Appl Pharmacol* 66(3):376-382.

Sjostrand M, Rylander R. 1997. Pulmonary cell infiltration after chronic exposure to (1->3)-beta-D-glucan and cigarette smoke. *Inflamm Res* 46(3):93-97.

Sopori ML, Goud NS, Kaplan AM. 1994. Effects of tobacco smoke on the immune system. Dean JH, Luster MI, Munson AE, Kimber I, eds. *Immunotoxicology and Immunopharmacology*. Second Edition ed. New York: Raven Press, Ltd. Pp. 413-434.

Subramaniam S, Srinivasan S, Bummer PM, Gairola CG. 1999. Perinatal sidestream cigarette smoke exposure and the developing pulmonary surfactant system in rats. *Hum Exp Toxicol* 18(4):206-211.

Subramaniam S, Whitsett JA, Hull W, Gairola CG. 1996. Alteration of pulmonary surfactant proteins in rats chronically exposed to cigarette smoke. *Toxicol Appl Pharmacol* 140(2):274-280.

Swafford DS, Middleton SK, Palmisano WA, et al. 1997. Frequent aberrant methylation of p16INK4a in primary rat lung tumors. *Mol Cell Biol* 17(3):1366-1374.

Thomas WR, Holt PG, Keast D. 1973. Cellular immunity in mice chronically exposed to fresh cigarette smoke. *Arch Environ Health* 27(6):372-375.

Thomas WR, Holt PG, Keast D. 1974a. Development of alterations in the primary immune response of mice by exposure to fresh cigarette smoke. *Int Arch Allergy Appl Immunol* 46(4):481-486.

Thomas WR, Holt PG, Keast D. 1975. Humoral immune response of mice with long-term exposure to cigarette smoke. *Arch Environ Health* 30(2):78-80.

Thomas WR, Holt PG, Papadimitriou JM, Keast D. 1974b. The growth of transplanted tumours in mice after chronic inhalation of fresh cigarette smoke. *Br J Cancer* 30(5):459-462.

Tice RR, Agurell E, Anderson D, et al. 2000. Single cell gel/comet assay: guidelines for in vitro and in vivo genetic toxicology testing. *Environ Mol Mutagen* 35(3):206-221.

Vial WC. 1986. Cigarette smoking and lung disease. *Am J Med Sci* 291(2):130-142.

Wahlberg I, Ringberger T. 1999. Smokeless tobacco. Davis DL, Nielsen MT, eds. *World Agriculture Series; Tobacco Production, Chemistry and Technology*. London: Blackwell Science. Pp 452-460.

Wang T, Hopkins D, Schmidt C, Silva S, Hougton R,Takita H, Repasky E, Reed SG. 2000. Identification of genes differentially over-expressed in lung squamous cell carcinomas using combination of cDNA subtraction and miroarray analysis. *Oncogene* 19:1519-1528.

Winn DM. 1997. Epidemiology of cancer and other systemic effects associated with the use of smokeless tobacco. *Adv Dent Res* 11(3):313-321.

Winn DM, Blot WJ, Shy CM, Pickle LW, Toledo A, Fraumeni JF. 1981. Snuff dipping and oral cancer among women in the southern United States. *N Engl J Med* 304(13):745-749.

Witschi H, Espiritu I, Maronpot RR, Pinkerton KE, Jones AD. 1997a. The carcinogenic potential of the gas phase of environmental tobacco smoke. *Carcinogenesis* 18(11):2035-2042.

Witschi H, Espiritu I, Peake JL, Wu K, Maronpot RR, Pinkerton KE. 1997b. The carcinogenicity of environmental tobacco smoke. *Carcinogenesis* 18(3):575-586.

Witschi H, Espiritu I, Uyeminami D. 1999. Chemoprevention of tobacco smoke-induced lung tumors in A/J strain mice with dietary myo-inositol and dexamethasone. *Carcinogenesis* 20(7):1375-1378.

Witschi H, Lundgaard SM, Rajini P, Hendrickx AG, Last JA. 1994. Effects of exposure to nicotine and to sidestream smoke on pregnancy outcome in rats. *Toxicol Lett* 71(3):279-286.

Witschi H, Uyeminami D, Moran D, Espiritu I. 2000. Chemoprevention of tobacco-smoke lung carcinogenesis in mice after cessation of smoke exposure. *Carcinogenesis* 21(5):977-982.

Zhu BQ, Sun YP, Sievers RE, Glantz SA, Parmley WW, Wolfe CL. 1994. Exposure to environmental tobacco smoke increases myocardial infarct size in rats. *Circulation* 89(3):1282-1290.

11

Exposure and
Biomarker Assessment in Humans

The evaluation of potential reduced-exposure agents (PREPs) has to be defined in the context of the outcome of interest (e.g., individual or population reduction in risk and disease type) and compared to an appropriate baseline (i.e., nonsmokers, former smokers, current smokers in the context of host susceptibility and previous level of smoke exposure). Tobacco exposure can be measured in the aggregate at the level of the entire population (e.g., through the measurement of tobacco sales or reported consumption in population-based surveys) and related to disease incidence or change in mortality rates. These methodologies are considered descriptive epidemiological tools that are useful in generating hypotheses and/or validating public health strategies, marketing programs, and so forth. Exposure can also be measured at the level of the individual through biomarker measurements. This type of assessment within epidemiological studies can be used for hypothesis generation or testing. A range of methodologies and assays can be used for assessing exposure, as well as a range of assays for assessing host susceptibilities to exposure.

The evaluation of a PREP can include four components: (1) external exposure measurements, (2) internal exposure measurements, (3) biomarkers estimating the biologically effective dose (Perera, 1987), and (4) biomarkers of potential harm (see Figure 11-1). The definitions of each are provided in Table 11-1 and explained further herein. There have been different definitions of types of exposure assessments used previously, but more recent understandings of biomarker uses and limitations, as

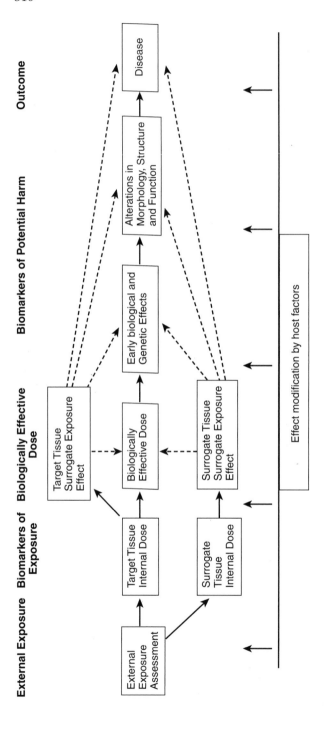

FIGURE 11-1 Assessing potential harm reduction products.

NOTE: Dashed lines indicate hypothetical indirect relationship. Solid lines indicate mechanistic direct relationship.

SOURCE: Modified with permission from Committee on Biological Markers of the National Research Council, 1987.

TABLE 11-1   Exposure and Biomarker Assessment Definitions

Exposure or Biomarker Assessment[a]	Definition
External exposure marker	A tobacco constituent or product that may reach or is at the portal of entry to the body
Biomarker of exposure	A tobacco constituent or metabolite that is measured in a biological fluid or tissue that has the potential to interact with a biological macromolecule; sometimes considered a measure of internal dose
Biologically Effective Dose (BED)	The amount that a tobacco constituent or metabolite binds to or alters a macromolecule; estimates of the BED might be performed in surrogate tissues
Biomarker of potential harm	A measurement of an effect due to exposure; these include early biological effects, alterations in morphology, structure, or function, and clinical symptoms consistent with harm; also includes "preclinical changes"

[a]Categories and definitions reflect concept that the critical exposure is at the level of a biological macromolecule, so that exposure for this discussion is not limited to a measurement at the portal of entry to the body.

well as different approaches needed for PREP evaluation lead to a need for clarification and redefinition. The latter three measurements improve upon the first by quantifying exposure at the cellular level to characterize low-dose exposures or low-risk populations, providing a relative contribution of individual chemical carcinogens from complex mixtures and estimating total burden of a particular exposure where there are many sources (Vineis and Porta, 1996). In assessing PREPs through biomarkers, understanding the biological effects of a wide range of exposures will be important. Within the context of this chapter, exposure at the level of the cell and critical macromolecules is considered with greater weight, rather than the traditional view of exposure at the portal of entry into a person.

Biomarkers are intuitively more informative and better disease risk markers when measured in the target tissue through biopsies (e.g., oral mucosa, lung, bladder). However, biomarker assays are technically limited, and target tissue can be difficult to obtain, especially in nondiseased smokers. Therefore, biomarker assays have been developed for surrogate tissues and fluids (e.g., expired breath, saliva, blood, urine). While these are technically simpler to use and easier to collect, the ability to prove a

TABLE 11-2   Measurements Used For Assessing Harm Reduction Products

Factor	Comment
Type of measurement	Types of measurements that can be used include external exposure assessment, biomarkers of exposure, biomarkers that represent the biologically effective dose, and biomarkers of potential harm. Depending on the context, the PREP, and the outcome of interest, different measurements might be more appropriate, although it is likely that a combination will be needed
Target tissue and outcome effect	Is the measurement used for detecting effects in target or surrogate tissues, and is this a measurement of pathogenesis?
Dose-response data	Measurements must have a dose-response relationship that is understood on a mechanistic basis. Biomarker should be able to demonstrate effects from exposure over the range of human experience, so that it can show exposure reduction from a PREP
Harm reduction in dose-response data	Biomarker should be able to predict a decrease in disease incidence after exposure is reduced
Specificity	Is the measurement specific for a tobacco product constituent, or does is also measure exposures from nontobacco products?
Sensitivity	Is the measurement sensitive enough to measure what it is supposed to measure in humans within the possible exposure ranges?
Validation	Are there sufficient data to show that the assay is reproducible?

predictive value for the potential harm reduction is more difficult. It should be noted that the biomarkers discussed in this chapter refer to either target or surrogate tissue or fluid assays, but that the biologically effective dose refers to the assessment of a mechanistically relevant biomarker only in the target organ.

The following factors should be considered when evaluating measurements for predicting or determining the effects of a PREP. Table 11-2 summarizes these factors and Table 11-3 provides an overview of available measures to predict the effects.

1. *Type of measurement.* Measurements are defined within four general categories—namely, external exposure, biomarkers of exposure, biomarkers estimating the biologically effective dose, and biomarkers of

potential harm. Placing a measurement solely within one category may not be possible. The external exposure assessment category is limited to those methods that are not detected by an assay using a body fluid or part. Although some external exposure methods might be poor predictors of disease risk and hence also poor measures for assessing a new product (e.g., the Federal Trade Commission [FTC] method described below), others might be strongly associated with disease risk and might therefore be better (e.g., cigarettes per day). While some external exposure assessments might be useful for harm reduction risk assessments (e.g., smoking history), they should not be used alone in assessing harm reduction because the predictive power for disease is not sufficient without corroborative biomarker data. Biomarkers of exposure are assayed in a body fluid (including exhaled air) or tissue that measures a constituent of tobacco smoke, tobacco-related products, or metabolites, where the constituent is not bound to a biomolecule. These biomarkers include unmetabolized compounds (e.g., carbon monoxide [CO], serum nicotine, carcinogen levels in serum or internal organs), biomarkers of exposure to individual cigarettes (e.g., incremental increases in exhaled CO or serum nicotine), and metabolites in any body fluid (e.g., cotinine in serum or urine, carcinogen metabolites in urine). Biomarkers assessing the biologically effective dose are those considered mechanistically related to disease outcomes (e.g., carcinogen-DNA adducts in the target tissue). Surrogate biologically effective doses, once validated, estimate a biological effect in a target organ (e.g., hemoglobin adducts or white blood cell carcinogen-DNA adducts). These biomarkers are in theory best able to link exposure (external and internal) to disease outcomes. Biomarkers of potential harm can reflect early or late damage (e.g., loss of heterozygosity in sputum, background mutations in nondiseased tissues, reactive airway disease, arrhythmias, premalignant lesions, mutations in premalignant lesions, chromosomal aberrations in smoking-damaged epithelium, hypermethylation of genes, atherosclerosis). In this context, potential harm implies that the assay might or might not reflect actual harm and that some change in physiological function, for example, might not represent a harmful effect.

2. *Target tissue and outcome relationship.* A biomarker assay should be shown to be relevant to the outcome of interest. Besides having a mechanistic relationship to pathogenesis, data should be available to determine the predictive capacity for disease and disease reduction. This validation includes supportive evidence that the assay reflects harm reduction, such as might be done in an experimental cell culture or animal study. Assays that measure the effects in target tissue would generally have the greatest weight to support the use of a PREP. Sometimes, the target tissue effect

TABLE 11-3 Methods for PREP Assessment

Category	Type of Measurement	Target vs. Surrogate	Examples
External exposure	External exposure assessment	Neither	Questionnaire data, FTC yield
Biomarker of exposure	Internal dose	Target tissue	Polycyclic aromatic hydrocarbon in lung tissue
		Surrogate tissue	Urinary measurement of tobacco constituent or metabolite, exhaled CO, carboxyhemoglobin, urinary mutagenicity
Biologically effective dose	Biologically effective dose	Target tissue	Carcinogen-DNA adducts in human lung tissue, exfoliated bladder cells, or oral mucosa
		Surrogate tissue	Carcinogen-DNA or hemoglobin adducts; DNA adducts; lipid peroxidation
Biomarker of potential harm	Early biological and genetic effects	Target tissue	Changes in RNA or protein expression, somatic mutations, and LOH in normally or abnormally appearing tissue; change in methylation or gene control; mitochondrial mutations, mRNA expression arrays, or proteomics
	Alterations in morphology, structure, or function	Target tissue	Osteoporosis, hypertension, cough, hyperplasia, dysplasia, lipids, blood coagulant pathways, mRNA expression arrays, or proteomics
	Surrogate assays	Surrogate tissue	Leukocytosis; HPRT mutations; chromosomal aberrations; circulating lymphocytes; mRNA or protein expression via microarrays in cultured blood cells
Effect modifiers	Measures of interindividual variation	Neither	Genetic polymorphisms for genes involved in disease pathways
		Target	Enzyme induction of metabolizing enzymes

NOTE: HPRT=hypoxanthine phosphoribosyltransferase; LOH=loss of heterozygosity.

Strengths	Limitations
Inexpensive	Does not reflect actual internal doses
Provides integrated measure of external exposure and smoking behavior	Expensive; may not be specific for tobacco products; does not necessarily reflect biologically effective dose; tissue may be difficult to access; may be difficult to validate as a risk marker for disease
Easily accessible; provides integrated measure of external exposure and smoking behavior; metabolites reflect host capacity for metabolism and clearance	May not be specific for tobacco products; does not necessarily reflect biologically effective dose; may be difficult to validate as a risk marker for disease
Reflects integrated measure of external exposure, smoking behavior, metabolic activation, DNA repair capacity, cell-cycle control, and capacity for apoptosis	Difficult to measure and validate as a disease risk marker, predictive value for disease risk is insufficiently studied, more commonly reflects internal dose to a target macromolecule rather than disease risk
Does not require invasive procedures, greater amount of tissue is generally available; more likely to be used in an epidemiological setting	Relationship to disease risk is not fully established
Assessment of mechanistic pathway leading to disease	Tissue difficult to obtain; technically difficult; relationship to disease risk difficult to establish; harmful effects may already be present; bioinformatics with which to process information not yet available
Greater ability to identify risk for disease with marker	Tissue difficult to obtain; late effects where harm has already occurred; bioinformatics with which to process information not yet available
Easily accessible; provides integrated measure of external exposure and smoking behavior; metabolites reflect host capacity for metabolism and clearance	Relationship to target organ effect is difficult to prove; specificity for tobacco product needs to be proved; bioinformatics not yet available
Reflects lifetime response to exposure; high throughput possible	Candidate gene approach will typically study many polymorphisms that are not related to disease risk
Integrated assessment of how prior exposures or genetic traits affect exposures and harm	Tissue technically difficult to obtain; laboratory validation difficult

might also be a surrogate for an effect in other tissues, and a surrogate tissue assay might reflect effects in multiple organs.

3. *Dose-response data for harm.* Assays that have a demonstrable dose-response relationship to actual disease outcomes is important for assessing a PREP and, if they do not, it should be shown to have a dose-response relationship to a biomarker of potential harm relevant to a disease pathway. The mechanistic basis for the relationship should be well understood in order to make meaningful interpretations of data used to assess a PREP. For example, assays that demonstrate a dose-response relationship between smoking and DNA damage in epithelial cells of a target organ could be useful. Methods assessing tobacco exposure as a complex mixture would have greater weight than a single component exposure.

4. *Dose-response data for harm reduction.* Assays that show a reduction in harm after reducing exposure to tobacco smoke or a tobacco product constituent would have the greatest weight, where the experimental design uses an initial dose level for a specific duration of time followed by exposure to a lower level at a later time. The intent is to simulate the effects of a person's switching from one level of exposure to another level of exposure. Importantly, the effects of the biomarker should be measurably different over the range of human exposures, so that the assessment can predictably measure the effects of exposure reduction from a PREP. Currently, there are some biomarker assays that have been assessed in former smokers or smoking cessation trials. These biomarker studies that indicate measurable decreases in effect can provide some information about the utility of markers for assessing exposure reduction. Included in this are half-life data, which must be measured and taken into account when evaluating a tobacco-related PREP. Methods assessing tobacco exposure as a complex mixture would have greater weight than a single component exposure.

5. *Specificity.* Consideration should be given to whether the effect is specific to a constituent of tobacco smoke or a tobacco product, or whether the method also measures exposure from other sources. Higher degrees of specificity are useful, although in some cases the method might be useful for assessing exposures from multiple sources other than tobacco in order to provide an understanding about relative contributions. Assays that are specific for tobacco's complex chemical mixture and those that are specific for a chemical or chemical class both have utility, but the former would have greater weight if appropriately validated, because persons are exposed simultaneously to all of the constituents. Validating assays for complex effects is more difficult because they may have less specificity for tobacco.

6. *Sensitivity.* The assay must be sufficiently sensitive to measure what it is supposed to measure in the human tissue of interest. This is especially problematic in measuring low-level effects, for example, in assessing the effects of environmental tobacco smoke (ETS) exposure.

7. *Validation.* It is critical that biomarkers for assessing PREPs be well validated in the laboratory. Validation includes proof that the assay measures what it claims to measure and that it is reproducible. Sensitivity, specificity, and predictive value are all important to consider.

EXTERNAL EXPOSURE ASSESSMENT: THE FTC METHOD AND QUESTIONNAIRE DATA

External exposure markers attempt to measure the amount of a tobacco smoke or tobacco product constituent that may enter at a portal to the body. However, these predictors generally do so without regard to most interindividual differences in smoking behavior and cellular processes. There are several types of external exposure assessment, some of which are listed in Table 11-4.

A common way to assess potential exposure to tobacco smoke is by measuring the yield of tobacco smoke constituents. One attempt to estimate delivered doses is the method adopted by the Federal Trade Commission in 1967. It was intended to provide a standardized estimate of tar and nicotine yield by cigarette brand, simulating a cursory observation of human smoking behavior. A cigarette is inserted into a smoking machine and lit, puffs are taken through a syringe (35 ml over 2 seconds, every 60 seconds) until the cigarette is "smoked" to a fixed length. Particulates are collected on a filter and weighed. Nicotine is assayed separately. Tar is measured as total particulate matter less nicotine, other alkaloids, and water. Although the machine provides yield data that can be used to compare one cigarette to another, this information has limited usefulness for understanding human exposure because people do not smoke cigarettes as the machine does due to different smoking behaviors. Smokers also can affect cigarette filter performance by covering ventilation holes in the filter with their lips or fingers, which would increase yields in vivo. Although FTC yields might define a comparative range of actual exposures, there is a wide overlap of actual to predicted yields among types of cigarettes (i.e., low, medium, and high yields), where smokers of low-nicotine cigarettes might have higher nicotine levels than those who smoke brands with higher FTC yields (Byrd et al., 1998, 1995). Altering the FTC method to simulate puffs and times for actual smokers results in

TABLE 11-4 External Exposure Assessment[a]

Category	Variables Used in Literature	Related to a Disease Outcome[b]	Strengths	Limitations
FTC machine method	Tar yield Nicotine yield Individual smoke constituent yield	Yes	Standardized method for yields	Little relationship to actual human experience
Subject smoking history	Cigarettes per day Years of smoking Age of initiation Recall of inhalation depth Usual type of cigarette smoked Quitting attempts Cumulative tar exposure	Yes	Inexpensive assessment; generally considered reliable, except in some circumstances listed in limitations	Recall is subject to self-perceptions of risk. Reporting is variable depending on context, such as smoking cessation program or where recall bias might exist in epidemiology studies. Known limitations for persons who are switching brands or altering smoking behavior; also not sufficiently reliable in smoking cessation studies. Thus, not sufficiently reliable in harm reduction studies
Smoking Topography	Puff duration Puffs per cigarette Interpuff interval Puff volume	No	Direct measure of inhalation exposure per cigarette. Can be used to assess effects of cigarette brand switching	Measurement performed in artificial environment

[a]References are not provided in this table but can be found in the text of this and disease-related chapters.
[b]Any report related to a disease outcome where the report is plausible but has not necessarily been replicated.

higher exposures to tar and specific carcinogens (e.g., tobacco-specific nitrosamines and benzo[a]pyrene) (Djordjevic et al., 2000; Fischer et al., 1989; Hoffman and Hoffman, 1997). For example, using modified protocols to stimulate human smoking behavior, the medium-yield (0.9-1.2 mg nicotine per cigarette) and low-yield (0.8 mg nicotine per cigarette) cigarettes deliver similar amounts of tar per day, although by FTC method measured per cigarette yields of tar, benzo[a]pyrene (BaP), and tobacco-specific nitrosamines (TSNAs) were higher in the former (Djordjevic et al., 2000). As cigarettes with different designs are developed and marketed, an assumption that the FTC method of estimating yields will be comparable to existing products is premature.

Over the last 30 years, data from surveys have been an important tool in the assessment of tobacco exposure among individuals and the population. They have been an effective means of tracking patterns of tobacco use and the societal perceptions that ultimately influence consumption. Individual exposure can be assessed through the measurement of the number of cigarettes smoked per day, duration of smoking, types or brands of cigarettes smoked (e.g., "tar" delivery, filter type, type of tobacco, mentholation), and age at initiation (IARC, 1986; Kaufman et al., 1989; La Vecchia et al., 1990; Lubin et al., 1984; Stellman and Garfinkel, 1989; U.S. DHHS, 1988; Vutuc and Kunze, 1983; Wilcox et al., 1988; Zang and Wynder, 1992). Lifetime exposures can be estimated by calculating pack-years (average packs per day multiplied by number of years smoked) or cumulative tar exposure (Zang and Wynder, 1992). A more detailed description of the most common surveys in use is presented in Table 11-5.

Most analyses indicate that self-report validity among adults is good (Patrick et al., 1994). Certain limitations, however, are evident in this type of exposure assessment (Giovino, 1999; U.S.DHHS, 1994). First, sampling errors may occur in any study in which generalizations are made from a selected population sample. One example is the over- or underrepresentation of certain groups, especially those that exhibit significant tobacco use or have differing smoking behavior. In fact, there is a built-in exclusion in many of the major surveillance tools of various segments of the population, such as the institutionalized mentally ill, prisoners, and those in areas of inadequate telephone coverage. Errors in response must be considered including memory errors, nonresponse errors, and misclassifications and inconsistencies in reporting. The validity of self-reported responses can be influenced by many factors (Velicer et al., 1992), particularly the respondent's perception of privacy (Giovino, 1999). This is especially a concern among adolescents in the home setting and among groups that have increased pressure to abstain or to quit, including pregnant women, adolescents, and patients with heart or lung disease. One

TABLE 11-5 Major Tobacco Use Surveys

Survey	Sponsor	Population
National Health Interview Survey (NHIS)	National Center for Health Statistics, Centers for Disease Control and Prevention (CDC)	Civilian, noninstitutionalized adults over age 18; children by proxy
Behavioral Risk Factor Surveillance System (BRFSS)	CDC and individual states	Noninstitutionalized adults over age 18
National Health and Nutrition Examination Survey (NHANES)	CDC	Age 2 and over
National Household Survey on Drug Abuse (NHSDA)	National Institute on Drug Abuse and Substance Abuse and Mental Health Services Administration	Noninstitutionalized civilian population over age 12
American Legacy Foundation Survey		Sixth to twelfth grade students
Monitoring the Future Survey (previously, the National High School Senior Survey)	University of Michigan Survey Research Center	Eighth, tenth, and twelfth grade students
Youth Risk Behavior Surveillance System (YRBSS)	CDC	Ninth to twelfth grade students

effort to validate self-report measures and to reveal any ETS exposure can be found in the National Health and Nutrition Examination Survey (NHANES; see Table 11-5), which collects serum cotinine levels of respondents (CDC, 2000; Giovino, 1999; Giovino et al., 1995; SAMHSA, 1998).

Population surveys have limited practicality in evaluating the consequence of tobacco exposure because of the relatively long time frame required. However, in context, population assessments have been studied extensively in relation to disease outcomes and thus can be considered a

Size	Setting	Comments
More than 38,000 families in 1998	Household interview with responses typed directly in laptop computer; annual	Excludes homeless not in shelters, military personnel, prisoners, hospital patients Data: cigarette, chewing tobacco, cigar, and pipe use since 1965 Oversampling of black American and Hispanic populations
	Computer-assisted telephone interviews; annual	Added smokeless tobacco use questions in 1987 State level
Approximately 40,000 participants between 1988 and 1994	Personal interview with physical exam and blood tests; periodic	Serum cotinine measurements Oversampling of children 1-5years, adults over age 60, black Americans, and Mexican Americans
Approximately 25,500 participants in 1998	Household interview; self-administered through a computer; annual	State level Oversampling of black Americans, Hispanic Americans, and youth
	School based	Evaluates knowledge of and attitudes towards all forms of tobacco, including bidis and Kreteks
Approximately 50,000 students from public and private high schools	Classroom based; self-administered; annual	Random sample from each senior class is followed after graduation for longitudinal data
	Classroom based; self-administered; biennial	Oversampling of black and Hispanic-American students. Combination of national, state, and local surveys

crude measurement of individual risk and a better measure of population risk. These surveys do provide insight into trends of tobacco product use within and across a variety of sociodemographic groups, including age, sex, race or ethnicity, educational status, and economic status. The data can be compared to morbidity and mortality registries to understand new or changing consequences of use patterns or specific products. In addition, these trends in prevalence, initiation, and cessation in turn aid in the evaluation of the effects of tobacco-related activities, policies, and interventions within the general population and its subgroups.

Methods for assessing external exposure (e.g., number of cigarettes per day) are widely used and relatively inexpensive but do not provide an assessment of how someone smokes cigarettes and how the body responds to exposure. Thus, these measures approximate the level of actual exposure and, as described below, become less reliable in assessing exposure reduction. Smoking topography is an additional method of assessing external exposure (e.g., how much smoke enters the lung, estimated by measuring puff volume, number of puffs per cigarette, puff duration, total inhalation time, and interpuff interval) (Bridges et al., 1990; Gritz et al., 1983; Herning et al., 1983; Hofer et al., 1992; Kolonen et al., 1992b). In the laboratory, if subjects smoke their own cigarettes, then it is presumed that the measurement reflects their usual smoking behavior. A limitation of smoking topography studies is that cigarettes are typically smoked via cigarette holders, which may influence puffing behaviors and prevent vent hole blocking that might normally occur when the cigarettes are smoked without the holder. Smoking topography studies have contributed to the findings that persons who switch from high-tar and nicotine to low-tar and nicotine cigarettes increase their intake of smoke per cigarette to compensate for a lower yield of nicotine (Benowitz et al., 1986a, b). It is well established that smokers self-titrate their blood nicotine levels, such that smokers of lower-nicotine cigarettes inhale more (Benowitz et al., 1983; Benowitz et al., 1986b; Benowitz et al., 1998; Ebert et al., 1983; Gritz et al., 1983; Hill and Marquardt, 1980), and altering topography leads to differences in nicotine absorption and CO boosts (Hofer et al., 1992; Kolonen et al., 1992a). Smoking lower-nicotine delivery cigarettes increases puff volume (Battig et al., 1982; Bridges et. al., 1986; Kolonen et al., 1992b) and, to a lesser extent, puff duration (Bridges et al., 1990). Using a multiple regression model for prediction of nicotine blood levels, the best-fit model incorporates interpuff interval, number of puffs per cigarette, puff volume, puff duration, inhaled volume, and inhalation duration (Herning et al, 1983). These studies are difficult to interpret, however, because cigarette and topographic parameters are interrelated (Bridges et al., 1990; Kolonen et al., 1992a; Nemeth-Coslett and Griffiths, 1984).

Different methods have been developed for the study of exposure to environmental tobacco smoke (EPA, 1992). Stationary and personal air monitors can be used to measure total particulates or individual constituents. Some measurements, such as nicotine, are more specific for ETS. Ambient air concentrations and personal exposures to polycyclic aromatic hydrocarbons (PAHs) and other tobacco constituents can be measured, but their relationship to disease risk has not been adequately studied.

BIOMARKERS OF EXPOSURE

Biomarkers of exposure, measured in a body fluid, tissue, or exhaled air, represent an internal dose of tobacco smoke or a tobacco product constituent that is either the parent compound or its metabolite. They are not measurements of how the constituents interact with body functions or macromolecules to cause harm. Some of these markers have been researched extensively, and they are more representative of actual human exposures to tobacco products than external measures of exposure. They are generally technically feasible and provide information about short-term (e.g., from a single cigarette) and long-term exposures. Examples are listed in Table 11-6, which gives a range of assays available but is not intended to be all inclusive. Because it has such a short half-life, carbon monoxide is best used for assessing recent exposures, although CO measurements also have been used to improve long-term exposure estimates of cigarette consumption (Law et al., 1997). The limitations of CO are that there are other sources of carbon monoxide, such as automobile exhaust and endogenous metabolism, and there is some variation with differences in physical activity, gender, and the presence of lung disease or other disease states. Nicotine blood levels are used and are helpful for assessing internal exposure primarily because it has a very short half-life. Serum, urinary, or salivary cotinine, which is a metabolite of nicotine with a longer half-life, however, has been extensively studied for confirmation of exposure in smokers, quitters, and persons exposed to ETS (Bono et al., 1996; Benowitz, 1999; Crawford et al., 1994). Cotinine levels are dependent on both the extent of formation from nicotine by cytochrome P450 (CYP) 2A6 and the rates of oxidation and glucuronidation of cotinine to 3-hydroxy-cotinine and glucuronide conjugates, respectively, which vary widely among individuals. Therefore, cotinine levels are only approximately correlated with the daily intake of nicotine. Carbon monoxide and nicotine boosts (i.e., the difference between levels before and after a single cigarette) reflect smoking topography and exposures from an individual cigarette.

Technologies exist for directly measuring internal exposure to tobacco smoke constituents in target organs through biopsies (e.g., PAHs in the lung) (Lodovici et al., 1998) and for measuring levels of metabolites of compounds (e.g., those from TSNAs in the urine) (Atawodi et al., 1998; Carmella et al., 1990, 1995). Tobacco smokers have higher levels of mutagens circulating in the body, which can be measured by using extracts of urine in the Ames *Salmonella* mutation assay (Jaffe et al., 1983; Mohtashamipur et al., 1985; Yamasaki and Ames, 1977). Levels have been found to decrease with some test cigarettes that heat, rather than burn, tobacco (Smith et al., 1996).

TABLE 11-6 Biomarkers of Exposure[a,b]

Category	Variables Used in Literature	Dose-Response Data Available	Associated with Cessation or Half-life	Chemical Specificity
Nicotine-related biomarkers	Nicotine	Yes	2 hr	Yes
	Nicotine boost (pre- and post-cigarette nicotine levels)	Yes	NA	Yes
	Cotinine	Yes	17 hr	Yes
	Other nicotine metabolites	Yes	Depends on metabolite	Yes
Minor tobacco alkaloids	Anatabine Anabasine	NDA	10-16 hr	Yes
Carbon monoxide	Exhaled CO	Yes	4-6 hr	Yes
	CO boost (pre- and post-cigarette levels)	Yes	NA	Yes
	Carboxyhemoglobin	Yes	Hours	Yes
Hydrogen cyanide	Thiocyanate	Yes	1-2 weeks	No

Specific to Tobacco	Related to a Disease Risk[c]	Strengths	Limitations
Yes (except when using nicotine replacement therapy [NRT])	Yes (addiction only)	Direct measure of exposure	Short half-life dependent on a person's ability to metabolize nicotine and time of sampling. Not useful with concurrent use of NRT
Yes	NDA	Measures exposure to single cigarette	Requires two blood draws. Short-term marker only
Yes (except when using NRT)	Yes (addiction only)	Well validated; can be measured easily in urine, plasma saliva, or hair. Useful for environmental tobacco smoke	Short-term marker only. At higher levels of smoking, dose-response relationship is less clear and there is wide overlap among smokers
Yes (except when using NRT)	NDA	Allows for assessment of nicotine metabolism	Low levels. No benefit over cotinine. Short term marker only
Yes	NDA	Useful when individuals are using NRT; may be precursors to nitrosamines	Short-term marker only
No	Yes	Easy to measure in exhaled air	Other sources exist, including endogenous processes. Short-term marker only.
Yes	NDA	Measures exposure to single cigarette	Short term marker only. Levels vary over the day
No	Yes	Measures cumulative, although short-term exposure to several cigarettes	Requires blood draw and special handling. Benefit above that for using exhaled CO not shown
No	NDA	Long-term marker. Can be measured in urine, saliva, and blood. Saliva easy to obtain	Many dietary sources. Dose-response curve flattens at higher smoking levels so cannot distinguish among heavy smokers

continues

TABLE 11-6 Continued

Category	Variables Used in Literature	Dose-Response Data Available	Associated with Cessation or Half-life	Chemical Specificity
Tobacco-specific nitrosamines	Urinary metabolites	Yes	45 d	Yes
Polycyclic aromatic hydrocarbons	Parent compounds	NDA	NDA	Yes
	Urinary 3-hydroxypyrene and 1-hydroxypyrene	Yes	NDA	Yes
Complex mixture assay	Urinary mutagenicity	Yes	Yes	No

NOTE: NA=not applicable; NDA=No data available.

The assessment of smoking exposure using nicotine or cotinine cannot be done in smokers who are concomitantly using nicotine replacement products. An alternative is to determine the levels of other tobacco alkaloids, such as anatabine or anabasine in the urine (Jacob et al., 1999).

BIOMARKERS ESTIMATING THE BIOLOGICALLY EFFECTIVE DOSE

The biologically effective dose (Perera, 1987) is the amount of a tobacco smoke or tobacco toxin that measurably binds to, or alters, a macromolecule (e.g., protein or DNA) in a cell. In some cases, the macromolecule may be a surrogate for a target molecule. The biologically effective dose represents the net effect of metabolic activation, decreased rate of detoxification, decreased repair capacity, loss of cell-cycle checkpoint control, and decreased rates of cell death. It should be noted that not all binding to, or alteration of, a macromolecule leads to an adverse health effect; so, often, what is really measured is the dose to a target macromolecule that estimates the biologically effective dose. Table 11-7 provides

Specific to Tobacco	Related to a Disease Risk[c]	Strengths	Limitations
Yes	NDA	May reflect biologically effective dose	Technically difficult to measure
No	Yes	Measured in organs where effect might occur	Technically difficult to obtain tissue and perform assay
No	NDA	Assay simple to perform	Other exposures can be substantial
No	NDA	May be related to in vivo mutagen exposure	Lack of specificity

[a]Selected examples; list is not all-inclusive.

[b]References are not provided in this table but can be found in the text of this and disease-related chapters.

[c]Any report related to a disease outcome associated with tobacco where the report is plausible but has not necessarily been replicated.

examples of biomarkers that estimate the biologically effective dose, but is not intended to be all inclusive.

Many tobacco-related toxins and chemical carcinogens are biologically inactive until transformed by cellular enzymes such as cytochrome-P450s into reactive intermediates. These reactive intermediates bind to macromolecules such as DNA and protein and disrupt their normal processes.

For cancer, a common assessment of the biologically effective dose is the measurement of carcinogen-DNA adduct levels. These are formed when carcinogen metabolites are alkylated to nucleotides, creating a promutagenic lesion. There are strong laboratory animal data and some human studies that indicate a relationship between tobacco smoke constituents, carcinogen–DNA adduct formation, and cancer (La and Swenberg, 1996). Laboratory animal studies have shown a correlation between cancer and increased adducts in target organs (Ashurst et. al., 1983; Nakayama et al., 1984; Pelkonen et al., 1980). In humans, tobacco smoking leads to increased adduct formation in target tissues such as the lung (Phillips et al., 1988; Schoket et al., 1998; Wiencke et al., 1995) and in

TABLE 11-7 Biomarkers Estimating the Biologically Effective Dose[a,b]

Category	Variables Used in Literature	Dose-Response Data	Associated with Cessation or Half-life	Target Tissue Assay Available
Carcinogen-DNA adducts	Nonidentified adducts/ ^{32}P-postlabeling	Yes	Yes	Yes
	PAH-DNA adducts	Yes	9-13 weeks (blood)	Yes
	4-Aminobiphenyl-DNA adducts	Yes	Yes	Yes
	NNK-DNA adducts	Yes	NDA	Yes
	8-hydroxydeoxy-guanosine	No	Yes	Yes
	5-(Hydroxy-methyl)uracil	No	NDA	No
	N-Nitrosamine-related-DNA adducts	NDA	26 hr (blood; O6-methyldeoxy-guanosine) and 60 hr (blood; 7-methyldeoxy-guanosine)	Yes

Chemical Specificity	Specific to Tobacco	Related to a Disease Risk[c]	Strengths	Limitations
No	No	Yes	Facile assay; does not require knowledge of specific adducts; blood may be surrogate for lung tissue. Adducts found in all tissues, including heart and blood vessels	Cannot identify adducts so mechanistic studies are problematic
Yes	No	Yes	Can be measured in any tissue and assays are available that are sufficiently sensitive	Low sensitivity and technical difficulties make assay use limited in large-scale studies. Diet might be greater contributor than smoking
Yes	No	NDA	Can be measured in any tissue; has some specificity for smoking if no known occupational exposure	Low sensitivity makes assay use limited in large-scale studies
Yes	Yes	NDA	Can be measured in any tissue, although methodology has low sensitivity. Highly specific for smoking	Low sensitivity makes assay use limited in large-scale studies
Yes	No	NDA	Can be measured in any tissue	Assay has large interlaboratory variation; it is easy to introduce oxidative damage into laboratory assay; low sensitivity makes assay use limited in large-scale studies
Yes	No	Not available	Sufficient sensitivity to use for ETS	Technically difficult
Yes	No	NDA	Can be measured in any tissue	Low sensitivity makes assay use limited in large-scale studies. Diet a common source

continues

TABLE 11-7 Continued

Category	Variables Used in Literature	Dose-Response Data	Associated with Cessation or Half-life	Target Tissue Assay Available
Carcinogen-hemoglobin (Hgb) adducts	PAH-Hgb adducts	Yes	NDA	No
	4-Aminobiphenyl-Hgb adducts	Yes	7-9 weeks	No
Carcinogen-protein adducts	PAH–albumen adducts	Yes	NDA	No
Carcinogen-DNA adduct antibodies	Anti-BPDE serum antibodies	NDA	NDA	No
	Adducts	Yes	Yes	Yes
Carbon monoxide	Carboxy-hemoglobin	Yes	Yes	No
Lipid peroxidation	F2-Isoprostanes	No	Yes	No

NOTE: NA=not applicable; NDA=no data available; NNK=nitrosonornicotine ketone; BPDE=benzo(a)pyrene-diol-epoxide

surrogate tissues such as blood (Tang et al., 1995; Vineis et al., 1994; Wiencke et al., 1995). Evidence exists that carcinogen-DNA adduct levels in target and nontarget organs are modulated by interindividual differences (Badawi et al., 1995; Grinberg-Funes et al., 1994; Kato et al., 1995;

Chemical Specificity	Specific to Tobacco	Related to a Disease Risk[c]	Strengths	Limitations
Yes	No	NDA	Large amount of adducts available in blood so method is facile	Surrogate assay not yet validated against target organ damage
Yes	No	NDA	Large amount of adducts available in blood so method is facile	Surrogate assay not yet validated against target organ damage
Yes	No	NDA	Large amount of adducts available in blood so method is facile	Surrogate assay not yet validated against target organ damage
No	NDA	NDA	May provide long-term marker of exposure	Doubtful that a dose-response relationship can be established due to complexity of immune response in individuals
Yes	No	NDA	Measured in organs where effect might occur	Technically difficult to obtain tissue and perform assay
Yes	No	Yes	Might also reflect a surrogate measure of biologically effective dose	Logistical problems in sample handling
Yes	No	NDA	Corroborative end point for oxidative damage without artifactual introduction of oxidative damage	Technically difficult

[a]Selected examples; list is not all-inclusive.

[b]References are not provided in this table but can be found in the text of this and disease-related chapters.

[c]Any report related to a disease outcome associated with tobacco where the report is plausible but has not necessarily been replicated.

Pastorelli et al., 1998; Rojas et al., 1998; Ryberg et al., 1997; Stern et al., 1993). Interestingly, in former smokers, age of initiation may influence lung adduct levels (Wiencke et al., 1999). In humans, only a few studies have investigated a link between carcinogen-DNA adducts and cancer risk. All data come from case-control studies of the lung and bladder, and almost all show a positive relationship (Dunn et al., 1991; Peluso et al., 1998; Tang et al., 1995; van Schooten et al., 1990). However, since no published prospective studies of tobacco smoking show a relationship of adducts to cancer, the case-control studies must be interpreted cautiously because there may be an effect due to differential metabolism or DNA repair. The utility of carcinogen-DNA adduct measurements in assessing harm reduction is suggested by studies showing that lung adduct levels are lower in persons who smoked filter cigarettes (van Schooten et al., 1990). Hemoglobin adducts, an estimate of the biologically effective dose, are higher in smokers than in nonsmokers (Bryant et al., 1987), and in those who smoke black rather than blond tobacco (Bryant et al, 1988). Snuff dipping may lead to even higher levels of some types of adduct than to smoking (Carmella et al., 1990).

A variety of assays are available to determine carcinogen–macro-molecular adducts in human tissues (Farmer and Shuker, 1999; Hecht, 1999; La and Swenberg, 1996; Lee et al., 1993; Wang et al., 2000). Although DNA adduct analysis is most commonly studied in relation to carcino-genesis, adducts also have been found in atherosclerotic lesions (Izzotti et al., 1995). Assay techniques include the phosphorus-32 (^{32}P)-postlabeling assay-nucleotide chromatography (Phillips, 1997; Randerath et al., 1981), immunoassays (Lee et al., 1993), fluorescence spectroscopy (Izzotti et al., 1991), gas chromatography-mass spectroscopy (GC-MS) (Farmer and Shuker, 1999; Hecht, 1999), and electrochemical detection (Helbock et al., 1998; Park et al., 1989). Each has its strengths and limitations, and almost all are challenged by low sensitivity and/or specificity. The less specific methods, such as the ^{32}P-postlabeling assay-nucleotide chromatography, when used as originally described (Randerath et al., 1981) or with modifi-cations (Reddy and Randerath, 1986), offer the benefit of assessing expo-sure to complex mixtures because multiple adducts are measured at the same time. However, because the assay does not identify the types of adducts, any interpretations of the results are limited. Chemical specificity is helpful in assessing harm reduction products when the adducts are specific for tobacco (e.g., TSNAs or 4-aminobiphenyl in the absence of occupational exposure), whereas adduct assays that determine levels from endogenous sources (e.g., oxidative damage, methylation) are more diffi-cult to use and interpret. The study of carcinogen-DNA adducts presents other challenges in interpretation; for example, carcinogen-DNA adduct levels are higher in the heart than in the lung (Randerath et al., 1989) while cancer is rare in the former. For the future, newer adduct methods may provide increased specificity and sensitivity, along with higher throughput.

The use of target organ biomarkers can provide specific information about potentially carcinogenic effects and will best represent the biologically effective dose. Target organs include lung for lung diseases, oral mucosa for oral cavity diseases, bladder mucosa for bladder disease, and so forth. Surrogate markers that estimate levels in target organs, such as carcinogen-DNA adducts in blood, have been partially studied, indicating that blood levels might reflect target organ levels (Mustonen and Hemminki, 1992; Mustonen et al., 1993; Tang et al., 1995; Wiencke et al., 1995), but this is not yet firmly established. Protein (Meyer and Bechtold, 1996) and hemoglobin (Wang et al., 2000) adducts also may estimate levels of exposure at the target organ and thus be surrogates. Such assays offer technological advantages because these macromolecules are more abundant in blood than DNA, but the relationship of these other macromolecular adducts to DNA levels has been insufficiently studied.

A few studies show the decline of adducts following short-term and long-term smoking cessation. Most studies will necessarily rely on blood levels, and the half-life of adducts in blood will depend on the life span of various blood cell types. In humans, the half-life for 4-aminobiphenyl-hemoglobin adducts is 7-9 weeks, which is shorter than the life span of a red blood cell (Jahnke et al., 1990). PAH-DNA adducts in white blood cells have a half-life of 9 to 13 weeks (Mooney et al., 1995). In human lungs, it was reported that adducts persist in the lungs of ex-smokers (Randerath et al., 1989), but it is not known whether this is truly persistence or the formation of new adducts from the continuing presence of tobacco constituents such as PAHs or from other exposures such as diet or air pollution (Rothman et al., 1990).

Carcinogen-DNA adduct data have essentially not been used for population risk assessments. In one example, it was considered that a doubling of PAH-DNA adduct levels would result in an additional 2,400 cancer cases per million persons (van Delft et al., 1998), but the model assumed linear dose-responses; was not adjusted for age, gender, or race; and was too simplistic.

BIOMARKERS OF POTENTIAL HARM

These biomarkers reflect changes in a cell and its macromolecules that result from tobacco. These can range from isolated changes, with or without effects on function, to events that clearly lead to illness or are symptoms of the illness (i.e., cough). Examples of biomarkers of effect are provided in Table 11-8, which gives the reader a range of assays available but is not intended to be all inclusive.

Among the most promising biomarkers of effect for assessing harm reduction claims for cancer are those that measure DNA damage or alterations of genetic function (mutations, gross chromosomal changes, DNA methylation of promoter regions, etc.). While these biomarkers are envi-

TABLE 11-8 Biomarkers of Potential Harmful Effects[a,b]

Category	Variables Used in Literature	Dose-Response Data	Associated with Cessation or Half-life	Target Tissue Assay Available	Chemical Specificity
Enzymatic induction	Aryl hydrocarbon hydroxylase	No	>30 d	Yes	Yes
	CYP1A2	No	NDA	Yes	Yes
	DNA repair enzymes	NDA	Yes	Yes	NA
	Microarray assays for mRNA expression and proteomics	NDA	NDA	Yes	NA
Chromosomal alterations	Chromosomal aberrations	Yes	Yes	Yes	No
	Micronuclei	Yes	Yes	Yes	No
	Sister chromatid exchanges	Yes	Yes	No	No
	Loss of heterozygosity	Yes	Yes	Yes	No
	Mutations in reporter genes (HPRT, GPA)	Yes	Yes	No	No

Specific to Tobacco	Related to a Disease Risk[c]	Strengths	Limitations
No	Yes	Indicates acquired changes in susceptibility; related to DNA-adduct levels	Technically difficult to assess in large epidemiological studies
No	Yes	Indicates acquired changes in susceptibility; related to DNA-adduct levels	Technically difficult to assess in large epidemiological studies
No	NDA	Indicates acquired changes in susceptibility; provides analysis of what is likely to be critical part of carcinogenesis	Technically difficult
No	NDA	Reflects integrated measure of multiple genotypes, provides complex data potentially usable for rapid identification of important risk factors	Difficult to perform; relationship to disease risk is technically difficult to prove; requires extensive laboratory validation; RNA and protein microarray assays are expensive; large-scale studies are needed; refined bioinformatic analysis required
No	Yes	Can be done in blood as surrogate tissue. Similar lesions observed in cancer. Can be measured in persons without cancer	Very nonspecific; relationship to target organ is not established; significant lack of specificity and wide overlap between smokers and nonsmokers
No	NDA	Facile assay	Lack of specificity
No	No	Easy to do in blood as surrogate tissue. Can be measured in persons without cancer	Very nonspecific; relationship to target organ is not established; predictivity for disease risk not established. Association with cancer in case-control studies may have case bias. Significant lack of specificity and wide overlap between smokers and nonsmokers
No	NDA	Similar lesions observed in cancer	Technically complex; relationship to cancer risk unknown
No	NDA	Facile assay in blood	Relationship to target tissue or blood unknown

continues

TABLE 11-8 Continued

Category	Variables Used in Literature	Dose-Response Data	Associated with Cessation or Half-life	Target Tissue Assay Available	Chemical Specificity
	Mutational load in target genes (p53, K-ras)	NA	NDA	Yes	No
Mitochondrial mutations	Deletions, insertions	NDA	NDA	Yes	No
Epigenetic cancer effects	Whole genome methylation	NDA	NDA	Yes	No
	Hypermethylation of promoter regions	NDA	NDA	Yes	No
Lipids	Blood lipids: HDL, LDL, oxidized LDL, triglycerides	Yes	NDA	Yes	Yes
Cardiovascular response	Heart rate, blood pressure	No	Yes	Yes	NA
Thrombosis	Bleeding time	No	NDA	Yes	No
	Fibrinogen	NDA	NDA	Yes	Yes
	Prothrombin time, partial thromboplastin time, plasminogen activator inhibitor, C-reactive protein	Yes	NDA	Yes	Yes
	Urinary thromboxane and prostacyclins	Yes	No	No	Yes

Specific to Tobacco	Related to a Disease Risk[c]	Strengths	Limitations
No	NDA	Target gene specificity	Very difficult to do in normal tissues
No	NDA	Provides corroborative marker	Relationship to disease not established
No	No	Facile assay	Relationship to disease unknown
No	No	Similar lesions observed in cancers	Technically difficult; relationship to risk unknown
No	Yes	May be directly related to disease risk	Levels among heavy smokers cannot be distinguished. Wide interindividual variation. Many individuals under medication therapy. Significant confounders exist
No	Yes	Easy to measure; intraindividual differences may be important for the individual	Both interindividual and intraindividual differences are significant. Substantial confounders exist, and many persons are on medications
No	No	Minimally invasive	Very nonspecific
No	NDA	Pathogenically related to disease	Does not distinguish levels of smoking. Nicotine might separately affect these parameters so limited use in persons using NRT
No	NDA	Leave a fingerprint at the site of their formation	
No	Yes	May be markers of platelet-vascular interactions; reflect chronic exposure	Technically difficult. Wide overlap of values due to individual differences in response

continues

TABLE 11-8   Continued

Category	Variables Used in Literature	Dose-Response Data	Associated with Cessation or Half-life	Target Tissue Assay Available	Chemical Specificity
	Platelet activation and survival	Yes	NDA	Yes	No
Blood cell parameters	White blood cell counts (i.e., lymphocytes, neutrophils, total counts)	Yes	Yes	Yes	Yes
	Hematocrit, hemoglobin, red blood cell mass	Yes	Yes	Yes	No
Bronchio-alveolar lavage response	Inflammatory cells, protein, cytokines	Yes	Yes	Yes	No
	Neutrophil elastase a1-antiprotease complex	Yes	Yes	Yes	No
	α1-antitrypsin	No	No	Yes	Yes
Inflammatory mediators of response	Leukotrienes	Yes	NDA	No	Yes
Pulmonary function tests	FEV1, FVC	Yes	Yes	Yes	No
Periodontal disease	Periodontal height	Yes	Yes	Yes	No
	Gum bleeding	Yes	Yes	Yes	No
Osteoporosis	Fractures	Yes	NDA	NA	No
	Bone density	NDA	NDA	Yes	No
Skin	Premature wrinkling	Yes	NDA	NA	No

Specific to Tobacco	Related to a Disease Risk[c]	Strengths	Limitations
No	No	Platelet activation in vivo might be pathophysiologically related to cardiac artery thrombosis	Technically difficult to use for large numbers of subjects. Significant number of confounding variables. Smoking increases platelet counts
No	Yes	Can be a surrogate marker for several processes including atherosclerosis and thrombosis	Relationship to disease uncertain, although alterations in levels are linked epidemiologically to disease. Wide interindividual and intraindividual variation and large number of confounders
No	No	Can reflect both cardiac and respiratory disease risk	Insensitive; wide interindividual differences
No	NDA	Provides different types of data with single procedure	Bronchoscopy is too invasive for large epidemiological studies
No	NDA	Provides different types of data with single procedure	Bronchoscopy is too invasive for large epidemiological studies
Yes	NDA	May be specific to tobacco smoke	Requires invasive test; short half-life
No	NDA	May be measured in urine, bronchioalveolar lavage, and serum	Substantial number of confounders
No	Yes	Widely available	Low sensitivity for mild disease. Decrease in function with aging. Large interindividual variation
No	Yes		
No	Yes		
No	Yes	Easily measured	Numerous confounders
No	Yes		
No	NA		Lack of specificity; involves subjective evaluation

continues

TABLE 11-8 Continued

Category	Variables Used in Literature	Dose-Response Data	Associated with Cessation or Half-life	Target Tissue Assay Available	Chemical Specificity
Fetal and neonatal effects	Birth weight	Yes	Yes	Yes	No
Weight	Weight loss and gain	Yes	Yes	Yes	No

NOTE: NA=not applicable; NDA=no data available; FVC=Forced vital capacity; FEV$_1$=forced expiratory volume in 1 sec; HDL=high-density lipoprotein; LDL=low-density lipoprotein.

sioned for use in developing a molecular fingerprint reflecting a particular exposure, this has not occurred for tobacco carcinogens, and measurable effects thus far are relatively nonspecific. Nonetheless, a reduction in the level of genetic damage would logically be required if a tobacco-related PREP were to be successful in reducing cancer risk, although how much reduction of genetic damage would be needed to derive a benefit in terms of disease risk is unknown. Several types of assays are available. The main limitation today is that no assays have been shown convincingly to be sufficiently predictive of cancer risk. Chromosomal damage can be measured through classical cytogenetic alterations (Bender et al., 1988; Obe et al., 1982; Ramsey et al., 1995), micronuclei formation (Thorne et al., 1998), COMET (Poli et al., 1999; Speit and Hartmann, 1999), fluorescent in situ hybridization (FISH) (Pressl et al., 1999; Ramsey et al., 1995; van Diemen et al., 1995), or polymerase chain reaction (PCR) methods assessing loss of heterozygosity (using tandem repeats or comparative genomic hybridization) (Mao et al., 1997), where the latter two methods can be used for morphologically normal-appearing cells. Mutations in reporter genes, such as hypoxanthine phosphoribosyltransferase (HPRT) (Ammenheuser et al., 1997; Hou et al., 1999; Jones et al., 1993) or glycophorin A (GPA), have been used in blood cells, but it is better to

Specific to Tobacco	Related to a Disease Risk[c]	Strengths	Limitations
No	Yes	Data collection is easy	Nonspecific; numerous confounders
No	Yes	Both a biomarker for metabolism and an important outcome for some people	Some people perceive weight loss as a benefit of smoking, despite significant adverse effects associated with smoking

[a]Selected examples; list is not all-inclusive.

[b]References are not provided in this table but can be found in the text of this and disease-related chapters.

[c]Any report related to a disease outcome associated with tobacco where the report is plausible but has not necessarily been replicated.

identify mutation rates for cancer genes in biopsies from target organs or in surrogate tissues, and for genes such as p53 (Greenblatt et al., 1994) or KRAS (Lehman et al., 1996; Mills et al., 1995; Scott et al., 1997; Yakubovskaya et al., 1995). Although these assays are available, current technology limits their use in large-scale epidemiological studies. The role of mitochondrial DNA lesions is receiving greater attention for cancer risk (Fliss et al., 2000), and the lesions associated with smoking might be useful (Liu et al., 1997). Among all the assays that have potential application to assessing harm reduction claims, only two studies have assessed prospectively the cancer predictive value of chromosomal aberrations (Bonassi et al., 1995; Hagmar et al., 1994), but they consisted of pooled heterogenous populations and were not focused on tobacco. Further studies are needed to indicate the value of these assays for determining harm reduction. Thus, none of these assays can be used today to allow claims of risk reduction, although in the proper setting they can suggest that such might occur.

Biomarkers of pathobiological effect include morphological markers of preneoplastic lesions (e.g., dysplasia), altered phenotypic expression of normal cellular functions (e.g., overexpression of the proto-oncogene Erb-B2), and mutations in cancer-related genes such as the p53 tumor suppressor gene. Some of these may be considered preclinical effects that are

occurring before diagnosis. The lesions demonstrate a person's phenotype for exposure and predisposition that persist following DNA damage. Recent advances have made it possible to measure background mutations in cancer-associated genes of noncancerous tissues (Aguilar et al.,
1994; Mao et al., 1997; Sidransky, 1997), which presumably are related to
future cancer risk.

The study of mutations in the p53 tumor suppressor gene is uniquely
suited for studying cancer etiology, exposure, and susceptibility (Harris
and Hollstein, 1993), because p53 is involved in many cellular processes
including maintenance of genomic stability, programmed cell death, DNA
repair, and others (Attardi and Jacks, 1999; Hollstein et al., 1999; Shimoda
et al., 1994; Soussi et al., 2000). The p53 gene, in particular, has a more
frequent spectrum of mutations in tobacco-associated lung cancers
(Bennett et al., 1999). An interactive effect of alcohol drinking and cigarette use in oral cavity and lung cancers leads to different types of p53
mutations (Ahrendt et al., 1999, 2000; Brennan et al., 1995). Interestingly,
given that the p53 mutational spectrum for lung cancer is similar worldwide (Hartmann et al., 1997), it is likely that tobacco smoke is the major
determinant of lung p53 mutations worldwide. Evidence for a relationship of gene-environment interactions and mutation risk in the p53 gene
can be found from a Japanese study of *CYP1A1* (Kawajiri et al., 1996),
where a fivefold increase in risk of p53 mutations was found for smokers
with lung cancer and the "at-risk" genetic variant. This risk increased
further for persons who also lacked the glutathione *S*-transferase (*GSTM1*)
gene. In one study from Norway, smokers with lung cancer who lacked
GSTM1 also had more p53 mutations, especially transversions (Ryberg et
al., 1994). For oropharyngeal tumors, the frequency of p53 mutations was
increased for the same *CYP1A1* variant allele (Lazarus et al., 1998). An
increased risk for p53 mutations in lung cancer also has been found in
Japanese persons with less common variants of *CYP2E1* (Oyama et al.,
1997).

Newly developed technologies allow for the detection of loss of heterozygosity (LOH) in small amounts of tissue. Losses at chromosome
3p14, 9p21, and 17p13 have been seen in the lungs of both smokers and
former smokers, where the first is less frequent in former smokers than
current smokers (Mao et al., 1997).

An important area that has not been well studied is the effect of
tobacco toxicants on the induction of enzymes that might affect cancer
risk. For example, cytochrome P-450 enzymes are induced with tobacco
smoking (e.g., arylhydrocarbon hydroxylases [AHHs]) (Bartsch et al.,
1995; Guengerich, 2000; McLemore et al., 1990; Nakajima et al., 1991, 1995;
Rojas et al., 1992). Induction is related to greater amounts of DNA damage
(Bartsch et al., 1991; Geneste et al., 1991). It remains to be tested whether a

tobacco-related PREP can reduce AHH exposure so that other carcinogenic exposures will be less harmful. Many proteins are induced in relation to DNA damage (e.g., p53) (Bjelogrlic et al., 1994). Whether higher levels of these proteins increase or decrease the risk of disease remains unknown.

Several biomarkers can be studied in relation to cardiovascular disease risk, but none of these are specific to tobacco smoking, such as blood lipid level (Cullen et al., 1997; Freeman et al., 1993; Hellerstein et al., 1994; Ludviksdottir et al., 1999; Stubbe et al., 1982; Wald et al., 1989), which changes with cessation (Green and Harari, 1995), or urinary excretion of thromboxane A_2 metabolites (Nowak et al., 1987; Lassila et al., 1988; Rangemark et al., 1992; Wennmalm et al., 1991). F_2-Isoprostanes in blood have a dose-response relationship to smoking (Morrow et al., 1995). Tobacco smoking is associated with decreased weight (Green and Harari, 1995) and therefore modifies the relationship of weight gain to increased risk of heart disease (Fulton and Shekelle, 1997). Blood pressure has been studied but is not clearly associated with smoking (Green and Harari, 1995). Other biomarkers that have been suggested to reflect an increased cardiac disease risk include reduced platelet survival (Fuster et al., 1981). Newer imaging methods such as electron-beam computed tomography (O'Malley et al., 2000; Raggi et al., 2000) are being used to assess heart disease risk, and these methods might be used to assess the decreasing rate of formation of atherosclerosis or calcium when using a PREP.

Biomarkers of developing respiratory illness have been assessed in different ways, and several studies have specifically assessed the effects of smoking reduction separately from cessation. Symptoms, albeit late effects, such as cough, chronic phlegm production, wheezing, and shortness of breath have been used and improve with smoking cessation (Buist et al., 1976; Kanner et al., 1999). Reducing smoking, without quitting, also is associated with a reduction in symptoms (Buist et al., 1976). There are many studies that explore decrements of pulmonary function related to cigarette smoking. While such decrements occur with aging independent of smoking, further decrements are induced by smoking (Lange et al., 1989; McCarthy et al., 1976). Declines in the forced expiratory volume at 1 second (FEV_1) are associated with increased disease and mortality, including nonpulmonary diseases (James et al., 1999). The decline in pulmonary function tests slows with complete cessation (Buist et al., 1976; Kanner et al., 1999; Lange et al., 1989; McCarthy et al., 1976) and with greater than 25% reduction in the number of cigarettes smoked per day (Buist et al., 1976; Lange et al., 1989; McCarthy et al., 1976). Smoking reduction in the elderly apparently showed no effect in slowing the rate of decline (Lange et al., 1989). Bronchioalveolar lavage has been used, although it is invasive, and different types of assays can assess inflammation, neutrophil elastase α_1-antiprotease complex, and α_1-antitrypsin (Rennard et al., 1990). Induction of these components reverses with smoking reduction (Rennard et al., 1990), and some markers such as alveolar neutrophils, neutrophil elastase α_1-antiprotease complexes, and alveolar macrophages decrease in smokers who reduce their amount of smoking when provided with nicotine replacement therapy (Rennard et al., 1990).

Several nonspecific biomarkers of effect are related to smoking, such as leukocyte count (Parry et al., 1997; Phillips et al., 1992; Sunyer et al., 1996; Wald et al., 1989), which reverses with cessation (Green and Harari, 1995; Sunyer et al., 1996) and then increases again with resumption of smoking (Sunyer et al., 1996). Levels remain increased to some extent in former smokers compared to never smokers (Parry et al., 1997). Whether these findings, however, are independent predictors of disease risk has had limited study (e.g., mortality) (James et al., 1999), and the differences that can be found may be due to disease unrelated to smoking (Wald et al., 1989). Some of these parameters are covariates (James et al., 1999). Thus, such markers would be less useful for assessing harm reduction claims but might be useful for assessing exposure reduction claims.

There are several short-term effects on the body that can be considered both from the perspective of disease and as a biomarker of effect. Examples include periodontal disease, abnormal glucose tolerance tests, and decreased birthweight of infants born to mothers who smoke. Also, changes in adult body weight can be measured in the context of harm reduction. It is well known that smoking increases metabolism and decreases appetite, while stopping smoking is associated with weight gain (O'Hara et al., 1998). This can be a very important marker of smoking effects since the consideration of weight is often a factor in persons' beginning smoking or resisting cessation.

HOST SUSCEPTIBILITY

Host susceptibility could modify the risk of tobacco-related disease and, therefore, the effects of PREPs. Host susceptibility can be influenced by genetic susceptibility, age, gender, ethnicity, health status, and so forth. These will not be discussed in detail except for genetic susceptibility, but any relevant potential modifying factor should be considered in the assessment of a PREP.

The study of genetic susceptibilities can improve the accuracy of estimates of disease associations (Khoury and Wagener, 1995). Tobacco toxicants affect people to variable degrees. It is therefore reasonable to assume that harm reduction strategies would affect people differently. There is large interindividual variation in cellular responses—for example, in metabolism and detoxification of toxicants and DNA repair. As other cellular responses to DNA damage are identified (e.g., cell-cycle delays, heat shock), interindividual variation in risk is likely to be discovered for these as well. Interindividual effects in cellular responses could be due to genetically determined enzyme expression, kinetics, or stability. Also, induction of enzymes from previous exposures or comorbidity also may contribute to cancer risk, and induction has a genetic component.

Susceptibility to disease from genetic variability can range from small to large, depending on the genetic penetrance. Highly-penetrant cancer susceptibility genes cause familial cancers but account for less than 1% of all cancers (Fearon, 1997). Low-penetrant genes cause common sporadic cancers and can have great public health consequences (Shields and Harris, 2000).

Genetic susceptibility can be assessed either phenotypically (measuring the resultant enzymatic function) or genotypically (determining the genetic code). Examples are provided in Table 11-9. Phenotypic assays may include determining enzymatic activity by administering probe

TABLE 11-9   Assays for Assessing Effect Modification by Heritable Traits

Assay Type	Example Used in Literature	Strengths	Limitations
Gene-based assays	Genetic polymorphisms for carcinogen metabolism and induction or DNA repair, smoking behavior	Inexpensive, simple to perform, specific gene effect when exists, high throughput available	Functional relationship of genotype to phenotype difficult to prove; disease risk for low-penetrant genes difficult to prove
Phenotypic assays	Mutagen sensitivity for DNA repair; host-reactivation assay for DNA repair; CYP450 metabolism and induction studies; RNA expression of specific genes; microarray RNA expression; proteomics	Reflects integrated measure of multiple genotypes; provides complex data potentially usable for rapid identification of important risk factors	Difficult to perform; relationship to disease risk technically difficult to prove; requires extensive laboratory validation; RNA and protein microarray assays are expensive; large-scale studies are needed; bioinformatics not available

drugs to individuals and measuring blood levels or urinary metabolites, assessing carcinogen metabolic capacity in cultured lymphocytes, or establishing the ratios of endogenously produced substances such as estrogen metabolite ratios. One extensively studied phenotype in relation to smoking risk is AHH activity (Kellermann et al., 1973; Kouri et al., 1982). In general, it is preferable to use a gene-based assay to assess disease risk because DNA is easier to obtain and the assays are technically simpler. However, phenotypes usually represent a multigenic trait, which may not be adequately characterized by only one genetic assay. Therefore, there is a role for both gene- and phenotype-based assays in research studies and PREP assessments. Examples of frequently studied genetic polymorphisms in tobacco-related cancers that have been shown in some studies to modify smoking-related disease risk include the N-acetyltransferase 2 (NAT2) (Brockmoller et al., 1996, 1998; Henning et al., 1999), glutathione S-transferase M1 (GSTM1) (Bell et al., 1993; Brockmoller et al., 1996, 1998; Cullen et al., 1997; Jourenkova et al., 1998; Jourenkova-Mironova et al., 1999; Rebbeck, 1997), cytochrome P-450 1A1 (CYP1A1) genes (Bishop, 1987; Ishibe et al., 1997), glutathione S-transferase Pi (Ryberg et al., 1997), and others (Jourenkova-Mironova et al., 1999; Rosvold et al., 1995; Wiencke et al., 1997). These and other genetic polymorphisms are believed to affect levels of biomarkers, such as DNA adducts (Kato et al., 1995; Pastorelli et al., 1998; Ryberg et al., 1997; Yu et al., 1995).

In the general population, DNA repair capacity decreases in humans with aging (Liu et al., 1994; Wei et al., 1993), which would make this an acquired risk factor for cancer and might explain a portion of the increased cancer risk in the elderly (Simpson, 1997). Both genotyping and phenotyping assays for DNA repair or cell-cycle control that affects DNA repair might be useful in identifying individuals who might benefit from harm reduction strategies. Tobacco toxicants can affect DNA repair (Grafstrom et al., 1994), so that the effects of both tobacco toxicants and heritable capacity on DNA repair can be considered in assessing harm reduction products. It should be noted that cigarette smoking induces levels of some repair enzymes (Drin et al., 1994; Hall et al., 1993; Slupphaug et al., 1992), so caution must be used for some phenotyping assays.

Inherited susceptibilities via specific genetic polymorphisms that affect the efficiency of DNA repair (e.g., for base excision repair) have been identified recently (Mohrenweiser and Jones, 1998). Studies now being completed indicate an effect of these genetic variants on tobacco-related cancer risk (Sumida et al., 1998), some of which have functional effects on DNA repair (Lunn et al., 1999, 2000). A nonspecific DNA repair assay, which measures chromosomal aberrations in human cultured lymphocytes after an in vitro challenge with a mutagen, has shown initial promise. In this case, an increased mutagen-related aberration rate has been observed in persons with primary and secondary upper aerodigestive tract cancers (Cloos et al., 1996), multiple primary cancers (Cloos et al., 1994), and lung cancer (Li et al., 1996; Spitz et al., 1995; Wei et al., 1996).

Genetic susceptibilities for genes other than those involved in carcinogen metabolism and DNA repair are also being investigated (Jin et al., 1995; Sjalander et al., 1996). There has been less study of genetic susceptibilities for coronary artery disease (Gealy et al, 1999). It is likely that these genes also will play a role in modifying disease risk (see Chapter 13).

GENETIC PREDISPOSITIONS TO SMOKING ADDICTION

The greatest contributors to smoking addiction are the availability of tobacco and cultural acceptance of tobacco smoking. Genetics plays a lesser role. The tobacco smoking epidemic has occurred only over the last 50 to 70 years, and it is unlikely that human genetics have evolved in that amount of time. Nonetheless, twin studies indicate a genetic role for both smoking initiation and smoking persistence (Carmelli et al., 1992; Heath et al., 1993a, b).

People smoke in ways that will maintain a desired blood nicotine level. Nicotine in turn stimulates reward mechanisms in the brain. Presynaptic nicotinic acetylcholine receptors stimulate the secretion of dopamine into neuronal synapses. There also are effects on other pathways, such as those that involve serotonin. For dopamine, synaptic dopamine stimulates dopamine receptors; five subtypes have been identified, which are considered to be D_1- or D_2-like. Synaptic dopamine levels are governed by presynaptic release and the presynaptic dopamine transporter protein. In humans, there are different types of data supporting the link between nicotine and dopamine. Nicotine self-administration through tobacco smoking may reduce the adverse consequences of Parkinson's disease, attention deficit disorder, and schizophrenia (Bannon et al., 1995; Olincy et al., 1997; Seeman, 1995), diseases thought to be related to dopamine abnormalities. Also, smoking probably relieves depression (Gilbert and Gilbert, 1995), and the dopamine transporter inhibitor antidepressants (e.g., bupropion SR) are now used to treat nicotine addiction (Hurt et al., 1997; Jorenby et al., 1999).

The genes that code for dopamine receptors (e.g., *DRD2*, *DRD4*), dopamine transporter reuptake (*SL6A3*), and dopamine synthesis (e.g., dopamine hydroxylase, tyrosine hydroxylase, tryptophan hydroxylase, catechol-*O*-methyltransferase, monoamine oxidase) are polymorphic. Some of the polymorphisms result in altered protein function. Persons with higher levels of synaptic dopamine, or "more stimulation" of dopamine receptors may have less rewarding effects of nicotine and so would be less likely to become smokers and would more easily quit. For example, in a study of 500 smokers and nonsmokers, several candidate genes have been implicated (Lerman et al., 1998, 1999; Shields et al., 1998), whereas other studies of candidate genes have yielded null results (Lerman et al., 1997). Other investigators also have reported supporting evidence (Comings et al., 1996; Noble et al., 1994; Spitz et al., 1998). Thus, it is likely that there is a genetic contribution to smoking addiction and behavior and there may also be a genetic influence on who benefits from PREPs.

BIOMARKER ASSESSMENT FOR ENVIRONMENTAL TOBACCO SMOKE EXPOSURE

Biomarker assessments in persons exposed to environmental tobacco smoke are problematic because exposures occur at much lower levels than in smokers, and therefore the level of detection is limiting (Benowitz, 1999). The most consistently used biomarkers are those that reflect exposures, namely cotinine (serum, plasma, or urine), rather than biologically effective doses or biomarkers of effect. Such biomarkers, for example, can show that adolescents are exposed to tobacco smoke through household smoking (Bono et al., 1996). Urinary metabolites of tobacco-specific nitrosamines also have been found in persons exposed passively to smoke (Atawodi et al, 1998; Hecht et al., 1993; Parsons et al., 1998). DNA adducts in the lung are also detected in persons who are thought to be nonsmokers (Kato et al., 1995). Children exposed to modest levels of ETS have been found to have increased concentrates of 4-aminobiohenyl adducts of PAH-albumin adducts (Tang et al., 1999). Although it may follow that proven methods to reduce harm in smokers would apply to nonsmokers with passive exposure, there are circumstances in which passive smoke exposure might be substantial (e.g., cigar smoking).

DEVELOPMENT AND VALIDATION OF BIOMARKER ASSAYS, INCLUDING QUALITY CONTROL

The use of biomarkers in assessing harm reduction can be helpful only when the assays have undergone rigorous development and validation. Reliance on insufficiently validated biomarkers becomes problematic because they are of uncertain value and so should not be used to support a claim of exposure or risk reduction. The design and development of a biomarker assay must conform to the original goals—that is, the assay should have sufficient specificity, it should be quantitatively reproducible in humans at the levels that occur when exposure reduction is achieved, and other assays should be available to corroborate the qualitative and quantitative results. Many pitfalls have already been found in biomarker development. There are examples of biomarker assays that are more difficult to perform at levels observed in humans compared to the use of higher-level laboratory chemical standards (e.g., immunoassays) (Santella et al., 1988). Some methodologies can artifactually affect assay results (e.g., introduction of oxidative damage) (Farmer and Shuker, 1999). In some cases, measurements of in vivo formation can be skewed by exogenous exposure to the biomarker (e.g., dietary ingestion of 3-alkyladenine) (Prevost and Shuker, 1996).

Validation of a biomarker assay includes a determination of replicability (e.g., coefficient of variation), interobserver and interlaboratory variability, intraindividual variation, and interindividual variation. These validation steps must be done using known controls that simulate human exposure levels and harm. Thus, the assay should be validated in light and heavy smokers, former smokers, and never smokers. Caution must be used in interpreting assay results in the context of certain study designs. For example, the reliability of biomarkers thought to be related to disease risk in case-control studies is problematic for markers that might be affected by disease status (differential case bias) (Wald et al., 1989).

Research laboratories providing data that can impact individual or public health should have adequate quality control and quality assurance procedures in place. The definition of adequate will depend on the population under study and the number of subjects. In clinical pathology laboratories, standards and protocols have been established by organizations such as the College of American Pathologists and the National Committee for Clinical Laboratory Standards. In a research laboratory that performs biomarkers studies assessing PREPs, there should be standards for proficiency testing, quality improvement, quality control, use of standards, methods for interpretation, specimen handling, specimen labeling, specimen processing, and reporting of results. There also should be criteria for facility and equipment maintenance.

CONCLUSIONS

The assessment of a PREP will have to consider external exposure and markers of internal exposure, estimates of the biologically effective dose, and biomarkers of potential harm. A risk reduction claim should be based on disease reduction, but time limitations mandate the use of biomarkers for both exposure and risk reduction assessments. Measurements of the number of cigarettes per day, smoking duration, estimated lifetime exposure, smoking topography, and so forth, provide an effective indicator of exposure that has been associated with risk. However, these measures may be insensitive to small changes in risk, are difficult to assess accurately over time, and have not been tested in the context of harm reduction. Also, because there is interindividual variation in how the body responds to these exposures, such measures might not be sufficiently accurate for new products intended to decrease exposure. The relationship of external exposure markers to disease risk might be less predictable for new products. Currently, there is sufficient evidence to show that biomarkers can provide better estimates of risk in the context of exposure, and therefore they will likely be able to provide improved assessments for harm reduction products. However, no single biomarker has been sufficiently validated and related to disease risk to be recommended as an intermediate biomarker of cancer risk. Thus, different types of biomarkers along the pathway from internal exposure to biologically effective dose, and to potential harm are needed, and additional research is necessary to identify the best combination of markers to be used. Experimental toxicity testing (in vitro and animal models) are not sufficient to support a tobacco-related PREP claim because only biomarkers can show that the PREP reduces exposure adequately enough to imply risk reduction. However, the use of intermediate biomarkers as surrogate risk factors for disease may overestimate the number of persons who actually develop disease because not all early changes in morphology or function progress to disease. On the other hand, it may underestimate if, as expected, other mechanisms are involved in the disease process that are not reflected by the biomarkers. Therefore, the implication of a potential benefit in a harm reduction strategy could also be an overestimate, but this limitation in the scientific methodology for identifying sufficiently specific biomarkers of risk requires acceptance at the current time.

Previously, the most common way in which exposure reduction has been inferred is through the use of methods that simulate human smoking behavior, such as the FTC method. Although they provide a standardized way to assess cigarettes, it is clear that these methods have limited usefulness because people smoke their cigarettes differently than the machine, with resultant differences in the types and amounts of exposure.

The use of biomarkers improves exposure assessments (e.g., characterizing low-dose exposures or low-risk populations), provides a relative contribution of individual chemical carcinogens from complex mixtures (e.g., TSNAs and PAHs in cigarette smoke), and estimates total burden of a particular exposure where there are numerous sources (e.g., BaP from air, tobacco, diet, and occupation) (Vineis and Porta, 1996). Biomarkers also can establish differences in individual susceptibilities and whether there are differences in response depending on dose. Thus, biomarkers that measure both complex exposures and single tobacco product constituents are needed and should be assessed for the range of possible human exposures and those that assess complex exposures should carry a greater weight. Also, some biomarkers should be used that are less specific for individual tobacco constituents in order to monitor for the introduction of new hazards from tobacco-related PREPs.

Today, there remain technical limitations to the use of biomarkers. Depending on the harmful effect, surrogate assays in nontarget fluids or organs that represent effects in target organs may be easier to perform in humans because the target tissues might not be easily accessible. However, if such is the case, the relevance of the surrogate biomarker to the effect in the target organ should be demonstrated.

The use of a biomarker for harm reduction assessment should include several considerations, including where it is along the pathway from exposure to disease, its specificity and sensitivity, available harm dose-response data, available reduction in harm dose-response data, target tissue effect, and how it is validated. The need for validation cannot be overemphasized. Each biomarker should be validated for its relationship to exposure and harm and also as a laboratory assay that provides reliable and reproducible data. Separately, the way a biomarker is affected by interindividual variation in response and by behavior should also be considered.

Assessment of harm and harm reduction should be made through direct human experience, as the products are used by the general population. Most of what is known about harmful tobacco products has resulted from epidemiology, supported by in vitro studies, laboratory animal studies, and human experiments. However, while epidemiological studies can provide the most definitive data about tobacco harm and harm reduction products, the study of diseases with long latency (e.g., cancer, heart disease, chronic obstructive pulmonary disease) is problematic because such studies require many years before they provide useful data. Thus, because definitive evidence for a new risk reduction product is not available short-term markers that reflect long-term outcomes are needed. If an approach for assessing risk reduction products required only epidemiological data measuring disease outcome prior to use by the public, then an opportunity to reduce morbidity and early mortality might be missed. However, the use of intermediate markers does not replace long-term follow-up and epidemiological surveillance, but allows judgments to be made until such data are forthcoming.

Biomarkers of internal exposure, biologically effective dose, or potential harm have been validated to different degrees. It is typically easier to show a direct relationship of external exposure to biomarkers in the following order: internal exposure, biologically effective dose, and potential harm. Conversely, it is typically easier to show a direct relationship of disease outcome to biomarkers in the following order: potential harm, biologically effective dose, and internal exposure. It might be acceptable to rely on external exposure measurements for considering risk and dose-response, but only with substantial corroborative biomarker data. The best strategy for assessing the claims for risk reduction methods is with several markers that range from exposure to outcome, one being linked to another, and at least one with which a dose-response risk assessment can be made.

The recommendation that harm reduction products should be assessed with the use of biomarkers reflects sufficient available data to show that the public is composed of individuals with different cultural and heritable traits that affect how people use tobacco products and respond to them. To achieve the best confidence that a PREP will reduce risks for persons who cannot stop smoking, both well-validated methods for predicting risk, including external exposure indicators, and the best available biomarker assays should be used.

RESEARCH AGENDA

There are currently different methodologies for assessing PREPs, but substantial research is needed to increase confidence in the application of these methods. Although it may be possible to improve external methods for assessing exposures, such as through modification of the FTC method or improving questionnaire assessments, there is so much variability in human smoking behavior that it is believed these methods could never be much more helpful than they already are. This recommendation does not imply that questionnaires and topography instruments are not helpful in assessing smoking behavior, because they are, but it is unlikely that the methodology can be improved substantially. Indeed, clinical epidemiological studies generally have to integrate more variables for smoking behavior (e.g., accurately documenting changes in smoking, brand switching).

The development and validation of biomarkers for assessing harm reduction must be accelerated for all diseases, especially for cardiovascular and respiratory diseases because less research has been conducted compared to cancer.

The use of a biomarker for assessing harm reduction should be considered using the criteria provided in this chapter. Dose-response relationships should be established, and the biomarker should be assessed for reversibility in smoking cessation trials. In all studies of biomarker validation, consideration should be made of what nontobacco exposures, if any, would influence the biomarker study results. Also, biomarkers have to be tested and validated in different populations, to determine whether they are affected by susceptible subpopulations, and within genders, races, or ethnicities. Research efforts should focus on biomarkers that might be used for existing cohort studies, where disease outcome already is known. For example, markers are needed that can be used in serum or small amounts of DNA from stored samples. This is the best way to identify a relationship between exposure, a biomarker, and disease risk. Substantial research is needed to identify the relationships between biomarkers to exposure, biologically effective doses, and biomarkers of harm. Study designs that can provide these linkages are needed, and the best evidence will come from cohort studies.

Internal biomarkers of exposure such as cotinine, nicotine boosts, CO, and CO boosts provide good information about exposure, including to environmental tobacco smoke, but additional markers, such as urinary anabasine and anatabine levels, have to be developed for use in persons who are concurrently using nicotine replacement therapy. Increased efforts to measure urinary excretion of carcinogen metabolites, which are currently showing promise for use in risk assessment of active smoking and ETS, are needed. Examples include urinary excretion of tobacco-specific nitrosamines and polycyclic aromatic hydrocarbons and urinary mutagenicity, where these reflect both single and complex markers of exposure, respectively. Also, markers with longer half-lives would be useful to avoid confounding by recent changes in smoking behavior.

Biomarkers that reflect the biologically effective doses of exposure to carcinogens must be improved and validated. Newer technologies are now available that are more sensitive (e.g., mass spectroscopy) and can provide more information, and these should be applied in experimental systems and human studies that were developed long before such methods were available. For example, the determination of carcinogen-DNA adducts might be useful where small amounts of tissue are available (e.g., buccal swabs, sputum, blood).

More biomarkers of potential harm are currently being developed than any other types. This is because pathobiological pathways are well understood and newer technologies are available to explore them. However, along with better technologies will come limitations in the interpretation of new data (e.g., mRNA expression assays, proteomics). As researchers explore greater numbers of gene-smoking interactions and accumulate data for numerous genes expressed in response to exposures, it is clear that there are insufficient methods to analyze data where there are a substantial number of predictor variables. Also, some data will have to be reduced to clusters or other smaller units that are understandable in the context of biological hypotheses. Increased research is needed in methodologies to interpret these types of data, to validate the new models in the context of disease outcome.

For cancer, increased efforts are needed to assess target organ assays, such as genetic damage in lung cells in sputum and exfoliated bladder cells in urine, in persons years before they have a clinically detectable cancer. Given that genetic damage is only one part of the carcinogenic process, additional efforts are necessary to develop biomarkers for other pathways, such as gene silencing through hypermethylation of promoter regions. For cardiac disease, additional studies are needed to validate biomarkers of platelet function, endothelial function, endothelial thickening, and plaque formation and thrombosis. For respiratory disease, better markers are needed to assess changes in lung function that predict chronic obstructive pulmonary disease and asthma, and to assess immunological changes that will increase risk of respiratory infections.

It would be optimal to identify biomarkers that can be used to assess risk for several diseases. For example, biomarkers of oxidative damage might identify risk for cardiac disease, cancer, and respiratory illness. However, because the relationship of oxidative damage to these diseases remains mostly an unproved hypothesis, research is needed in this area.

Biomarkers will have to assess PREPs for single tobacco constituents and complex mixtures. The use of biomarkers that can assess multiple exposures from complex mixtures is critical because new tobacco-related PREPs might include compounds that are not present in existing tobacco constituents, or the ratio of exposures to individual constituents might change. A committee of experts should be convened to consider and identify those biomarkers that have the most promise and to determine what combination of biomarkers should be part of a panel for assessing PREPs.

To identify those biomarkers most useful for assessing harm reduction products, current efforts have to be focused on clinical trials that assess the effects of switching brands, using new products, and reducing daily consumption of tobacco through the concomitant use of nicotine replace therapy or other aids used for smoking cessation.

There are unique opportunities in epidemiological studies to validate biomarkers for use in assessing harm reduction strategies. Specifically, cohorts of participants in smoking cessation programs and former smokers should be established because these individuals represent the best possible reduction in the risk due to smoking. The collection of tissues and fluids from persons who have quit smoking and comparisons of persons who do and do not develop disease would be very helpful in determining which biomarkers have the most predictive value. This should be done in the context of previous smoking history to identify which persons would obtain the greatest benefit from cessation and how biomarkers might be able to identify individuals at greatest risk within these groups. Some diseases, such as cardiovascular disease, have a relatively rapid decline in risk following cessation so it would be quicker to validate cardiac disease risk factors. For cancer, the studies will take much longer. Monitoring populations that are at the highest risk of cancer, such as persons with resected early-stage lung cancer or bladder cancer, might be useful in this context. If a biomarker cannot predict increased risk in former smokers, it is unlikely to be useful in assessing PREPs.

REFERENCES

Aguilar F, Harris CC, Sun T, Hollstein M, Cerutti P. 1994. Geographic variation of p53 mutational profile in nonmalignant human liver. *Science* 264(5163):1317-1319.

Ahrendt SA, Chow JT, Yang SC, et al. 2000. Alcohol consumption and cigarette smoking increase the frequency of p53 mutations in non-small cell lung cancer. *Cancer Res* 60(12):3155-3159.

Ahrendt SA, Halachmi S, Chow JT, et al. 1999. Rapid p53 sequence analysis in primary lung cancer using an oligonucleotide probe array. *Proc Natl Acad Sci U S A* 96(13):7382-7387.

Ammenheuser MM, Hastings DA, Whorton EB, Ward JB. 1997. Frequencies of hprt mutant lymphocytes in smokers, non-smokers, and former smokers. *Environ Mol Mutagen* 30(2):131-138.

Ashurst SW, Cohen GM, Nesnow S, DiGiovanni J, Slaga TJ. 1983. Formation of benzo(a)-pyrene/DNA adducts and their relationship to tumor initiation in mouse epidermis. *Cancer Res* 43(3):1024-1029.

Atawodi SE, Lea S, Nyberg F, et al. 1998. 4-Hydroxy-1-(3-pyridyl)-1-butanone-hemoglobin adducts as biomarkers of exposure to tobacco smoke: validation of a method to be used in multicenter studies. *Cancer Epidemiol Biomarkers Prev* 7(9):817-821.

Attardi LD, Jacks T. 1999. The role of p53 in tumour suppression: lessons from mouse models. *Cell Mol Life Sci* 55(1):48-63.

Badawi AF, Hirvonen A, Bell DA, Lang NP, Kadlubar FF. 1995. Role of aromatic amine acetyltransferases, NAT1 and NAT2, in carcinogen-DNA adduct formation in the human urinary bladder. *Cancer Res* 55(22):5230-5237.

Bannon MJ, Granneman JG, Kapatos G. 1995. The dopamine transporter. Bloom FE, Kupfer DL, ed. *Psychopharmacology: The 4th Generation of Progress.* New York: Raven Press. Pp. 179-188.

Bartsch H, Petruzzelli S, De Flora S, et al. 1991. Carcinogen metabolism and DNA adducts in human lung tissues as affected by tobacco smoking or metabolic phenotype: a case-control study on lung cancer patients. *Mutat Res* 250(1-2):103-114.

Bartsch H, Rojas M, Alexandrov K, et al. 1995. Metabolic polymorphism affecting DNA binding and excretion of carcinogens in humans. *Pharmacogenetics* 5 Spec No:S84-90.

Battig K, Buzzi R, Nil R. 1982. Smoke yield of cigarettes and puffing behavior in men and women. *Psychopharmacology (Berl)* 76(2):139-148.

Bell DA, Taylor JA, Paulson DF, Robertson CN, Mohler JL, Lucier GW. 1993. Genetic risk and carcinogen exposure: a common inherited defect of the carcinogen-metabolism gene glutathione S-transferase M1 (GSTM1) that increases susceptibility to bladder cancer. *J Natl Cancer Inst* 85(14):1159-1164.

Bender MA, Awa AA, Brooks AL, et al. 1988. Current status of cytogenetic procedures to detect and quantify previous exposures to radiation. *Mutat Res* 196(2):103-159.

Bennett WP, Hussain SP, Vahakangas KH, Khan MA, Shields PG, Harris CC. 1999. Molecular epidemiology of human cancer risk: gene-environment interactions and p53 mutation spectrum in human lung cancer. *J Pathol* 187(1):8-18.

Benowitz NL. 1999. Biomarkers of environmental tobacco smoke exposure. *Environ Health Perspect* 107 Suppl 2:349-355.

Benowitz NL, Hall SM, Herning RI, Jacob P 3d, Jones RT, Osman AL. 1983. Smokers of low-yield cigarettes do not consume less nicotine. *N Engl J Med* 309(3):139-142.

Benowitz NL, Jacob P 3d, Kozlowski LT, Yu L. 1986a. Influence of smoking fewer cigarettes on exposure to tar, nicotine, and carbon monoxide. *N Engl J Med* 315(21):1310-1313.

Benowitz NL, Jacob P 3d, Yu L, Talcott R, Hall S, Jones RT. 1986b. Reduced tar, nicotine, and carbon monoxide exposure while smoking ultralow- but not low-yield cigarettes. *JAMA* 256(2):241-246.

Benowitz NL, Zevin S, Jacob P 3rd. 1998. Suppression of nicotine intake during ad libitum cigarette smoking by high-dose transdermal nicotine. *J Pharmacol Exp Ther* 287(3):958-962.

Bishop JM. 1987. The molecular genetics of cancer. *Science* 235:305-311.

Bjelogrlic NM, Makinen M, Stenback F, Vahakangas K. 1994. Benzo[a]pyrene-7,8-diol-9,10-epoxide-DNA adducts and increased p53 protein in mouse skin. *Carcinogenesis* 15(4):771-774.

Bonassi S, Abbondandolo A, Camurri L, et al. 1995. Are chromosome aberrations in circulating lymphocytes predictive of future cancer onset in humans? Preliminary results of an Italian cohort study. *Cancer Genet Cytogenet* 79(2):133-135.

Bono R, Russo R, Arossa W, Scursatone E, Gilli G. 1996. Involuntary exposure to tobacco smoke in adolescents: urinary cotinine and environmental factors. *Arch Environ Health* 51(2):127-131.

Brennan JA, Boyle JO, Koch WM, et al. 1995. Association between cigarette smoking and mutation of the p53 gene in squamous-cell carcinoma of the head and neck. *N Engl J Med* 332(11):712-717.

Bridges RB, Combs JG, Humble JW, Turbek JA, Rehm SR, Haley NJ. 1990. Puffing topography as a determinant of smoke exposure. *Pharmacol Biochem Behav* 37(1):29-39.

Bridges RB, Humble JW, Turbek JA, Rehm SR. 1986. Smoking history, cigarette yield and smoking behavior as determinants of smoke exposure. *Eur J Respir Dis Suppl* 146:129-137.

Brockmoller J, Cascorbi I, Kerb R, Roots I. 1996. Combined analysis of inherited polymorphisms in arylamine N-acetyltransferase 2, glutathione S-transferases M1 and T1, microsomal epoxide hydrolase, and cytochrome P450 enzymes as modulators of bladder cancer risk. *Cancer Res* 56(17):3915-3925.

Brockmoller J, Cascorbi I, Kerb R, Sachse C, Roots I. 1998. Polymorphisms in xenobiotic conjugation and disease predisposition. *Toxicol Lett* 102-103:173-183.

Bryant MS, Skipper PL, Tannenbaum SR, Maclure M. 1987. Hemoglobin adducts of 4-aminobiphenyl in smokers and nonsmokers. *Cancer Res* 47(2):602-608.

Bryant MS, Vineis P, Skipper PL, Tannenbaum SR. 1988. Hemoglobin adducts of aromatic amines: associations with smoking status and type of tobacco. *Proc Natl Acad Sci U S A* 85(24):9788-9791.

Buist AS, Sexton GJ, Nagy JM, Ross BB. 1976. The effect of smoking cessation and modification on lung function. *Am Rev Respir Dis* 114(1):115-122.

Byrd GD, Davis RA, Caldwell WS, Robinson JH, deBethizy JD. 1998. A further study of FTC yield and nicotine absorption in smokers. *Psychopharmacology (Berl)* 139(4):291-299.

Byrd GD, Robinson JH, Caldwell WS, deBethizy JD. 1995. Comparison of measured and FTC-predicted nicotine uptake in smokers. *Psychopharmacology (Berl)* 122(2):95-103.

Carmella SG, Akerkar SA, Richie JP, Hecht SS. 1995. Intraindividual and interindividual differences in metabolites of the tobacco-specific lung carcinogen 4-(methylnitrosamino)-1-(3-pyridyl)-1-butanone (NNK) in smokers' urine. *Cancer Epidemiol Biomarkers Prev* 4(6):635-642.

Carmella SG, Kagan SS, Kagan M, et al. 1990. Mass spectrometric analysis of tobacco-specific nitrosamine hemoglobin adducts in snuff dippers, smokers, and nonsmokers. *Cancer Res* 50(17):5438-5445.

Carmelli D, Swan GE, Robinette D, Fabsitz R. 1992. Genetic influence on smoking—a study of male twins. *N Engl J Med* 327(12):829-833.

CDC (Centers for Disease Control and Prevention). 2000. National Health and Nutrition Examination Survey. [Online]. Available: http://www.cdc.gov/nchs/nhanes.htm [accessed 2001].

Cloos J, Braakhuis BJ, Steen I, et al. 1994. Increased mutagen sensitivity in head-and-neck squamous-cell carcinoma patients, particularly those with multiple primary tumors. *Int J Cancer* 56(6):816-819.

Cloos J, Spitz MR, Schantz SP, et al. 1996. Genetic susceptibility to head and neck squamous cell carcinoma. *J Natl Cancer Inst* 88(8):530-535.

Comings DE, Ferry L, Bradshaw-Robinson S, Burchette R, Chiu C, Muhleman D. 1996. The dopamine D2 receptor (DRD2) gene: a genetic risk factor in smoking. *Pharmacogenetics* 6(1):73-79.

Committee on Biological Markers of the National Research Council. 1987. Biological markers in environmental health research. *Environ Health Perspect* 74:3-9.

Crawford FG, Mayer J, Santella RM, et al. 1994. Biomarkers of environmental tobacco smoke in preschool children and their mothers. *J Natl Cancer Inst* 86(18):1398-1402.

Cullen P, Schulte H, Assmann G. 1997. The Munster Heart Study (PROCAM): total mortality in middle-aged men is increased at low total and LDL cholesterol concentrations in smokers but not in nonsmokers. *Circulation* 96(7):2128-2136.

Djordjevic MV, Stellman SD, Zang E. 2000. Doses of nicotine and lung carcinogens delivered to cigarette smokers. *J Natl Cancer Inst* 92(2):106-111.

Drin I, Schoket B, Kostic S, Vincze I. 1994. Smoking-related increase in O6-alkylguanine-DNA alkyltransferase activity in human lung tissue. *Carcinogenesis* 15(8):1535-1539.

Dunn BP, Vedal S, San RH, et al. 1991. DNA adducts in bronchial biopsies. *Int J Cancer* 48(4):485-492.

Ebert RV, McNabb ME, McCusker KT, Snow SL. 1983. Amount of nicotine and carbon monoxide inhaled by smokers of low-tar, low-nicotine cigarettes. *JAMA* 250(20):2840-2842.

EPA (Environmental Protection Agency). 1992. *Respiratory Health Effects of Passive Smoking; Lung Cancer and Other Disorders.* Washington, DC: EPA.

Farmer PB, Shuker DE. 1999. What is the significance of increases in background levels of carcinogen-derived protein and DNA adducts? Some considerations for incremental risk assessment. *Mutat Res* 424(1-2):275-286.

Fearon ER. 1997. Human cancer syndromes: clues to the origin and nature of cancer. *Science* 278(5340):1043-1050.

Fischer S, Spiegelhalder B, Preussmann R. 1989. Influence of smoking parameters on the delivery of tobacco-specific nitrosamines in cigarette smoke—a contribution to relative risk evaluation. *Carcinogenesis* 10(6):1059-1066.

Fliss MS, Usadel H, Caballero OL, et al. 2000. Facile detection of mitochondrial DNA mutations in tumors and bodily fluids. *Science* 287(5460):2017-2019.

Freeman DJ, Griffin BA, Murray E, et al. 1993. Smoking and plasma lipoproteins in man: effects on low-density lipoprotein cholesterol levels and high-density lipoprotein subfraction distribution. *Eur J Clin Invest* 23(10):630-640.

Fulton JE, Shekelle RB. 1997. Cigarette smoking, weight gain, and coronary mortality: results from the Chicago Western Electric Study. *Circulation* 96(5):1438-1444.

Fuster V, Chesebro JH, Frye RL, Elveback LR. 1981. Platelet survival and the development of coronary artery disease in the young adult: effects of cigarette smoking, strong family history and medical therapy. *Circulation* 63(3):546-551.

Gealy R, Zhang L, Siegfried JM, Luketich JD, Keohavong P. 1999. Comparison of mutations in the p53 and K-ras genes in lung carcinomas from smoking and nonsmoking women. *Cancer Epidemiol Biomarkers Prev* 8(4 Pt 1):297-302.

Geneste O, Camus AM, Castegnaro M, et al. 1991. Comparison of pulmonary DNA adduct levels, measured by 32P-postlabelling and aryl hydrocarbon hydroxylase activity in lung parenchyma of smokers and ex-smokers. *Carcinogenesis* 12(7):1301-1305.

Gilbert DG, Gilbert BO. 1995. Personality, psychopathology, and nicotine response as mediators of the genetics of smoking. *Behav Genet* 25(2):133-147.

Giovino GA. 1999. Epidemiology of tobacco use among US adolescents. *Nicotine Tob Res* 1(Suppl 1):S31-40.

Giovino GA, Henningfield JE, Tomar SL, Escobedo LG, Slade J. 1995. Epidemiology of tobacco use and dependence. *Epidemiol Rev* 17(1):48-65.

Grafstrom RC, Dypbukt JM, Sundqvist K, et al. 1994. Pathobiological effects of acetaldehyde in cultured human epithelial cells and fibroblasts. *Carcinogenesis* 15(5):985-990.

Green MS, Harari G. 1995. A prospective study of the effects of changes in smoking habits on blood count, serum lipids and lipoproteins, body weight and blood pressure in occupationally active men. The Israeli CORDIS Study. *J Clin Epidemiol* 48(9):1159-1166.

Greenblatt MS, Bennett WP, Hollstein M, Harris CC. 1994. Mutations in the p53 tumor suppressor gene: clues to cancer etiology and molecular pathogenesis. *Cancer Res* 54(18):4855-4878.

Grinberg-Funes RA, Singh VN, Perera FP, et al. 1994. Polycyclic aromatic hydrocarbon-DNA adducts in smokers and their relationship to micronutrient levels and the glutathione-S-transferase M1 genotype. *Carcinogenesis* 15(11):2449-2454.

Gritz ER, Rose JE, Jarvik ME. 1983. Regulation of tobacco smoke intake with paced cigarette presentation. *Pharmacol Biochem Behav* 18(3):457-462.

Guengerich FP. 2000. Metabolism of chemical carcinogens. *Carcinogenesis* 21(3):345-351.

Hagmar L, Brogger A, Hansteen IL, et al. 1994. Cancer risk in humans predicted by increased levels of chromosomal aberrations in lymphocytes: Nordic study group on the health risk of chromosome damage. *Cancer Res* 54(11):2919-2922.

Hall J, Bresil H, Donato F, et al. 1993. Alkylation and oxidative-DNA damage repair activity in blood leukocytes of smokers and non-smokers. *Int J Cancer* 54(5):728-733.

Harris CC, Hollstein M. 1993. Clinical implications of the p53 tumor-suppressor gene. *N Engl J Med* 329(18):1318-1327.

Hartmann A, Blaszyk H, Kovach JS, Sommer SS. 1997. The molecular epidemiology of p53 gene mutations in human breast cancer. *Trends Genet* 13(1):27-33.

Heath AC, Cates R, Martin NG, et al. 1993a. Genetic contribution to risk of smoking initiation: comparisons across birth cohorts and across cultures. *J Subst Abuse* 5(3):221-246.

Heath AC, Martin NG. 1993b. Genetic models for the natural history of smoking: evidence for a genetic influence on smoking persistence. *Addict Behav* 18(1):19-34.

Hecht SS. 1999. DNA adduct formation from tobacco-specific N-nitrosamines. *Mutat Res* 424(1-2):127-142.

Hecht SS, Carmella SG, Murphy SE, Akerkar S, Brunnemann KD, Hoffmann D. 1993. A tobacco-specific lung carcinogen in the urine of men exposed to cigarette smoke. *N Engl J Med* 329(21):1543-1546.

Helbock HJ, Beckman KB, Shigenaga MK, et al. 1998. DNA oxidation matters: the HPLC-electrochemical detection assay of 8-oxo-deoxyguanosine and 8-oxo-guanine. *Proc Natl Acad Sci U S A* 95(1):288-293.

Hellerstein MK, Benowitz NL, Neese RA, et al. 1994. Effects of cigarette smoking and its cessation on lipid metabolism and energy expenditure in heavy smokers. *J Clin Invest* 93(1):265-272.

Henning S, Cascorbi I, Munchow B, Jahnke V, Roots I. 1999. Association of arylamine N-acetyltransferases NAT1 and NAT2 genotypes to laryngeal cancer risk. *Pharmacogenetics* 9(1):103-111.

Herning RI, Jones RT, Benowitz NL, Mines AH. 1983. How a cigarette is smoked determines blood nicotine levels. *Clin Pharmacol Ther* 33(1):84-90.

Hill P, Marquardt H. 1980. Plasma and urine changes after smoking different brands of cigarettes. *Clin Pharmacol Ther* 27(5):652-658.

Hofer I, Nil R, Wyss F, Battig K. 1992. The contributions of cigarette yield, consumption, inhalation and puffing behaviour to the prediction of smoke exposure. *Clin Investig* 70(3-4):343-351.

Hoffmann D, Hoffmann I. 1997. The changing cigarette, 1950-1995. *J Toxicol Environ Health* 50(4):307-364.

Hollstein M, Hergenhahn M, Yang Q, Bartsch H, Wang ZQ, Hainaut P. 1999. New approaches to understanding p53 gene tumor mutation spectra. *Mutat Res* 431(2):199-209.

Hou SM, Yang K, Nyberg F, Hemminki K, Pershagen G, Lambert B. 1999. Hprt mutant frequency and aromatic DNA adduct level in non-smoking and smoking lung cancer patients and population controls. *Carcinogenesis* 20(3):437-444.

Hurt RD, Sachs DP, Glover ED, et al. 1997. A comparison of sustained-release bupropion and placebo for smoking cessation. *N Engl J Med* 337(17):1195-1202.

IARC (International Agency on the Research of Cancer). 1986. *IARC Monographs on the Evaluation of the Carcinogenic Risk of Chemicals to Humans: Tobacco Smoking.* Vol. 38 ed. Lyon, France: IARC.

Ishibe N, Wiencke JK, Zuo ZF, McMillan A, Spitz M, Kelsey KT. 1997. Susceptibility to lung cancer in light smokers associated with CYP1A1 polymorphisms in Mexican- and African-Americans. *Cancer Epidemiol Biomarkers Prev* 6(12):1075-1080.

Izzotti A, De Flora S, Petrilli GL, et al. 1995. Cancer biomarkers in human atherosclerotic lesions: detection of DNA adducts. *Cancer Epidemiol Biomarkers Prev* 4(2):105-110.

Izzotti A, Rossi GA, Bagnasco M, De Flora S. 1991. Benzo[a]pyrene diolepoxide-DNA adducts in alveolar macrophages of smokers. *Carcinogenesis* 12(7):1281-1285.

Jacob P, Yu L, Shulgin AT, Benowitz NL. 1999. Minor tobacco alkaloids as biomarkers for tobacco use: comparison of users of cigarettes, smokeless tobacco, cigars, and pipes. *Am J Public Health* 89(5):731-736.

Jaffe RL, Nicholson WJ, Garro AJ. 1983. Urinary mutagen levels in smokers. *Cancer Lett* 20(1):37-42.

Jahnke GD, Thompson CL, Walker MP, Gallagher JE, Lucier GW, DiAugustine RP. 1990. Multiple DNA adducts in lymphocytes of smokers and nonsmokers determined by 32P-postlabeling analysis. *Carcinogenesis* 11(2):205-211.

James AL, Knuiman MW, Divitini ML, Musk AW, Ryan G, Bartholomew HC. 1999. Associations between white blood cell count, lung function, respiratory illness and mortality: the Busselton Health Study. *Eur Respir J* 13(5):1115-1119.

Jin X, Wu X, Roth JA, et al. 1995. Higher lung cancer risk for younger African-Americans with the Pro/Pro p53 genotype. *Carcinogenesis* 16(9):2205-2208.

Jones IM, Moore DH, Thomas CB, Thompson CL, Strout CL, Burkhart-Schultz K. 1993. Factors affecting HPRT mutant frequency in T-lymphocytes of smokers and nonsmokers. *Cancer Epidemiol Biomarkers Prev* 2(3):249-260.

Jorenby DE, Leischow SJ, Nides MA, et al. 1999. A controlled trial of sustained-release bupropion, a nicotine patch, or both for smoking cessation. *N Engl J Med* 340(9):685-691.

Jourenkova-Mironova N, Voho A, Bouchardy C, et al. 1999. Glutathione S-transferase GSTM1, GSTM3, GSTP1 and GSTT1 genotypes and the risk of smoking-related oral and pharyngeal cancers. *Int J Cancer* 81(1):44-48.

Jourenkova N, Reinikainen M, Bouchardy C, Dayer P, Benhamou S, Hirvonen A. 1998. Larynx cancer risk in relation to glutathione S-transferase M1 and T1 genotypes and tobacco smoking. *Cancer Epidemiol Biomarkers Prev* 7(1):19-23.

Kanner RE, Connett JE, Williams DE, Buist AS. 1999. Effects of randomized assignment to a smoking cessation intervention and changes in smoking habits on respiratory symptoms in smokers with early chronic obstructive pulmonary disease: the Lung Health Study. *Am J Med* 106(4):410-416.

Kato S, Bowman ED, Harrington AM, Blomeke B, Shields PG. 1995. Human lung carcinogen-DNA adduct levels mediated by genetic polymorphisms in vivo. *J Natl Cancer Inst* 87(12):902-907.

Kaufman DW, Palmer JR, Rosenberg L, Stolley P, Warshauer E, Shapiro S. 1989. Tar content of cigarettes in relation to lung cancer. *Am J Epidemiol* 129(4):703-711.

Kawajiri K, Eguchi H, Nakachi K, Sekiya T, Yamamoto M. 1996. Association of CYP1A1 germ line polymorphisms with mutations of the p53 gene in lung cancer. *Cancer Res* 56(1):72-76.

Kellermann G, Shaw CR, Luyten-Kellerman M. 1973. Aryl hydrocarbon hydroxylase inducibility and bronchogenic carcinoma. *N Engl J Med* 289(18):934-937.

Khoury MJ, Wagener DK. 1995. Epidemiological evaluation of the use of genetics to improve the predictive value of disease risk factors. *Am J Hum Genet* 56(4):835-844.

Kolonen S, Tuomisto J, Puustinen P, Airaksinen MM. 1992a. Effects of smoking abstinence and chain-smoking on puffing topography and diurnal nicotine exposure. *Pharmacol Biochem Behav* 42(2):327-332.

Kolonen S, Tuomisto J, Puustinen P, Airaksinen MM. 1992b. Puffing behavior during the smoking of a single cigarette in a naturalistic environment. *Pharmacol Biochem Behav* 41(4):701-706.

Kouri RE, McKinney CE, Slomiany DJ, Snodgrass DR, Wray NP, McLemore TL. 1982. Positive correlation between high aryl hydrocarbon hydroxylase activity and primary lung cancer as analyzed in cryopreserved lymphocytes. *Cancer Res* 42(12):5030-5037.

La DK, Swenberg JA. 1996. DNA adducts: biological markers of exposure and potential applications to risk assessment. *Mutat Res* 365(1-3):129-146.

La Vecchia C, Bidoli E, Barra S, et al. 1990. Type of cigarettes and cancers of the upper digestive and respiratory tract. *Cancer Causes Control* 1(1):69-74.

Lange P, Groth S, Nyboe GJ, et al. 1989. Effects of smoking and changes in smoking habits on the decline of FEV1. *Eur Respir J* 2(9):811-816.

Lassila R, Seyberth HW, Haapanen A, Schweer H, Koskenvuo M, Laustiola KE. 1988. Vasoactive and atherogenic effects of cigarette smoking: a study of monozygotic twins discordant for smoking. *BMJ* 297(6654):955-957.

Law MR, Morris JK, Watt HC, Wald NJ. 1997. The dose-response relationship between cigarette consumption, biochemical markers and risk of lung cancer. *Br J Cancer* 75(11):1690-1693.

Lazarus P, Sheikh SN, Ren Q, et al. 1998. p53, but not p16 mutations in oral squamous cell carcinomas are associated with specific CYP1A1 and GSTM1 polymorphic genotypes and patient tobacco use. *Carcinogenesis* 19(3):509-514.

Lee CK, Brown BG, Reed EA, Coggins CR, Doolittle DJ, Hayes AW. 1993. Ninety-day inhalation study in rats, using aged and diluted sidestream smoke from a reference cigarette: DNA adducts and alveolar macrophage cytogenetics. *Fundam Appl Toxicol* 20(4):393-401.

Lehman TA, Scott F, Seddon M, et al. 1996. Detection of K-ras oncogene mutations by polymerase chain reaction-based ligase chain reaction. *Anal Biochem* 239(2):153-159.

Lerman C, Caporaso N, Audrain J, Main D, Bowman ED, Lockshin B, Boyd NR, Shields PG. 1999. Evidence suggesting the role of specific genetic factors in cigarette smoking. *Health Psychol* 18(1):14-20.

Lerman C, Caporaso N, Main D, Audrain J, Boyd NR, Bowman ED, Shields PG. 1998. Depression and self-medication with nicotine: the modifying influence of the dopa receptor gene. *Health Psychol* 17(1):56-62.

Lerman C, Shields PG, Main D, et al. 1997. Lack of association of tyrosine hydroxylase genetic polymorphism with cigarette smoking. *Pharmacogenetics* 7(6):521-524.

Li D, Wang M, Cheng L, Spitz MR, Hittelman WN, Wei Q. 1996. In vitro induction of benzo(a)pyrene diol epoxide-DNA adducts in peripheral lymphocytes as a susceptibility marker for human lung cancer. *Cancer Res* 56(16):3638-3641.

Liu CS, Kao SH, Wei YH. 1997. Smoking-associated mitochondrial DNA mutations in human hair follicles. *Environ Mol Mutagen* 30(1):47-55.

Liu Y, Hernandez AM, Shibata D, Cortopassi GA. 1994. BCL2 translocation frequency rises with age in humans. *Proc Natl Acad Sci U S A* 91(19):8910-8914.

Lodovici M, Akpan V, Giovannini L, Migliani F, Dolara P. 1998. Benzo[a]pyrene diol-epoxide DNA adducts and levels of polycyclic aromatic hydrocarbons in autoptic samples from human lungs. *Chem Biol Interact* 116(3):199-212.

Lubin JH, Blot WJ, Berrino F, et al. 1984. Patterns of lung cancer risk according to type of cigarette smoked. *Int J Cancer* 33(5):569-576.

Ludviksdottir D, Blondal T, Franzon M, Gudmundsson TV, Sawe U. 1999. Effects of nicotine nasal spray on atherogenic and thrombogenic factors during smoking cessation. *J Intern Med* 246(1):61-66.

Lunn RM, Helzlsouer KJ, Parshad R, et al. 2000. XPD polymorphisms: effects on DNA repair proficiency. *Carcinogenesis* 21(4):551-555.

Lunn RM, Langlois RG, Hsieh LL, Thompson CL, Bell DA. 1999. XRCC1 polymorphisms: effects on aflatoxin B1-DNA adducts and glycophorin A variant frequency. *Cancer Res* 59(11):2557-2561.

Mao L, Lee JS, Kurie JM, et al. 1997. Clonal genetic alterations in the lungs of current and former smokers. *J Natl Cancer Inst* 89(12):857-862.

McCarthy DS, Craig DB, Cherniack RM. 1976. Effect of modification of the smoking habit on lung function. *Am Rev Respir Dis* 114(1):103-113.

McLemore TL, Adelberg S, Liu MC, et al. 1990. Expression of CYP1A1 gene in patients with lung cancer: evidence for cigarette smoke-induced gene expression in normal lung tissue and for altered gene regulation in primary pulmonary carcinomas. *J Natl Cancer Inst* 82(16):1333-1339.

Meyer MJ, Bechtold WE. 1996. Protein adduct biomarkers: state of the art. *Environ Health Perspect* 104 Suppl 5:879-882.

Mills NE, Fishman CL, Scholes J, Anderson SE, Rom WN, Jacobson DR. 1995. Detection of K-ras oncogene mutations in bronchoalveolar lavage fluid for lung cancer diagnosis. *J Natl Cancer Inst* 87(14):1056-1060.

Mohrenweiser HW, Jones IM. 1998. Variation in DNA repair is a factor in cancer susceptibility: a paradigm for the promises and perils of individual and population risk estimation? *Mutat Res* 400(1-2):15-24.

Mohtashamipur E, Norpoth K, Lieder F. 1985. Isolation of frameshift mutagens from smokers' urine: experiences with three concentration methods. *Carcinogenesis* 6(5):783-788.

Mooney LA, Santella RM, Covey L, et al. 1995. Decline of DNA damage and other biomarkers in peripheral blood following smoking cessation. *Cancer Epidemiol Biomarkers Prev* 4(6):627-634.

Morrow JD, Frei B, Longmire AW, et al. 1995. Increase in circulating products of lipid peroxidation (F2-isoprostanes) in smokers. Smoking as a cause of oxidative damage. *N Engl J Med* 332(18):1198-1203.

Mustonen R, Hemminki K. 1992. 7-Methylguanine levels in DNA of smokers' and nonsmokers' total white blood cells, granulocytes and lymphocytes. *Carcinogenesis* 13(11):1951-1955.

Mustonen R, Schoket B, Hemminki K. 1993. Smoking-related DNA adducts: 32P-postlabeling analysis of 7-methylguanine in human bronchial and lymphocyte DNA. *Carcinogenesis* 14(1):151-154.

Nakajima T, Elovaara E, Anttila S, et al. 1995. Expression and polymorphism of glutathione S-transferase in human lungs: risk factors in smoking-related lung cancer. *Carcinogenesis* 16(4):707-711.

Nakayama J, Yuspa SH, Poirier MC. 1984. Benzo(a)pyrene-DNA adduct formation and removal in mouse epidermis in vivo and in vitro: relationship of DNA binding to initiation of skin carcinogenesis. *Cancer Res* 44(9):4087-4095.

Nemeth-Coslett R, Griffiths RR. 1984. Determinants of puff duration in cigarette smokers: I. *Pharmacol Biochem Behav* 20(6):965-971.

Noble EP, St Jeor ST, Ritchie T, et al. 1994. D2 dopamine receptor gene and cigarette smoking: a reward gene? *Med Hypotheses* 42(4):257-260.

Nowak J, Murray JJ, Oates JA, FitzGerald GA. 1987. Biochemical evidence of a chronic abnormality in platelet and vascular function in healthy individuals who smoke cigarettes. *Circulation* 76(1):6-14.

O'Hara P, Connett JE, Lee WW, Nides M, Murray R, Wise R. 1998. Early and late weight gain following smoking cessation in the Lung Health Study. *Am J Epidemiol* 148(9):821-830.

O'Malley PG, Taylor AJ, Jackson JL, Doherty TM, Detrano RC. 2000. Prognostic value of coronary electron-beam computed tomography for coronary heart disease events in asymptomatic populations. *Am J Cardiol* 85(8):945-948.

Obe G, Vogt HJ, Madle S, Fahning A, Heller WD. 1982. Double-blind study on the effect of cigarette smoking on the chromosomes of human peripheral blood lymphocytes in vivo. *Mutat Res* 92(1-2):309-319.

Olincy A, Young DA, Freedman R. 1997. Increased levels of the nicotine metabolite cotinine in schizophrenic smokers compared to other smokers. *Biol Psychiatry* 42(1):1-5.

Oyama T, Kawamoto T, Mizoue T, et al. 1997. Cytochrome P450 2E1 polymorphism as a risk factor for lung cancer: in relation to p53 gene mutation. *Anticancer Res* 17(1B):583-587.

Park JW, Cundy KC, Ames BN. 1989. Detection of DNA adducts by high-performance liquid chromatography with electrochemical detection. *Carcinogenesis* 10(5):827-832.

Parry H, Cohen S, Schlarb JE, et al. 1997. Smoking, alcohol consumption, and leukocyte counts. *Am J Clin Pathol* 107(1):64-67.

Parsons WD, Carmella SG, Akerkar S, Bonilla LE, Hecht SS. 1998. A metabolite of the tobacco-specific lung carcinogen 4-(methylnitrosamino)-1-(3-pyridyl)-1-butanone in the urine of hospital workers exposed to environmental tobacco smoke. *Cancer Epidemiol Biomarkers Prev* 7(3):257-260.

Pastorelli R, Guanci M, Cerri A, et al. 1998. Impact of inherited polymorphisms in glutathione S-transferase M1, microsomal epoxide hydrolase, cytochrome P450 enzymes on DNA, and blood protein adducts of benzo(a)pyrene-diolepoxide. *Cancer Epidemiol Biomarkers Prev* 7(8):703-709.

Patrick DL, Cheadle A, Thompson DC, Diehr P, Koepsell T, Kinne S. 1994. Validity of self-reported smoking: a review and meta-analysis. *American Journal of Public Health* 84(7): 1086-1093.

Pelkonen O, Vahakangas K, Nebert DW. 1980. Binding of polycyclic aromatic hydrocarbons to DNA: comparison with mutagenesis and tumorigenesis. *J Toxicol Environ Health* 6(5-6):1009-1020.

Peluso M, Airoldi L, Armelle M, et al. 1998. White blood cell DNA adducts, smoking, and NAT2 and GSTM1 genotypes in bladder cancer: a case-control study. *Cancer Epidemiol Biomarkers Prev* 7(4):341-346.

Perera FP. 1987. Molecular cancer epidemiology: a new tool in cancer prevention. *J Natl Cancer Inst* 78(5):887-898.

Phillips AN, Neaton JD, Cook DG, Grimm RH, Shaper AG. 1992. The leukocyte count and risk of lung cancer. *Cancer* 69(3):680-684.

Phillips DH. 1997. Detection of DNA modifications by the 32P-postlabelling assay. *Mutat Res* 378(1-2):1-12.

Phillips DH, Hewer A, Martin CN, Garner RC, King MM. 1988. Correlation of DNA adduct levels in human lung with cigarette smoking. *Nature* 336(6201):790-792.

Poli P, Buschini A, Spaggiari A, Rizzoli V, Carlo-Stella C, Rossi C. 1999. DNA damage by tobacco smoke and some antiblastic drugs evaluated using the Comet assay. *Toxicol Lett* 108(2-3):267-276.

Pressl S, Edwards A, Stephan G. 1999. The influence of age, sex and smoking habits on the background level of fish-detected translocations. *Mutat Res* 442(2):89-95.

Prevost V, Shuker DE. 1996. Cigarette smoking and urinary 3-alkyladenine excretion in man. *Chem Res Toxicol* 9(2):439-444.

Raggi P, Callister TQ, Cooil B, et al. 2000. Identification of patients at increased risk of first unheralded acute myocardial infarction by electron-beam computed tomography. *Circulation* 101(8):850-855.

Ramsey MJ, Moore DH, Briner JF, et al. 1995. The effects of age and lifestyle factors on the accumulation of cytogenetic damage as measured by chromosome painting. *Mutat Res* 338(1-6):95-106.

Randerath E, Miller RH, Mittal D, Avitts TA, Dunsford HA, Randerath K. 1989. Covalent DNA damage in tissues of cigarette smokers as determined by 32P-postlabeling assay. *J Natl Cancer Inst* 81(5):341-347.

Randerath K, Reddy MV, Gupta RC. 1981. 32P-labeling test for DNA damage. *Proc Natl Acad Sci U S A* 78(10):6126-6129.

Rangemark C, Benthin G, Granstrom EF, Persson L, Winell S, Wennmalm A. 1992. Tobacco use and urinary excretion of thromboxane A2 and prostacyclin metabolites in women stratified by age. *Circulation* 86(5):1495-1500.

Rebbeck TR. 1997. Molecular epidemiology of the human glutathione S-transferase genotypes GSTM1 and GSTT1 in cancer susceptibility. *Cancer Epidemiol Biomarkers Prev* 6(9):733-743.

Reddy MV, Randerath K. 1986. Nuclease P1-mediated enhancement of sensitivity of 32P-postlabeling test for structurally diverse DNA adducts. *Carcinogenesis* 7(9):1543-1551.

Rennard SI, Daughton D, Fujita J, et al. 1990. Short-term smoking reduction is associated with reduction in measures of lower respiratory tract inflammation in heavy smokers. *Eur Respir J* 3(7):752-759.

Rojas M, Alexandrov K, Cascorbi I, et al. 1998. High benzo[a]pyrene diol-epoxide DNA adduct levels in lung and blood cells from individuals with combined CYP1A1 MspI/Msp-GSTM1*0/*0 genotypes. *Pharmacogenetics* 8(2):109-118.

Rojas M, Camus AM, Alexandrov K, et al. 1992. Stereoselective metabolism of (-)-benzo[a]pyrene-7,8-diol by human lung microsomes and peripheral blood lymphocytes: effect of smoking. *Carcinogenesis* 13(6):929-933.

Rosvold EA, McGlynn KA, Lustbader ED, Buetow KH. 1995. Identification of an NAD(P)H:quinone oxidoreductase polymorphism and its association with lung cancer and smoking. *Pharmacogenetics* 5(4):199-206.

Rothman N, Poirier MC, Baser ME, et al. 1990. Formation of polycyclic aromatic hydrocarbon-DNA adducts in peripheral white blood cells during consumption of charcoal-broiled beef. *Carcinogenesis* 11(7):1241-1243.

Ryberg D, Kure E, Lystad S, et al. 1994. p53 mutations in lung tumors: relationship to putative susceptibility markers for cancer. *Cancer Res* 54(6):1551-1555.

Ryberg D, Skaug V, Hewer A, et al. 1997. Genotypes of glutathione transferase M1 and P1 and their significance for lung DNA adduct levels and cancer risk. *Carcinogenesis* 18(7):1285-1289.

SAMHSA (Substance Abuse and Mental Health Services Administration). 1998. National Household Survey on Drug Abuse: Population Estimates 1998. [Online]. Available: http//:www.samhsa.gov/oasNHSDA/Pe1998/Pop98web1.htm#TopOfPage [accessed 2001].

Santella RM, Weston A, Perera FP, et al. 1988. Interlaboratory comparison of antisera and immunoassays for benzo[a]pyrene-diol-epoxide-I-modified DNA. *Carcinogenesis* 9(7):1265-1269.

Schoket B, Phillips DH, Kostic S, Vincze I. 1998. Smoking-associated bulky DNA adducts in bronchial tissue related to CYP1A1 MspI and GSTM1 genotypes in lung patients. *Carcinogenesis* 19(5):841-846.

Scott FM, Modali R, Lehman TA, et al. 1997. High frequency of K-ras codon 12 mutations in bronchoalveolar lavage fluid of patients at high risk for second primary lung cancer. *Clin Cancer Res* 3(3):479-482.

Seeman P. 1995. Dopamine receptors. Bloom FE, Lupfer DJ, ed. *Psycopharmacology; 4th Generation of Progress.* New York: Raven Press. Pp. 295-302.

Shields PG, Harris CC. 2000. Cancer risk and low-penetrance susceptibility genes in gene-environment interactions. *J Clin Oncol* 18(11):2309-2315.

Shields PG, Lerman C, Audrain J, et al. 1998. Dopamine D4 receptors and the risk of cigarette smoking in African-Americans and Caucasians. *Cancer Epidemiol Biomarkers Prev* 7(6):453-458.

Shimoda R, Nagashima M, Sakamoto M, et al. 1994. Increased formation of oxidative DNA damage, 8-hydroxydeoxyguanosine, in human livers with chronic hepatitis. *Cancer Res* 54(12):3171-3172.

Sidransky D. 1997. Nucleic acid-based methods for the detection of cancer. *Science* 278(5340):1054-1059.

Simpson AJ. 1997. The natural somatic mutation frequency and human carcinogenesis. *Adv Cancer Res* 71:209-240.

Sjalander A, Birgander R, Rannug A, Alexandrie AK, Tornling G, Beckman G. 1996. Association between the p21 codon 31 A1 (arg) allele and lung cancer. *Hum Hered* 46(4):221-225.

Slupphaug G, Lettrem I, Myrnes B, Krokan HE. 1992. Expression of O6-methylguanine-DNA methyltransferase and uracil-DNA glycosylase in human placentae from smokers and non-smokers. *Carcinogenesis* 13(10):1769-1773.

Smith CJ, McKarns SC, Davis RA, et al. 1996. Human urine mutagenicity study comparing cigarettes which burn or primarily heat tobacco. *Mutat Res* 361(1):1-9.

Soussi T, Dehouche K, Beroud C. 2000. p53 website and analysis of p53 gene mutations in human cancer: forging a link between epidemiology and carcinogenesis. *Hum Mutat* 15(1):105-113.

Speit G, Hartmann A. 1999. The comet assay (single-cell gel test). A sensitive genotoxicity test for the detection of DNA damage and repair. *Methods Mol Biol* 113:203-212.

Spitz MR, Hsu TC, Wu X, Fueger JJ, Amos CI, Roth JA. 1995. Mutagen sensitivity as a biological marker of lung cancer risk in African Americans. *Cancer Epidemiol Biomarkers Prev* 4(2):99-103.

Spitz MR, Shi H, Yang F, et al. 1998. Case-control study of the D2 dopamine receptor gene and smoking status in lung cancer patients. *J Natl Cancer Inst* 90(5):358-363.

Stellman SD, Garfinkel L. 1989. Lung cancer risk is proportional to cigarette tar yield: evidence from a prospective study. *Prev Med* 18(4):518-525.

Stern SJ, Degawa M, Martin MV, et al. 1993. Metabolic activation, DNA adducts, and H-ras mutations in human neoplastic and non-neoplastic laryngeal tissue. *J Cell Biochem Suppl* 17F:129-137.

Stubbe I, Eskilsson J, Nilsson-Ehle P. 1982. High-density lipoprotein concentrations increase after stopping smoking. *Br Med J (Clin Res Ed)* 284(6328):1511-1513.

Sumida H, Watanabe H, Kugiyama K, Ohgushi M, Matsumura T, Yasue H. 1998. Does passive smoking impair endothelium-dependent coronary artery dilation in women? *J Am Coll Cardiol* 31(4):811-815.

Sunyer J, Munoz A, Peng Y, et al. 1996. Longitudinal relation between smoking and white blood cells. *Am J Epidemiol* 144(8):734-741.

Tang D, Santella RM, Blackwood AM, et al. 1995. A molecular epidemiological case-control study of lung cancer. *Cancer Epidemiol Biomarkers Prev* 4(4):341-346.

Tang D, Warburton D, Tannenbaum SR, Skipper P, Santella RM, Cereijido GS, Crawford FG, Perera FP. 1999. Molecular and genetic damage from environmental tobacco smoke in young children. *Cancer Epidemiol Biomarkers Prev* 8(5)427-431.

Thorne S, Mullen MJ, Clarkson P, Donald AE, Deanfield JE. 1998. Early endothelial dysfunction in adults at risk from atherosclerosis: different responses to L-arginine. *J Am Coll Cardiol* 32(1):110-116.

U.S. DHHS (U.S. Department of Health and Human Services). 1988. *Smoking and Health; A Report of the Surgeon General.* Washington, DC: U.S. Department of Health and Human Services, Centers for Disease Control and Prevention.

U.S. DHHS (U.S. Department of Health and Human Services). 1994. *Preventing Tobacco Use Among Young People: A Report of the Surgeon General.* Washington, DC: U.S. Department of Health and Human Services, Centers for Disease Control and Prevention.

van Delft JH, Baan RA, Roza L. 1998. Biological effect markers for exposure to carcinogenic compound and their relevance for risk assessment. *Crit Rev Toxicol* 28(5):477-510.

van Diemen PC, Maasdam D, Vermeulen S, Darroudi F, Natarajan AT. 1995. Influence of smoking habits on the frequencies of structural and numerical chromosomal aberrations in human peripheral blood lymphocytes using the fluorescence in situ hybridization (FISH) technique. *Mutagenesis* 10(6):487-495.

van Schooten FJ, Hillebrand MJ, van Leeuwen FE, et al. 1990. Polycyclic aromatic hydrocarbon-DNA adducts in lung tissue from lung cancer patients. *Carcinogenesis* 11(9):1677-1681.

Velicer WF, Prochaska, JO, Rossi, JS, Snow, MG. 1992. Assessing outcome in smoking cessation studies. *Psychological Bulletin* 111(1): 23-41.

Vineis P, Bartsch H, Caporaso N, et al. 1994. Genetically based N-acetyltransferase metabolic polymorphism and low-level environmental exposure to carcinogens. *Nature* 369(6476):154-156.

Vineis P, Porta M. 1996. Causal thinking, biomarkers, and mechanisms of carcinogenesis. *J Clin Epidemiol* 49(9):951-956.

Vutuc C, Kunze M. 1983. Tar yields of cigarettes and male lung cancer risk. *J Natl Cancer Inst* 71(3):435-437.

Wald NJ, Thompson SG, Law MR, Densem JW, Bailey A. 1989. Serum cholesterol and subsequent risk of cancer: results from the BUPA study. *Br J Cancer* 59(6):936-938.

Wang LE, Bondy ML, de Andrade M, et al. 2000. Gender difference in smoking effect on chromosome sensitivity to gamma radiation in a healthy population. *Radiat Res* 154(1):20-27.

Wei Q, Gu J, Cheng L, et al. 1996. Benzo(a)pyrene diol epoxide-induced chromosomal aberrations and risk of lung cancer. *Cancer Res* 56(17):3975-3979.

Wei Q, Matanoski GM, Farmer ER, Hedayati MA, Grossman L. 1993. DNA repair and aging in basal cell carcinoma: a molecular epidemiology study. *Proc Natl Acad Sci U S A* 90(4):1614-1618.

Wennmalm A, Benthin G, Granstrom EF, Persson L, Petersson AS, Winell S. 1991. Relation between tobacco use and urinary excretion of thromboxane A2 and prostacyclin metabolites in young men. *Circulation* 83(5):1698-1704.

Wiencke JK, Kelsey KT, Varkonyi A, et al. 1995. Correlation of DNA adducts in blood mononuclear cells with tobacco carcinogen-induced damage in human lung. *Cancer Res* 55(21):4910-4914.

Wiencke JK, Spitz MR, McMillan A, Kelsey KT. 1997. Lung cancer in Mexican-Americans and African-Americans is associated with the wild-type genotype of the NAD(P)H: quinone oxidoreductase polymorphism. *Cancer Epidemiol Biomarkers Prev* 6(2):87-92.

Wiencke JK, Thurston SW, Kelsey KT, et al. 1999. Early age at smoking initiation and tobacco carcinogen DNA damage in the lung. *J Natl Cancer Inst* 91(7):614-619.

Wilcox HB, Schoenberg JB, Mason TJ, Bill JS, Stemhagen A. 1988. Smoking and lung cancer: risk as a function of cigarette tar content. *Prev Med* 17(3):263-272.

Yakubovskaya MS, Spiegelman V, Luo FC, et al. 1995. High frequency of K-ras mutations in normal appearing lung tissues and sputum of patients with lung cancer. *Int J Cancer* 63(6):810-814.

Yamasaki E, Ames BN. 1977. Concentration of mutagens from urine by absorption with the nonpolar resin XAD-2: cigarette smokers have mutagenic urine. *Proc Natl Acad Sci U S A* 74(8):3555-3559.

Yu MC, Ross RK, Chan KK, et al. 1995. Glutathione S-transferase M1 genotype affects aminobiphenyl-hemoglobin adduct levels in white, black and Asian smokers and nonsmokers. *Cancer Epidemiol Biomarkers Prev* 4(8):861-864.

Zang EA, Wynder EL. 1992. Cumulative tar exposure. A new index for estimating lung cancer risk among cigarette smokers. *Cancer* 70(1):69-76.

12

Cancer

Around 1950, Doll and Hill (Doll and Hill, 1950), Wynder and Graham (Wynder and Graham, 1950), and others reported the extremely high incidence of smoking in lung cancer patients. In fact, lung cancer was a rare disease before smoking (Doll and Hill, 1950). If one employs almost any method to assess causality, such as that proposed in the first Surgeon General's report on smoking (U.S. PHS, 1964) and later articulated in more detail by Sir Austin Bradford Hill (Hill, 1965), then clearly the use of tobacco products causes cancer. This conclusion comes from substantial epidemiology, laboratory animal, and in vitro studies. Tobacco smoke contains more than 100 carcinogens and mutagens, many of which are classified as carcinogens based upon human and animal studies (IARC, 1986), the latter include lung tumors in the same organs as cancers occur in humans. It is estimated that 20% of all cancers worldwide are attributable to smoking (Parkin et al., 1999).

If a regular smoker successfully quits, then the risk of cancer decreases, but the risk of cancer in former smokers does not decrease to the level of "never smokers." Thus, the concept of harm reduction by reducing exposure to tobacco carcinogens might be plausible if the exposure is significantly reduced, but the reduction in risk could not be more than that for a former smoker and would probably be less. Therefore, the most beneficial harm reduction strategy in smokers is to stop smoking.

The assessment of cancer risk from potential reduced-exposure agents (PREPs) must consider mechanisms of mutagenesis and carcinogenesis. This chapter will focus on only four types of cancer caused by cigarettes and tobacco-containing products, although smoking causes other cancers as well (Doll, 1996); two of these are examples of the most common cancers related to tobacco (lung and oropharyngeal), and one is an example of a cancer that occurs remotely from the site of entry of the carcinogen into the body (bladder). The fourth cancer is one in which tobacco is believed to reduce risk (endometrial). In this chapter, a mutagen is defined as a compound that causes DNA damage of any sort. A carcinogen is defined as a compound that contributes to cancer, independent of the mechanism. A tobacco constituent is any compound from a tobacco-containing product, used in an intended or unintended fashion, which results in human exposure.

MUTAGENESIS AND DNA DAMAGE

Cancers result from an accumulated amount of mutations (changes in nucleotide sequence) or gross chromosomal damage. There are several pathways to such DNA damage. Genetic damage occurs because a mutagen, or its activated metabolite, binds to or otherwise interacts with DNA. This mutagen can then cause a promutagenic lesion or in some other way perturb the genetic structure resulting in a gross chromosomal alteration (aneuploidy, break, translocation, amplification, deletion). The genetic damage follows a failure of several protective mechanisms. The first line of defense against chemical insult involves metabolizing enzymes that are intended to aid excretion of potentially damaging chemicals in the body (produced endogenously or coming from exogenous sources; Guengerich, 2000). For tobacco constituents, this "excretory" process gone wrong is a multistep pathway simplistically described as (1) entry of the mutagen into the body (i.e., oral, respiratory, and gastrointestinal mucosa) and its distribution throughout the body; followed by (2) recognition by an organ that this is a foreign substance in need of excretion (e.g., lung, liver, bladder); (3) use of enzymes for metabolic conversion of the chemical so that it can be bound to an excretory conjugate (Guengerich, 2000); (4) binding to DNA rather than an excretory conjugate; and (5) formation of a DNA adduct or a lesion that then results in DNA damage. A mutagen might be made more water soluble or able to bind an excretory conjugate (e.g., glutathione) through several chemical reactions catalyzed by cytochrome P-450 (CYP) and other enzymes, followed by conjugation catalyzed by enzymes (e.g., glutathione S-transferases, glucuronyl transferases, sulfuronyl transferases). Every one of these steps can influence cancer risk (Hecht, 1999a; Perera, 1997; Van Delft et al., 1998), where greater activity increases the risk of DNA damage, while greater conjugation and excretory capacity could reduce risk. Metabolic conversion and conjugate binding is a complex pathway that differs for different classes of mutagens, and there are redundant pathways (Anttila et al., 1992, 1993, 1995; Brennan, 1998; Grundy et al., 1998; Guengerich, 2000; Nakajima et al., 1995). Different parts of an organ such as the lung may have different capacities for activation and detoxification (Anttila et al., 1993; Bartsch et al., 1991; Geneste, 1991; Petruzzelli et al., 1989; Rojas et al., 1992; Shimada et al., 1996b). Enzymes that are responsible for metabolic activation and detoxification can be induced by exposures, which could further affect the level of subsequent DNA damage (Bartsch et al., 1995; Ciruzzi et al., 1998; Guengerich, 2000; McLemore et al., 1990; Nakajima et al., 1995). Thus, when manipulating the levels of carcinogens in tobacco products, it is important to consider how these changes might affect any of the above steps. Separately, it is well known that people have different heritable abilities for these steps, so manipulating the level of one or more tobacco product constituents might affect people differently.

If a mutagen binds to DNA, additional processes must fail before a mutagenic event occurs, and this takes place more often for some mutagens than for others. Thus, not all mutagens are human carcinogens. One form of DNA damage is a DNA adduct, which is a nucleotide with a chemically bound mutagen or some part of the mutagen. There may be some specificity for the sites of DNA adducts to occur within the genome, but adducts can form anywhere in the genome (La and Swenberg, 1996). Importantly, for the DNA adduct to contribute to the carcinogenic process, it must lead to a mutation and that mutation has to occur in a critical part of a critical gene.

Chemicals within the same class can have different capacities to form reactive intermediates and cause DNA-adduct formation in different parts of a gene. Therefore, simply altering the levels of a specific class in a harm reduction strategy might not affect the important chemical within that class, and the formation of a new adduct due to changes in chemical constituents might result in greater degrees of mutagenicity.

The chemical binding to DNA through the formation of adducts, for example, can lead to nucleotide sequence changes (insertions, deletions, or substitutions). It also can lead to gross chromosomal aberrations such as breaks, deletions, or translocations. These events occur when the mutagen causes errors during DNA replication or mitosis. However, there are protective mechanisms should any of these types of DNA damage occur and cancer develop due to the imbalance of DNA damage and DNA repair (Loeb and Loeb, 2000). Individual adducts may be repaired by excision repair pathways, while chromosomal aberrations are repaired by recombination repair pathways. In addition to DNA repair, other protective mechanisms can reduce the harmful effects of DNA damage, such as lengthening the G1 or G2 checkpoints to allow more time for DNA repair or triggering cell death (apoptosis). Unfortunately, some mutations might block the entry into these checkpoints or evade cell death processes. In addition to repair pathways that remove adducts, there are other control methods if these mechanisms or pathways fail. There are DNA repair enzymes that recognize and repair mismatch damage, and if this does not occur, cell death may be triggered. The combination of repair and cell turnover leads to a half-life of carcinogen-DNA adducts. To date, the effects of chemicals on these repair and control pathways, and interindividual differences in DNA repair, cell-cycle control, and cell death have only recently received some attention (Sumida et al., 1998).

The relationship between a mutagen and mutation is complex and may depend on the dose of the mutagen (La and Swenberg 1996; Van Delft et al., 1998). Low-dose exposures are often difficult to evaluate in vitro or in vivo because of mutational background rates, and extrapolation of mutation rates from high-dose to low-dose exposures depends on assumptions that may not be true (Liber et al., 1985). Mutation rates for exposures that switch from high dose to low dose and how the mutational spectra changes have not been studied.

In summary, for a tobacco constituent to cause a DNA lesion that confers a selective clonal advantage on a cell that ultimately becomes cancerous, the constituent must be absorbed and metabolically activated; it has to damage DNA, which evades repair; it has to occur in a critical part of a critical gene; and finally, the cell must evade cell death. Moreover, this has to occur in the target organ. Internal exposure can be affected by smoking behavior or storage depots in the body (e.g., adipose tissue). These pathways are important to consider for harm reduction strategies because altering the levels of different tobacco constituents or complex mixtures might affect these pathways differently, so that the net effect on carcinogenicity may not be predictable a priori, either for an individual or for the population.

CARCINOGENESIS

Carcinogenesis is a multistage process involving many different genes (Devereux et al., 1999). DNA damage is necessary, but not sufficient, to cause tumors as evidenced in experimental models (Pledger et al., 1977). One conceptual approach to understanding carcinogenesis is to consider that cancer is driven by defects in either caretaker, gatekeeper, or landscaper genes (Kinzler and Vogelstein, 1997, 1998). Caretaker genes are those responsible for maintaining genomic integrity, such as DNA repair and metabolism. Mutations or inherited variants in these genes increase the risk of mutations in other genes. Gatekeeper genes are those involved in controlling cell cycle, and replication of the genome, triggering apoptosis, and assisting caretaker genes in maintaining genomic integrity. Mutations in gatekeeper genes increase the risk for a cell to replicate uncontrollably and increase the likelihood of permanently establishing mutations. Landscaper genes are those that affect the external environment around the cells and thus control adjacent cells. Current data do not exist to indicate that tobacco-related harm preferentially affects any classes of genes, but an effect of tobacco carcinogens on all classes of genes is plausible and suggested by the observed complex genetic alterations in lung and other cancers.

Another way to classify cancer genes is to consider them as oncogenes or tumor suppressor genes. This classification is based on studies showing that overexpression or mutation of the former increases proliferative potential, while loss of the latter stimulates proliferative potential. For oncogenes, only one allele has to be activated, whereas for tumor suppressor genes, both alleles are usually inactivated. Thus, the former is a dominant trait, while the latter is recessive.

Oncogenes occur when proto-oncogenes, responsible for normal cellular processes, are mutated. Multiple oncogenes are involved in the pathogenesis of solid tumors including lung cancer (Fong et al., 1999; Kohno and Yokota, 1999; Rom et al., 2000). Only a few oncogenes, such as K-ras and c-MYC have been identified as playing crucial roles in the pathogenesis of several tobacco-related tumors (Reynolds et al., 1991; Rodenhuis et al., 1987; Slebos et al., 1991, 1992). However, most oncogenes remain to be discovered. The ras gene family consists of three members (K, N, and H). They are membrane-bound proteins that bind to guanosine 5'-triphosphate (GTP) when activated, and to guanosine 5-diphosphate (GDP) when inactivated. Activation sends a signal to the nucleus, via a cascade of kinases, that eventually results in the activation of transcription factors. Activating ras mutations occur at codons 12, 13, and 61 and result in loss of intrinsic guanosine triphosphates (GTPase) activity, locking in the activated form (Bos, 1988); experimental studies support the role of tobacco carcinogens that affect ras in lung cancers (Ronai et al., 1993). It has been reported that ras mutations are present usually only in smoke-related lung cancers (Gealy et al., 1999; Slebos et al., 1991). Such mutations also can be observed in smokers without lung cancer, suggesting that they can be early markers of smoking-related damage (Lehman et al., 1996; Slebos et al., 1991; Scott et al., 1997; Yakubovskaya et al., 1995).

Tumor suppressor genes in tobacco-related cancers include p53 (also known as TP53), p16[INK4A] (p16), retinoblastoma (RB), and fragile histidine triad (FHIT) genes. A frequent method of inactivation of one allele of recessive oncogenes is by allelic deletion (i.e., loss of DNA material on one of the alleles). Often, this deletion is extensive and involves not only the gene of interest, but adjacent genes as well. The p53 gene is a tumor suppressor gene, and mutations of this gene may represent the most common genetic abnormality discovered to date in tumors, being present in about 50% of human carcinomas (Hollstein et al., 1996). It plays a central role in the balance between gene transcription, cell proliferation, and apoptosis. DNA damage results in the induction of genes upstream of p53 (Oren, 1999), which then stimulates p53 induction and stability through posttranslational modifications. This in turn affects *p21*, *MDM2*, *GADD45*, *BAX*, and other genes responsible for DNA repair, and delay of the cell cycle to allow additional time for DNA repair, or triggers cell death when DNA repair is not possible. The (p16) gene is located on chromosome 9p, and its protein plays a crucial role (along with the retinoblastoma gene product and the cyclins) in regulating the cell cycle. It is inactivated in many smoking-related cancers including non-small-cell lung, head and neck, and pancreatic cancer and squamous carcinomas of the esophagus (Geradts et al., 1999; Liu et al., 1995; Lydiatt et al., 1995), and occasionally bladder tumors (Orlow et al., 1999). Inactivation occurs by many mechanisms including hemizygous or homozygous deletions, point mutations, or aberrant methylation of the promoter region. The latter is an example of tobacco smoke constituents affecting genetic function without causing a mutation (i.e., an epigenetic change; Belinsky et al., 1998). The FHIT gene is a putative tumor suppressor gene located on chromosome 3p14 (Sozzi et al., 1998a, b). Inactivation of the gene product has been described in many tumors including lung, head and neck, and esophagus.

It might be possible to consider mutations in different genes as "molecular fingerprints" of causation by tobacco smoke for an individual. This could be helpful in considering the effects of different types of tobacco products and changes in tobacco constituents over time. It might also be possible to identify which tumors in an individual were caused by tobacco versus some other agent. However, no such "fingerprints" have been identified for tobacco smoking, although some types of lesions occur more frequently (Kondo et al., 1996). This may be due to the numerous carcinogens in tobacco, which cause many types of DNA damage. New microarray technologies will provide sequence data for many genes, RNA expression profiles or protein expression profiles that—with sufficient bioinformatic support—a characteristic profile could enable tobacco effects to be discerned.

Tobacco smoke exposes the entire respiratory and upper gastrointestinal mucosa to carcinogens, whereas smokeless tobacco exposes only the oral cavity and the gastrointestinal mucosa. Thus, these entire "fields" are at risk for the development of preneoplastic and neoplastic lesions (Slaughter et al., 1953; Strong et al., 1984). A field effect for cancer has been demonstrated on a molecular basis, where different molecular lesions were found in persons with multiple synchronous lesions (Sozzi et al., 1995).

TOBACCO MUTAGENS AND CARCINOGENS

The use of tobacco products, as they are intended to be used, results in exposure to more than 100 mutagens and carcinogens (Hoffman and Hoffman,1997; Zaridze et al., 1991) that have different potencies and effects. Mainstream smoke consists of particulate and vapor phases. Although carcinogens have been identified in both the vapor and the particulate phase, the latter shows more overall carcinogenic activity. The particulate phase contains more than 3,500 compounds, of which at least 55 have been identified as possible human carcinogens (Hoffman and Hoffman, 1997). The vapor phase contains more than 500 compounds (Hoffman and Hoffman, 1997). A list of some of these constituents is provided in Table 12-1, which is not all inclusive.

Tobacco mutagens and carcinogens have different potencies and target organ specificities. A recent critical review summarizing data for tobacco constituents proposed that tobacco-specific nitrosamines (TSNAs) and polycyclic aromatic hydrocarbons (PAHs) are classes of compounds that most affect human cancer risk (Hecht, 1999b). Although this may be true, it is currently difficult to prove in human cancer, especially because these exposures are mixed with others. Other compounds also are likely to be important. The existing data are not sufficient to determine whether some compounds are clearly more carcinogenic in humans than others when delivered through the use of tobacco products, and whether there is a synergistic effect of coexposure. Therefore, the assessment of a harm reduction strategy for cancer must consider these constituents individually and as part of a complex mixture since the former can provide mechanistic information but only the latter can be used to fully understand the effect of PREPs on carcinogenesis.

Tobacco and tobacco products have changed over time, with resultant differences in predicted exposure using the Federal Trade Commission (FTC) method for the measurement of "tar" and "nicotine" (Hoffman and Hoffman, 1997). It is known that the FTC method for estimating tar exposure underestimates actual human exposure because it does not sufficiently mimic human smoking behavior (Hoffman and Hoffman, 1997). Specifically, using a protocol that mimics actual human smoking behavior shows that the FTC method substantially underestimates the exposure to TSNAs and benzo[a]pyrene [BaP] (Djordjevic et al., 2000). While smokers of low-nicotine cigarettes have somewhat lower delivered levels of BaP and TSNAs, the daily amount of tar delivered is similar (Djordjevic et

TABLE 12-1 List of Selected Tobacco Mutagens and Carcinogens[a]

Constituent Class	Phase	IARC Evaluation	Examples
N-Nitrosamines	Particulate	Sufficient in animals	Tobacco-specific nitrosamines (NNK, NNN), dimethylnitrosamine, diethylnitrosamine
Polycyclic aromatic hydrocarbons	Particulate	Probable in humans	Benzo[a]pyrene, benzo[a]anthracene, benzo[b]fluoranthene, 5-methylchrysene
Arylamines	Particulate	Sufficient in humans	4-Aminobiphenyl, 2-toluidine, 2-naphthylamine
Heterocyclic amines	Particulate	Probable in humans	2-Amino-3-methylimidazo[4,5-[b]quino- lone (IQ)
Organic solvents	Vapor	Sufficient in humans	Benzene, methanol, toluene, styrene
Aldehydes	Vapor	Limited in humans	Acetaldehyde, formaldehyde
Volatile organic compounds	Vapor	Probable	1,3-Butadiene, isoprene
Inorganic compounds	Particulate	Sufficient in humans	Arsenic, nickel, chromium, polonium-210

[a]This list is intended to provide a conceptual overview of the complexity of tobacco product exposures. It is not all inclusive, but is included to allow the reader to understand the number of considerations that must be made in assessing PREPs.

[b]International Agency for Research on Cancer: The classifications here refer to evaluations of the compound from any exposure, not just tobacco. Not all chemicals within the class are considered carcinogenic in humans. There is no consideration in this table of delivered dose or route of exposure.

NOTE: NNK=nitrosonornicotine ketone; NNN=N-nitrosonornicotine.

al., 2000). Therefore, in this report so-called tar yields do not imply actual tar exposure.

Although it is important to understand the differences in risks by chemical class, in order to assess PREPs, it must be realized that affecting the exposure to one compound or class might not account for similar proportional decreases of other compounds, and we do not know if removing one compound or even a whole class will reduce unless other classes of compounds are also decreased. Further, the study of mixtures (i.e., the real-life scenario of simultaneous exposure to many chemicals and classes) has received insufficient attention, and exposure to tobacco constituents as complex mixtures would provide the most compelling evidence for prediction of a successful PREP. Cigarette smoke condensate is mutagenic in bacterial and human cell lines (Matsukura et al., 1991) and can cause a malignant transformation in human bronchial epithelial cells (Klein-Szanto et al., 1992). Whole smoke, which is also mutagenic, can be used as well (Bombick et al., 1997). Both the vapor phase of environmental tobacco smoke (ETS) and unfiltered ETS exposure causes lung cancer in laboratory animals (Witschi et al., 1997a). There is some evidence to suggest that the mutational spectra of a complex mixture reflects mostly that of the dominant chemicals in the mixture (DeMarini, 1998), although experimental animal studies of DNA adducts from benzo[a]pyrene and coal tars indicate that total adduct levels are not related to BaP content alone (Goldstein et al., 1998), suggesting that studying single chemicals is not sufficient to represent the effects of complex mixtures. Further, various polycyclic aromatic hydrocarbons cause different hotspots in p53 (Smith et al., 2000), and different dose levels of complex mixtures might have additive or synergistic effects (Hecht et al., 1999; Poirier and Beland, 1992). Thus, more studies are needed to determine the best approaches to assess the mutagenicity and carcinogenicity of complex mixtures (Guengerich, 2000).

Several constituents of tobacco are considered likely agents of human carcinogenesis. Some of these are reviewed here to highlight the considerations needed in considering harm reduction strategies.

People are commonly exposed to PAHs through tobacco products, diet, occupation, and consumption of fossil fuels (i.e., burning coal or wood). These compounds are formed from the incomplete combustion of tobacco leaves, and many types of PAHs are present in tobacco smoke as a complex mixture. Parent PAHs can be detected in human lung tissue (Lodovici et al., 1998; Seto et al., 1993). It is estimated that smokers are exposed to 2-5µg of PAHs per day per pack of cigarettes, and our diet provides PAHs of 3µg per day (Hoffman and Hoffman, 1997; Lioy and Greenberg, 1990; Waldman et al., 1991). As a class, they are mutagenic and carcinogenic in organs of laboratory animals (including the lung) and humans (Hoffman and Hoffman, 1997; IARC, 1986; Van Delft et al., 1998). PAHs have different potencies, which are thought to be related to metabolic activation of a compound that leads to either a bay region diol-epoxide (potent), or a fjord region diol-epoxide (nonpotent) compound. PAHs are metabolically activated in humans through *CYP1A1*, *CYP1B1*, and *CYP3A4* (Kim et al., 1998; Shimada et al., 1996a). They are conjugated for excretion by glutathione *S*-transferases, sulfuronyl transferases, and glucuronyl transferases (Robertson et al., 1988), and the lack of such activity increases mutagenic potential (Romert et al., 1989). In laboratory animals treated with benzo[a]pyrene, the half-life of DNA adducts following a single dose is 15 days in the liver, 17 days in peripheral blood lymphocytes, and 22 days in lung (Ross et al., 1990). In humans, there is more than a hundredfold variation in the resultant capacity for DNA-adduct formation (Harris et al., 1974) due to variation in induction and activity for these activating and detoxifying enzymes. PAH-related DNA adducts have been demonstrated in human lung (Kato et al., 1995), while the presence of hemoglobin and albumin adducts also shows that these compounds circulate in human blood (Day et al., 1990; Kriek et al., 1998). In vitro studies indicate that PAHs can cause the same types of p53 mutations observed in human tumors (Denissenko et al., 1996; Smith et al., 2000). DNA damage from PAHs is repaired by both excision and recombination pathways, and while there is clearly interindividual variation in the DNA repair capacity of these pathways, such variation have received little attention for PAHs (Xu et al., 1998).

Tobacco products and smoke contain N-nitrosamines (Brunneman et al., 1996; Fischer et al., 1989b, 1990; Tricker et al., 1991), which are among the most potent rodent carcinogens (Lewis et al., 1997). Some of the N-nitrosamines in tobacco smoke are specific for tobacco, whereas others are the same types formed from dietary exposures. N-nitrosamines cause cancer in more than 40 animal species, and there is target organ specificity, including for TSNAs and lung tumors (Lewis et al., 1997; Rivenson et al., 1989), where there is a biphasic response in experimental animals indicating both a high affinity response at lower exposure levels and a saturation effect at higher levels (Belinsky et al., 1990; Pledger et al., 1977). Experimental animal studies also show that higher doses of exposure cause tumors in less time, suggesting that intensity and duration are equally important (La and Swenberg, 1996; Lewis et al., 1997). Mutations in K-ras have been found in lung tumors of experimentally exposed animals (Chen et al., 1993). TSNAs can transform human bronchial epithelial cells (Klein-Szanto et al., 1992). The same type of adducts that occur from TSNAs in experimental animals also have been detected in humans, including in lung tissue (Hecht et al., 1994). Different types of tobacco have different TSNA yields (Brunnemann et al., 1996). In humans, metabolites of TSNAs are found in urine (Carmella et al., 1993, 1995; Hecht et al., 1994), and adducts are detected in blood, so TSNAs circulate through the body, including in persons who are passively exposed (Atawodi et al., 1998; Hecht et al., 1993; Parsons et al., 1998). The elimination half-life of several TSNAs through the urine is estimated to be 40-45 days (Hecht et al., 1999). There is no mutational specificity for N-nitrosamines in several genes studied to date, although there is a propensity for G→A (guanine to adenine) transitions in experimental models (Chen et al., 1993; Ronai et al., 1993;

Tiano et al., 1994). N-nitrosamines undergo metabolic activation by human cytochrome P-450s located in the lung, buccal mucosa, and other tissues (e.g., CYP2E1 and CYP2A6; Hecht, 1998; Patten et al., 1997; Smith et al., 1995, 1992). Ethanol induces CYP2E1 which may have implications for oropharyngeal and esophageal cancer (Garro et al., 1981; Ma et al. 1991). Cigarette smoking and exposure to other tobacco products increase endogenous nitrosation, so that there are additional exposures to nitroso compounds (Nair et al., 1996). The metabolic activation of TSNAs and other tobacco N-nitrosamines leads to the formation of DNA adducts in target tissues or is associated with specific cancers (Chang et al., 1990; La and Swenberg, 1996; Liu et al., 1993; Nesnow et al., 1994; Tiano et al., 1993; Yang et al., 1990; Zhang et al., 1991). TSNAs form three different classes of DNA adducts (Hecht, 1999a). The first involves methylation of different nucleotides, which also are formed by other N-nitrosamines, and some of these adducts are more promutagenic than others (O^6-methylguanine more readily causes mutations than N^7-methylguanine), and there are different repair enzymes for each. The O^6-methylguanine is repaired by O^6-alkyl-alkyltransferase. The activity of this enzyme varies among people and can be reduced in smokers, because once methylated it becomes inactive (Liu et al., 1997). Other classes of adducts formed by TSNAs, which are bulky (Atawodi et al., 1998; Hecht, 1999a), are probably repaired by nucleotide excision and recombination repair, similar to PAHs. The activity of these repair pathways also varies widely among individuals. Different tobacco products contain widely differing amounts of TSNAs (Fischer et al., 1989b). Snuff use can lead to higher levels of TSNAs than smoking (Hecht et al., 1994), and changing smoking patterns can result in higher delivery of TSNAs (Fischer et al. 1989a). For example, Swedish snuff products contain substantially fewer TSNAs than snuff sold in the United States. Lower-tar and nicotine cigarettes result in greater exposure to TSNAs than high-tar and nicotine cigarettes (Brunnemann et al., 1996; Hoffman and Hoffman, 1997). The third type of DNA adducts formed from TSNAs is related to oxidative damage (Hecht, 1999a).

Aromatic amines are another class of compounds present in tobacco smoke; these consist of aryl amines and heterocyclic amines. The latter are not reviewed here. There are substantial data to implicate aryl amines and their metabolism in human carcinogenesis (Vineis and Pirastu, 1997), especially bladder cancer in occupationally exposed cohorts (e.g., dye workers; Cartwright et al., 1982). In experimental animals, 4-aminobiphenyl (4-ABP) adduct levels increase in both liver and bladder tissues, but the rise in the bladder as a target organ is substantial and correlated with tumor incidence (Poirier and Beland, 1992). Saturation pathways might occur in female mice at lower doses than male mice, and saturation effects at higher levels of smoking have been reported in humans (Dallinga et al., 1998). Aromatic amines are thought to contribute to bladder carcinogenesis in smokers too, so this is an example of a tobacco carcinogen that affects an organ distant to the route of entry. Aromatic amines are initially activated by CYP1A2, which ultimately leads to the formation of nitreunium ion that then forms a DNA adduct. *N-Acetyltransferases* (NAT1 and NAT2) play an activating or detoxifying role, depending on the arylamine (Windmill et al., 1997). These compounds are metabolically activated in the liver and transported to the kidney. Upon excretion in the urine, the bladder mucosa can further activate and detoxify the conjugates. Both NAT1 and NAT2 are present in the bladder mucosa (Badawi et al., 1995; Kloth et al., 1994). Aromatic amine biomarker studies have generally focused on hemoglobin rather than DNA adducts. Levels are higher in smokers than nonsmokers, and different types of tobacco can lead to higher adduct levels (Bryant et al., 1988; Carmella et al., 1990; Dallinga et al., 1998). Different types of adducts also have been detected in urinary bladder (Badawi et al., 1996; Bryant et al., 1988; Carmella et al., 1990; Kadlubar, 1994).

Both the gaseous and the particulate phases of cigarette smoke contain free radicals (e.g., nitric oxides in the gaseous phase) that induce oxidative damage (Hoffman and Hoffman, 1997; Pryor et al., 1998). Many components of cigarette smoke can individually cause oxidative damage (Leanderson and Tagesson, 1990). Nitric oxides may act synergistically with the particulate phase to induce DNA breaks. Although free radicals cause DNA damage in experimental systems and are suspected to be involved in carcinogenesis (Floyd, 1990), a direct relationship to human carcinogenesis has been suspected but not proven (Loft and Poulsen, 1996; Marnett, 2000; Poulsen et al., 1998). It is difficult to measure free radicals and oxidative damage in humans from tobacco smoke or any other source (endogenous or exogenous), because it is impossible to distinguish free radical sources and biomarker methods can artifactually induce oxidative damage (Loft and Poulsen, 1996; Marnett, 2000). Among the most common methods is to measure 8-oxodeoxyguanosine, one of the most abundant products of oxidative damage and can be seen in human lung (Inoue et al., 1998), using high-performance liquid chromatography (HPLC) and electrochemical detection (Helbock et al., 1998; Loft and Poulsen, 1996; Park et al., 1989). Levels are generally higher in leucocytes and in the urine of smokers compared to nonsmokers (Asami et al., 1996; Loft et al., 1992). In a comparison of 100 smokers in a smoking cessation program and 82 smokers who were not quitting, the cessation group lowered urinary excretion levels of 8-oxodeoxyguanosine by 21%, while there was no effect in smokers (Prieme et al., 1998). While levels of 8-oxoguanine are elevated in smokers, so is the capacity to repair these lesions (Asami et al., 1996; Hall, 1993). Separately, free F2-isoprostanes, a marker of lipid peroxidation, were measured in the plasma of ten smokers and ten non-smokers. Levels were higher in smokers than in nonsmokers and decreased with abstinence (Morrow et al., 1995).

Other constituents of tobacco smoke can also play a role in human carcinogenesis. For example, although 1,3-butadiene is present in cigarette smoke at "low" levels, it is considered a potent carcinogen in experimental animal studies that specifically affects the lung (Huff et al., 1985; Melnick and Huff, 1992; Owen et al., 1987). Heterocyclic amines, are emerging as an important human carcinogen, which are formed from the pyrolysis of creatines (Knize et al., 1999). These compounds are found in smoke, and the removal of proteins from tobacco before making cigarette smoke condensate substantially reduces mutagenicity (Clapp et al., 1999).

Although it is conceivable that some harm reduction strategies might decrease exposures to PAHs, TSNAs, or aromatic amines, singly or combined, the assessment of real benefit must consider the effects of altering these carcinogens in the context of the complex exposures resulting from tobacco products (Krewski and Thomas, 1992). Thus, it is critical that PREP assessment include consideration of complex mixtures. Assays are available that can do this, as reviewed in Chapter 10. Separately, decreasing one or more tobacco constituents might not affect cancer risk if other compounds such as 1,3-butadiene, benzene, aldehydes, and acrolein are not affected, or the risks from additives or fibers might be comparatively more important. There are insufficient data to currently conclude that overall cancer rates would decrease proportionally to the reduction of a carcinogen, because although one type of cancer might decrease, others might increase. Alternatively, the substitution of one lung carcinogen for another might not allow for a sufficient benefit from harm reduction strategies.

SCIENTIFIC METHODS FOR ASSESSING
HARM REDUCTION STRATEGIES

The actual exposure to tobacco mutagens is dependent on the type of cigarette and how it is smoked. Over time, manufactured cigarettes and consumer choices have changed substantially. Prior to the 1950s, most manufactured cigarettes did not have filters, but more than half of all cigarettes had filters by the beginning of the 1960s (Cummings, 1984). Although filtered cigarettes were available earlier, about the time of the first Surgeon General's report in 1964, many people began switching to filtered cigarettes (Hoffman and Hoffman, 1997; Stellman and Garfinkel, 1986). Today, more than 98% of cigarettes have filters (NCI, 1996). The so-called tar content also has declined since the 1950s, from about 37 mg to less than 15 mg. (Cummings, 1984; Hoffman and Hoffman, 1997). Most studies published through the 1980s included subjects who probably had smoked nonfiltered cigarettes in their lifetime, and this could make smoking studies from that time more difficult to interpret.

The introduction of low-tar and nicotine cigarettes was conceptualized to make cigarettes "safer," but currently available scientific data suggest that potential benefits may not have been realized for some or most persons (see discussion of lung cancer). In actuality, many persons who smoke low-tar and nicotine cigarettes compensate for lower nicotine delivery by smoking more (Benowitz et al., 1983, 1986b, 1998; Ebert et al., 1983; Gritz et al., 1983). Levels of TSNAs and benzo[a]pyrene in tobacco smoke can be similar for low- and high-tar cigarettes when people oversmoke their cigarettes (Djordjevic et al., 1995, 1997). It is unknown, however, what happens to carcinogenic biologically effective doses. While switching results in a higher peak nicotine level per cigarette, there is a lesser proportional reduction in urinary mutagenicity (Benowitz et al., 1986a, b; Sorsa et al., 1984). In one study, sister chromatid exchange levels, which have not been validated as a surrogate marker for the harm, do not change when persons switch from high- to low-tar cigarettes (Djordjevic et al., 1997), but the relation of this marker to short-term and long-term exposures, half-lives, and other factors that might confound such studies is not known.

Smokers have learned to block filter ventilation holes, with concomitant increase in tar exposure of more than tenfold compared to the standard FTC method (Kozlowski et al., 1982). Blocking can occur with the fingers, lips, or tape, and can be intentional or unintentional.

To consider the value of a harm reduction strategy, one must consider the effects on cancer risk of the targeted reduction in exposure and then place these risks in the context of reducing exposures by any means. This process can be used for individual constituents of exposure or in the aggregate. Through this process, several predictive models should be developed that are based on adequate scientific data, not speculation. The methods listed below are used if the proposed reduction in exposure is for carcinogens. Importantly, an assessment of a PREP can only be made in the context of risk due to conventional tobacco products. Following a risk assessment model typically used by regulatory agencies, the following three steps are proposed:

1. Cancer linkage assessment (linking exposure to cancer)

Causality assessment. Does the compound targeted for reduced exposure cause cancer, and what do we know about the biological mechanisms for that compound? The criteria proposed in the first Surgeon General's report on smoking and health (U.S. Public Health Service, 1964), and later extended by Sir Austin Bradford Hill, or their equivalents, proposed by others, should be applied, using experimental data and human experience.

Dose-response assessment. What is the dose-response relationship between targeted carcinogens and cancer risk? Where is the risk to persons from environmental tobacco smoke within the shape of the dose-response curve? What is the optimal way to consider dose?

Causality in context assessment. How do the risks of harmful effects from the targeted compounds compare with other carcinogens in tobacco smoke? Are there data to indicate how these compounds can affect risks for former smokers or persons with ETS.

Individual susceptibility and attributable risk assessment. Are there individual susceptibilities (e.g., age, gender, race, heritable traits, prior tobacco exposure) that can modify the carcinogenic risks of these targeted compounds, and what is the frequency of these traits in different smoker populations?

2. Cancer reduction assessment (decreasing exposure to reduce cancer risk from a higher base line level)

Reduction in carcinogenicity assessment. What is the experimental evidence that reducing exposure will reverse or halt carcinogenic processes that have already begun?

Decreasing dose-response assessment. What happens to cancer risk if doses are reduced, and how much does exposure have to be reduced to result in a meaningful benefit? A model should be developed that predicts the cancer risk or rate over time for the consistent use of the harm reduction method. This model will include several scenarios in which different categories of smoking (current smoking, lifetime smoking, and age at initiation) at the time a PREP is used are studied and, for example, assume a 10, 25, 50, 75, and 90% reduction in exposure. (Response is measured by a biologically effective dose, a biomarker of potential harm, or a disease outcome, not what is measured in tobacco smoke or delivered to the oral and respiratory mucosa.)

Adverse effects of harm reduction strategies. A model should be developed that predicts the numbers of individuals who begin smoking, do not quit, or resume smoking after quitting due to the availability and knowledge of a PREP, or belief that such products are safe. For new products that contain tobacco or tobacco constituents, a dose-response model should be developed for cancer risk that indicates risk for use by "never smokers" who initiate regular use with the PREP.

Harm reduction method in context assessment. An assessment is needed of the predicted reduction in risk versus the remaining risk of other tobacco constituents and consideration of new or altered tobacco constituents.

Individual susceptibility and population-attributable risk for successful harm reduction. Are there individual susceptibilities (e.g., age, gender, race, heritable traits, previous tobacco exposure) that can modify the success of a PREP (increased, decreased, or no change in risk), and what is the frequency of these traits in different smoker populations?

3. Integrated harm reduction method assessment

Final considerations. Consideration of the above models should provide a summary statement and prediction of the numbers of cancers that are avoided due to the method. If the reduction in exposure is proposed only for carcinogens, possible effects on other tobacco-related illnesses and total mortality must be considered.

The causality assessments (linkage and reduction) can be provided in the format of Sir Austin Bradford Hill (Hill, 1965), which was originally proposed in the first Surgeon General's report (U.S. PHS, 1964; Table 12-2). In the evaluation of a PREP, it must be decided which disease outcome or outcomes are to be targeted. The role for all cancers could be evaluated with this model, or perhaps only lung cancer, because the lung is the most sensitive organ to tobacco smoking. For smokeless tobacco products, oral

TABLE 12-2 Causality Assessment[a]

Criteria	Comment
Strength of association	What is magnitude of risk?
Consistency	Are there repeated observations by multiple investigators in different populations?
Specificity	Is the effect specific or are there other known causes?
Temporality	Does exposure precede effect?
Biological gradient	Is there a dose-response relation?
Biological plausibility	Is the effect predictable?
Coherence	Is the effect consistent with other scientific data?
Analogy	Do similar agents act similarly?

[a]These criteria should serve as guidelines for assessment of PREPs. They rely on human studies (and are applicable to biomarker studies) and use experimental data to provide supportive evidence. The criteria were originally published in the first Surgeon General's report and later expanded by Sir Bradford Austin-Hill.

cancers can be used. These targeted outcomes might change because new PREPs may increase risk in non-lung or non-oral cavity cancers, or though exposures to some carcinogens decrease, the risk from other tobacco product constituents (i.e., additives, fibers) might increase.

Various scientific methods are used to assess carcinogenicity and carcinogenic risk. These appear in Table 12-3, along with their strengths and limitations. Most of these methods, however, have limited use currently for harm reduction evaluations. The assessment of carcinogenicity in the laboratory is focused on identifying potential human carcinogens, and fewer data are available for effects at different doses in ranges to which humans are exposed. Also, these studies are designed to identify risks using continuous exposures, and there are few studies that consider intermittent exposures or exposures that begin at high dose and then change

TABLE 12-3 Scientific Methods for Carcinogenicity Testing

Method	Strengths	Limitations
In vitro cell culture models	Rapid, inexpensive. Can be used to study mechanisms in human cells. Can be used to prioritize other studies	Quantitative extrapolation to human risk cannot be done, so relevance to human experience is questionable
Laboratory animal models	Provides *in vivo* experience. Dosing protocols can be modified to assess changes in PREPs. Can be used to prioritize human studies. Can be used to develop and validate biomarkers. Transgenic mice available	Quantitative extrapolation to human risk cannot be done and so the relevance to human experience is questionable. Expensive
Human experimental models	Can provide data about short-term changes in exposure and resultant effects on biomarkers. Can assess complex mixture	Best used to assess effects on exposure rather than outcome (i.e., cancer). Expensive
Epidemiology	Provides direct experience and assesses tobacco use in the context of individual susceptibilities and complex exposures. Case-control studies provide rapid data. Prospective studies provide best evidence	Use of cancer end point means long latency periods and adverse effects on many people. Expensive

to low dose. These types of studies are likely more relevant to predicting the effects of PREPs.

The use of in vitro cell culture models and DNA-binding studies that focus on mutagenesis is attractive because they can provide toxicity data rapidly. Many models are available and have helped elucidate various mechanisms of carcinogenesis (Balmain and Harris, 2000; Taningher et al., 1990). Combinations of assays are traditionally used to better identify genotoxic compounds and to assess the genotoxic potential of different cigarettes (Bombick et al., 1997). Although in vitro models that provide positive results for genotoxic damage best predict carcinogenesis in experimental animal studies qualitatively or semiquantitatively (Taningher et al., 1990), mutagenicity in a cell culture is difficult to extrapolate to the in vivo models, whether animal or human (Rosenkranz and Klopman, 1993). Also, a negative result in cell culture systems is less reliable (Thilly, 1985; Zeiger, 1987). Mutations found in the *Salmonella* assay predicted carcinogenicity in laboratory animals with an overall accuracy of 59.5% (Lee et al., 1996). However, simple concordance analysis of predicitivity (yes or no) may not be as useful as considering predictivity in relation to mechanistic pathways (Butterworth, 1990). Some limitations for dosing in cell culture studies are related to the balance between mutation and cell survival (Thilly, 1985). The limitations for extrapolation to animal studies are described below. Thus, in vitro models might be useful for suggesting which tobacco constituents should receive the highest priority in evaluating PREPs, but they cannot be used to quantitatively support such a claim.

The use of laboratory animal models can provide an estimate of the effects of PREPs. Typically, studies used for risk assessment have followed protocols established by the National Toxicology Program to prioritize compounds for study as potential human carcinogens (Boorman et al., 1994; Fung et al., 1995; Goldstein, 1994). However, these studies use maximally tolerated doses (MTDs) over the animal's lifetime. Definition of the MTD is based on a subchronic, 90-day study, in which the highest dose that does not affect overt toxicity or growth is chosen. A consideration of the various limitations of MTDs for extrapolation to low-level exposures would lead to the conclusion that risk assessments based on MTD tumor incidences can overestimate human risk (Portier et al., 1994). In trying to understand the prediction of risk based on the MTD, there is a correlation for a dose that induces tumors at 50% of the MTD (Krewski et al., 1993) and at 50-75% of the MTD (Haseman and Lockhart, 1994), but it is unclear whether this information validates the use of the MTD or suggests that even 50% dose levels still cause significant physiological perturbations. Thus, the extrapolation of animal studies to humans remains questionable for harm reduction evaluation under these experimental designs. The use of animal studies might be rational, however, if

the doses approximate human levels of exposure and if the experimental
animal model does not possess a sufficiently different physiological response to carcinogen exposure compared to humans. Transgenic mice are
now available with human genes that may screen potentially carcinogenic agents more rapidly than the two-year bioassay (Tennant et al.,
1995), but multiple dosing protocols are needed (Schmezer et al., 1998).
However, their use for harm reduction has not been considered previously, and doses affecting toxicity will be different than those for other
animal models.

The use of experimental animal studies to predict cancer risk is more
qualitative than quantitative for a number of reasons usually related to
physiological differences and expanded protocols (Int. Panel, 1996;
Swenberg et al., 1987). Mice and humans have many similar genetic alterations, but there are some important differences such as telomerase activity. Target organ specificity is very different under typical experimental
conditions (Benigni and Pino, 1998). There are large quantitative differences in cancer risks among any organ in experimental animals. For a
variety of tissue sites, including lung, liver, breast, and skin, pairs of
inbred mice differ by a hundredfold in risk for tumor development. Detailed analyses of these differences using back-cross, recombinant inbred,
and recombinant congenic breeding protocols have shown specific determinants for initiation, promotion, premalignant progression, and metastatic stages. There are even larger quantitative differences in cancer risk
by organ among experimental animals. In most cases, susceptibility or
resistance is a property of the target tissue, not of the host, for genotoxic
substances such as N-nitrosamines. A direct relationship between in vitro
experimental studies (i.e., metabolic activation of chemicals by cytochrome P-450s), in vitro mutagenicity tests, experimental animal studies,
and human epidemiology has not been proven. Although there tends to
be concordance among experimental studies, this is not 100% true, and
extrapolation to humans is far from concordant. For the prediction of
carcinogenicity in humans, the sensitivity of experimental studies is high
and the specificity is very low. Some studies suggest that consideration of
multiple assays yields greater productivity (Tennant and Ashby, 1991),
but only if these compounds are mutagenic (Cunningham et al., 1998;
Gold et al., 1993; Int. Panel, 1996; Tennant, 1993; Tennant and Ashby,
1991). Chemicals that are highly reactive toward DNA are commonly
mutagenic; however, in experimental animals, 84% of tested carcinogens
and 66% of noncarcinogens are mutagenic. Chemicals that are not predictive of reactivity and are nonmutagenic are carcinogenic less than 5% of
the time and so animal testing of nongenotoxic chemicals is less reliable
without understanding the mechanistic etiology or carcinogenesis (Cohen,
1995). Concordance among animal species for the same organ site is less

than 50%, almost that predicted by chance (Haseman and Lockhart, 1993; Gold et al., 1992). Importantly, the actual concordance might vary between 20 and 100%, where the currently observed concordance among rats and mice might be estimated wrongly due to measurement error (Freedman et al., 1996; Haseman and Seilkop, 1992). Other studies indicate that concordance from rat to mouse occurred more commonly for chemicals that induce tumors at different sites, chemicals that induce tumors in both sexes, those that have a reduced latency, and those that increase the rates of rare tumors (Gray et al., 1995; Tennant and Spalding, 1996). Thus, if these same criteria are considered to apply in humans, it is less likely that there would be concordance for studies assessing changing levels of tobacco constituents as a way to assess PREPs. In general, the concordance between animal and human experience also is highly variable (Gold et al., 1992; Lutz, 1999; Monro, 1994), although carcinogens that are more potent in one species tend to be more potent in others, including humans. The major reason for differences in extrapolation from experimental systems to humans is that in vitro cell cultures and experimental animals from different species handle chemical exposures, DNA damage, and stress differently.

The inclusion of biomarkers in experimental animal studies with analogy to human exposures would be helpful (Guengerich, 2000). Carcinogen-DNA adduct levels correlate with tumor incidence for the target organ of these animals (Pledger et al., 1977; Ross et al., 1990). In some cases, target organ specificity is only approximate; for example, in benzo[a]pyrene-treated animals, levels are almost equal in liver and lung, but there is a higher cancer incidence in the former (Ross et al., 1990). Adducts of different classes have relatively similar potency (i.e., the number of adducts correlated with tumor induction is within one to two orders of magnitude; Otteneder and Lutz, 1999). One limitation is that adduct levels associated with tumor incidence can vary widely among species. The relative potency of carcinogens may be assessed by normalizing adduct levels for doses that induce 50% of tumors at the MTD and then considering the dose required to generate that number of adducts (Otteneder and Lutz, 1999). Adduct measurements also can estimate the occurrence of saturation pathways and where dose-response relationships become nonlinear for both adduct formation and tumor incidence. Different tumor-adduct profiles have been shown, depending on the carcinogens and the dose level (Pledger et al., 1977). Figure 12-1 summarizes different types of data for DNA-adduct levels and tumor formation (Poirier and Beland, 1992). It shows that in some cases, the relationship of adduct levels to tumor occurrence is linear, but in some cases, the slope of the effect can be different and saturation pathways occur. Thus, prediction of the effects of individual tobacco constituents in the context of other

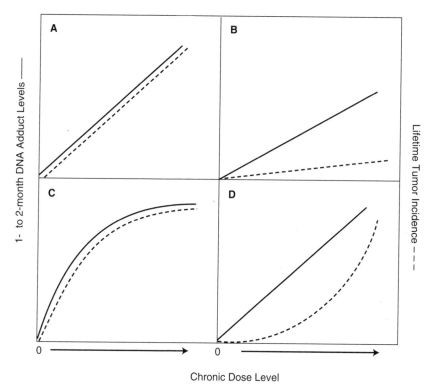

FIGURE 12-1   Lifetime tumor incidence.

NOTE: "Schematic diagrams for the different patterns of lifetime tumor incidence
(—) and DNA adducts at 4-8 weeks of administration (-), observed as a function
of chronic carcinogen dosing" (Poirier and Beland, 1992).
SOURCE: Reprinted with permission from Poirier and Beland, 1992. Copyright
(1992) American Chemical Society.

constituents is complex. It is unknown whether the cancer effect of coexposures is synergistic or additive, but at different dose levels, either can
be observed (Poirier and Beland, 1992). Extrapolation of animal studies to
humans also can be strengthened by demonstrating the same occurrence
of carcinogen-DNA adducts in experimental animals and humans. Unfortunately, most studies have used only single-dose or short-term exposure
(Otteneder and Lutz, 1999), so it is unknown how this relates to chronic
exposure. Most of these studies also investigate levels in the experimental
animal liver rather than in human target organs.

Different methods are available for assessing tobacco-related cancer
risk in humans. External exposure markers attempt to predict exposure

without regard to interindividual differences in smoking behavior and cellular processes. Biomarker assays can assess internal exposure, the biologically effective dose, and potential harm. Examples of how biomarkers might be applied to the assessment of PREPs are provided in Table 12-4. Epidemiological studies of lung cancer as the outcome would provide the best evidence for the use of a PREP. However, these studies have many pitfalls (Table 12-5), and more importantly, the assessment of a particular product would be difficult because of the short duration of use and changes in these products as technology develops, which would make it difficult to assess the risk of any individual product among all the others. (See Chapter 11 for a detailed discussion of biomarkers.)

The biologically effective dose (Perera, 1987) is the amount of a tobacco smoke or tobacco toxin that measurably binds to, or alters, a macromolecule (e.g., protein or DNA) in a cell. The biologically effective dose represents the net effect of metabolic activation, decreased rate of detoxification, decreased repair capacity, loss of cell-cycle checkpoint control, and decreased rates of cell death. It should be noted that not all binding to, or alteration of, a macromolecule leads to an adverse health effect, and so often we are really measuring the dose to a target macromolecule that estimates a biologically effective dose.

For cancer, a common assessment of the biologically effective dose is the measurement of carcinogen-DNA adduct levels (Farmer and Shuker, 1999). These are formed when carcinogen metabolites are alkylated to nucleotides, creating a promutagenic lesion. There are strong laboratory animal data and some human studies that prove a relationship between tobacco smoke constituents, carcinogen-DNA adduct formation, and cancer (Farmer and Shuker, 1999; La and Swenberg, 1996). Laboratory animal studies have shown a cancer correlation with increased adducts in target organs (Ashurst et al., 1983; Nakayama et al., 1984; Pelkonen et al., 1980). In humans, tobacco smoking leads to increased adduct formation in target tissues such as the lung (Bartsch, 1991; Phillips, 1996; Phillips et al., 1988; Schoket et al., 1998; Wiencke et al., 1995) and in surrogate tissues such as the blood (Bartsch, 1991; Hou et al., 1999; Phillips, 1996; Tang et al., 1995a; Vineis et al., 1994; Wiencke et al., 1995). Evidence exists that carcinogen-DNA adducts levels in target and nontarget organs are modulated by interindividual differences (Badawi et al., 1995; Bartsch, 1991; Grinberg-Funes et al., 1994; Kato et al., 1995; Pastorelli et al., 1998; Rojas et al., 1998; Ryberg et al., 1997; Stern et al., 1993;).

In humans, only a few studies have investigated a link between carcinogen-DNA adducts and cancer risk. All data come from case-control studies of the lung and bladder, and all show a positive relationship (Dunn et al., 1991; Peluso et al., 1998; Tang et al., 1995a; Van Schooten et al., 1990). However, since no published prospective studies shows a rela-

TABLE 12-4 Examples of Variables Affecting the Success or Failure of PREPs

Variable	Questions
Effect on metabolic enzymes	1. Does a change in one or more tobacco constituent affect enzymatic induction or activity? 2. Does a change in enzymatic induction or activity result in a change in resultant DNA damage by the agent and, if so, by how much? 3. Does a change in enzymatic induction or activity result in a change in resultant DNA damage by other agents that are substrates in the same metabolic pathway and, if so, by how much? 4. Will people be affected differently because of interindividual differences in enzymatic induction or activity, and if so, what is the proportion of persons affected in a population (race, gender, age, etc.)? 5. Does a change in one or more tobacco constituents affect smoking behavior (e.g., absorption of nicotine, irritation, mucosal damage)?
Effect on DNA repair, cell-cycle control, and programmed cell death	1. Does a change in one or more tobacco constituents affect the induction or activity of DNA repair enzymes? 2. Does lowering exposure to one or more tobacco constituents result in less DNA damage from these constituents but also result in less time for cell-cycle arrest or less cell death, so that other carcinogens can cause more DNA damage or cell proliferation is enhanced? 3. Will individuals be affected differently because of interindividual differences in DNA repair, cell-cycle control, or programmed cell death, and if so, what is the proportion of persons affected in a population (race, gender, age, etc.)?
Effect on mutational spectra of gatekeeper and caretaker genes or tumor suppressor genes and oncogenes	1. Does a change in exposure to tobacco smoke carcinogens result in a change in the mutational spectra of one or several genes or in different RNA and protein expression profiles? 2. Do different levels of exposure result in different mutational spectra or RNA and protein expression profiles? 3. Does the mutational spectra or expression profile differ in persons who have changed their exposures compared to those who have not? 4. Does a tobacco constituent cause DNA damage through the formation of DNA adducts, free-radical damage, and/or gross chromosomal aberrations, (each of these mechanisms might affect the success of a harm reduction strategy and all might have to be considered)?

TABLE 12-5 Competing Risk Factors, Confounders, and Sources of Error in Tobacco-Cancer Associations

Problem	Examples
Competing risk factors and confounders	Type of tobacco product
	Variability in tobacco constituents
	Duration of smoking
	Dose estimates: daily vs. cumulative, packs per day, pack-years, etc.
	Depth of inhalation
	Smoking behavior following a change in brand
	Tumor histology
	Age-related factors: starting, quitting, age at diagnosis
	ETS (duration, intensity, dose)
	Occupational exposure to carcinogens
	Air pollution
	Dietary factors: carcinogens, preventive substances
	Genetic predisposition
	Social and cultural factors
	Gender, race, or ethnicity
	Education
Study design errors	Selection bias
	Recall bias
	Information bias (especially use of proxies in interviews)
	Poor choice of control subjects
	Inadequate matching, overmatching
	Lack of control for confounders
External factors	Publication bias

tionship of adducts to cancer, the case control studies must be interpreted cautiously because they may suffer from differential metabolism or DNA repair due to case status. A variety of assays are available to determine carcinogen-macromolecular adducts in human tissues (Farmer and Shuker, 1999; Hecht, 1999a; La and Swenberg, 1996; Lee et al., 1993; Wang et al., 2000). These are reviewed in Chapter 11, along with data indicating which markers are useful for assessment in target organs.

Biomarkers of potential harm can range from isolated early changes with or without effects on function to events that clearly lead to carcinogenesis and can be observed in cancer cells. Assessing PREPs through clinical and epidemiological studies would consider this full range of effects. These studies are better focused on the earliest events that have been linked to disease, so that the adversity of disease is not a consequence. One goal has been to develop a molecular fingerprint of genetic damage that reflects a particular exposure in persons without cancer, although this has not happened for tobacco carcinogens, and any measurable effects are nonspecific. Nonetheless, a reduction in the level of genetic damage would logically be required if a PREP were to be successful, although the amount of reduction needed to derive a benefit in terms of disease risk is unknown. Several types of assays are available. The main limitation today is that no assays have convincingly been shown to be sufficiently predictive of cancer risk, so they can not be used singly to predict harm reduction. Chromosomal damage can be measured using classical cytogenetic methods (Obe et al., 1982), micronuclei formation (including in bronchial mucosa; Lippman et al., 1990; Schmid, 1975; Xue et al., 1992), COMET (Poli et al., 1999; Speit and Hartmann, 1999), fluorescent in situ hybridization (FISH), or polymerase chain reaction (PCR) methods assessing loss of heterozygosity (using tandem repeats or comparative genomic hybridization), where the latter two methods can be used for morphologically appearing cells. Use of mutations in reporter genes, such as HPRT (Ammenheuser et al., 1997; Bailar, 1999; Duthie et al., 1995; Hou et al., 1999; Jones et al., 1993) or glycophorin A (GPA) have been used, but it is better to identify mutation rates in cancer susceptibility genes such as p53 (Brennan et al., 1995; Ciruzzi et al., 1998; Kure et al., 1996; Yang et al., 1990) or K-ras (Gealy et al., 1999; Mills et al., 1995; Scott et al., 1997; Slebos et al., 1991; Valkonen and Kuusi, 1998; Yakubovskaya et al., 1995).

Biomarkers of potential harm that reflect later stages of carcinogenesis include morphological markers of preneoplastic lesions (e.g., dysplasia), altered phenotypic expression of normal cellular function (e.g., overexpression of the proto-oncogene Erb-B2), and mutations in cancer-related genes such as the p53 tumor suppressor gene. It is possible to measure p53 mutation rates in normal tissues (Hussain and Harris, 1999) of persons without cancer and to measure mutations in sputum for persons with cancer (Sidransky, 1997b). Although these assays are available, current technology limits their use for large-scale epidemiological studies. It also has been found that measuring loss of heterozygosity (Mao et al., 1997) or hypermethylation of genes involved in neoplasia (Belinsky et al., 1998) might be useful for assessing the effects of tobacco smoke.

The study of p53 tumor suppressor genes in tumors might be helpful in determining which tumors were related to specific etiologies. It has been reported that there is a dose-response relationship between tobacco smoking and p53 mutations in general (Kondo et al., 1996) and for G→T (guanine-to-thymine) transversions in particular (Kure et al., 1996; Takagi et al., 1998). Women have more G→T transversions than men for similar levels of smoking, even though men have p53 mutations more commonly (Kondo et al., 1996; Kure et al., 1996). In vitro studies show that BaP and other PAHs cause the same types of lesions, but in different parts of p53

(Smith et al., 2000). An interactive effect of alcohol drinking and cigarette use in oral cavity and lung cancers leads to different types of p53 mutations (Ahrendt, 2000; Brennan et al., 1995). Interestingly, given that the p53 mutational spectrum for lung cancer is similar worldwide (Hartmann et al., 1997), tobacco smoke is likely the major determinant of lung p53 mutations worldwide.

LUNG CANCER

In this country, there were about 171,000 newly diagnosed lung cancer cases in 1999; 92.6% of these were incurable (Landis et al., 1999). Lung cancer consists of four major histological types, namely squamous cell cancer (SCC), small-cell lung cancer (SCLC), adenocarcinoma (AD), and large-cell carcinoma (LCC; Travis et al., 1995). The first two types tend to arise from the large or medium-sized bronchi ("central tumors"), while the latter two tend to develop from the small bronchi, bronchioles, and alveoli ("peripheral tumors"). There has been a shift in the prevalence of histology types over time, in which AD has been increasing relative to SCC (Charloux et al., 1997; Thun et al., 1997; Travis et al., 1995). In Connecticut from 1959 through 1991, associations between cigarette smoking and death from AD versus SCC increased nearly seventeenfold in women and nearly tenfold in men, while smoking-related lung cancer risk increased from 4.6 to 19 in men and 1.5 to 8.1 in women (Thun et al., 1997). This is presumably due to the use of lower-nicotine cigarettes, increased exposures to TSNAs, and greater depths of inhalation.

Lung cancer is preceded by a series of histopathological changes. These changes have been identified for SCC and consist of both reactive changes (hyperplasia, metaplasia) and preneoplastic changes (dysplasia, carcinoma in situ [CIS]; Auerbach et al., 1961). These histologic changes occur far less frequently in never smokers than in cigarette smokers and increase in frequency with the amount of smoking, adjusted for age (Auerbach et al., 1979). Advanced histologic changes are rarely seen in nonsmokers, but occur in 2.6% of those who smoked 1 to 19 cigarettes a day, 13.2% of those who smoked 20 to 39 per day, and 22.5% of those who smoked 40 cigarettes or more a day. Advanced bronchial preoplastic changes (moderate to severe dysplasia and CIS) do not regress on smoking cessation, persist for many years, and possibly persist for life. Compared to men, women have a lower prevalence of high-grade preinvasive lesions in the observed airways (14% vs. 31%; odds ratio [OR]=0.18; 95% confidence interval [CI]=0.04, 0.88; Lam et al., 1999). Bronchial preoplastic lesions are difficult to identify by routine white-light bronchoscopy but may be visualized by fluorescence bronchoscopy (Lam et al., 1998). Preneoplastic lesions preceding SCLC and AD are not well defined. AD has been associated with the presence of peripheral lesions known as atypical adenomatous hyperplasia (Kerr et al., 1994), while SCLC may arise directly from histologically normal or reactive epithelium (Wistuba et al., 2000).

Extensive molecular changes already are present in the bronchi of smokers before any morphological changes can be discerned (Mao, 1996; Wistuba et al., 1997). Allelic losses at chromosome regions 3p and 9p are present in about 80% of smokers, but are rarely present in those who have never smoked. Approximately one-third of the bronchial epithelium of smokers with lung cancer has sustained molecular damage (Park et al., 1999). These changes are more frequent and extensive in SCLC, intermediate in SCC, and much less frequent in AD (Wistuba et al., 2000).

Allelotyping of human lung cancers indicates multiple sites of frequent allelic (>30%; Virmani et al., 1998) loss, which may represent the sites of recessive oncogenes. Loss of heterozygosity (LOH) on chromosomes 11 and 17 occurs more frequently with increased smoking (Mao et al., 1997; Schreiber et al., 1997). A molecular technique known as comparative genomic hybridization (CGH) permits identification of genomic sites of allelic loss as well as sites of amplification. CGH studies of human lung cancers and cell lines confirm allelotyping data and indicate that multiple sites of increased copy number are present in lung cancers, including c-MYC (Levin et al., 1995; Petersen et al., 1997).

Oncogene mutation frequency in lung cancers varies with the histological type. Ras gene mutations are present in about 30% of adenocarcinomas but are relatively rare in other lung cancers (Gealy et al., 1999; Slebos and Rodenhuis, 1992). It has been reported that ras mutations are present only in smoking-related cancers and that these mutations are associated with cigarette smoking (Reynolds et al., 1987; Slebos et al., 1991). Such mutations also can be observed in persons without lung cancer, suggesting that they may be an early marker of smoking-related damage (Lehman et al., 1996; Scott et al., 1997; Slebos et al., 1991; Valkonen and Kuusi, 1998; Yakubovskaya et al., 1995). Interestingly, the K-RAS mutations are G→T transversions, typical of PAH exposure (Gealy et al., 1999; Hutchison et al., 1997; Slebos et al., 1991). LOH affecting at least one locus of the FHIT gene was observed in 80% of lung cancers in smokers, but in only 22% of cancers in nonsmokers (Sozzi et al., 1998a). These findings suggest that FHIT is a candidate molecular target of carcinogens contained in tobacco smoke. Many other molecular changes are present in lung cancers, such as the ERB-B family and MYC (Fong et al., 1999; Hecht, 1999b; Kanetsky et al., 1998; Sekido et al., 1998), but because they have not been directly linked to smoking, they are not discussed here.

Overall allelic losses (Wistuba et al., 2000) and mutations of the p53 gene (Chiba et al., 1990; D'Amico et al., 1992) are more common in SCLC and SCC than in AD. For p53, a positive relationship exists between life-time cigarette consumption and the frequency of mutations (Ahrendt et al., 2000), and of G→T transversions in particular (Kondo et al., 1996), and these lesions are more common than G→A transitions. Supporting the role of PAHs in p53 mutations, benzo[a]pyrene-diol epoxide and other PAHs bind to guanine at codons 157, 248, and 273 of the p53 gene (Denissenko et al., 1996; Smith et al., 2000), which are hotspots for the G→T transversions (Denissenko et al., 1997). Never smokers who develop lung cancer have a completely different, almost random grouping of p53 mutations (Ahrendt et al., 2000; Rom et al., 2000). Alcohol consumption might further increase the frequency of p53 mutations (Ahrendt et al., 2000).

The contribution of DNA repair pathways to lung carcinogenesis has had limited study. Increased frequencies of microsatellite alterations (single shifts of nucleotides), rather than microsatellite instability (multiple shifts of multiple nucleotides), are seen in 20-30% of lung cancers (Fong et al., 1999), suggesting defective DNA repair through a mechanism similar to that seen for colon cancer and defective mismatch repair. Separately, mutations in the mitochondrial base excision repair enzyme, *OGG1* have been seen (Chevillard et al., 1998). Also reported is O^6-alkyl-alkylguanine transferase mutations (et al., 1997), which might increase susceptibility to *N*-nitrosamines.

There are nonmutational effects of smoking in lung carcinogenesis. For example, telomerase expression is very frequent in lung cancer (Fong et al., 1999), which is increased in smokers compared to nonsmokers and former smokers, suggesting that this is reversible (Xinarianos et al., 1999). Expression of gastrin-releasing peptide is seen in lung cancer in response to tobacco smoking (Shriver et al., 2000), and might contribute to bronchial epithelial cell proliferation before lung cancer develops. Some data exist to implicate nicotine as the inducer of gastrin-releasing peptide (Shriver et al., 2000). Hypermethylation of promoter regions, such as that for p16, death-associated protein kinase, glutathione *S*-transferase P1, estrogen receptors and O^6-alkyl-alkylguanine transferase have frequently been observed in lung cancer (Fong et al., 1999), and smoking in particular has been associated with an increased frequency of hypermethylation of p16 (Belinsky et al., 1998). Reactive oxygen species, produced by tobacco smoke constituents, as well as through inflammatory responses, can affect many different protein kinases and transcription factors (Minamoto et al., 1999). How these effects influence carcinogenesis is currently unknown, but perturbations in the balance between oxidative stress and response clearly affect cell survival.

Lung cancer survival after surgical resection and for more advanced stages might be affected by smoking history (Hendriks et al., 1996; Hinds et al., 1982; Sekine et al., 1999; Xavier, 1996), especially in the earlier years after surgical resection (Sobue et al., 1991). While women who smoke might be diagnosed at later stages because their symptoms have been ignored, the early-stage data supports a relationship between smoking history and survival of lung cancer patients. Smoking also increases the risk of lung metastases in breast cancer patients (Scanlon et al., 1995) and of second primary cancers after the diagnosis of lung cancer (Levi et al., 1999). Thus, a PREP in a person with lung cancer or breast cancer might be beneficial, in addition to reducing risks for secondary cancers.

Dose-Response for Smoking and Lung Cancer

The evaluation of a PREP would include an assessment of how the dose-response curve for smoking is altered as exposure is reduced. Today, however there are limited data on the effects of decreasing exposure short of complete cessation. When people stop smoking, their risk of lung cancer decreases over time. Thus, it is plausible that exposure reduction short of cessation may also reduce risk. One might be able to predict the effects of a particular harm reduction method by assuming that the risk for a smoker who achieves exposure reduction will drop to that of a continuous lower-level smoker. This assumption would be speculative at present and it must be evaluated through an understanding of the dose-response relationship of smoking to lung cancer. The modeling for a PREP, however, might be more complicated and analogous to intermittent exposure situations (Murdoch et al., 1992).

A dose-response relationship between cigarette smoking and lung cancer has been established in cohort studies of both men and women (Chyou et al., 1992; Doll and Peto, 1976; Engeland, 1996; Friedman et al., 1979; Nordlund et al., 1997; Shaten et al., 1997; Thun et al., 1995; Tverdal et al., 1993; Winter et al., 1985). These studies show remarkable consistency. Both daily smoking amounts and duration of smoking are important contributors to risk, although the lung cancer risk ratios for daily smoking are higher than for duration of smoking. An earlier age at initiation is a separate lung cancer risk factor (Benhamou et al., 1994; Benhamou et al., 1985; Hegmann et al., 1993; Khuder et al., 1998). Zang and Wynder had proposed to estimate cumulative tar exposure by determining all brands used for different periods of life and the quantity per day for a person, using milligram yields calculated by the FTC method (Zang and Wynder, 1992). The reported effect of how deeply someone inhales also has been associated with an increased risk (Agudo, 1994; Benhamou et al., 1994; Joly et al., 1983; Khuder et al., 1998). There is some data to indicate that smoking more cigarettes per day increases the risk for small-cell carcinoma (Weiss et al., 1977).

The slope of a dose-response curve may provide an indication of the success of a PREP in smokers. Several groups of investigators have modeled different cohort study sets for lung cancer and have found different relationships. Using data from single studies, Armitage (1985) and Gaffney and Altshuler (1988) provided evidence that smoking risk fits either a multistage or a two-stage model for initiation and cell proliferation. Doll and Peto (1978) found that their data on British doctors fit a second-order polynomial equation for number of cigarettes per day and a quadrate equation for duration, although this was only true for smoking less than 40 cigarettes per day. Reanalysis of their data using different approaches confirmed the quadratic function where any two of three dose-dependent parameters (cigarettes per day, duration, age) were considered and the greatest predictor was cigarettes per day (Moolgavkar et al., 1989). A linear relationship was reported in Japanese men (Mizuno and Akiba, 1989), with duration identical to the estimate of Doll and Peto (1978). Puntoni and coworkers summarized data for nine different cohort studies and found that a multistage model best fit the data ($R^2=0.67$, which increased to 0.80 when one study was eliminated; Figure 12-2; Puntoni et al., 1995). Logit, Probit, and Weibull models also provided a good fit (all $R^2=0.61$), but a one-hit model provided a poor fit ($R^2=0.36$). As a validation step, the authors included data for passive smoking and the fit was acceptable. Importantly, these models all predict a greater slope for increasing risk at lower doses and a plateau at higher doses. The multistage model showed that risk increased to a plateau at about 20 cigarettes per day, while the Weibull model showed a decrease in the rate of increasing risk at about 5 cigarettes per day. The latter model, however, never clearly reaches a plateau so that any decrease in smoking might have some benefit. These data imply that small amounts of exposure reduction at higher levels of smoking may not be sufficient to achieve harm reduction. In contrast, Law and coworkers (1997) argued that if smoking data in cohort studies were adjusted for actual internal exposure, in their case predicted by carboxyhemoglobin levels, then the shape of the curve assumes a quadratic relationship, where the slope is higher at more than 20 cigarettes per day; these authors argue that a quadratic relationship is more biologically plausible (Figure 12-3). This data imply that the greatest effect of harm reduction would occur in smokers with the highest levels of baseline smoking. Because of these conflicting models, it is impossible at this time to conclude a lesser or greater benefit from exposure reduction at higher or lower levels of smoking. However, the argument by Law and coworkers for including an internal exposure assessment are compelling (Law et al., 1997). Therefore it is likely that the use of biomarkers would provide better estimates of dose-response relationships and harm reduction. It also should be noted that the above analyses that derive

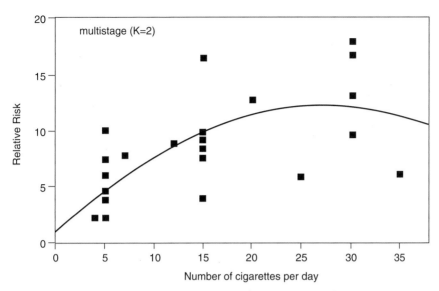

FIGURE 12-2   Relative risk verses number of cigarettes per day.

SOURCE: Reprinted with permission from Puntoni et al., 1995. Copyright (1995) Oxford University Press.

models for dose-response relationships mostly do not consider duration of smoking. It is possible that the daily consumption of cigarettes and lung cancer risk have different relationships based on the number of years smoked (Mizuno and Akiba, 1989; Rylander et al., 1996).

The available cohort studies do not provide data that allows for modeling at smoking levels of less than ten cigarettes per day, which is important for understanding the effects of PREPs that result in low exposure levels. Models developed by Puntoni et al. (1995) for smoking less than an average of 1.4 cigarettes per day do not indicate the existence of a threshold, and there are no current data today to show where one might exist. It should be noted that thresholds are difficult to demonstrate (Purchase and Auton, 1995). Puntoni and colleagues (1995) predict an increased risk for smoking at levels as low as 0.25 cigarette per day. Depending on the model however, there are different slopes for low-dose exposures, where a Weibull model predicts a rapid rise to a relative risk (RR) greater than 2 at 0.2 cigarette per day. A multistage model, however, shows a linear response at low levels of exposure. Either model, however, indicates that the best goal for harm reduction is complete cessation because there are

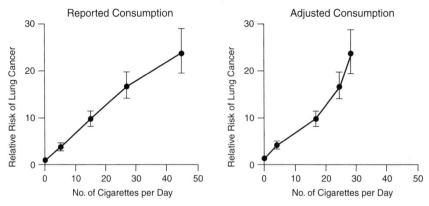

FIGURE 12-3   Relative risk of lung cancer verses number of cigarettes per day.

NOTE: Estimates (with 95% confidence limits) from the U.S. veterans study (Kahn, 1996) of lung cancer mortality in current smokers relative to that in never smokers, according to reported cigarette consumption and adjusted cigarette consumption.
SOURCE: Reprinted from Law et al. (1997) by permission of the publisher Churchill Livingstone.

risks at the lowest levels of smoking, and no safe level can be discerned from the literature. It has been estimated that ETS exposure is equivalent to approximately one cigarette per day (Vutuc, 1984), and although this seems to be a small amount, the risk will depend on duration of exposure as well. Thus while hormesis is thought to exist (Teeguarden et al., 1988), it has not been shown for tobacco exposure.

Biomarker studies have attempted to establish a dose-response relationship for smoking, but the relationship has been variable. Most studies crudely examine smokers versus nonsmokers, rather than reporting a relationship by levels of smoking. Dallinga and coworkers (Dallinga et al., 1998) determined 4-ABP-hemoglobin adducts and "total" phosphorus-32 postlabeled (^{32}P-postlabeled) DNA adducts in peripheral lymphocytes in 55 smokers. The slope of the response was greater for the "total" ^{32}P-postlabeled DNA adducts than for the 4-ABP-hemoglobin adducts in relation to cigarettes per day, and a third-order polynomial curve best fit the data. A plateau began at about 10 cigarettes per day for the former and 20 cigarettes for the latter. In this same study, so-called tar consumption per day had a similar effect. The correlation coefficients for tar were higher than for cigarettes per day, but no models were presented that would allow one to assess the relative contribution of tar versus cigarettes smoked per day, and topography was not considered. The data suggest that for a given level of cigarettes per day, the type of cigarette affects the adduct levels, but that at certain higher levels of cigarettes per day, there is less adduct formation per cigarette. This might be due to differences in smoking topography at higher levels of smoking or saturation of metabolic, repair, and apoptotic capacities. All of these factors probably play a role and have led to differences in adduct formation rates in hemoglobin and in lymphocytes.

A PREP would have to work by reducing exposure, but there are few data today to provide an estimate of how much reduction would have a measurable effect in disease outcome. Studies that examine switching to lower amounts of smoking are few. Benhamou and coworkers reported that compared to 1,503 controls, 1,057 lung cancer patients who reported decreasing daily cigarette use by more than 25% had a 20% reduction in risk, although this was not statistically significant (Benhamou et al., 1989). Lubin, et al. reported that reducing daily consumption by more than 50% reduced risk by about 16%, which was barely statistically significant (Lubin et al., 1984a). Thus, the data available is not sufficient to suggest how much decreased consumption must occur to show a measurable benefit. Wald and Watt (1997) reported that persons who switched from cigarettes to pipes or cigars had a higher risk of lung cancer than persons who had never smoked cigarettes. They also found that these individuals inhaled more of their cigars and pipes; so while there was a reduction in risk compared to continuous cigarette smoking, there was still a persistent effect (Wald and Watt, 1997). Graham and Levin found a sixfold increased risk in persons who switched from cigarettes to other tobacco products, compared to an 8.8-fold risk of continued cigarette use and a 2.6-fold risk for persons who had only smoked other tobacco products and continued to smoke them (Graham and Levin, 1971). This represented a 27% reduction in risk compared to continued smoking.

Lung Cancer Risk and Cigarette Type

An important question is whether smoking low-tar and nicotine cigarettes is associated with lower lung cancer risk and whether switching to such cigarettes has shown a benefit. The available data can be summarized as those that examine risk in relation to tar content or to the use of filtered cigarettes. It is important to understand that there are few differences in type of tobacco used among commercial cigarettes and that most changes in tar and nicotine delivery are achieved mainly through the use of a filter, which affects the absorption of carcinogens as well as diluting the smoke with air through ventilation holes. Thus, under similar smoking circumstances (which does not happen in humans), the amount of tar that would be delivered from a cigarette varies in the following order from the highest to the lowest: nonfilter, filter high tar and filter low tar. Confounding this relationship, however, is the fact that over the past 30 years, the amount of tar yield per cigarette type has also decreased.

While the studies described below indicate a lesser risk in lung cancer for low-tar and nicotine cigarettes, there are outstanding questions about smoking behavior that might affect the interpretation of epidemiological studies. While it is possible to examine risks for persons who smoke low-tar and nicotine cigarettes, the perceived decreased risk may be due to several confounding variables that have not yet been measured. For example, such individuals might be smoking these cigarettes because of a desire to quit or reduce their smoking, and so might smoke less of their cigarettes, have more quitting attempts, or underreport what they smoke. They also might have some illness (i.e., respiratory problems) that may lead them to reduce the amount they smoke. Finally, these may be individuals who have "healthier" life styles (e.g., diets that reduce lung cancer risk).

Studies Assessing Tar Content. Several large studies have suggested that higher-tar cigarettes are associated with increased lung cancer risk or that the risk is less for smokers of low-tar cigarettes compared to high-tar or mixed exposures (Benhamou et al., 1985; Kaufman et al., 1989; Lubin et al., 1984b; Stellman and Garfinkel, 1989; Vutuc and Kunze, 1983). Other studies have not shown a reduced risk due to low-tar cigarettes (Kuller et al., 1991; Lee and Garfinkel, 1981; Sidney et al., 1993; Wilcox et al., 1988). The American Cancer Society Cancer Prevention Study (CPS-1) Cohort study had a lower standardized mortality ratio (SMR) for persons who smoked lower-tar cigarettes within smoking amount categories (Stellman and Garfinkel, 1989). The SMRs increased from 841 to 1,236 in smokers using cigarettes with less than 17.6 mg for the lowest-tar category and more than 25.7 mg for the highest-tar category. The greatest proportional increase was in persons who smoked the highest number of cigarettes per day. An important limitation of this study was that no adjustment was made for smoking duration, with the claim that an age adjustment approximated this need, although models of lung cancer risk do not support this assumption (Moolgavkar et al., 1989). It also classified tar exposure based on current cigarette use, rather than usual brand smoked or switching. A hospital-based case-control study of 881 cases and 2,570 controls indicated that there was a threefold increased risk from smoking 22-28 mg tar cigarettes compared to less than 22-mg tar cigarettes, which increased to fourfold for more than 29 mg (Kaufman et al., 1989). Lubin and coworkers (1984) studied 7,804 lung cancer cases and 15,207 hospital-based controls in five different European cities. This study collected data for previously used cigarette brands, up to a maximum of four, and considered cigarettes per day and duration for each of those brands. So-called total tar exposure was estimated, and a statistically significant trend for increasing risk with highest-tar exposure was observed in both men and women. Compared to persons who only used low-tar brands, a person who used other brands less than 25% of the time had a risk of 1.2, using other brands more than 25% of the time led to a risk of 1.5; using high-tar brands more than 75% of the time led to a risk of 1.8; and exclusive use of high-tar brands produced risk of 1.7. By contrast in another study (Benhamou et al., 1985), the risk of mixing cigarette types was no different than exclusive use of nonfiltered or dark-tobacco cigarettes (Benhamou et al., 1994), where data were provided for smoking duration and cigarettes smoked per day, and there was no clear subgroup (high or low levels of smoking) that had a particularly higher or lower risk compared to low-tar or filtered cigarettes. In a hospital-based case-control study by Alderson and coworkers, reported switching to filtered cigarettes for either less than or more than ten years before lung cancer diagnosis did not show a statistically significant difference in difference risk (Alderson et al., 1985). The potential for increasing consumption when switching brands is important and can modify the possible benefits of presumed lower-tar exposure, as shown in a case-control study of 763 cases and 900 controls by Wilcox and coworkers, in which increasing weighted average tar yields crudely predicted increased risk until exposures were controlled for by cigarettes per day, duration, or pack-years (Wilcox et al., 1988). The Wilcox group, interestingly, concluded that there was more compensation by increasing cigarettes per day in cases compared to controls that switched to lower-tar and nicotine cigarettes. Data pooled from four cohorts failed to show a statistically significant benefit for low-tar cigarettes in terms of lung cancer risk, even among different levels of smoking (Tang et al., 1995b), as did another large cohort study (Sidney et al., 1993). Lee and Garfinkel provided a summary of lung cancer risk and type of cigarette smoked (Lee and Garfinkel, 1981) and were unable to demonstrate a significant decrease in risk based on tar content. There was some observed benefit in terms of tar content when comparing high- to low-tar cigarettes but not for medium versus high tar. Thiocyanate levels, which reflect quantity of cigarettes per day and smoking topography, are associated with increased lung cancer risk, whereas the tar yields of cigarettes were not (Kuller et al., 1991).

Case-control studies have provided evidence that black tobacco carries almost a fourfold risk higher risk tan blond tobacco (Benhamou et al., 1985, 1989; Joly et al., 1983), and a 6.6-fold higher risk in women (Agudo et al., 1994).

In summary, while some studies suggest that there is a lower risk of lung cancer with lower-tar cigarettes, many do not, especially when risk is considered along with biomarkers in relation to smoking behavior.

Studies Assessing Filter Cigarette Use. One of the earliest reports suggesting that filtered cigarettes were less hazardous than nonfiltered cigarettes was published by Bross and Gibson, who estimated a 60% decrease in risk in 265 cases compared to 214 controls (Bross and Gibson, 1968). Lubin and coworkers (1984b) reported that nonfilter cigarette use compared to filtered cigarettes consistently gave higher RR estimates for lung cancer, no matter the history of cigarette smoking duration or cigarettes per day. Interestingly, the magnitude of the risk for cigarettes per day and years of use increased substantially more for persons using filtered compared to nonfiltered cigarettes. This suggests that decreasing cigarette use in persons using PREPs might have more easily demonstrable benefits. In other studies of filtered versus nonfiltered cigarettes, a decreased lung cancer risk of 30% was found in a French study of 1,625 lung cancer cases and 3,091 controls (Benhamou et al., 1985, 1989), a twofold lower risk in a Philadelphia study (Khuder et al., 1998), a 3.5 fold decrease in an Argentinian study (Pezzotto et al., 1993), and a fourfold decrease for women in a Spanish study (Agudo et al., 1994). Wynder and Stellman (Wynder and Stellman, 1979) reported that among 684 cases and 9,547 controls, there was a reduced risk for those who smoked nonfiltered cigarettes for ten years or more, although the results were not statistically significant. However, when they later reported data for 1,242 lung cancer cases compared to 2,300 controls and accounted for increasing smoking per day after switching to lower-tar cigarettes, they found that lung cancer risk was not reduced and even increased in the highest levels of compensation (Figure 12-4; Augustine et al., 1989). Other studies also have reported a reduced risk for filtered cigarettes (Rimington, 1981; Pathak et al., 1986), but a dose-response relationship for persons who mix their brands was harder to demonstrate (Lubin et al., 1984b; Pathak et al., 1986). Lee and Garfinkel provided a summary of lung cancer risk and type of cigarette smoked (Lee and Garfinkel, 1981) and concluded that there was about a 25% reduction in lung cancer mortality for filtered cigarettes. In this review as in others, there was no consideration of biomarkers, actual smoking consumption, or analysis for persons who switched cigarette types. Several studies do not support a decreased risk for filtered cigarettes. In a population-based case-control study, when amount of smoking was considered, there was no benefit from filtered cigarettes (Wilcox et al., 1988). Data pooled from four cohorts failed to show a statistically significant benefit for filtered cigarettes and lung cancer risk, even among different levels of smoking (Tang et al., 1995b), as did another large cohort of 79,946 members of Kaiser Permanente (RR=1.03 for men and 0.65 for women; neither

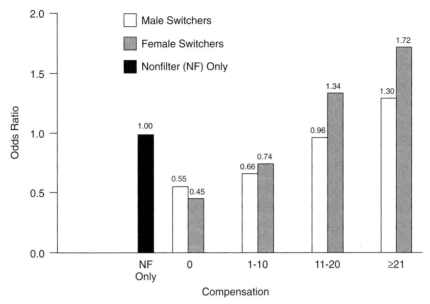

FIGURE 12-4 Risks among switchers relative to nonfilter only smokers according to difference levels of compensation.

NOTE: Compensation is the increase in cigarettes per day after switching to filter cigarettes. Computed from the result of Augutine, Harris and Wynder study (1989) and the results of the study by Wynder and Kabat, 1988.
SOURCE: Reprinted from Augustine et al., 1989. Copyright by the American Public Health Association.

was statistically significant), although women who used filtered cigarettes for more than 20 years had a risk of 0.36 (95% CI=0.18, 0.75; Sidney et al., 1993).

In comparison to nonfiltered cigarettes, filtered cigarettes are more closely associated with AD rather than SCC (Pezzotto et al., 1993; Stellman et al., 1997a, b), although this was observed mostly in women smokers in another study (Lubin and Blot, 1984; Stellman, 1997a, b). Hand rolled cigarettes had an increased risk for SCC and not AD (Engeland et al., 1996). This association is thought to be due to greater depths of inhalation when the filter is in place to compensate for lower nicotine amounts, as well as increased delivery of TSNAs. Evidence for this includes self-report of frequency and depth of inhalation associated with AD rather than SCC (Lubin and Blot, 1984). When considering persons who switched from nonfiltered to filtered cigarettes, compared to lifetime nonfilter cigarette smokers, there was a decrease in risk for SCC but not AD in men and women (Stellman et al., 1997a, b).

Most of the above studies suggest that the use of filtered or low-tar cigarettes was associated with lower lung cancer risk, even though there is clearly an increased risk of all types of lung cancer with all types of cigarettes. Ecological studies also support such a conclusion when smoking rates and lung cancer mortality over time are compared and the slopes for the decline in smoking are less steep than the decline in mortality (Peto et al., 2000). However, many significant considerations that make such data difficult to interpret. These studies assess smoking at a specific point in time, which prevents the collection of useful data later in prospective studies and is subject to recall bias in case-control studies. Indeed, assessing smoking prospectively at multiple time points or considering individual smoking improves estimates (Lee, 1998; Akiba, 1994). Separately, these studies do not account for cohort effects where the public's overall lung cancer risk may have changed due to diet and lifestyle or for the possibility that persons who smoke low-tar cigarettes otherwise differ from those who smoke high-tar cigarettes. Additionally, it is important to note that these studies do not provide an assessment of what happens to persons who switch cigarette type, which is directly relevant to assessing PREPs.

Gender Differences in Lung Cancer Risk

The prevalence of smoking among women is less than among men and consequently have overall lower rates of lung cancer (Shopland et al., 1991) and preneoplastic lesions (Lam et al., 1999). However, lung cancer rates recently have been decreasing for men but not for women (Wingo et al., 1999). The rates of increase since the 1950s for lung cancer in general, and AD in particular, are higher for women than men, because women have increased the amounts of cigarettes smoked per day, start smoking earlier, and smoke different types of cigarettes (Haldorsen and Grimsrud, 1999; Levi et al., 1987; Shopland et al., 1991; Thun et al., 1997). Women more commonly have AD than SCC, even after controlling for smoking status (Ernster, 1996). Several studies have provided evidence that women have a higher risk of lung cancer for a given level of smoking. In a study of 1,108 males and 781 females with lung cancer, compared to 1,122 male and 948 female controls, women were found to have a 1.2- to 1.7-fold higher risk, which was limited to AD and SCLC cancers, rather than SCC (Zang and Wynder, 1996). These data provided similar results when examined by estimated tar yields according to the FTC method, pack-years, and recent smoking per day. Other studies have provided similar findings (Brownson et al., 1992; Cohn et al., 1996; Engeland, 1996; Lubin and

Blot, 1984; Osann et al., 1993; Risch et al., 1993), although some have not (Doll and Peto, 1976; Halpern et al., 1993). While some might hypothesize that the difference in cancer risks between men and women are due to differing baseline nonsmoking rates (Prescott et al., 1998), this was found not to be the case using summary statistics from several large cohort studies (Risch et al., 1994). An increased risk in women is also evidenced by data showing that there is a higher risk for lung cancer in women at similar ages of initiation and the risks are the same for women who start smoking over age 25 as for men over age 20 (Hegmann et al., 1993). In a study of lung cancer risk among persons who switched from filtered to nonfiltered cigarettes, both women and men had a lower lung cancer risk if they smoked similar amounts of cigarettes per day before and after switching (Augustine et al., 1989). However, women had a greater likeli-hood of compensation and smoking more cigarettes, especially at lower doses. For similar levels of increased cigarettes per day, women had a much higher risk of lung cancer. There are several plausible explanations for this increased risk that bear directly on the effects of PREPs. The increased risk observed in women might be due to smoking behavior and/or biological differences. For example, women may be at higher risk because they have begun smoking in recent times with lower-tar and nicotine cigarettes (Thun et al., 1997), which can deliver greater amounts of TSNAs (Hoffman and Hoffman, 1997). There is a lack of evidence, however, showing consistent differences in greater smoking topography, but this may be because both men and women were smoking similar types of cigarettes.

Another explanation for a higher lung cancer risk in women might be related to biological differences between men and women. There might be a hormonal relationship, because women more commonly have estro-gen or progesterone receptors in lung cancer (Kaiser et al., 1996). Two studies found a high abundance in both males and females, but a differ-ence between the two could not be discerned (Canver et al., 1994; Su et al., 1996). Also, because women suffer more tobacco withdrawal symptoms during menses, they might have greater lifetime exposure (O'Hara et al., 1989). Women have higher levels of carcinogen-DNA adducts in lung tissues, even though they have the same or lower levels of smoking (Ryberg et al., 1994a), which supports the latter hypothesis. Separately, Chinese women have a higher risk of lung cancer if they have more and shorter menstrual cycles (Liao et al., 1996; Gao et al., 1987). The frequency of p53 mutations is higher in men, consistent with the greater amounts of smoking, but women more commonly have G→T transversions (Kure et al., 1996), suggesting a particular susceptibility to tobacco smoke carcino-gens. Women might also be more susceptible if they have particular meta-bolic polymorphisms that affect carcinogen detoxification (Mollerup et al., 1999; Ryberg et al., 1997; Tang et al., 1998). Separately, gastrin-releasing peptide has been shown to be more highly expressed in women than men for the same level of smoking (Shriver et al., 2000). The gene for this peptide is located on the X chromosome, so a double copy might result in increased levels that in turn trigger growth stimulation. Therefore, assessment of PREPs also should consider possible effects on hormonal status and differences in the effects of individual tobacco constituents in women.

Racial and Ethnic Differences in Lung Cancer Risk

Lung cancer rates differ by race and ethnicity (U.S. DHHS, 1998). Lung cancer incidence rates are highest among African-American males (112.3 per 100,000), followed by Caucasian males (73.1 per 100,000). African-American and Caucasian females have similar rates (46.2 and 43.3 per 100,000, respectively). Asian-American males and females have relatively lower rates (52.4 and 22.5, respectively), while Hispanics have the lowest rates (38.8 and 19.6, respectively). In the United States, smoking rates differ; African Americans, Hispanics, Chinese Americans, and Hawaiians tend to smoke fewer cigarettes per day than European Americans (Le Marchand et al., 1992). In the United States, Hispanic males and females tend to smoke less than nonHispanic whites, but the risks within smoking levels are similar (Humble et al., 1985). Hawaiians have a greater risk of smoking compared to Filipinos and Caucasians living in Hawaii (Le Marchand et al., 1992). The differences in lung cancer rates within smoking categories may be due to smoking topography, types of cigarettes smoked, differences in the frequencies of heritable traits, and/or environmental, lifestyle, and dietary differences. For example, Japanese have lower rates of lung cancer than persons from other countries, which may be due to the use of tobacco with lower TSNAs, more frequent use of charcoal filters, or lifestyle differences. However, the effects of these factors on lung cancer rates are small compared to the overall increased risk for use of cigarettes.

There are some data to suggest that lung cancer risk is higher in African Americans than Caucasians for a given level of smoking (Harris et al., 1993; Schwartz and Swanson, 1997). For example, Harris et al. reported an RR of 1.8 for African Americans compared to Caucasians at the same level of tobacco consumption calculated as cumulative tar intake (Harris et al., 1993). African Americans tend to smoke menthol cigarettes, while the opposite is true for Caucasians (Cummings et al., 1987). Menthol cigarettes provide cooler smoke that helps anesthetize airways (Sant'Ambrogio et al., 1991), so smoking topography might be affected (Orani et al., 1991). The greater use of mentholated cigarettes among African Americans (Cummings et al., 1987; Wagenknecht et al., 1990) is associated with a higher lung cancer risk (Sidney et al., 1995), although not in other studies of menthol cigarettes (Kabat et al., 1991). In two studies the carbon monoxide (CO) boost was shown to be higher among users of menthol cigarettes (Clark et al., 1996; Jarvik et al., 1994), but not in another study that examined women (Ahijevych et al., 1996). Another mechanism for potential harm from mentholation is pyrolysis of menthol, which leads to benzo[a]pyrene production (Schmeltz and Schlotzhauer, 1968). Socioeconomic status may also contribute to racial differences in lung cancer incidence (McWhorter et al., 1989).

African Americans tend to smoke less than Caucasians (Hahn et al., 1990; Kabat et al., 1991; Royce et al., 1993; Vander Martin et al., 1990), but may be more highly nicotine dependent (Royce et al., 1993; Vander Martin et al., 1990). They also tend to smoke higher-tar and nicotine cigarettes, resulting in higher nicotine levels and greater tar yields (Hahn et al., 1990; Perez-Stable et al., 1998). This is consistent with repeated findings of higher cotinine levels in African Americans compared to Caucasians (Caraballo et al., 1998; Wagenknecht et al., 1990) and higher urinary TSNAs in African Americans (Richie et al., 1997). Even though it was recently shown that African Americans have decreased cotinine clearance, rather than different nicotine metabolism (Perez-Stable et al., 1998), there is still the possibility of higher relative carcinogen exposure. For example, this study did not consider smoking topography. There are data showing that African Americans smoke more of their cigarettes (Clark et al., 1996) and have a higher CO boost per millimeter of cigarette smoked (Ahijevych et al., 1996; Clark et al., 1996). They also have higher nicotine intake per cigarette (Perez-Stable et al., 1998).

When considering a PREP, we must consider different races within the United States and other countries, where more environmental exposures are shared. In general, within levels of smoking, the risks of lung cancer around the world are similar, with some exceptions. In China, a combined analysis of studies that appeared in the Chinese literature reportedly demonstrated a dose-response relationship for men and women (Liu, 1992). However, the slope of the dose-response relationship was less than that in Western studies and similar to that for Japan. Importantly, the attributable risk for smoking was only 57% in men and 26% in women, where 88% and 46% of male and females smoked, respectively. A substantial number of Chinese also smoked pipes of various types, and among these individuals, the risk of lung cancer was lower than among persons who smoked only cigarettes (Lubin et al., 1992). Whether this represents a method of harm reduction remains unknown since that was not specifically studied, and the lower risk might have been observed for many reasons. Caution must be used in attempting to implicate any particular factor for differences in lung cancer rates in different geographical areas. These differences might be due to uses of different tobacco products, smoking topography, modifying effects of tobacco (e.g., diet), or increased frequencies of genetic traits. Separately, differences in health system access, reporting methods, or diagnostic procedures also may substantially affect the accuracy of risk estimates.

In summary, there are sufficient data to show that the use of tobacco products, exposures, and outcomes can vary in different racial and ethnic groups. Thus, PREPs must be assessed in the context of the ethnicity and race of intended users.

Factors Modifying Lung Cancer Risk

Although this report focuses on the evaluation of PREPs, it also is recognized that other factors affect lung cancer risk, both positively and negatively (Lee and Forey, 1998; Lutz et al., 1999). These factors are not detailed here, but evidence follows to show that the evaluation of PREPs should consider other exposures and host susceptibilities that might affect an individual's risk related to PREPs. Many cancer risk factors covary with smoking consumption and thus can confound some studies (Thornton et al., 1994). Factors shown to modify cancer risk may include diet (Breslow et al., 2000; Grinberg-Funes et al., 1994; Mendilaharsu et al., 1998; Voorrips et al., 2000; Yong et al., 1997), vitamin use, and chemopreventive agents (Clapp et al., 1999; Omenn et al., 1996), although whether these factors actually modify cancer risk has not been conclusively demonstrated (Koo, 1997). While the latter might have sufficient impact because of dose, diet and vitamin supplementation would unlikely provide a significant benefit to proven harm reduction methods. This report does not consider these areas as potential harm reduction strategies.

Coexposures to other lung carcinogens from nonsmoking sources can lead to a multiplicative risk effect. These exposures include occupational asbestos exposure, occupational radiation exposure, radon, and therapeutic radiation exposure (Brownson et al., 1993; Carstensen et al., 1988; Inskip and Boice, 1994; Jockel et al., 1992; Moolgavkar et al., 1993; Neugut et al., 1994; Osann et al., 2000; Qiao et al., 1989; Tokarskaya et al., 1995; Torkarskaya et al., 1997). This report does not consider these coexposures in detail, but it should be recognized that they might affect the efficacy of PREPs. Prior lung disease history might increase lung cancer risk in smokers and persons exposed to tobacco smoke (Mayne et al., 1999b).

Heritable susceptibilities can affect tobacco-related cancer risks (Brockmoller et al., 1998; Shields, 2000; Shields and Harris, 2000). The use of genetic susceptibilities in the context of a given type of exposure can increase risk assessment prediction (Khoury and Wagener, 1995). Evidence for familial transmission of risks comes from analyses of lung cancer patients and their parents (Amos et al., 1999; Sellers et al., 1992a, b), although some studies have disagreed (Braun, 1994; Mayne et al., 1999a). While these types of studies tend to implicate high-penetrance genes, there also is evidence that low-penetrance genes in carcinogen metabolism can modify cancer risks, and it is likely that other types of genes will also (e.g., DNA repair, cell-cycle control, apoptosis, signal transduction). Examples of frequently studied genetic polymorphisms in tobacco-related cancers that have been shown in some studies to modify smoking-related disease risk include the glutathione *S*-transferase M1 (GSTM1; Bell et al., 1993; Brockmoller et al., 1996, 1998; Jourenkova-Mironova et al., 1998, 1999; Kihara et al., 1995; Lehman et al., 1996; Rebbeck, 1997), cytochrome P-450 1A1 (CYP1A1) genes (Ishibe et al., 1997; Kihara et al., 1995), glutathione *S*-transferase Pi (Ryberg et al., 1997), and others (Bouchardy, 1998; Jourenkova-Mironova et al., 1999; Rosvold et al., 1995; Sjalander et al., 1996; Wiencke et al., 1997). These genetic polymorphisms and others are believed to affect levels of biomarkers, such as DNA adducts (Kato et al., 1995; Pastorelli et al., 1998; Ryberg et al., 1997; Yu et al., 1995). Interestingly, in Japanese, the risk for the GSTM1 null genotypes increases with increasing levels of smoking (Kihara and Noda, 1994), but the opposite is true for CYP1A1 (Nakachi et al., 1991), indicating a saturation effect. Also, several biomarker phenotypes representing carcinogen metabolism and DNA repair have been shown to modify the effects of smoking-related risks (Li et al., 1996; Spitz et al., 1995; Wei et al., 1996). More specific evidence for a relationship between gene-environment interactions and mutations in the p53 gene can be found from Japanese studies of CYP1A1 (Kawajiri et al., 1996), where a fivefold increase in risk was found for smokers with lung cancer. This risk increased further for persons who also lacked GSTM1. In one study from Norway, smokers who lacked GSTM1 also had an increased risk of lung cancer from p53 mutations, especially transversions (Ryberg et al., 1994b).

Thus, it is important to study the differences in responses to PREPs by individual susceptibilities, in addition to gender, race, and diet, because they might increase or decrease product effectiveness in specific persons. In addition to carcinogen metabolism and DNA repair, genetic traits for other aspects of carcinogenesis and smoking behavior will undoubtedly be identified.

Former Smokers and Lung Cancer

Central to the issue of decreasing the harm from tobacco use in the general population is the evidence for lung cancer risk reduction among former smokers. The risk that is observed in former smokers can reasonably be predicted to be the lowest risk achievable by a PREP. This section describes studies of persons who quit smoking cigarettes and were not reported to have switched to other tobacco products.

The risk of lung cancer in former smokers is less than in current smokers, as demonstrated by both case-control and prospective studies. Case-control studies among former smokers are numerous and most have reported statistically significant reductions in the odds ratio of lung cancer relative to smokers. Some individual studies have been limited by the use of hospital controls, lack of adequate age adjustment, reliance on proxy interviews for information about the smoking behavior of deceased cases, and lack of information on potential risk reduction variability by histological type of lung cancer. A study conducted in Germany (Pohlabeln et al., 1997) addressed many of these issues. That study compared 839 lung cancer cases with an equal number of population-based controls matched on age, gender, and region and examined ORs by years since quitting and by histological type of lung cancer. Relative to current smokers, the lung cancer ORs among former smokers were 0.97 for those who had quit 10 years ago, 0.55 for 11-20 years since quitting, and 0.25 for those who had quit more than 20 years ago. The same pattern of OR reduction was observed among all histological types. In a separate study by Muscat and Wynder (Muscat and Wynder, 1995), the frequency of AD compared with SCC after 25 years of cessation appeared more like that in those who have never smoked than in current smokers. This was true for both men and women and is in agreement with other studies (Tong et al., 1996). Other studies indicate that risk differs depending on previous smoking history and duration of abstinence (Ben-Shlomo, 1994; Graham and Levin, 1971).

One of the difficulties in obtaining reliable estimates of the reduction in lung cancer risk among former smokers in case-control studies is the variability in the ages at which people take up smoking. In the study conducted by Sobue et al. in Japan (Sobue et al., 1993), all of the male former smokers had started smoking cigarettes between the ages of 18 and 22, a far narrower range than reported in many other case-control studies. In that study, the OR reduction for lung cancer among former smokers ranged from 50 to 65%, depending on the age at smoking cessation and the age at admission, with stronger benefits accruing to men who quit at younger ages and who were diagnosed at younger ages. This study also is noteworthy because it documented a hazard reduction even among men in their seventies who had stopped smoking for just a few years: the decrease in cancer risk among male ex-smokers of this age group is significant due to the higher incidence of lung cancer among men this age than in the general population. While the greatest absolute reduction in risk occurred among older smokers, the greatest rate of reduction in risk occurred among younger smokers.

Alavanja et al. (1995) used case-control study data from 618 female lung cancer patients and 1,402 population-based age-matched controls to estimate the population attributable risks (PARs) of lung cancer among nonsmokers and ex-smokers. By far, the most important factor for female ex-smokers was their history of active smoking, explaining 56% of lung cancer incidence in the population of female ex-smokers and far outweighing factors such as environmental tobacco smoke (PAR=1.7), occupational risk factors (PAR=4), and family history of lung cancer (PAR=14).

Unlike case-control studies, prospective studies can directly compute the relative risk of lung cancer among former smokers, and several large cohort studies have examined this in detail. (Doll and Peto, 1976) followed 34,440 British male physicians for 20 years and observed an age-adjusted annual death rate from lung cancer of 43 per 100,000 among ex-smokers, compared to 10 in nonsmokers and 140 in men smoking cigarettes exclusively. Thus, the rate was clearly lower than that in smokers but remained higher than that of nonsmokers. Results from the American Cancer Society's CPS-II of nearly 900,000 men and women (reported by Halpern et al., 1993), demonstrated dose-response curves for the association of lung cancer mortality with years since quitting. For both men and women, the RR of lung cancer death by age 75 was <0.05 among lifelong nonsmokers compared to current smokers, while among ex-smokers who had quit in their thirties and forties, the RRs were in the range of 0.07-0.15. Even those who had quit at ages 60-64 experienced a reduction in risk of lung cancer mortality at age 75 (RR=0.45 for men, 0.49 for women). The 20-year follow-up report on mortality among 12,866 participants in the Multiple Risk Factor Intervention (MRFIT) study observed a 60% reduction in deaths from lung cancer among ex-smokers compared to men who continued to smoke (Kuller et al., 1991).

Figure 12-5 shows the reduction in the RR of death from lung cancer among former smokers compared to those who continued to smoke from the large prospective cohort study of (Halpern et al., 1993). The amount of harm reduction depends not only on the length of time since quitting but also on the person's current age. Nonetheless, while the curves for each cohort of former smokers show downward trends for fatal lung cancer, the risk ratios never reach the low level experienced by persons who have never smoked cigarettes. Quitting at an earlier age provided the greatest risk reduction. In a logistic regression model, significant independent variables included gender, education, age (β=0.085), number of cigarettes per day (β=0.025), years smoked (β=0.055), and years quitting (β depends on quit cohort). Enstrom and Heath (1999) compared lung cancer mortality among smokers and quitters in a cohort of 118,000 men and women in

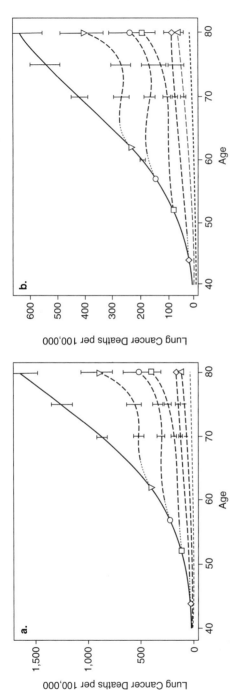

FIGURE 12-5 Lung cancer death rates of women and men.

NOTE: Figure on the left: Model estimates of lung cancer death rates by age for female current, former, and never smokers, based on smokers who started at age 18.5 and smoked 22 cigarettes/day. Estimates are plotted for current smokers (solid line), never smokers (dotted line), and former smokers (dashed lines). The five age-at-quitting cohorts are distinguished by the following symbols on the graph at the age of quitting and also at age 80: △ 30-39, ◇ 40-49, □ 50-54, ○ 55-59, ▽ 60-64.

Figure on the right: Model estimates of lung cancer death rates by age for male current, former and never smokers, based on smokers who started at age 17.5 and smoked 26 cigarettes/day. Estimates are plotted for current smokers (solid line), never smokers (dotted line), and former smokers (dashed lines). The five age-at-quitting cohorts are distinguished by the following symbols on the graph at the age of quitting and also at age 80: △ 30-39, ◇ 40-49, □ 50-54, ○ 55-59, ▽ 60-64.
SOURCE: Halpern et al., 1993. Copyright by the Journal of the National Cancer Institute. Reprinted with permission.

California born between 1900 and 1929 and traced from 1960 through 1997. Lung cancer mortality declined among ex-smokers, but never reached the rate experienced by people who had never smoked and was still twice as high after 20 years of quitting. This result is in agreement with the lung cancer mortality RR of 1.73 for people who had stopped smoking at least 15 years ago observed by Enstrom (1999) in the analysis of both National Health and Nutrition Examination Survey (NHANES I) data (N=4,900) and a large veterans study (N=284,000). Figure 12-6 shows (Enstrom, 1999) results on the RR of fatal lung cancer according to the number of years since quitting, relative to men who never smoked. Note that the RR does not decline all the way to unity but remains modestly elevated (RR=1.8) even after 15 or more years since smoking cessation. Such results (Enstrom, 1999) suggest the possibility that many former smokers are in poor health around the time of quitting, with substantial mortality with the first few years. This effect diminishes with time, and after five years the benefits become increasingly substantial among survivors. Other data show that risks of lung cancer shortly after quitting do not change much, but decrease to a 30% increased risk over nonsmokers after ten years of cessation (Graham and Levin, 1971). No data are available to suggest differences in risk by cigarette type among former smokers. Some data do not indicate a greater benefit for quitting in women

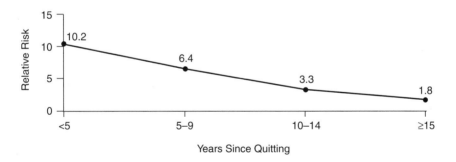

FIGURE 12-6 Relative risk of lung cancer death among former male smokers, relative to never smokers.

NOTE: Data from Enstrom's (1999) prospective cohort study of U.S. male veterans, comparing 43,559 former smokers to 59,351 men who never smoked cigarettes.
SOURCE: Adapted from Enstrom, J.E., 1999. Copyright by Elsevier Science. Reprinted with permission.

compared to men (Halpern et al., 1993), although other studies do (Risch et al., 1993).

An overall challenge in assessing this cancer risk reduction is the question: Are quitters systematically different from current smokers and never smokers in ways that would explain the differences in cancer risks among these groups? A large, cross-sectional survey of nearly 9,000 adults in Britain (Thornton et al., 1994) attempted to answer this question by randomly sampling nonsmokers, ex-smokers, and current smokers in the population. The study collected data on a wide range of risk factors for poor overall health, including dietary, lifestyle, medical, and socioeconomic factors. Of 33 risk factors assessed, 27 were most prevalent among current smokers, and most risk factors decreased in prevalence with the amount of time since quitting. The data allowed estimation of the degree of confounding in smoking studies due to multiple risk factors and, importantly, called attention to the likely impact of these factors on epidemiological studies where weak associations are detected. In other words, the potential effects of confounding variables may be profound for weak associations between cancer and smoking variables, including those that attempt to estimate risk reductions among former smokers (where RRs have approached 1.5 or less and have been much debated). Finally, greater smoking and lower tendency to quit have been observed among socioeconomically disadvantaged groups relative to others, suggesting an increasing burden of tobacco-related cancers on persons of lower socioeconomic status. It should be noted that the MRFIT prospective trial found a higher lung cancer rate in the group with smoking cessation counseling versus usual care (Shaten et al., 1997). While this is likely due to chance alone, it may suggest that the greatest decrease in risk occurs in persons who are able to quit without counseling or that those who continue to smoke after counseling have high levels of exposure.

Many studies indicate that lung cancer mortality is increased for the first five years after quitting (Alderson et al., 1985; Enstrom, 1999; Graham and Levin, 1971; Halpern et al., 1993; Higgins and Wynder, 1988; Pohlabeln et al., 1997), which has been called the "quitting-ill" effect. It is presumed that this occurs because people who are ill are induced to quit. Although that is probably the explanation for this phenomena, it is not known if there is also some chronic induction of cytotoxicity or other mechanism in the lungs induced by tobacco smoking. When smoking ceases, this effect diminishes and allows already present neoplastic cells to replicate, so persons who would have developed lung cancer do so at a more rapid rate. This is currently speculative. The issue of biological evidence for tobacco harm among former smokers has been addressed in studies examining lung tissue biopsies. Mao et al. (1997) compared LOH at several chromosomal loci in nontumor lung tissue samples from smokers, ex-smokers, and nonsmokers. LOH at 3p14 was found in 88% of smokers, compared to 45% in ex-smokers (p=.01) and 20% in nonsmokers. A similar pattern of increasing levels of genetic damage in current, former, and nonsmokers has also been reported with regard to p53 mutations in bladder tumors (Djordjevic et al., 2000). Witsuba et al., on the other hand, did not detect such differences in LOH at the loci they surveyed, but ex-smokers were just as likely as current smokers to have genetic changes typical of lung tumors, and these changes persisted many years after quitting (Wistuba et al., 1997). Thus, there do appear to be irreversible sequellae of past tobacco use even among people who have abstained for many years and are not currently diagnosed as having lung cancer.

In summary, stopping smoking decreases the risk of lung cancer, and the earlier that an individual stops, the greater is the reduction in risk. Data are not sufficient to determine whether certain levels of prior smoking result in proportionately greater or lesser risk reduction, although clearly, a greater smoking history carries a greater lung cancer risk (Graham and Levin, 1971). The data supports the conclusion that after about 20 years of quitting, the risk reduction plateaus and remains slightly above never smokers. Thus, it is unlikely that a PREP would achieve a greater level of risk reduction than the level at 20 years, and at any time point before that, the risk is likely to be greater than that of someone who quit.

OROPHARYNGEAL CANCERS

Oropharyngeal cancers include cancers arising in the oral cavity, tongue, pharynx, and larynx. Almost all are squamous cell carcinomas. Their incidence is about 40,000 cases annually, of which about 12,000 will eventually die from the disease (Greenlee et al., 2000). There is a male-female ratio of about 2:1.

Preneoplastic lesions, which include keratosis, dysplasia, carcinoma in situ, and microinvasive cancer, are considered a sequential continuum (Gillis et al., 1983). Keratosis is the most common oral lesion, occurring as white (leukoplakia) or red (erythroplakia) patches, and is present in 1-10% of adults (Mao and El-Naggar, 1999). Some molecular evidence exists that premalignant lesions are the direct precursors of invasive lesions (Califano et al., 2000). Cessation of smoking does not remove the potential for progression of the disease and all patients must be followed indefinitely (Gillis et al., 1983).

The major risk factors for oropharyngeal cancers are tobacco and alcohol use. The role of smokeless tobacco is discussed below. There is a dose-response for both smoking and alcohol use; together the two agents act synergistically (Ahrens, 1991; Barasch et al., 1994; Blot et al., 1988; Hayes et al., 1999; Iribarren et al., 1999; Keller and Terris, 1965; La Vecchia et al., 1990; Lewin et al., 1998; Macfarlane et al., 1995; Mashberg et al., 1993; Muscat et al., 1996; Sanderson et al., 1997; Schildt et al., 1998; Schlecht et al., 1999; Takezaki et al., 1996; Talamini et al., 1998). The attributable risk for alcohol and/or tobacco use is about 75-80% for males and 52-61% for females (Blot et al., 1988; Hayes et al., 1999). There is some evidence for a weak familial association in smokers (Goldstein et al., 1994). Some studies suggest that tobacco consumption is more likely than alcohol consumption to cause precursor lesions (Jaber et al., 1999; Kulasegaram et al., 1995) and cancer (Elwood et al., 1984; Macfarlane et al., 1995). Another study reported the converse in women (Schildt et al., 1998), although this was a small study of Swedish women who may have been snuff users (see section on Smokeless Tobacco; Sanderson et al., 1997). Actual consideration of the relative carcinogenicity of the two agents depends on the level of consumption for each. Talamini and coworkers studied 60 nonsmoking drinkers and 32 nondrinking smokers and compared them to controls (Talamini et al., 1998). Depending on the amount of drinks per week, the OR reached 5.3 (95% CI=1.1, 24.8) in the nonsmokers and 7.2 (95% CI=1.1, –46) in smokers. Thus, the dose–response curves overlapped. In a pooled analysis of three studies form New York, Italy, and China, the OR for males with greater than 33 pack-years was 1.3 (95% CI=0.6, 3.1) and for females who smoked more than 18 pack-years was 4.6 (95% CI=1.9, 10.9). Three published studies that report data by gender all indicate an increased risk for women compared to men, especially at the highest levels of smoking (Blot et al., 1988; Hayes et al., 1999; Muscat et al., 1996).

Cigarette type and oropharyngeal cancer risk have not been extensively studied. Three reports have not shown a difference between filter and nonfilter cigarettes (Blot et al., 1988; Hayes et al., 1999). Hand-rolled cigarettes appear to carry greater risks than manufactured cigarettes (De Stefani et al., 1998). Only one study could be identified that examined so-called tar content for cigarettes, and a lower risk was associated with low-tar cigarettes (La Vecchia et al., 1990). Black tobacco carried about a five-fold higher risk than blond (De Stefani et al., 1998). Where studies are available, there are no differences in risk for similar levels of smoking in Caucasian Americans compared to African Americans (Blot et al., 1988; Day et al., 1993), although one study suggested that African Americans were at a lower risk, but there was no breakdown by smoking and drinking categories.

Smoking cessation changes the risk of oropharyngeal cancers. Cancer of the larynx has been found to be markedly less likely among ex-smokers than among current cigarette smokers (U.S. PHS, 1964). In a relatively large case-control study in Brazil (Schlecht et al., 1999), 784 cases of cancer of the mouth, pharynx, and larynx were compared to 1,578 noncancer controls, compared to never smokers, the ex-smokers of >20 years had an

OR=2.0 (95% CI=1.0, -3.8) for all types combined, lower risks for mouth (OR=1.6) and pharyngeal cancer (OR=1.5), and a high risk for laryngeal cancer (OR=3.6). The benefits of quitting were greatest for cigarettes and lesser for cigars and pipes.

Excellent reviews have been published of the molecular changes present in oropharyngeal cancer (Mao and El-Nagger, 1999; Sidransky, 1997b). Many of the molecular changes in smoking-related upper aerodigestive tract tumors, including lung and oropharyngeal cancers, are similar and commence during multistage pathogenesis (Mao and El-Nagger, 1999; Sidransky, 1997b). Changes include frequent losses at chromosome arms 3p, 9p, 17p, 5q, and 8p, aneuploidy, p53 gene mutations and expression abnormalities of the TGF-b signaling pathway, activation of telomerase, downregulation of RAR-a, and inactivation of the p16 gene (Brennan et al., 1995; Field et al., 1995; Izzo et al., 1998; Picard et al., 1999). Deregulation of the cell cycle is related to the degree of tobacco exposure (Davidson et al., 1996; Gallo et al., 1995).

Oropharyngeal tissues clearly have the capacity to metabolically activate tobacco smoke carcinogens and cause DNA damage (Badawi et al., 1996; Degawa et al., 1994; Kabat et al., 1991; Liu et al., 1993; Matthias et al., 1998). Among the highest levels of CYP1A1 have been reported in these tissues compared to others (Kabat et al., 1991). NAT1, but not NAT2, activity is present, and there is some evidence that CYP2C plays an important role in these tissues. Aromatic DNA and 4-ABP adducts have been detected in laryngeal tissues; these were higher in smokers than in nonsmokers (Flamini et al., 1998; Szyfter et al., 1994). Adduct levels in oral mucosa are correlated with biopsy levels indicating that mucosa can be used as a surrogate marker (Besarati et al., 2000; Jones et al., 1993).

Several studies have indicated an increased risk of oropharyngeal cancers in those who have a heritable trait demonstrated by genetic polymorphisms, although which markers play the greatest role is not yet known (Cullen et al., 1997; Helbock et al., 1998; Henning et al., 1999; LeVois, 1997; Morita et al.,1999; Rebbeck, 1997; Sturgis et al., 1999; Sumida et al., 1998; Trizna, 1995), and there is some evidence for a greater effect in persons with lower levels of smoking (Jourenkova et al., 1998). In one study, heritable traits in carcinogen metabolism increased the frequency of p53 mutations (Lazarus et al., 1998). When cultured lymphocytes are exposed to mutagens and resultant chromosomal breaks are counted, there is a greater mutagen sensitivity in cases, especially smokers (Cheng et al., 1998; Cloos et al., 1996; Schantz et al., 1997; Spitz et al., 1993). This trait also predicts the risk of secondary cancers in persons with oropharyngeal cancer (Spitz et al., 1994). There are some data to suggest that smoking might increase mutagen sensitivity, so there might be an inductive effect in this assay (Wang et al., 2000).

The mutational spectrum of p53 in oropharyngeal cancers is similar to that in lung (Liloglou et al., 1997), although some studies disagree (Olshan et al., 1997). Mutations occur more commonly in smokers than nonsmokers (Brennan et al., 1995; Field et al., 1994; Lazarus et al., 1996b; Liloglou et al., 1997). In a study by Brennan and coworkers, the frequency of p53 mutations for tobacco and alcohol users was higher than for either of these exposures alone (Brennan et al., 1995).

The above information suggests that assessments for oropharyngeal and lung cancer risk related to the use of potential inhaled (i.e., tobacco smoke) PREPs are similar. It can be inferred that a suggested benefit for lung cancer would also benefit oropharyngeal cancer. However, these studies cannot imply that the quantitative benefits might be similar or even measurable in persons who continue to drink alcoholic beverages because of the synergistic effect of tobacco smoking and alcohol. The study of persons with oropharyngeal neoplasms provides some opportunities because of the accessibility of tissue and the occurrence of preneoplastic lesions.

BLADDER CANCER

More than 53,000 cases of bladder cancer will occur in the United States in the year 2000, and approximately 12,000 persons will eventually die from this disease (Greenlee et al., 2000). The male-female ratio is about 2.6:1. About 70% of bladder cancers are superficial at the time of presentation (i.e., confined to the mucosa or submucosa), while the rest are deeply invasive (Soloway and Perito, 1992). Most bladder cancers in this country are transitional cell carcinomas, arising from the normal transitional epithelium after multistage progression (hyperplasia, dysplasia, carcinoma in situ, superficial invasion). However, in some parts of the world such as Egypt, where schistosome infection of the bladder is common, squamous cell carcinomas are associated with chronic inflammation and squamous metaplastic changes. The chemicals most commonly implicated in bladder cancer in humans are aromatic amines, although other compounds such as PAHs might also play a role (Ross et al., 1996; Vineis and Pirastu, 1997).

Patients with cancer of the urinary bladder often present with metachronous tumors, appearing at different times and at different sites in the bladder. This observation has been attributed to a "field defect" in the bladder that allows the independent transformation of epithelial cells at a number of sites. Analyses of clonality indicate that a number of bladder tumors can arise from the uncontrolled spread of a single transformed clonal population (Sidransky et al., 1992). The molecular pathology and development of bladder cancer have been reviewed recently (Rao et al., 1999). As with other cancers, many molecular changes have been described in bladder cancers, including p53 gene mutations, p16 and retinoblastoma gene silencing, LOH at various chromosomal regions, aberrant methylation, and the presence of microsatellite alterations.

Many studies have shown a dose-response effect of smoking on bladder cancer risk in both men and women (Hartge et al., 1987; Slattery et al., 1988; Vineis et al., 1983, 1984). A recent report summarized a combined analysis of 11 case-control studies (Brennan et al., 2000). The authors found a linear increasing risk of bladder cancer with increasing duration of smoking, ranging from an OR of 1.96 after 20 years of smoking (95% CI=1.48, –2.61) to 5.57 after 60 years (95% CI=4.18, –7.44). A dose-response relationship was observed between number of cigarettes smoked per day and bladder cancer up to a limit of 15-20 cigarettes per day (OR=4.50, 95% CI=3.81, –5. 33), after which no increased risk was observed. An immediate decrease in risk of bladder cancer was observed for those who gave up smoking. This decrease amounted to more than 30% after 1-4 years, (OR=0.65; 95% CI=0. 53, –0.79) and was more than 60% after 25 years of cessation (OR=0.37; 95% CI=0.30, –0.45). However, even after 25 years, the decrease in risk did not reach the level of the never smokers (OR=0.20; 95% CI=0.17, –0.24). The proportion of bladder cancer cases attributable to ever smoking was 0.66 (95% CI=0.61, –0.70) for all men and 0.73 (95% CI=0.66, –0.79) for men younger than 60. These estimates are higher than previously calculated. Using a modeling approach to mortality data, the RR for 20 cigarettes per day for 20 years was 2.9 for men and women in England and Wales (Stevens and Moolgavkar, 1979). Another important bladder cancer risk factor is occupational exposure to aromatic amines (Ross et al., 1996). Several studies report an interactive effect for increasing risk in such workers who smoke (D'Avanzo et al., 1990; Vineis et al., 1984; Vineis and Martone, 1996). For bladder cancer, PREPs for individuals must be considered in the context of the workplace.

A recent study focused on the relationship of smoking and the progression of superficial cancers (Fleshner et al., 1999). Continued smokers experience worse disease-associated outcomes than patients who quit smoking. The authors recommended that smoking cessation be employed as a tertiary prevention strategy for patients with superficial cancers.

Cigarette type can influence bladder cancer risk. There is a higher risk with black tobacco than with blond tobacco (D'Avanzo et al., 1990; Vineis et al., 1984; Vineis and Martone, 1996). Filter-tip cigarettes pose a lower risk (Vineis et al., 1983, 1984), although this finding is not consistent (Burch et al., 1989). Importantly, only one study of which the committee is aware collected data for switching from nonfilter to filter cigarettes conflicts (Burch et al., 1989; Hartge et al., 1987). In this, there was a small benefit from switching, but this was more pronounced in persons who had switched more than 15 years prior to diagnosis, and there was no benefit when the data were examined for persons aged 21-64 years rather than 21-84 years (Anwar et al., 1993; Hartge et al., 1987). The data were adjusted for smoking duration and cigarettes per day, but compensation was not specifically queried. Moreover, although there was a decreased risk with filtered cigarettes, there was no difference between smokers who smoked only nonfiltered cigarettes and those who switched. Increasing depth of inhalation has been reported as a separate risk factor (Burch et al., 1989; Slattery et al., 1988).

Doll and Peto (1976) found that among male British physicians followed for 20 years since an initial survey of smoking habits, the annual age-adjusted rate of bladder cancer deaths was 11 per 100,000 among men who had quit smoking, compared to 9 among nonsmokers and 19 among men who smoked cigarettes exclusively. Benefits from giving up cigarette smoking were quantified in a large cohort study of Kaiser Permanente Medical Care Program members in the United States (Habel et al., 1998). Among current smokers, former smokers, and never smokers the standardized bladder cancer incidence ratios were 0.56, 0.68, and 1.04, respectively.

Chromosome 9 alterations and TP53 mutations are among the most frequent events in bladder cancer. Several studies have explored the relationships between epidemiological risk factors (especially smoking) and these genetic alterations. Elevated odds ratios were found for chromosome 9 alterations in smokers compared to nonsmokers (OR=4.2, 95% CI=1.02, –17.0) after controlling for age, sex, race, occupational history, and stage of disease. For chromosome 9 alterations, the OR was 3.6 for those smoking 20 cigarettes per day (Zhang et al., 1997). One study reported an association of smoking status and p53 mutations (Zhang et al., 1994), although others disagree (Spruck et al., 1993; Xu et al., 1997). For p53, a significant association between the number of cigarettes smoked per day and p53 protein nuclear overexpression was found (p=.02; Zhang et al., 1994). The odds ratios were 2.3 for those smoking one to two packs per day and 8.4 for those smoking more than two packs a day. In addition, a distinct mutational spectrum for the p53 tumor suppressor gene in bladder carcinomas was reported in patients with known exposures to cigarette smoke (Spruck et al., 1993; Xu et al., 1997).The P53 mutations in bladder cancers from workers with aromatic amine exposure have the same spectra (Djordjevic et al., 2000). These data support the hypothesis that certain carcinogens derived from cigarette smoking and occupation may induce p53 mutations, which in turn are involved in early steps of bladder carcinogenesis.

Aromatic amines are metabolically conjugated in the liver, excreted in the urine, and then metabolically activated in the bladder (Ross et al., 1996). DNA adducts have been described in the bladder epithelium of smokers and nonsmokers (Phillips and Hewer, 1993; Talaska et al., 1991). Although most of the DNA binding appears not to be smoking related, the levels of several specific adducts were found to be significantly elevated in DNA samples of current smokers, as opposed to never smokers or former smokers (five years' abstinence). Detection of DNA adducts in cells in voided urine may be a noninvasive method for following subjects at increased risk (Talaska et al., 1993).

Studies have shown that smoking-related bladder cancer risk and survival increases with genetic susceptibilities for carcinogen metabolism and detoxification, mostly for GSTM1 and NAT2 (Bell et al., 1992; Brockmoller et al., 1996, 1998; Katoh et al., 1995, 1998; Mommsen and Aagaard, 1986; Okkels et al., 1996, 1997; Rebbeck, 1997; Risch et al., 1995; Taylor et al., 1995). Persons with low activity of CYP3A were associated with higher p53 overexpression (Romkes et al., 1996). Only one study relates adduct levels to bladder cancer risk (Peluso et al., 1998), but because this was a case-control study, conclusions are limited. However, in a small group of patients *(N=45)*, adduct levels were not related to p53 mutations in tumors, but this was not a prospective study (Martone et al., 1998).

In summary, the bladder is a remote site from carcinogen entry into the body. Because urine is easily accessible, there are unique opportunities to study the effects of PREPs on bladder epithelial cells, especially in the context of genetic susceptibilities. However, although there are data showing that some cigarettes produce lesser risks if they are filtered, there also are data to indicate that changing to lower-tar cigarettes is not beneficial. If these data can be replicated, then it is suggested that the potential benefits of PREPs will be difficult to measure. In persons with occupational exposures that increase risk of bladder cancer, the potential benefit of a PREP may be minimized.

ENDOMETRIAL CANCER

Several studies have found that cigarette smoking reduces the risk of endometrial cancer. Although there are no prospective studies, case-control studies are fairly consistent. For example, in a study by Lesko and coworkers, 510 women with endometrial cancer were compared with 727 women with other types of cancers; the RR for current smokers was 0.7 (95% CI=0.5, 1.0), and a dose-response effect was noted (Lesko et al., 1985). The effect occurred predominantly in postmenopausal women. A study by Brinton and coworkers reported a RR of 0.6 (95% CI=0.4, –0.9) in postmenopausal women, where current smokers had the lowest risk and former smokers had an intermediate risk (Brinton et al., 1993). Other studies are in agreement (Austin et al., 1993; Levi et al., 1987; Parazzini et al.,

1995). This reduced risk is thought to be related to an effect of smoking on circulating estrogens and androgens, which are also affected by other factors such as increased weight (Austin et al., 1993). In a study of post-menopausal women, smoking was associated with a decreased risk in women who did and did not use estrogens, although the effect was greater in the former (Franks et al., 1987). Smoking was found to modify the association of increased weight and endometrial cancer, where there was no increased risk in smokers (Lawrence et al., 1987; Parazzini et al., 1995). No studies have reported data for the effects by cigarette type.

ENVIRONMENTAL TOBACCO SMOKE

Environmental tobacco smoke (ETS), also termed passive smoking or exposure to secondhand smoke, has been estimated to cause 2,600 to 7,400 lung cancer deaths per year among nonsmokers in the United States, according to a review of nine studies of lung cancer mortality (Repace and Lowrey, 1990). Animal models have established the carcinogenicity of ETS (Witschi, 1997a; Witschi et al., 1997b) Despite widespread workplace restrictions on smoking and public education about the dangers of second-hand smoke to adults and children, millions of people continue to be exposed to ETS. Data from NHANES III, a representative sample of the health of the U.S. population, show that 43% of children were living in a house with one or more smokers, and 37% of adult nonsmokers reported either having one or more smokers living in the same house or being exposed to tobacco smoke in the workplace (Pirkle et al., 1996). Also, 88% of nonsmokers tested positive for serum cotinine. Workplace bans on smoking have been highly effective in reducing ETS, but lesser workplace restrictions have been shown to be much less effective and many non-smokers continue to be exposed to ETS on the job (Hammond, 1999). Any future strategies to reduce the harm from smoked tobacco products must therefore consider the potential effects on persons exposed to ETS.

Many studies have focused on indirect markers of ETS exposure, such as the presence in the home of a spouse who smokes or the number of years exposed to ETS. Direct measurements of ETS biomarkers (e.g., cotinine in urine, blood, or saliva) have also been widely implemented. Cotinine, the direct metabolic breakdown product of nicotine, with a bio-logical half-life of 20 hours in urine, has been shown to meet all of the criteria for a highly sensitive and specific marker of ETS (Benowitz, 1999). Cotinine levels in children are highly correlated with adult cotinine levels (Crawford et al., 1994) and with the number of adult smokers in the household and the number of cigarettes smoked by the adults (Bono et al., 1996). ETS results in increased adduct levels and carcinogen metabolites in humans (Hecht et al., 1993; Maclure et al., 1989).

The initial evidence linking ETS with increased risks of lung cancer came from studies in Japan and other countries in which smoking among women is rare. The conclusion that ETS is a cause of lung cancer has been opined by several reviewers and persons conducting meta-analysis (Brownson et al., 1997, 1998). In many studies, the risk of lung cancer among nonsmoking women was evaluated in relation to the presence or absence of a husband who smokes. For example, Fontham et. al. (1991) reported an OR of 1.5 for the association of lung cancer among lifetime nonsmoking women who lived with a spouse who smoked. In that study, there was no significant association with childhood ETS exposures. On the other hand, (Janerich et al., 1990) found no association with ETS exposure in adulthood, but an OR of 2.0 was found for high levels of household tobacco smoke in childhood. (Stockwell et al., 1992) compared 210 women with lung cancer who were lifetime nonsmokers with 301 controls assembled by random-digit dialing. The maximum effect detected was an OR of 2.4 (95% CI=1.1, –5.3) for more than 40 smoke-years of exposure (with p=0.004 for trend). Childhood ETS exposures yielded ORs and trends very similar to those associated with adult ETS exposures. Numerous other studies support the conclusion that ETS exposure increases lung cancer risk (Brownson et al., 1992a, 1998; Darby and Pike, 1988; Hirayama, 1981; Stockwell et al., 1992; Tweedie and Mengersen, 1992).

A recent review by Lee of 44 ETS studies revealed that the RR of lung cancer among nonsmokers is between 1.16 and 1.24 for women having a husband who smokes, relative to nonsmokers whose husbands are also nonsmokers (Lee et al., 1998). Furthermore, this report assessed the impact of a number of potential covariates on the magnitude of the ETS association. This is a critical question when the RR is weak (i.e., <1.5) because the impact of statistical confounding can obscure the true level of risk in such situations. Although Lee concluded that it was impossible to determine that ETS exposure is linked to increased lung cancer risk because of potential biases, his own estimates examining the available literature in many different ways still lead to the conclusion of an association. For example, providing a summary estimate where dose-response data are available, thereby reducing some biases, a risk estimate of 1.24 (95% CI=1.15, 1.35) was found. In that review, Lee reported evidence for confounding by a number of factors: (1) continent—with RRs in European studies exceeding those in Asia and the United States; (2) publication date—with earlier publications (1981-1989) tending to report higher risks than more recent publications; (3) histology—with RRs in studies that confirmed primary lung tumors tending to be greater than those using unconfirmed cases; (4) dose-response data—with studies providing such data more likely to detect significant effects of ETS than those that did not provide such data; (5) type of control—with RRs from studies using disease controls tending to be higher than studies using healthy controls; (6) assessment of confounders, which tends to reduce RRs compared to studies that did not adjust for confounders; and (7) age matching, which tends to produce lower risk estimates than unmatched studies. Additionally, Lee noted that many studies of ETS did not match on marital status, most only adjusted for one or two confounders (leaving a high potential for uncontrolled confounding), and most considered only a single source of ETS rather than taking multiple sources into account. Still another factor that should be mentioned is the method of interviewing proxies when a case is deceased, which tends to introduce information bias or at the least nondifferential errors.

Considering other sources of bias, one study examined recall bias and misclassification of smoking status by examining smoking histories reported by spouses and concluded that there was no recall bias (Nyberg et al., 1998). Examining studies that use cotinine to classify ETS exposures, Tweedie and Mengersen used a meta-analysis approach and concluded that ETS risk was 1.17 (95% CI=1.06, 1.28; Tweedie and Mengersen, 1992).

Epidemiological studies of ETS risks that incorporate cotinine measurements are able to validate the classification of subjects according to self-reported levels of ETS. For studies lacking this biomarker, a major issue in estimating the cancer risks associated with ETS is the potential misclassification of former smokers as nonsmokers, which would tend to inflate the true risk ratios. A study of two large cohorts in Sweden, one involving twins and the other of randomly surveyed adults in the population, estimated that about 5% of former smokers were misclassified as never smokers, with roughly equal proportions in men and women. The RR for lung cancer among misclassified men was 1.9 (95% CI=0.5, –9.1), indicating no statistical association, compared to 4.5 for correctly classified former smokers and 13.3 for current smokers (Nyberg et al., 1997). The authors of that study concluded that misclassification occurs mainly among very light smokers and long-term ex-smokers. Future studies of ETS should use cotinine measurements to estimate the impacts of such classification errors.

It was not easy to show that ETS affects biomarkers of cancer risk (Scherer et al., 1992). However, improved methodologies now show that ETS-exposed persons have elevated levels of TSNA metabolites in their urine (Hecht, 1999b). Other studies have reported an increase of aryl amine-related adducts (Maclure et al., 1989).

CIGAR SMOKING

Cigar smoking has increased tremendously in the United States in recent years, with sales increasing as much as 50% between 1993 and 1998 according to a recent commentary by the Surgeon General (Satcher, 1999). In that report, Dr. David Satcher notes that the popularity of cigar smoking has been especially pronounced among well-educated people. During this same period, cigarette sales fell by 3%, and taxes on cigarettes, but not cigars, were increased nationwide. Regular cigar smoking, however, is not a safe substitute for cigarette smoking. Cigar smoking is associated with increased risks of oral, esophageal, laryngeal, and lung cancers and with coronary heart disease and chronic obstructive pulmonary disease. Accordingly, the Surgeon General warned that cigars should not be viewed by the public as a safe and lower-cost alternative to cigarette smoking and called for warning labels, increased public awareness, and youth education efforts about the risks of cigars.

One of the limitations in studying the effects of cigar smoking on lung cancer risk is the relative rarity of cigar smokers compared to cigarette smokers in the population, and the fact that some smokers tend to mix tobacco products presents further challenges to disentangling potential cigar-related effects from cigarette-related effects. Many case-control studies have not had sufficient numbers of cigar smokers to analyze the risks.

Boffeta et al. (1999) Pooled data from seven large case-control studies in Germany, Italy, and Sweden. All except one of the studies used population-based controls in comparison to lung cancer cases, and proxies provided interview data for deceased cases. In Europe, small cigars (cigarillos) are popular, and these were analyzed separately from large cigars. Relative to lifelong nonsmokers, age- and study-adjusted analyses revealed that the risk of lung cancer was elevated among people who smoked only large cigars (OR=5.6; 95% CI=2.9, 10.6), and among those who smoked only cigarillos (OR=12.7; 95% CI=6.9, –23.7). The OR was 14.9 (95% CI=12.3, –18.1) among exclusive cigarette smokers and 12.7 (95% CI=10.3, –15.6) among smokers of mixed tobacco products. The dose-response relationship between cigar and cigarillo use and lung cancer risk was strong, whether analyzed by years of use (P=.0003 for trend), grams of tobacco per day (P=.01), or age at smoking initiation (P=.002).

The above risk ratios have been described independently and consistently in several other European studies, among them the Seven Areas Study in which (Lubin et al., 1984b) reported an OR of 5.6 for large cigars and 11.6 for cigarettes. Boffeta et al. (1999) suggested that the lower OR for cigars compared to cigarettes is not due to lower carcinogenic potential, but rather to lower cumulative lifelong consumption and later age at smoking initiation of cigars.

In comparison, a review of 14 American studies (Shanks and Burns, 1998) documented cigar-associated lung cancer risks as lower than those of cigarette users, but the risk estimates for cigars were only modestly elevated. Reasons for the risk-level discrepancies in European and American case-control studies of lung cancer and cigar smoking may include several factors: (1) Americans overall prefer large types of cigars, whereas Europeans favor cigarillos; (2) differences in the constituents of the products; (3) differences in inhalation and other behavioral parameters of smoking; (4) consistent misclassification of cigar smokers in American studies; (5) differences in the proportions of histological types of lung tumors; and (6) differences in age at taking up smoking. Clearly, more research is needed to resolve these issues.

Several large, prospective studies of cigar smokers have documented significantly increased risks of lung cancer during long periods of follow-up compared to nonusers of tobacco. For example, (Iribarren et al., 1999) followed a cohort of nearly 18,000 men enrolled in the Kaiser Permanente Health Plan, among whom 1,546 cigar smokers were studied for several decades. The RR for lung cancer, adjusted for age and other covariates, was 2.14 (95% CI=1.12, –4.11). Shapiro and coworkers (2000) studied the risk of lung cancer in cigar smokers who never used cigarettes. These individuals had an RR of 5.1 (95% CI=4.0, 6.6).

In addition to the increased risk of lung cancer among cigar smokers in the Kaiser Permanente study Iribarren et al. (1999), a synergistic effect of cigars and alcohol consumption was observed for oropharyngeal cancer, in which the RR for this cancer due to cigars and alcohol combined was much greater than for the independent effects of each substance alone.

Relatively few studies have examined the association of cigar smoking with bladder cancer. In their 20-year mortality follow-up study of 34,440 male British physicians who answered a questionnaire about smoking habits, Doll and Peto (1976) documented a significantly higher rate of bladder cancer deaths among men who smoked only pipes and cigars (14 per 100,000, age standardized) than among nonsmokers (9 per 100,000), compared to 19 per 100,000 among men who smoked only cigarettes. In the CPS-II cohort of male cigar smokers who never smoked cigarettes Shapiro also found an association between cigar smoking and bladder cancer (Shapiro et al., 2000).

SMOKELESS TOBACCO PRODUCTS

Smokeless tobacco is consumed in a variety of different ways in various cultures around the world. Examples of smokeless tobacco products include chewing tobacco, dry snuff (used in the nasal cavity), wet snuff (a moist wad of tobacco, usually placed between the lips and gums), and nass (a mixture of tobacco, lime, ash, and cotton oil), with many local variations in Asia, the Middle East, and Africa. Large geographical differences in the prevalence of smokeless tobacco consumption are evident, with particularly high consumption in Scandinavia (where a popular form of snuff is known as snus), India, Southeast Asia, Sudan, and parts of the United States. Smokeless tobacco products from these different regions are produced differently and have different levels of carcinogens (Gupta et al., 1996; Hoffmann and Djordjevic, 1997). The popularity of smokeless tobacco increased sharply in the 1980s among young men in the United States, particularly among athletes and high school or college students (Christen, 1980). Data from the 1986-1987 National Survey of Oral Health in U.S. School Children examined relationships between smokeless tobacco, alcohol, and the presence of oral soft-tissue lesions (Tomar et al., 1997). In the study sample of more than 17,000 children between the ages of 12 and 17, 1.5% had mouth lesions from smokeless tobacco. Factors associated with these lesions included male gender, white race, current snuff use, and current chewing tobacco use, with snuff having the highest OR (18.4). There was little evidence for risk modification by the use of alcohol.

Persistent use of snuff in the oral cavity causes a characteristic lesion, "snuff pouch keratosis," with a prevalence of 1.6 per 1,000 adults in a population-based study of 23,616 white American adults over age 35 (Bouquot, 1986). Almost 7% of the leukoplakias examined in that study were either carcinomas or severely dysplastic lesions. Early lesions can commonly be found in adolescent snuff users (Tomar et al., 1997). In India, chewing tobacco is associated with erythroplakia (Hashibe et al., 2000), and oral dysplasia (Kulasegaram et al., 1995). Stopping the use of smokeless tobacco results in the disappearance of oral leukoplakia (Martin et al., 1999).

Smokeless tobacco products are associated with cancers of the head and neck, depending on the type of tobacco used (Brinton et al., 1984; Jacob III et al., 1999; Rao et al., 1994; Winn et al., 1981; Winn, 1997; Wynder et al.,1957), although not all studies are supportive of this conclusion. A large number of studies in India, including cohort, case-control, and intervention studies, support the association between oral cancer and smokeless tobacco, and these studies are consistent, strong, coherent, and temporally plausible (Idris et al., 1998; Nandakumar et al., 1990; Sankaranarayanan et al., 1990; Wasnik et al.,1998). Idris et al., compared oral cancer risks in the Sudan, where there is an extremely high consumption of smokeless tobacco and found that users of toombak had an extremely high RR for such tumors (Idris et al., 1998). Research from other parts of the world is far less complete, so it is not known at this point whether ethnic or cultural differences in susceptibility explain any of the geographic variability. Studies conducted in the United States are more conflicting. In some of these studies it is difficult to separate the effects of chewing tobacco from alcohol drinking because of the few nondrinkers. In the United States, Winn and coworkers reported a 4.2-fold increased risk (95% CI=2.6, 6.7) in southern white women who exclusively use snuff (Winn et al., 1981). In contrast, an analysis of the relationship between smokeless tobacco and cancer of the oral cavity in the National Mortality Followback Study did not detect increased risk (Sterling et al., 1992). In spite of the conflicting U.S. data, it can be concluded that snuff use in the United States also increases the risk of oropharyngeal cancers.

In Sweden, there is a very high rate of Swedish snuff (snus) use. But, the use of snus in Sweden has generally not been associated with oral cavity cancer (Idris et al., 1998; Kresty et al., 1996; Lewin et al., 1998; Nilsson, 1998; Schildt et al., 1998). Snus is not fermented and so has a much lower level of N-nitrosamines (Nilsson, 1998) and has a lower genotoxic potential (Jansson et al., 1991), which might be related to the lack of increased risk.

The risks of other types of head and neck cancers from smokeless tobacco products have not been studied as extensively. Evidence for an elevated risk of nasal cancer in association with the use of snuff was reported in a case control study in North Carolina and Virginia (Brinton et al., 1984). Chewing tobacco was not associated with salivary gland cancer in a recent study of cases and controls in the United States (Muscat and Wynder, 1998).

Some authors have concluded that there is no association between smokeless tobacco and bladder cancer risk, but few studies have examined this association and the results are inconsistent (Burch et al., 1989). One of the few studies to report an increased risk of bladder cancer among snuff users was a very small study of 76 female cases and 254 controls, among whom 3 cases and 1 control reported snuff use (Kabat et al., 1986). In a larger case-control study of 332 white men with bladder cancer and 686 population-based controls, Slattery et al. (1988) reported an increased but statistically nonsignificant risk among nonsmokers who used snuff or chewing tobacco. In another population-based study, Burch et al. (1989) found no association of bladder cancer with chewing tobacco or snuff among men or women, but neither of these studies incorporated any exposure biomarkers and neither was specifically designed to test the hypothesis of smokeless tobacco risk.

The p53 gene is commonly mutated in cancers associated with smokeless tobacco products, and while some differences in the spectra have been reported for different regions of the world, no hotspots or patterns have been consistently shown compared to oral cavity cancers related to smoking (Ibrahim et al., 1999; Kannan, 1999; Lazarus et al., 1996 a, b; Saranath et al., 1999; Xu et al., 1998).

A major concern of health professionals is the presence of carcinogenic N-nitroso compounds in smokeless tobacco, which have been demonstrated to cause cancers of the mouth and lip, nasal cavity, esophagus, stomach, and lungs in laboratory animals. Hemoglobin adducts to these carcinogens are measurable in the blood of smokeless tobacco users (Carmella, et al., 1990) and, thus, may be useful biomarkers for measuring exposure levels among users. Urinary metabolites of TSNAs have been measured in persons using smokeless tobacco products, and higher levels were associated with oral leukoplakia, indicating greater use of the products (Kresty et al., 1996). Also, levels in snuff users were higher than those of chewing tobacco users. Hemoglobin adduct levels have been found to be higher in snuff users than in nonusers (Carmella et al., 1990). However, such markers have been used only rarely in epidemiological studies to date and have not been used frequently in studies of human cancer risk. Users of smokeless tobacco products have higher rates of endogenous nitrosation, as well (Nair et al., 1996). Other exposures that occur with the use of smokeless tobacco products include compounds that cause oxidative DNA damage (e.g., polyphenols), where the amount may be related to pH (Nair et al., 1996). Genetic susceptibilities, namely GSTM1 null, is associated with an increased risk of oral leukoplakia in India (Nair et al., 1999).

STUDIES OF NICOTINE MUTAGENICITY AND CARCINOGENICITY

Several studies have been conducted to determine if nicotine is genotoxic. Almost all studies that could be identified failed to find increased genotoxicity (Doolittle et al., 1991, 1995; Mizusaki et al., 1977; Trivedi et al., 1990, 1993; Yim and Hee, 1995), although there are conflicting data about the potential mutagenic activity of cotinine (Yim and Hee, 1995; Doolittle et al., 1995). Urine from rats exposed to nicotine was not mutagenic (Doolittle et al., 1991). The effects of coculture of nicotine and known genotoxic substances indicated an increased rate of mutations for some compounds and a decrease for others (Yim and Hee, 1995). Experimental animal studies using nicotine alone have not found that nicotine is carcinogenic to the exposed animal (Martin et al., 1979; Schuller et al., 1995) or to offspring of animals treated with nicotine (Martin et al., 1979). However, in experimental animals, nicotine can increase the frequency of tumors induced by other agents such as 7,12-dimethylbenz[a]anthracene (Chen and Squier, 1990), N-nitrosamines (Chen et al, 1994; Gurkalo and

Volfson, 1982), hyperoxemia (Schuller et al., 1995), and N-[4-(5-nitro-2-furyl)-2-thiazolyl]formamide (LaVoie et al., 1985), although there was no effect for other N-nitrosamines (Habs and Schmahl, 1984) and an antitumor effect in some cases (Zeller and Berger, 1989). Nicotine also is reported to reduce apoptosis in normal and transformed cells (Wright et al., 1993), but not in lung cancer cell lines (Maneckjee and Minna, 1994). Nicotine was shown not to induce growth of lung cancer cell lines (Pratesi et al., 1996). Long-term studies of persons treated with nicotine replacement therapy are not yet possible because of the short time during which such products have been available. Even though it is possible that nicotine might increase tumor occurrence due to other agents (e.g., have a promotional effect), the risk from co-treatment with nicotine replacement therapy in persons who continue to smoke is likely to be small compared to continued use of tobacco products at a higher rate. The amount of nicotine replaced is less than that available from cigarettes, and it does not have the spectrum of carcinogens present in tobacco products. Human studies have shown that the use of nicotine replacement products does not result in the formation of TSNAs such as NNK (Hecht et al., 1999), although no human liver microsomes in vitro increase the formation of NNK precursors (Hecht et al., 2000).

CONCLUSIONS

There are sufficient laboratory and human data to suggest that harm reduction might be an achievable goal for persons who cannot stop smoking, because there is evidence of decreasing risk for persons who quit smoking, and there are differences in risk for persons who use different tobacco products and differ in the way they use them. Clearly, quitting smoking is the most effective method of reducing cancer risk, and the cancer risk to former smokers is the lowest-level risk that might occur from the use of any PREP. Importantly, it must be recognized that the use of any PREP will likely increase the risk of cancer at some level as long as there is exposure to tobacco carcinogens, in contrast to quitting, which stops exposure to all tobacco constituents. Nonetheless, reduction of exposures to tobacco smoke and tobacco products to the lowest possible levels may provide some benefit to individual users and to the general population. However, data are insufficient to conclude how much reduction in exposure would yield a measurable benefit for whom. Currently, methods that would reduce exposure to tobacco constituents to the greatest extent would likely provide the greatest benefit, but this remains to be proven.

A thoughtful approach to the assessment of PREPs and cancer risk requires consideration of many factors obtained from laboratory and human studies. Data are sufficient to conclude that a dose-response relationship exists for the use of tobacco products and cancer risk. In laboratory animals, the shape of the dose-response curves differs for different tobacco constituents, so that understanding the relationship to tobacco smoke as a complex mixture is difficult. In humans, there are sufficient data for different cancer types and tobacco products, although more data exist for tobacco smoking and lung cancer risk. These studies suggest a curvilinear response that increases most around 5 cigarettes per day and plateaus at 20 cigarettes per day. However, these studies do not consider actual smoking behavior and exposure, so the shape of this curve is not certain. There is some evidence to indicate that when internal exposure is considered through biomarkers, the shape of this curve follows a quadratic equation. Thus, data are insufficient to conclude a greater or lesser effect for a PREP at particular levels of smoking history. There are sufficient data to suggest that different populations have different dose-response relationships depending on gender, race, age, and ethnicity, although the actual risk levels have not been sufficiently defined to draw definitive conclusions about risks among groups. Based on these type of data and possible modifiers of cancer risk (e.g., genetic susceptibilities, diet, lifestyle, occupation), it should be considered that PREPs might benefit people differently or not at all.

There is no evidence of a threshold for tobacco smoking and cancer risk. This conclusion is consistent with the knowledge that there are many carcinogens in tobacco smoke, the aggregate would work to increase risk at any level. Modeling for low-dose exposures indicates an increased risk with less than one cigarette per day. Thus, persons who initiate smoking with PREPs that contain tobacco would increase their risk for cancer, and there is unlikely to be a "safe" cigarette. Former smokers who resume smoking with such products would increase their risk further.

It is possible to conceptually extrapolate the regression of risk using PREPs and assume that the use of such products would bring a smoker to a risk equal to some lower level of lifetime exposure. However, it must be acknowledged that there are insufficient data to validate this assumption or indicate that a decrease in risk would be measurable for all or some smokers. There are insufficient data to indicate what the shape of the curve for regression of risk would look like.

Data are sufficient to conclude, with some caveats, that filtered cigarettes pose a lower risk than nonfilter cigarettes for lung cancer and possibly other cancers. The caveats are that this only occurs in persons who do not substantially increase the number of cigarettes they smoke per day or otherwise compensate for their smoking behavior due to lower delivered levels of nicotine. Also, these studies may be confounded by diet, lifestyle, or other characteristics of people who choose filtered cigarettes.

The available data are suggestive, but not sufficient, to conclude that smokers of so-called low-tar cigarettes have a lower cancer risk compared to those who smoke higher tar cigarettes, with the same caveats as for filter smoking studies. However, there are insufficient data to assess the differences in risk for "ultralow-", low- and high-tar cigarettes that are filtered. This is because these cigarettes became available at a later date, so there was not enough latency in the general population to assess them until recently. There are insufficient data to adequately consider how risk changes from switching types of cigarettes.

This chapter has not reviewed potential cancer risks due to fibers released from cigarette filters or tobacco additives, because it is thought that the risk from these exposures is substantially less than the risk from tobacco smoke constituents. However, there are no existing data to prove this assumption. Importantly, as PREPs are developed that substantially reduce exposure to tobacco constituents, the role of fibers and additives in carcinogenesis might become more important. Thus, fiber and additive exposure should be considered when assessing PREPs.

There are some experimental models (e.g., in vitro cell culture and laboratory animal) that are useful for the assessment of PREPs. Although there are many reasonable models for assessing individual tobacco smoke products, better models are needed to assess exposures to complex mixtures. Such studies are not sufficient alone to support claims of potential harm reduction, and no claim of potential harm reduction should be allowed without adequate human clinical and epidemiological studies. These studies, however, are very important for (1) determining those products that are not likely to result in measurable harm reduction (e.g., if a product results in exposures that increase genotoxicity, there would be less enthusiasm for it, while the converse indicates only that further testing should be considered in humans) and so should not be tested in a human clinical study in anticipation and should not be introduced into the marketplace; (2) identifying unforeseen reactions (e.g., if a product reduces exposure but does not decrease tumors then there might be some constituent or combination of constituents that are either new or more important than those targeted for reduction in the product); (3) providing supportive evidence for the use of a particular bioassay in humans (e.g., if the same biomarker predicts cancer risk in experimental animals) and; (4) assessing the dose-response and the shape of the regression of risk for the PREP, although the data should be considered qualitative or semi-quantitative and cannot be extrapolated directly to human smoking risk. Both in vitro cell culture and experimental animal studies should be used in assessing PREPs, where both can assess genotoxic and nongenotoxic end points, and chronic animal bioassays are needed to assess the end point of cancer risk. It is beyond the scope of this committee to recommend a specific panel of assays, but such a panel needs to be developed. Also, these studies should assess changes both for specific carcinogens and for complex mixtures in which the latter should be mandatory.

There is sufficient evidence to conclude that human experimental studies and short-term clinical studies can indicate the harmful effects of tobacco products. Thus, such studies can be used to assess harm reduction. Through the use of biomarkers and surrogate indicators of cancer risk these studies can evaluate the manipulation of carcinogens and nicotine to reduce exposures and how these changes might affect smoking behavior, metabolic activation, enzymatic induction, conjugation, excretion, biologically effective doses (or their validated surrogates), and biomarkers of potential harm. Separately, these studies can assess differences in risk and provide evidence for modifying effects by genetic susceptibility, diet, lifestyle, occupation, and so forth. However, at the current time, there is no single biomarker or panel of biomarkers that can be considered adequate indicators of cancer risk by themselves because most have not been sufficiently validated. New technologies are offering new opportunities for biomarkers. Thus, experts will be needed to devise a panel of biomarkers that reflect different exposures, biologically effective doses, and pathways for potential harm.

It is clearly possible to assess the effects of PREPs on cancer as the ultimate outcome, and only such studies can provide definitive evidence for the success of a product. However, the long latency for cancer renders these studies infeasible for making such claims today or in the near future. Preneoplastic lesions or the identification of harmful effects in single cells might be used as indicators of the carcinogenetic pathway, but the technology to identify these in the general population or in large epidemiological studies is not yet available. In these studies, the characterization of smoking history and behavior is well validated for recent exposures but problematic for assessing lifetime exposure. Also, self-reported smoking history is insufficient to adequately assess risk in the context of PREP assessments, so biomarkers also are needed to assess exposure, biologically effective dose, and potential harm.

Currently, the best means of assessing PREPs and cancer risk is to focus on lung cancer because this is the most common cancer and so will provide studies with the greatest statistical power. However, data are sufficient to conclude that a risk exists that the widespread use of PREPs will shift the burden of cancer in the population from one type of cancer to another or from cancer to a different disease. Thus, a particular cancer type cannot be the sole indicator for the success of a PREP, and other cancers, diseases, and overall mortality must be evaluated.

There has been sufficient study of nicotine in the laboratory to conclude that it is unlikely to be a cancer-causing agent, although there are no studies of long-term human exposure. The consideration of nicotine as a carcinogenic agent is trivial compared to the risk from other tobacco constituents.

Smokeless tobacco products are associated with oral cavity cancers, and a dose-response relationship exists. However, the overall risk is lower than for cigarette smoking, and some products such as Swedish snus may have no increased risk. It may be considered that such products could be used as a PREP for persons addicted to nicotine, but these products must undergo testing as PREPs using the guidelines and research agenda contained herein.

Studying of the effects of PREPs on cancer risk from ETS is problematic because of the difficulties in measuring reductions in exposure. Also, while there is clearly an increased risk of lung cancer from ETS, the determination of changes in risk from the use of PREPs will require studies of large numbers of people, and smoking currently is prohibited in many places where ETS might have occurred.

RESEARCH AGENDA

Use of Laboratory Studies to Predict Effects of Potential Reduced-Exposure Products

Both in vitro cell culture and in vivo experimental animal studies are needed to identify the possible benefits of a PREP. It is beyond the scope of this report to recommend specific assays or animal models, but it is clear that no single model is sufficiently validated to be solely relied upon. Given that there is insufficient confidence in extrapolating results from laboratory studies to human risk, such studies should not be used alone to support a claim of possible or potential harm reduction. However, these studies are considered important because they can (1) reduce the enthusiasm for a product if there is an increased amount of harm compared to current marketed products; (2) can be used to identify mechanisms of carcinogenesis and how the PREP affects these mechanisms; and (3) identify unanticipated affects such as exposures to new or different carcinogens, different carcinogenic effects, or other unanticipated toxic effects.

The design of experimental studies that assess PREPs should be done in two ways. The first would focus on exposure to the components of the PREP as the sole exposure (i.e., tar fraction or a mixture of carcinogens that simulates yields from the product), preferably as a complex mixture. The main value of these studies would be not to identify the potential benefits of the product, but to assess its impact on persons who initiate smoking with this product and then stay with it. The second experimental design would be to use initial exposures that simulate those from reference cigarettes at different levels and times, and then change the exposure to simulate the PREP at different levels and times. This would better mimic the human scenario where persons who cannot quit will switch to such a product. These models would provide supportive evidence to identify the range of exposure situations that might provide the most benefits and whether some circumstances might result in absolutely no benefit.

There has been difficulty in developing animal models of tobacco-related cancers because animals breathe smoke differently than humans. Thus, additional efforts are needed in developing such models. Separately, genetically altered animals and cell lines can be used in the context of harm reduction in order to elucidate effects on different carcinogenic pathways.

It is important that laboratory experimental studies incorporate biomarkers along the range from exposure to potential harm and that molecular analysis of tumors be undertaken to provide data that would prioritize biomarker use in humans.

Development and Validation of Biomarkers for Cancer Risk Assessment

The following research agenda is intended to provide sufficient data to develop methods that can assess PREPs. It does not describe the methods for assessing a PREP.

Currently, the best biomarkers assess internal exposure, such as the use of exhaled carbon monoxide. There has been significant progress in developing biomarkers that measure the biologically effective dose, but these generally are not sufficiently sensitive or are too labor intensive to be used in large epidemiological studies. Thus, there are some biomarkers that can be used in smaller studies for assessing PREPs, such as the measurement of carcinogen-DNA adducts, where there are some data to support a relationship to cancer risk. There is less enthusiasm for markers where less specificity exists (i.e., chromosomal aberrations and sister chromatid exchanges).

Great emphasis is needed on developing more sensitive technologies for biomarkers of harm in normal-appearing tissues. For example, mutational load studies measuring p53 or K-ras mutations in normal tissues, LOH in sputum and urine cells, and so forth, have the potential to identify effects very early in the carcinogenic process. These biomarkers would have more importance than biomarkers that reflect mostly exposure.

It is clear that human studies and surrogate biomarkers will have the greatest applicability for assessing PREPs. However, although there are some data to show that the use of surrogate biomarkers can reflect the effects in a target organ, a need clearly exists for more studies of both existing and new biomarkers.

Biomarkers are needed that will assess the effects of specific carcinogens as well as complex mixtures. They also should consider exposures from particulate and gaseous phases, as well as oxidative damage.

Many new technologies are being used in studies of carcinogenesis and cancer diagnosis, such as proteomics and expression arrays. These technologies have the potential to provide important information about the use of a PREP and how the product would affect the carcinogenic process. Currently, the use of these technologies in cancer risk studies is embryonic, but such studies should begin for PREPs at the present time, notwithstanding the significant bioinformatic and data interpretation issues. If, for example, it is shown that enzyme induction in human lymphocytes is not substantially different between product types for key genetic pathways, then there would be less enthusiasm for the PREP.

Current and new biomarkers should be tested in different populations in order to understand the range of results and how they relate to different populations. It is important to understand the utility of biomarkers in relation to conventional tobacco products in order to evaluate difference among PREPs. For example, dose-response relationships can be established in ETS-exposed persons, former smokers, low- and high-level smokers, women and men, and different races and ethnic groups. Many biomarker studies classify persons only as current, former, and never smokers. However, this classification is inadequate to provide supporting evidence for use of these biomarkers in the assessment for PREPs. When a dose-response relationship does not exist, modifying factors have to be identified, and until then, the enthusiasm for a particular biomarker is lessened because its laboratory validity would be questioned.

Determine Whether There Are Susceptible Subpopulations That Might Benefit More or Less from a Potential Reduced-Exposure Product

Data are sufficient to conclude that certain subpopulations are susceptible to tobacco-related harm, although there are insufficient data to conclude that any specific group carries different risks based on genetic susceptibilities, race, gender, et cetera. Human studies must consider different subpopulations and assess whether some who would benefit differently from a product.

Determine How Potential Reduced-Exposure Products Affect Human Exposure and Harm in Short-Term Human Clinical and Epidemiological Studies

Clinical trials should be used. These studies would focus on biomarkers of exposure, biologically effective dose, and potential harm. Studies should be designed for smokers willing to try PREPs continuously for a specified period (e.g., six months). Accurate assessment of smoking behavior, such as cigarettes per day and smoking topography, and effects on biomarkers should be determined. Baseline specimens should be collected (sputum, buccal swabs, blood, and urine and, for some studies, possibly bronchoalveolar lavage or biopsies of internal organs) and then are followed frequently throughout the study. The choice of the biomarkers would depend on the criteria described in this report. The study of smoking behavior alone is insufficient because persons who increase use of a PREP might still have reduced exposure and harm. Biomarkers of internal exposure, biologically effective dose, and harm should be used.

In the context of harm reduction, developing fluorescent bronchoscopic techniques, coupled with molecular analysis, has the potential to identify smokers who have already sustained molecular damage and, depending on the type of damage, to indicate who might benefit most from harm reduction strategies.

The design of studies should allow for sufficient time to acclimate the smoker to the new PREP.

Determine How Potential Reduced-Exposure Products Affect Human Exposure and Harm in Long-term Epidemiological Studies

These studies are the most problematic in early assessments of PREPs, because the latency period for cancers is too long to provide the needed results. Also, the technology for PREPs will likely change quickly, and an individual may use multiple types over a period of time. However, such studies should be started immediately in order to provide the definitive proof of success. Specifically, the emphasis in long-term epidemiological research should be on prospective cohort studies. Either existing cohorts should be modified, or new cohorts should be established. While it is recognized that such studies are expensive, additional costs will have to be added. These studies must follow subjects closely to document smoking behavior, use of all types of tobacco, modifying factors, and so forth. Smoking topography after switching brands should be assessed in at least a subset of individuals. Specimens should be collected at regular intervals, including buccal swabs, sputum, blood, and urine.

Given the long time for such cohorts to mature, emphasis might also be placed on establishing a complementary cohort of former smokers, because these persons will have the highest risk of recurrence in shorter periods of time. Biomarkers used in the former-smoker cohort that are found to be predictive of risk could then be used sooner in short-term epidemiological studies of PREPs.

Another group of subjects that might provide useful information about PREPs would be persons with surgically resectable cancers who will remain at high risk for developing new primary tumors or experience recurrence. These products might reduce the risk of either, so cohorts should be established to study these persons. Such studies would also provide valuable data for the worthiness of biomarkers.

SPECIAL ISSUES IN STUDY DESIGNS

Harm Reduction Compared to What Exposures?

Depending on study design, different exposure comparisons can be made, some of which are more applicable than others. For the development of biomarkers that are tested first in the laboratory setting (in vitro and in vivo animal studies), authentically synthesized tobacco carcinogens, or components of a reference cigarette (i.e., the tar fraction or cigarette smoke condensate from a Kentucky reference cigarette), would be preferable. Levels of exposure should be quantitated accurately so that the results can be interpreted in relation to the range of human exposures.

Experimental human studies in which the product is initially tested would optimally be compared to both reference cigarettes and separately to the smokers' usual cigarettes, where a range of smokers and cigarette types are used (i.e., low- and high-tar contents, menthol). The decreasing use of nonfilter cigarettes would make comparison to nonfilter cigarettes a low-priority consideration.

The comparison group for short- and long-term epidemiological studies would have to be usual cigarettes, because it is not feasible to ask persons to smoke reference cigarettes, which are not designed for appeal. Thus, these studies must carefully characterize smoking behavior, and a range of cigarette types and smoking behaviors must be included.

Harm Reduction Compared to What Risks?

It is clear that any PREP will likely bring some risk of cancer to the user. Comparison groups could be either never smokers, former, smokers or continuous smokers. Preferably, a PREP should reduce risk for the individual compared to the risk for that individual from the usual smoking behavior before switching to the PREP.

Is There a Threshold for Cancer Risk and Does It Matter?

It is important to consider whether a threshold exists for risk of cancer from tobacco products. This is important for new smokers who might initiate smoking with a PREP. It is possible that the delivered exposures from a PREP would be below the threshold, or at least this should be the goal of such a product. It is recognized that ETS data would suggest that the threshold, if any, is very low.

Do All New Products Need to Be Tested or Only Some That Would Serve as Indicators for Similar Products?

It would be efficient to have data that are representative of a class or type of product, where some screening tests could be used to ensure that a potential product fits within that class or type. However, data are needed to validate the assumption that representative products actually are representative.

Which Carcinogens Should Be Prioritized for Study, Given the Large Number Delivered By Tobacco Products?

While there is a need to study both individual carcinogens and tobacco exposure as a complex mixture, it will not be possible to evaluate all possible carcinogens. Thus, additional studies are needed that can compare the harmful effects of tobacco constituents, and models should be developed to prioritize their study. Further data are needed to achieve confidence that only a few key, but high-priority, carcinogens are sufficient to evaluate a PREP.

REFERENCES

Agudo A, Barnadas A, Pallares C, et al. 1994. Lung cancer and cigarette smoking in women: a case-control study in Barcelona (Spain). *Int J Cancer* 59(2):165-169.

Ahijevych K, Gillespie J, Demirci M, Jagadeesh J. 1996. Menthol and nonmenthol cigarettes and smoke exposure in black and white women. *Pharmacol Biochem Behav* 53(2):355-360.

Ahrendt SA, Chow JT, Yang SC, et al. 2000. Alcohol consumption and cigarette smoking increase the frequency of p53 mutations in non-small cell lung cancer. *Cancer Res* 60(12):3155-3159.

Ahrens W, Jockel KH, Patzak W, Elsner G. 1991. Alcohol, smoking, and occupational factors in cancer of the larynx: a case-control study. *Am J Ind Med* 20:477-493.

Akiba S. 1994. Analysis of cancer risk related to longitudinal information on smoking habits. *Environ Health Perspect* 102(Suppl 8):15-19.

Alavanja MC, Brownson RC, Benichou J, Swanson C, Boice JD Jr. 1995. Attributable risk of lung cancer in lifetime nonsmokers and long-term ex-smokers (Missouri, United States). *Cancer Causes Control* 6(3):209-216.

Alderson MR, Lee PN, Wang R. 1985. Risks of lung cancer, chronic bronchitis, ischaemic heart disease, and stroke in relation to type of cigarette smoked. *J Epidemiol Community Health* 39(4):286-293.

Ammenheuser MM, Hastings DA, Whorton EB Jr., Ward JB Jr. 1997. Frequencies of hprt mutant lymphocytes in smokers, non-smokers, and former smokers. *Environ. Mol. Mutagen* 30:131-138.

Amos CI, Xu W, Spitz MR. 1999. Is there a genetic basis for lung cancer susceptibility? *Recent Results Cancer Res* 151:3-12.

Anttila S, Hirvonen A, Vainio H, Husgafvel-Pursiainen K, Hayes JD, Ketterer B. 1993. Immunohistochemical localization of glutathione S-transferases in human lung. *Cancer Res* 53(23):5643-5648.

Anttila S, Luostarinen L, Hirvonen A, et al. 1995. Pulmonary expression of glutathione S-transferase M3 in lung cancer patients: association with GSTM1 polymorphism, smoking, and asbestos exposure. *Cancer Res* 55(15):3305-3309.

Anttila S, Vainio H, Hietanen E, et al. 1992. Immunohistochemical detection of pulmonary cytochrome P450IA and metabolic activities associated with P450IA1 and P450IA2 isozymes in lung cancer patients. *Environ Health Perspect* 98:179-182.

Anwar K, Nakakuki K, Imai H, Naiki H, Inuzuka M. 1993. Over-expression of p53 protein in human laryngeal carcinoma. *Int J Cancer* 53(6):952-956.

Armitage P. 1985. Multistage models of carcinogenesis. *Environ Health Perspect* 63:195-201.

Asami S, Hirano T, Yamaguchi R, Tomioka Y, Itoh H, Kasai H. 1996. Increase of a type of oxidative DNA damage, 8-hydroxyguanine, and its repair activity in human leukocytes by cigarette smoking. *Cancer Res* 56(11):2546-2549.

Ashurst SW, Cohen GM, Nesnow S, DiGiovanni J, Slaga TJ. 1983. Formation of benzo(a)pyrene/DNA adducts and their relationship to tumor initiation in mouse epidermis. *Cancer Res* 43:1024-1029.

Atawodi SE, Lea S, Nyberg F, et al. 1998. 4-Hydroxy-1-(3-pyridyl)-1-butanone-hemoglobin adducts as biomarkers of exposure to tobacco smoke: validation of a method to be used in multicenter studies. *Cancer Epidemiol Biomarkers Prev* 7(9):817-821.

Auerbach O, Hammond EC, Garfinkel L. 1979. Changes in bronchial epithelium in relation to cigarette smoking, 1955-1960 vs. 1970-1977. *N Engl J Med* 300(8):381-385.

Auerbach O, Stout AP, Hammond E.C. 1961. Changes in bronchial epithelium in relation to cigarette smoking. *NEJM* 256:252-256.

Augustine A, Harris RE, Wynder EL. 1989. Compensation as a risk factor for lung cancer in smokers who switch from nonfilter to filter cigarettes. *Am J Public Health* 79(2):188-191.

Austin H, Drews C, Partridge EE. 1993. A case-control study of endometrial cancer in relation to cigarette smoking, serum estrogen levels, and alcohol use. *Am J Obstet Gynecol* 169(5):1086-1091.

Badawi AF, Hirvonen A, Bell DA, Lang NP, Kadlubar FF. 1995. Role of aromatic amine acetyltransferases, NAT1 and NAT2, in carcinogen-DNA adduct formation in the human urinary bladder. *Cancer Res* 55(22):5230-5237.

Badawi AF, Stern SJ, Lang NP, Kadlubar FF. 1996. Cytochrome P-450 and acetyltransferase expression as biomarkers of carcinogen-DNA adduct levels and human cancer susceptibility. *Prog Clin Biol Res* 395:109-140.

Bailar JC III. 1999. Passive smoking, coronary heart disease, and meta-analysis. *N Engl J Med* 340(12):958-959.

Balmain A, Harris CC. 2000. Carcinogenesis in mouse and human cells: parallels and paradoxes. *Carcinogenesis* 21(3):371-377.

Barasch A, Morse DE, Krutchkoff DJ, Eisenberg E. 1994. Smoking, gender, and age as risk factors for site-specific intraoral squamous cell carcinoma. A case-series analysis. *Cancer* 73:509-513.

Bartsch H, Petruzzelli S, De Flora S, Hietanen E, Camus AM, Castegnaro M, Geneste O, Camoirano A, Saracci R, Giuntini C. 1991. Carcinogen metabolism and DNA adducts in human lung tissues as affected by tobacco smoking or metabolic phenotype: a case-control study on lung cancer patients. *Mutat Res* 250:103-114.

Bartsch H, Rojas M, Alexandrov K, et al. 1995. Metabolic polymorphism affecting DNA binding and excretion of carcinogens in humans. *Pharmacogenetics* 5 Spec No:S84-90.

Belinsky SA, Foley JF, White CM, Anderson MW, Maronpot RR. 1990. Dose-response relationship between O6-methylguanine formation in Clara cells and induction of pulmonary neoplasia in the rat by 4-(methylnitrosamino)-1-(3-pyridyl)-1-butanone. *Cancer Res* 50(12):3772-3780.

Belinsky SA, Nikula KJ, Palmisano WA, et al. 1998. Aberrant methylation of p16(INK4a) is an early event in lung cancer and a potential biomarker for early diagnosis. *Proc Natl Acad Sci U S A* 95(20):11891-11896.

Bell DA, Taylor JA, Paulson DF, Robertson CN, Mohler JL, Lucier GW. 1993. Genetic risk and carcinogen exposure: a common inherited defect of the carcinogen-metabolism gene glutathione S-transferase M1 (GSTM1) that increases susceptibility to bladder cancer. *J Natl Cancer Inst* 85(14):1159-1164.

Bell D, Taylor J, Paulson D, et al. 1992. Glutathione transferase u gene deletion is associated with increased risk for bladder cancer. *Proc Am Assoc Cancer Res.* 33:291.

Benhamou E, Benhamou S, Auquier A, Flamant R. 1989. Changes in patterns of cigarette smoking and lung cancer risk: results of a case-control study. *Br J Cancer* 60(4):601-604.

Benhamou S, Benhamou E, Auquier A, Flamant R. 1994. Differential effects of tar content, type of tobacco and use of a filter on lung cancer risk in male cigarette smokers. *Int J Epidemiol* 23(3):437-443.

Benhamou S, Benhamou E, Tirmarche M, Flamant R. 1985. Lung cancer and use of cigarettes: a French case-control study. *J Natl Cancer Inst* 74(6):1169-1175.

Benigni R and Pino A. 1998. Profiles of chemically-induced tumors in rodents: quantitative relationships. *Mutat Res* 421:93-107.

Benowitz NL. 1999. Biomarkers of environmental tobacco smoke exposure. *Environ Health Perspect* 107 Suppl 2:349-355.

Benowitz NL, Hall SM, Herning RI, Jacob P III, Jones RT, Osman AL. 1983. Smokers of low-yield cigarettes do not consume less nicotine. *N Engl J Med* 309(3):139-142.

Benowitz NL, Jacob P III, Kozlowski LT, Yu L. 1986a. Influence of smoking fewer cigarettes on exposure to tar, nicotine, and carbon monoxide. *N Engl J Med* 315(21):1310-1313.

Benowitz NL, Jacob P III, Yu L, Talcott R, Hall S, Jones RT. 1986b. Reduced tar, nicotine, and carbon monoxide exposure while smoking ultralow- but not low-yield cigarettes. *JAMA* 256(2):241-246.

Benowitz NL, Zevin S, Jacob P III. 1998. Suppression of nicotine intake during ad libitum cigarette smoking by high-dose transdermal nicotine. *J Pharmacol Exp Ther* 287(3):958-962.

Ben-Shlomo Y, Smith GD, Shipley MJ, Marmot MG. 1994. What determines mortality risk in male former cigarette smokers? *Am J Public Health* 84:1235-1242.

Besarati Nia A, Van Straaten HW, Godschalk RW, Van Zandwijk N, Balm AJ, Kleinjans JC, Van Schooten FJ. 2000. Immunoperoxidase detection of polycyclic aromatic hydrocarbon-DNA adducts in mouth floor and buccal mucosa cells of smokers and nonsmokers. *Environ Mol Mutagen* 36(2):127-133.

Blot WJ, McLaughlin JK, Winn DM, et al. 1988. Smoking and drinking in relation to oral and pharyngeal cancer. *Cancer Res* 48(11):3282-3287.

Boffetta P, Pershagen G, Jockel KH, et al. 1999. Cigar and pipe smoking and lung cancer risk: a multicenter study from Europe. *J Natl Cancer Inst* 91(8):697-701.

Bombick DW, Bombick BR, Ayres PH, et al. 1997. Evaluation of the genotoxic and cytotoxic potential of mainstream whole smoke and smoke condensate from a cigarette containing a novel carbon filter. *Fundam Appl Toxicol* 39(1):11-17.

Bono R, Russo R, Arossa W, Scursatone E, Gilli G. 1996. Involuntary exposure to tobacco smoke in adolescents: urinary cotinine and environmental factors. *Arch Environ Health* 51(2):127-131.

Boorman GA, Maronpot RR, Eustis SL. 1994. Rodent carcinogenicity bioassay: past, present, and future. *Toxicol Pathol* 22(2):105-111.

Bos JL. 1988. The ras gene family and human carcinogenesis. *Mutat Res* 195(3):255-271.

Bouchardy C, Mitrunen K, Wikman H, Husgafvel-Pursiainen K, Dayer P, Benhamou S, Hirvonen A. 1998. N-acetyltransferase NAT1 and NAT2 genotypes and lung cancer risk. *Pharmacogen* 8:291-298.

Bouquot JE, Gorlin RJ. 1986. Leukoplakia, lichen planus, and other oral keratoses in 23,616 white Americans over the age of 35 years. *Oral Surg Oral Med Oral Pathol* 61(4):373-81.

Braun MM, Caporaso NE, Page WF, Hoover RN. 1994. Genetic component of lung cancer: cohort study of twins. *Lancet* 344:440-443.

Brennan FJ. 1998. Passive smoking and coronary heart disease. *Circulation* 97(18):1871, discussion 1872-1873.

Brennan JA, Boyle JO, Koch WM, et al. 1995. Association between cigarette smoking and mutation of the p53 gene in squamous-cell carcinoma of the head and neck. *N Engl J Med* 332(11):712-717.

Brennan P, Bogillot O, Cordier S, et al. 2000. Cigarette smoking and bladder cancer in men: a pooled analysis of 11 case-control studies. *Int J Cancer* 86(2):289-294.

Breslow RA, Graubard BI, Sinha R, Subar AF. 2000. Diet and lung cancer mortality: a 1987 National Health Interview Survey cohort study. *Cancer Causes Control* 11(5):419-431.

Brinton LA, Barrett RJ, Berman ML, Mortel R, Twiggs LB, Wilbanks GD. 1993. Cigarette smoking and the risk of endometrial cancer. *Am J Epidemiol* 137(3):281-291.

Brinton LA, Blot WJ, Becker JA, Winn DM, Browder JP, Farmer JC, Fraumeni JF. 1984. A case-control study of cancers of the nasal cavity and paranasal sinuses. *Am J Epidemiol* 119(6):896-906.

Brockmoller J, Cascorbi I, Kerb R, Roots I. 1996. Combined analysis of inherited polymorphisms in arylamine N-acetyltransferase 2, glutathione S-transferases M1 and T1, microsomal epoxide hydrolase, and cytochrome P450 enzymes as modulators of bladder cancer risk. *Cancer Res* 56(17):3915-3925.

Brockmoller J, Cascorbi I, Kerb R, Sachse C, Roots I. 1998. Polymorphisms in xenobiotic conjugation and disease predisposition. *Toxicol Lett* 102-103:173-183.

Bross ID, Gibson R. 1968. Risks of lung cancer in smokers who switch to filter cigarettes. *Am J Public Health Nations Health* 58(8):1396-1403.

Brownson RC, Alavanja MC, Caporaso N, Simoes EJ, Chang JC. 1998. Epidemiology and prevention of lung cancer in nonsmokers. *Epidemiol Rev* 20(2):218-236.

Brownson RC, Alavanja MC, Chang JC. 1993. Occupational risk factors for lung cancer among nonsmoking women: a case-control study in Missouri (United States). *Cancer Causes.Control* 4:449-454.

Brownson RC, Alavanja MC, Hock ET, Loy TS. 1992a. Passive smoking and lung cancer in nonsmoking women. *Am J Public Health* 82(11):1525-1530.

Brownson RC, Chang JC, Davis JR. 1992b. Gender and histologic type variations in smoking-related risk of lung cancer. *Epidemiology* 3(1):61-64.

Brownson RC, Eriksen MP, Davis RM, Warner KE. 1997. Environmental tobacco smoke: health effects and policies to reduce exposure. *Annu Rev Public Health* 18:163-185.

Brunnemann KD, Prokopczyk B, Djordjevic MV, Hoffmann D. 1996. Formation and analysis of tobacco-specific N-nitrosamines. *Crit Rev Toxicol* 26(2):121-137.

Bryant MS, Vineis P, Skipper PL, Tannenbaum SR. 1988. Hemoglobin adducts of aromatic amines: associations with smoking status and type of tobacco. *Proc Natl Acad Sci U S A* 85(24):9788-9791.

Burch JD, Rohan TE, Howe GR, et al. 1989. Risk of bladder cancer by source and type of tobacco exposure: a case-control study. *Int J Cancer* 44(4):622-628.

Butterworth BE. 1990. Consideration of both genotoxic and nongenotoxic mechanisms in predicting carcinogenic potential. *Mutat Res* 239:117-132.

Califano J, Westra WH, Meininger G, Corio R, Koch WM, Sidransky D. 2000. Genetic progression and clonal relationship of recurrent premalignant head and neck lesions. *Clin Cancer Res* 6(2):347-352.

Canver CC, Memoli VA, Vanderveer PL, Dingivan CA, Mentzer RM Jr. 1994. Sex hormone receptors in non-small-cell lung cancer in human beings. *J Thorac Cardiovasc Surg* 108(1):153-157.

Caraballo RS, Giovino GA, Pechacek TF, et al. 1998. Racial and ethnic differences in serum cotinine levels of cigarette smokers: Third National Health and Nutrition Examination Survey, 1988-1991. *JAMA* 280(2):135-139.

Carmella SG, Akerkar S, Hecht SS. 1993. Metabolites of the tobacco-specific nitrosamine 4-(methylnitrosamino)-1-(3-pyridyl)-1-butanone in smokers' urine. *Cancer Res* 53:721-724.

Carmella SG, Akerkar SA, Richie JP Jr, Hecht SS. 1995. Intraindividual and interindividual differences in metabolites of the tobacco-specific lung carcinogen 4-(methylnitrosamino)-1-(3-pyridyl)-1-butanone (NNK) in smokers' urine. *Cancer Epidemiol Biomarkers Prev* 4(6):635-642.

Carmella SG, Kagan SS, Kagan M, et al. 1990. Mass spectrometric analysis of tobacco-specific nitrosamine hemoglobin adducts in snuff dippers, smokers, and nonsmokers. *Cancer Res* 50(17):5438-5445.

Carstensen JM, Pershagen G, Eklund G. 1988. Smoking-adjusted incidence of lung cancer among Swedish men in different occupations. *Int J Epidemiol* 17:753-758.

Cartwright RA, Glashan RW, Rogers HJ, et al. 1982. Role of N-acetyltransferase phenotypes in bladder carcinogenesis: a pharmacogenetic epidemiological approach to bladder cancer. *Lancet* 2(8303):842-845.

Chang F, Syrjanen S, Shen Q, Ji HX, Syrjanen K. 1990. Human papillomavirus (HPV) DNA in esophageal precancer lesions and squamous cell carcinomas from China. *Int J Cancer* 45(1):21-25.

Charloux A, Quoix E, Wolkove N, Small D, Pauli G, Kreisman H. 1997. The increasing incidence of lung adenocarcinoma: reality or artefact? A review of the epidemiology of lung adenocarcinoma. *Int J Epidemiol* 26(1):14-23.

Chen B, Liu L, Castonguay A, Maronpot RR, Anderson MW, You M. 1993. Dose-dependent ras mutation spectra in N-nitrosodiethylamine induced mouse liver tumors and 4-(methylnitrosamino)-1-(3-pyridyl)-1-butanone induced mouse lung tumors. *Carcinogenesis* 14(8):1603-1608.

Chen YP, Johnson GK, Squier CA. 1994. Effects of nicotine and tobacco-specific nitrosamines on hamster cheek pouch and gastric mucosa. *J Oral Pathol Med* 23(6):251-255.

Chen YP, Squier CA. 1990. Effect of nicotine on 7,12-dimethylbenz[a]anthracene carcinogenesis in hamster cheek pouch. *J Natl Cancer Inst* 82(10):861-864.

Cheng L, Eicher SA, Guo Z, Hong WK, Spitz MR, Wei Q. 1998. Reduced DNA repair capacity in head and neck cancer patients. *Cancer Epidemiol Biomarkers Prev* 7(6):465-468.

Chevillard S, Radicella JP, Levalois C, et al. 1998. Mutations in OGG1, a gene involved in the repair of oxidative DNA damage, are found in human lung and kidney tumours. *Oncogene* 16(23):3083-3086.

Chiba I, Takahashi T, Nau MM, et al. 1990. Mutations in the p53 gene are frequent in primary, resected non-small cell lung cancer. Lung Cancer Study Group. *Oncogene* 5(10):1603-1610.

Christen AG. 1980. Tobacco chewing and snuff dipping. *N Engl J Med* 302(14):818.

Chyou PH, Nomura AM, Stemmermann GN. 1992. A prospective study of the attributable risk of cancer due to cigarette smoking. *Am J Public Health* 82:37-40.

Ciruzzi M, Pramparo P, Esteban O, et al. 1998. Case-control study of passive smoking at home and risk of acute myocardial infarction. Argentine FRICAS Investigators. Factores de Riesgo Coronario en America del Sur. *J Am Coll Cardiol* 31(4):797-803.

Clapp WL, Fagg BS, Smith CJ. 1999. Reduction in Ames Salmonella mutagenicity of mainstream cigarette smoke condensate by tobacco protein removal. *Mutat Res* 446(2):167-174.

Clark PI, Gautam S, Gerson LW. 1996. Effect of menthol cigarettes on biochemical markers of smoke exposure among black and white smokers. *Chest* 110(5):1194-1198.

Cloos J, Spitz MR, Schantz SP, et al. 1996. Genetic susceptibility to head and neck squamous cell carcinoma. *J Natl Cancer Inst* 88(8):530-535.

Cohen SM. 1995. Human relevance of animal carcinogenicity studies. *Regul Toxicol Pharmacol* 21:75-80.

Cohn BA, Wingard DL, Cirillo PM, Cohen RD, Reynolds P, Kaplan GA. 1996. Re: Differences in lung cancer risk between men and women: examination of the evidence. *J Natl Cancer Inst* 88(24):1867-1868.

Crawford FG, Mayer J, Santella RM, et al. 1994. Biomarkers of environmental tobacco smoke in preschool children and their mothers. *J Natl Cancer Inst* 86(18):1398-1402.

Cullen P, Schulte H, Assmann G. 1997. The Munster Heart Study (PROCAM): total mortality in middle-aged men is increased at low total and LDL cholesterol concentrations in smokers but not in nonsmokers. *Circulation* 96(7):2128-2136.

Cummings KM. 1984. Changes in the smoking habits of adults in the United States and recent trends in lung cancer mortality. *Cancer Detect Prev* 7(3):125-134.

Cummings KM, Giovino G, Mendicino AJ. 1987. Cigarette advertising and black-white differences in brand preference. *Public Health Rep* 102(6):698-701.

Cunningham AR, Rosenkranz HS, Zhang YP, Klopman G. 1998. Identification of 'genotoxic' and 'non-genotoxic' alerts for cancer in mice: the carcinogenic potency database. *Mutat Res* 398(1-2):1-17.

D'Amico D, Carbone D, Mitsudomi T, et al. 1992. High frequency of somatically acquired p53 mutations in small-cell lung cancer cell lines and tumors. *Oncogene* 7(2):339-346.

D'Avanzo B, Negri E, La Vecchia C, et al. 1990. Cigarette smoking and bladder cancer. *Eur J Cancer* 26(6):714-718.

Dallinga JW, Pachen DM, Wijnhoven SW, et al. 1998. The use of 4-aminobiphenyl hemoglobin adducts and aromatic DNA adducts in lymphocytes of smokers as biomarkers of exposure. *Cancer Epidemiol Biomarkers Prev* 7(7):571-577.

Darby SC, Pike MC. 1988. Lung cancer and passive smoking: predicted effects from a mathematical model for cigarette smoking and lung cancer. *Br J Cancer* 58(6):825-831.

Davidson BJ, Lydiatt WM, Abate MP, Schantz SP, Chaganti RS. 1996. Cyclin D1 abnormalities and tobacco exposure in head and neck squamous cell carcinoma. *Head Neck* 18(6):512-521.

Day BW, Naylor S, Gan LS, et al. 1990. Molecular dosimetry of polycyclic aromatic hydrocarbon epoxides and diol epoxides via hemoglobin adducts. *Cancer Res* 50(15):4611-4618.

Day GL, Blot WJ, Austin DF, et al. 1993. Racial differences in risk of oral and pharyngeal cancer: alcohol, tobacco, and other determinants. *J Natl Cancer Inst* 85(6):465-473.

Degawa M, Stern,SJ, Martin MV, Guengerich FP, Fu PP, Ilett KF, Kaderlik RK, Kadlubar FF. 1994. Metabolic activation and carcinogen-DNA adduct detection in human larynx. *Cancer Res* 54:4915-4919.

De Stefani E, Boffetta P, Oreggia F, Mendilaharsu M, Deneo-Pellegrini H. 1998. Smoking patterns and cancer of the oral cavity and pharynx: a case-control study in Uruguay. *Oral Oncol* 34(5):340-346.

DeMarini DM. 1998. Mutation spectra of complex mixtures. *Mutat Res* 411(1):11-18.

Denissenko MF, Chen JX, Tang MS, Pfeifer GP. 1997. Cytosine methylation determines hot spots of DNA damage in the human P53 gene. *Proc Natl Acad Sci U S A* 94(8):3893-3898.

Denissenko MF, Pao A, Tang M, Pfeifer GP. 1996. Preferential formation of benzo[a]pyrene adducts at lung cancer mutational hotspots in P53. *Science* 274(5286):430-432.

Devereux TR, Risinger JI, Barrett JC. 1999. Mutations and altered expression of the human cancer genes: what they tell us about causes. *IARC Sci Publ* (146):19-42.

Djordjevic MV, Fan J, Ferguson S, Hoffmann D. 1995. Self-regulation of smoking intensity. Smoke yields of the low-nicotine, low-'tar' cigarettes. *Carcinogenesis* 16(9):2015-2021.

Djordjevic MV, Hoffmann D, Hoffmann I. 1997. Nicotine regulates smoking patterns. *Prev Med* 26(4):435-440.

Djordjevic MV, Stellman SD, Zang E. 2000. Doses of nicotine and lung carcinogens delivered to cigarette smokers. *J Natl Cancer Inst* 92(2):106-111.

Doll, R. 1996. Cancers weakly related to smoking. *Br Med Bull* 52:35-49.

Doll R, Hill AB. 1950. Smoking and carcinoma of the lung. Preliminary report. *Bull World Health Organ* 2: 739-748.

Doll R, Peto R. 1976. Mortality in relation to smoking: 20 years' observations on male British doctors. *Br Med J* 2(6051):1525-1536.

Doll R, Peto R. 1978. Cigarette smoking and bronchial carcinoma: dose and time relationships among regular smokers and lifelong non-smokers. *J Epidemiol Community Health* 32(4):303-313.

Doolittle DJ, Rahn CA, Lee CK. 1991. The effect of exposure to nicotine, carbon monoxide, cigarette smoke or cigarette smoke condensate on the mutagenicity of rat urine. *Mutat Res* 260(1):9-18.

Doolittle DJ, Winegar R, Lee CK, Caldwell WS, Hayes AW, de Bethizy JD. 1995. The genotoxic potential of nicotine and its major metabolites. *Mutat Res* 344(3-4):95-102.

Dunn BP, Vedal S, San RH, Kwan WF, Nelems B, Enarson DA, Stich HF. 1991. DNA adducts in bronchial biopsies. *Int J Cancer* 48(4):485-492.

Duthie SJ, Ross M, Collins AR. 1995. The influence of smoking and diet on the hypoxanthine phosphoribosyltransferase (hprt) mutant frequency in circulating T lymphocytes from a normal human population. *Mutat Res* 331:55-64.

Ebert RV, McNabb ME, McCusker KT, Snow SL. 1983. Amount of nicotine and carbon monoxide inhaled by smokers of low-tar, low-nicotine cigarettes. *JAMA* 250(20):2840-2842.

Elwood JM, Pearson JC, Skippen DH, Jackson SM. 1984. Alcohol, smoking, social and occupational factors in the aetiology of cancer of the oral cavity, pharynx and larynx. *Int J Cancer* 34(5):603-612.

Engeland A. 1996. Trends in the incidence of smoking-associated cancers in Norway, 1954-93. *Int J Cancer* 68:39-46.

Engeland A, Andersen A, Haldorsen T, Tretli S. 1996. Smoking habits and risk of cancers other than lung cancer: 28 years' follow-up of 26,000 Norwegian men and women. *Cancer Causes Control* 7(5):497-506.

Enstrom JE. 1999. Smoking cessation and mortality trends among two United States populations. *J Clin Epidemiol* 52(9):813-825.

Enstrom JE, Heath CW Jr. 1999. Smoking cessation and mortality trends among 118,000 Californians, 1960-1997. *Epidemiology* 10(5):500-512.

Ernster VL. 1996. Female lung cancer. *Annu Rev Public Health* 17:97-114.

Farmer PB, Shuker DE. 1999. What is the significance of increases in background levels of carcinogen-derived protein and DNA adducts? Some considerations for incremental risk assessment. *Mutat Res* 424:275-286.

Field JK, Kiaris H, Risk JM, et al. 1995. Allelotype of squamous cell carcinoma of the head and neck: fractional allele loss correlates with survival. *Br J Cancer* 72(5):1180-1188.

Field JK, Zoumpourlis V, Spandidos DA, Jones AS. 1994. p53 expression and mutations in squamous cell carcinoma of the head and neck: expression correlates with the patients' use of tobacco and alcohol. *Cancer Detect Prev* 18(3):197-208.

Fischer S, Spiegelhalder B, Eisenbarth J, Preussmann R. 1990. Investigations on the origin of tobacco-specific nitrosamines in mainstream smoke of cigarettes. *Carcinogenesis* 11(5):723-730.

Fischer S, Spiegelhalder B, Preussmann R. 1989a. Influence of smoking parameters on the delivery of tobacco-specific nitrosamines in cigarette smoke—a contribution to relative risk evaluation. *Carcinogenesis* 10(6):1059-1066.

Fischer S, Spiegelhalder B, Preussmann R. 1989b. Preformed tobacco-specific nitrosamines in tobacco—role of nitrate and influence of tobacco type. *Carcinogenesis* 10(8):1511-1517.

Flamini G, Romano G, Curigliano G, et al. 1998. 4-Aminobiphenyl-DNA adducts in laryngeal tissue and smoking habits: an immunohistochemical study. *Carcinogenesis* 19(2):353-357.

Fleshner N, Garland J, Moadel A, et al. 1999. Influence of smoking status on the disease-related outcomes of patients with tobacco-associated superficial transitional cell carcinoma of the bladder. *Cancer* 86(11):2337-2345.

Floyd RA. 1990. The role of 8-hydroxyguanine in carcinogenesis. *Carcinogenesis* 11(9):1447-1450.

Fong KM, Sekido Y, Minna JD. 1999. Molecular pathogenesis of lung cancer. *J Thorac Cardiovasc Surg* 118(6):1136-1152.

Fontham ET, Correa P, WuWilliams A, et al. 1991. Lung cancer in nonsmoking women: a multicenter case-control study. *Cancer Epidemiol Biomarkers Prev* 1(1):35-43.

Franks AL, Kendrick JS, Tyler CW Jr. 1987. Postmenopausal smoking, estrogen replacement therapy, and the risk of endometrial cancer. *Am J Obstet Gynecol* 156(1):20-23.

Freedman DA, Gold LS, Lin TH. 1996. Concordance between rats and mice in bioassays for carcinogenesis. *Regul Toxicol Pharmacol* 23(3):225-232.

Friedman GD, Dales LG, Ury HK. 1979. Mortality in middle-aged smokers and nonsmokers. *NEJM* 300:213-217.

Fung VA, Barrett JC, Huff J. 1995. The carcinogenesis bioassay in perspective: application in identifying human cancer hazards. *Environ Health Perspect* 103(7-8):680-683.

Gaffney M, Altshuler B. 1988. Examination of the role of cigarette smoke in lung carcinogenesis using multistage models. *J Natl Cancer Inst* 80(12):925-931.

Gallo O, Bianchi S, Porfirio B. 1995. Bcl-2 overexpression and smoking history in head and neck cancer. *J Natl Cancer Inst* 87(13):1024-1025.

Gao YT, Blot WJ, Zheng W, et al. 1987. Lung cancer among Chinese women. *Int J Cancer* 40(5):604-609.

Garro AJ, Seitz HK, Lieber CS. 1981. Enhancement of dimethylnitrosamine metabolism and activation to a mutagen following chronic ethanol consumption. *Cancer Res* 41:120-124.

Gealy R, Zhang L, Siegfried JM, Luketich JD, Keohavong P. 1999. Comparison of mutations in the p53 and K-ras genes in lung carcinomas from smoking and nonsmoking women. *Cancer Epidemiol Biomarkers Prev* 8:297-302.

Geneste O, Camus AM, Castegnaro M, Petruzzelli S, Macchiarini P, Angeletti CA, Giuntini C, Bartsch H. 1991. Comparison of pulmonary DNA adduct levels, measured by 32P-postlabelling and aryl hydrocarbon hydroxylase activity in lung parenchyma of smokers and ex-smokers. *Carcinogenesis* 12:1301-1305.

Geradts J, Fong KM, Zimmerman PV, Maynard R, Minna JD. 1999. Correlation of abnormal RB, p16ink4a, and p53 expression with 3p loss of heterozygosity, other genetic abnormalities, and clinical features in 103 primary non-small cell lung cancers. *Clin Cancer Res* 5(4):791-800.

Gillis TM, Incze J, Strong MS, Vaughan CW, Simpson GT. 1983. Natural history and management of keratosis, atypia, carcinoma-in situ, and microinvasive cancer of the larynx. *Am J Surg* 146(4):512-516.

Gold LS, Manley NB, Ames BN. 1992. Extrapolation of carcinogenicity between species: qualitative and quantitative factors. *Risk Anal* 12(4):579-588.

Gold LS, Slone TH, Stern BR, Bernstein L. 1993. Comparison of target organs of carcinogenicity for mutagenic and non-mutagenic chemicals. *Mutat Res* 286(1):75-100.

Goldstein AM, Blot WJ, Greenberg RS, et al. 1994. Familial risk in oral and pharyngeal cancer. *Eur J Cancer B Oral Oncol* 30B(5):319-322.

Goldstein BD. 1994. Risk assessment methodology: maximum tolerated dose and two-stage carcinogenesis models. *Toxicol Pathol* 22(2):194-197.

Goldstein LS, Weyand EH, Safe S, et al. 1998. Tumors and DNA adducts in mice exposed to benzo[a]pyrene and coal tars: implications for risk assessment. *Environ Health Perspect* 106(Suppl 6):1325-1330.

Graham S, Levin ML. 1971. Smoking withdrawal in the reduction of risk of lung cancer. *Cancer* 27(4):865-871.

Gray GM, Li P, Shlyakhter I, Wilson R. 1995. An empirical examination of factors influencing prediction of carcinogenic hazard across species. *Regul Toxicol Pharmacol* 22(3):283-291.

Greenlee RT, Murray T, Bolden S, Wingo PA. 2000. Cancer statistics, 2000. *CA Cancer J Clin* 50(1):7-33.

Grinberg-Funes RA, Singh VN, Perera FP, et al. 1994. Polycyclic aromatic hydrocarbon-DNA adducts in smokers and their relationship to micronutrient levels and the glutathione-S-transferase M1 genotype. *Carcinogenesis* 15(11):2449-2454.

Gritz ER, Rose JE, Jarvik ME. 1983. Regulation of tobacco smoke intake with paced cigarette presentation. *Pharmacol Biochem Behav* 18(3):457-462.

Grundy SM, Balady GJ, Criqui MH, et al. 1998. Primary prevention of coronary heart disease: guidance from Framingham: a statement for healthcare professionals from the AHA Task Force on Risk Reduction. American Heart Association. *Circulation* 97(18):1876-1887.

Guengerich FP. 2000. Metabolism of chemical carcinogens. *Carcinogenesis* 21(3):345-351.

Gupta PC, Murti PR, Bhonsle RB. 1996. Epidemiology of cancer by tobacco products and the significance of TSNA. *Crit Rev Toxicol* 26(2):183-198.

Gurkalo VK, Volfson NI. 1982. Nicotine influence upon the development of experimental stomach tumors. *Arch Geschwulstforsch* 52(4):259-265.

Habel LA, Bull SA, Friedman GD. 1998. Barbiturates, smoking, and bladder cancer risk. *Cancer Epidemiol Biomarkers Prev* 7(11):1049-1050.

Habs M, Schmahl D. 1984. Influence of nicotine on N-nitrosomethylurea-induced mammary tumors in rats. *Klin Wochenschr* 62 (Suppl 2):105-108.

Hahn LP, Folsom AR, Sprafka JM, Norsted SW. 1990. Cigarette smoking and cessation behaviors among urban blacks and whites. *Public Health Rep* 105(3):290-295.

Haldorsen T, Grimsrud TK. 1999. Cohort analysis of cigarette smoking and lung cancer incidence among Norwegian women. *Int J Epidemiol* 28(6):1032-1036.

Hall J, Bresil H, Donato F, Wild CP, Loktionova NA, Kazanova OI, Komyakov IP, Lemekhov VG, Likhachev AJ, Montesano R. 1993. Alkylation and oxidative-DNA damage repair activity in blood leukocytes of smokers and non-smokers. *Int J Cancer* 54:728-733.

Halpern MT, Gillespie BW, Warner KE. 1993. Patterns of absolute risk of lung cancer mortality in former smokers. *J Natl Cancer Inst* 85(6):457-464.

Hammond SK. 1999. Exposure of U.S. workers to environmental tobacco smoke. *Environ Health Perspect* 107 Suppl 2:329-340.

Harris CC, Genta VM, Frank AL, et al. 1974. Carcinogenic polynuclear hydrocarbons bind to macromolecules in cultured human bronchi. *Nature* 252(5478):68-69.

Harris RE, Zang EA, Anderson JI, Wynder EL. 1993. Race and sex differences in lung cancer risk associated with cigarette smoking. *Int J Epidemiol* 22(4):592-599.

Hartge P, Silverman D, Hoover R, Schairer C, Altman R, Austin D, Cantor K, Child M, Key C, Marrett LD, et al. 1987. Changing cigarette habits and bladder cancer risk: a case-control study. *J Natl Cancer Inst* 78:1119-1125.

Hartmann A, Blaszyk H, Kovach JS, Sommer SS. 1997. The molecular epidemiology of p53 gene mutations in human breast cancer. *Trends Genet* 13(1):27-33.

Haseman JK, Lockhart AM. 1993. Correlations between chemically related site-specific carcinogenic effects in long-term studies in rats and mice. *Environ Health Perspect* 101(1):50-54.

Haseman JK, Lockhart A. 1994. The relationship between use of the maximum tolerated dose and study sensitivity for detecting rodent carcinogenicity. *Fundam Appl Toxicol* 22(3):382-391.

Haseman JK, Seilkop SK. 1992. An examination of the association between maximum-tolerated dose and carcinogenicity in 326 long-term studies in rats and mice. *Fundam Appl Toxicol* 19:207-213.

Hashibe M, Mathew B, Kuruvilla B, Thomas G, Sankaranarayanan R, Parkin DM, Zhang ZF. 2000. Chewing tobacco, alcohol, and the risk of erythroplakia. *Cancer Epidemiol Biomarkers Prev* 9(7):639-645.

Hayes RB, Bravo-Otero E, Kleinman DV, et al. 1999. Tobacco and alcohol use and oral cancer in Puerto Rico. *Cancer Causes Control* 10(1):27-33.

Hecht SS. 1998. Biochemistry, biology, and carcinogenicity of tobacco-specific N-nitrosamines. *Chem Res Toxicol* 11(6):559-603.

Hecht SS. 1999a. DNA adduct formation from tobacco-specific N-nitrosamines. *Mutat Res* 424(1-2):127-142.

Hecht SS. 1999b. Tobacco smoke carcinogens and lung cancer. *J Natl Cancer Inst* 91(14):1194-1210.

Hecht SS, Carmella SG, Chen M, et al. 1999. Quantitation of urinary metabolites of a tobacco-specific lung carcinogen after smoking cessation. *Cancer Res* 59(3):590-596.

Hecht, SS, Carmella, SG, Foiles, PG, Murphy, SE. 1994. Biomarkers for human uptake and metabolic activation of tobacco-specific nitrosamines. *Cancer Res* 54:1912s-1917s.

Hecht SS, Carmella SG, Murphy SE, Akerkar S, Brunnemann KD, Hoffmann D. 1993. A tobacco-specific lung carcinogen in the urine of men exposed to cigarette smoke. *N Engl J Med* 329(21):1543-1546.

Hecht SS, Hochalter JB, Villalta PW, Murphy SE. 2000. 2'-Hydroxylation of nicotine by cytochrome P450 2A6 and human liver microsomes: formation of a lung carcinogen precursor. *Proc Natl Acad Sci USA* 97:12493-12497.

Hegmann KT, Fraser AM, Keaney RP, et al. 1993. The effect of age at smoking initiation on lung cancer risk. *Epidemiology* 4(5):444-448.

Helbock HJ, Beckman KB, Shigenaga MK, et al. 1998. DNA oxidation matters: the HPLC-electrochemical detection assay of 8-oxo-deoxyguanosine and 8-oxo-guanine. *Proc Natl Acad Sci U S A* 95(1):288-293.

Hendriks J, Van Schil P, Van Meerbeeck J, et al. 1996. Short-term survival after major pulmonary resections for bronchogenic carcinoma. *Acta Chir Belg* 96(6):273-279.

Henning S, Cascorbi I, Munchow B, Jahnke V, Roots I. 1999. Association of arylamine N-acetyltransferases NAT1 and NAT2 genotypes to laryngeal cancer risk. *Pharmacogen* 9:103-111.

Higgins IT, Wynder EL. 1988. Reduction in risk of lung cancer among ex-smokers with particular reference to histologic type. *Cancer* 62(11):2397-2401.

Hill AB. 1965. The environment and disease: association or causation. *Proc Royal Soc Med* 58:295-300.

Hinds MW, Yang HY, Stemmermann G, Lee J, Kolonel LN. 1982. Smoking history and lung cancer survival in women. *J Natl Cancer Inst* 68(3):395-399.

Hirayama T. 1981. Non-smoking wives of heavy smokers have a higher risk of lung cancer: a study from Japan. *Br Med J (Clin.Res.Ed)* 282:183-185.

Hoffmann D, Djordjevic MV. 1997. Chemical composition and carcinogenicity of smokeless tobacco. *Adv Dent Res* 11(3):322-329.

Hoffmann D, Hoffmann I. 1997. The changing cigarette, 1950-1995. *J Toxicol Environ Health* 50(4):307-364.

Hollstein M, Shomer B, Greenblatt M, et al. 1996. Somatic point mutations in the p53 gene of human tumors and cell lines: updated compilation. *Nucleic Acids Res* 24(1):141-146.

Hou SM, Yang K, Nyberg F, Hemminki K, Pershagen G, Lambert B. 1999. Hprt mutant frequency and aromatic DNA adduct level in non-smoking and smoking lung cancer patients and population controls. *Carcinogenesis* 20(3):437-444.

Huff JE, Melnick RL, Solleveld HA, Haseman JK, Powers M, Miller RA. 1985. Multiple organ carcinogenicity of 1,3-butadiene in B6C3F1 mice after 60 weeks of inhalation exposure. *Science* 227(4686):548-549.

Humble CG, Samet JM, Pathak DR, Skipper BJ. 1985. Cigarette smoking and lung cancer in 'Hispanic' whites and other whites in New Mexico. *Am J Public Health* 75(2):145-148.

Hussain SP, Harris CC. 1999. p53 mutation spectrum and load: the generation of hypotheses linking the exposure of endogenous or exogenous carcinogens to human cancer. *Mutat Res* 428(1-2):23-32.

Hutchison SJ, Sudhir K, Chou TM, et al. 1997. Testosterone worsens endothelial dysfunction associated with hypercholesterolemia and environmental tobacco smoke exposure in male rabbit aorta. *J Am Coll Cardiol* 29(4):800-807.

IARC (International Agency on the Research of Cancer). 1986. *IARC Monographs on the Evaluation of the Carcinogenic Risk of Chemicals to Humans: Tobacco Smoking.* Lyon, France: IARC.

Ibrahim SO, Vasstrand EN, Johannessen AC, Idris AM, Magnusson B, Nilsen R, Lillehaug JR. 1999. Mutations of the p53 gene in oral squamous-cell carcinomas from Sudanese dippers of nitrosamine-rich toombak and non-snuff-dippers from the Sudan and Scandinavia. *Int J Cancer* 81(4):527-534.

Idris AM, Ibrahim SO, Vasstrand EN, Johannessen AC, Lillehaug JR, Magnusson B, Wallstrom M, Hirsch JM, Nilsen R. 1998. The Swedish snus and the Sudanese toombak: are they different? *Oral Oncol* 34(6):558-566.

Inoue M, Osaki T, Noguchi M, Hirohashi S, Yasumoto K, Kasai H. 1998. Lung cancer patients have increased 8-hydroxydeoxyguanosine levels in peripheral lung tissue DNA. *Jpn J Cancer Res* 89:691-695.

Inskip PD, Boice JD, Jr. 1994. Radiotherapy-induced lung cancer among women who smoke. *Cancer* 73:1541-1543.

International Expert Panel on Carcinogen Risk Assessment. 1996. The use of mechanistic data in the risk assessments of ten chemicals: an introduction to the chemical-specific reviews. *Pharmacol Ther* 71:1-5.

Iribarren C, Tekawa IS, Sidney S, Friedman GD. 1999. Effect of cigar smoking on the risk of cardiovascular disease, chronic obstructive pulmonary disease, and cancer in men. *N Engl J Med* 340(23):1773-1780.

Ishibe N, Wiencke JK, Zuo ZF, McMillan A, Spitz M, Kelsey KT. 1997. Susceptibility to lung cancer in light smokers associated with CYP1A1 polymorphisms in Mexican- and African-Americans. *Cancer Epidemiol Biomarkers Prev* 6(12):1075-1080.

Izzo JG, Papadimitrakopoulou VA, Li XQ, et al. 1998. Dysregulated cyclin D1 expression early in head and neck tumorigenesis: in vivo evidence for an association with subsequent gene amplification. *Oncogene* 17(18):2313-2322.

Jaber MA, Porter SR, Gilthorpe MS, Bedi R, Scully C. 1999. Risk factors for oral epithelial dysplasia—the role of smoking and alcohol. *Oral Oncol* 35(2):151-156.

Jacob P III, Yu L, Shulgin AT, Benowitz NL. 1999. Minor tobacco alkaloids as biomarkers for tobacco use: comparison of users of cigarettes, smokeless tobacco, cigars, and pipes. *Am J Public Health* 89(5):731-736.

Janerich DT, Thompson WD, Varela LR, et al. 1990. Lung cancer and exposure to tobacco smoke in the household. *N Engl J Med* 323(10):632-636.

Jansson T, Romert L, Magnusson J, and Jenssen D. 1991. Genotoxicity testing of extracts of a Swedish moist oral snuff. *Mutat Res* 261:101-115.

Jarvik ME, Tashkin DP, Caskey NH, McCarthy WJ, Rosenblatt MR. 1994. Mentholated cigarettes decrease puff volume of smoke and increase carbon monoxide absorption. *Physiol Behav* 56(3):563-570.

Jockel KH, Ahrens W, Wichmann HE, Becher H, Bolm-Audorff U, Jahn I, Molik B, Greiser E, Timm J 1992. Occupational and environmental hazards associated with lung cancer. *Int J Epidemiol* 21:202-213.

Joly OG, Lubin JH, Caraballoso M. 1983. Dark tobacco and lung cancer in Cuba. *J Natl Cancer Inst* 70:1033-1039.

Jones IM, Moore DH, Thomas CB, Thompson CL, Strout CL, Burkhart-Schultz K. 1993. Factors affecting HPRT mutant frequency in T-lymphocytes of smokers and nonsmokers. *Cancer Epidemiol Biomarkers Prev* 2:249-260.

Jones NJ, McGregor AD, Waters R. 1993. Detection of DNA adducts in human oral tissue: correlation of adduct levels with tobacco smoking and differential enhancement of adducts using the butanol extraction and nuclease P1 versions of 32P postlabeling. *Cancer Res* 53(7):1522-1528.

Jourenkova-Mironova N, Voho A, Bouchardy C, et al. 1999. Glutathione S-transferase GSTM1, GSTM3, GSTP1 and GSTT1 genotypes and the risk of smoking-related oral and pharyngeal cancers. *Int J Cancer* 81(1):44-48.

Jourenkova-Mironova N, Reinikainen M, Bouchardy C, Dayer P, Benhamou S, Hirvonen A. 1998. Larynx cancer risk in relation to glutathione S-transferase M1 and T1 genotypes and tobacco smoking. *Cancer Epidemiol Biomarkers Prev* 7(1):19-23.

Kabat GC, Dieck GS, Wynder EL. 1986. Bladder cancer in nonsmokers. *Cancer* 57(2):362-367.

Kabat GC, Morabia A, Wynder EL. 1991. Comparison of smoking habits of blacks and whites in a case-control study. *Am J Public Health* 81(11):1483-1486.

Kadlubar FF. 1994. DNA adducts of carcinogenic aromatic amines. *IARC Sci Publ* (125):199-216.

Kaiser U, Hofmann J, Schilli M, et al. 1996. Steroid-hormone receptors in cell lines and tumor biopsies of human lung cancer. *Int J Cancer* 67(3):357-364.

Kanetsky PA, Gammon MD, Mandelblatt J, et al. 1998. Cigarette smoking and cervical dysplasia among non-Hispanic black women. *Cancer Detect Prev* 22(2):109-119.

Kannan K, Munirajan AK, Krishnamurthy J, Bhuvarahamurthy V, Mohanprasad BK, Panishankar KH, Tsuchida N, Shanmugam G. 1999. Low incidence of p53 mutations in betel quid and tobacco chewing-associated oral squamous carcinoma from India. *Int J Oncol* 15(6):1133-1136.

Kato S, Bowman ED, Harrington AM, Blomeke B, Shields PG. 1995. Human lung carcinogen-DNA adduct levels mediated by genetic polymorphisms in vivo. *J Natl Cancer Inst* 87(12):902-907.

Katoh T, Inatomi H, Nagaoka A, Sugita A. 1995. Cytochrome P4501A1 gene polymorphism and homozygous deletion of the glutathione S-transferase M1 gene in urothelial cancer patients. *Carcinogenesis* 16:655-657.

Katoh T, Kaneko S, Boissy R, Watson M, Ikemura K, Bell DA. 1998. A pilot study testing the association between N-acetyltransferases 1 and 2 and risk of oral squamous cell carcinoma in Japanese people. *Carcinogenesis* 19(10):1803-1807.

Kaufman DW, Palmer JR, Rosenberg L, Stolley P, Warshauer E, Shapiro S. 1989. Tar content of cigarettes in relation to lung cancer. *Am J Epidemiol* 129(4):703-711.

Kawajiri K, Eguchi H, Nakachi K, Sekiya T, Yamamoto M. 1996. Association of CYP1A1 germ line polymorphisms with mutations of the p53 gene in lung cancer. *Cancer Res* 56(1):72-76.

Keller AZ, Terris M. 1965. The association of alcohol and tobacco with cancer of the mouth and pharynx. *Am J Public Health Nations Health* 55(10):1578-1585.

Kerr KM, Carey FA, King G, Lamb D. 1994. Atypical alveolar hyperplasia: relationship with pulmonary adenocarcinoma, p53, and c-erbB-2 expression. *J Pathol* 174(4):249-256.

Khoury MJ, Wagener DK. 1995. Epidemiological evaluation of the use of genetics to improve the predictive value of disease risk factors. *Am J Hum Genet* 56(4):835-844.

Khuder SA, Dayal HH, Mutgi AB, Willey JC, Dayal G. 1998. Effect of cigarette smoking on major histological types of lung cancer in men. *Lung Cancer* 22(1):15-21.

Kihara M, Noda K. 1995. Risk of smoking for squamous and small cell carcinomas of the lung modulated by combinations of CYP1A1 and GSTM1 gene polymorphisms in a Japanese population. *Carcinogenesis* 16:2331-2336.

Kihara M, Noda K, Kihara M. 1995. Distribution of GSTM1 null genotype in relation to gender, age and smoking status in Japanese lung cancer patients. *Pharmacogen* 5 Spec No:S74-S79.

Kim JH, Stansbury KH, Walker NJ, Trush MA, Strickland PT, Sutter TR. 1998. Metabolism of benzo[a]pyrene and benzo[a]pyrene-7,8-diol by human cytochrome P450 1B1. *Carcinogenesis* 19(10):1847-1853.

Kinzler KW, Vogelstein B. 1997. Cancer-susceptibility genes. Gatekeepers and caretakers. *Nature* 386(6627):761, 763.

Kinzler KW, Vogelstein B. 1998. Landscaping the cancer terrain. *Science* 280(5366):1036-1037.

Klein-Szanto AJ, Iizasa T, Momiki S, et al. 1992. A tobacco-specific N-nitrosamine or cigarette smoke condensate causes neoplastic transformation of xenotransplanted human bronchial epithelial cells. *Proc Natl Acad Sci U S A* 89(15):6693-6697.

Kloth MT, Gee RL, Messing EM, Swaminathan S. 1994. Expression of N-acetyltransferase (NAT) in cultured human uroepithelial cells. *Carcinogenesis* 15(12):2781-2787.

Knize MG, Salmon CP, Pais P, Felton JS. 1999. Food heating and the formation of heterocyclic aromatic amine and polycyclic aromatic hydrocarbon mutagens/carcinogens. *Adv Exp Med Biol* 459:179-193.

Kohno T, Yokota J. 1999. How many tumor suppressor genes are involved in human lung carcinogenesis? *Carcinogenesis* 20(8):1403-1410.

Kondo K, Tsuzuki H, Sasa M, Sumitomo M, Uyama T, Monden Y. 1996. A dose-response relationship between the frequency of p53 mutations and tobacco consumption in lung cancer patients. *J Surg Oncol* 61(1):20-26.

Koo LC. 1997. Diet and lung cancer 20+ years later: more questions than answers? *Int J Cancer* Suppl 10:22-29.

Kozlowski LT, Rickert WS, Pope MA, Robinson JC, Frecker RC. 1982. Estimating the yield to smokers of tar, nicotine, and carbon monoxide from the 'lowest yield' ventilated filter-cigarettes. *Br J Addict* 77(2):159-165.

Kresty LA, Carmella SG, Borukhova A, Akerkar SA, Gopalakrishnan R, Harris RE, Stoner GD, Hecht SS. 1996. Metabolites of a tobacco-specific nitrosamine, 4-(methylnitros-amino)- 1-(3-pyridyl)-1-butanone (NNK), in the urine of smokeless tobacco users: relationship between urinary biomarkers and oral leukoplakia. *Cancer Epidemiol Biomarkers Prev* 5(7):521-525.

Krewski D, Gaylor DW, Soms AP, Szyszkowicz M. 1993. An overview of the report: correlation between carcinogenic potency and the maximum tolerated dose: implications for risk assessment. *Risk Anal* 13(4):383-398.

Krewski D, Thomas RD. 1992. Carcinogenic mixtures. *Risk Anal* 12:105-113.

Kriek E, Rojas M, Alexandrov K, Bartsch H. 1998. Polycyclic aromatic hydrocarbon-DNA adducts in humans: relevance as biomarkers for exposure and cancer risk. *Mutat Res* 400(1-2):215-231.

Kulasegaram R, Downer MC, Jullien JA, Zakrzewska JM, Speight PM. 1995. Case-control study of oral dysplasia and risk habits among patients of a dental hospital. *Eur J Cancer B Oral Oncol* 31B(4):227-231.

Kuller LH, Ockene JK, Meilahn E, Wentworth DN, Svendsen KH, Neaton JD. 1991. Cigarette smoking and mortality. MRFIT Research Group. *Prev Med* 20(5):638-654.

Kure EH, Ryberg D, Hewer A, et al. 1996. p53 mutations in lung tumours: relationship to gender and lung DNA adduct levels. *Carcinogenesis* 17(10):2201-2205.

La DK, Swenberg JA. 1996. DNA adducts: biological markers of exposure and potential applications to risk assessment. *Mutat Res* 365(1-3):129-146.

La Vecchia C, Bidoli E, Barra S, et al. 1990. Type of cigarettes and cancers of the upper digestive and respiratory tract. *Cancer Causes Control* 1(1):69-74.

Lam S, leRiche JC, Zheng Y, et al. 1999. Sex-related differences in bronchial epithelial changes associated with tobacco smoking. *J Natl Cancer Inst* 91(8):691-696.

Lam S, Kennedy T, Unger M, et al. 1998. Localization of bronchial intraepithelial neoplastic lesions by fluorescence bronchoscopy. *Chest* 113(3):696-702.

Landis SH, Murray T, Bolden S, Wingo PA. 1999. Cancer statistics, 1999. *CA Cancer J Clin* 49(1):8-31, 1.

LaVoie EJ, Shigematsu A, Rivenson A, Mu B, Hoffmann D. 1985. Evaluation of the effects of cotinine and nicotine-N'-oxides on the development of tumors in rats initiated with N-[4-(5-nitro-2-furyl)-2-thiazolyl]formamide. *J Natl Cancer Inst* 75(6):1075-1081.

Law MR, Morris JK, Watt HC, Wald NJ. 1997. The dose-response relationship between cigarette consumption, biochemical markers and risk of lung cancer. *Br J Cancer* 75(11):1690-1693.

Lawrence C, Tessaro I, Durgerian S, et al. 1987. Smoking, body weight, and early-stage endometrial cancer. *Cancer* 59(9):1665-1669.

Lazarus P, Idris AM, Kim J, Calcagnotto A, Hoffmann D. 1996a. p53 mutations in head and neck squamous cell carcinomas from Sudanese snuff (toombak) users. *Cancer Detect Prev* 20(4):270-278.

Lazarus P, Sheikh SN, Ren Q, et al. 1998. p53, but not p16 mutations in oral squamous cell carcinomas are associated with specific CYP1A1 and GSTM1 polymorphic genotypes and patient tobacco use. *Carcinogenesis* 19(3):509-514.

Lazarus P, Stern J, Zwiebel N, Fair A, Richie JP Jr, Schantz S. 1996b. Relationship between p53 mutation incidence in oral cavity squamous cell carcinomas and patient tobacco use. *Carcinogenesis* 17: 733-739.

Le Marchand L, Wilkens LR, Kolonel LN. 1992. Ethnic differences in the lung cancer risk associated with smoking. *Cancer Epidemiol Biomarkers Prev* 1(2):103-107.

Leanderson P, Tagesson C. 1990. Cigarette smoke-induced DNA-damage: role of hydroquinone and catechol in the formation of the oxidative DNA-adduct, 8-hydroxydeoxyguanosine. *Chem Biol Interact* 75(1):71-81.

Lee CK, Brown BG, Reed EA, Coggins CR, Doolittle DJ, Hayes AW. 1993. Ninety-day inhalation study in rats, using aged and diluted sidestream smoke from a reference cigarette: DNA adducts and alveolar macrophage cytogenetics. *Fundam Appl Toxicol* 20:393-401.

Lee PN. 1998. Difficulties in assessing the relationship between passive smoking and lung cancer. *Stat Methods Med Res* 7(2):137-163.

Lee PN, Forey BA. 1998. Trends in cigarette consumption cannot fully explain trends in British lung cancer rates. *J Epidemiol Community Health* 52:82-92.

Lee PN, Garfinkel L. 1981. Mortality and type of cigarette smoked. *J Epidemiol Community Health* 35(1):16-22.

Lee Y, Buchanan BG, Klopman G, Dimayuga M, Rosenkranz HS. 1996. The potential of organ specific toxicity for predicting rodent carcinogenicity. *Mutat Res* 358:37-62.

Lehman TA, Scott F, Seddon M, et al. 1996. Detection of K-ras oncogene mutations by polymerase chain reaction-based ligase chain reaction. *Anal Biochem* 239(2):153-159.

Lesko SM, Rosenberg L, Kaufman DW, et al. 1985. Cigarette smoking and the risk of endometrial cancer. *N Engl J Med* 313(10):593-596.

Levi F, la Vecchia C, Decarli A. 1987. Cigarette smoking and the risk of endometrial cancer. *Eur J Cancer Clin Oncol* 23(7):1025-1029.

Levi F, Randimbison L, Te VC, La Vecchia C. 1999. Second primary cancers in patients with lung carcinoma. *Cancer* 86(1):186-190.

Levin NA, Brzoska PM, Warnock ML, Gray JW, Christman MF. 1995. Identification of novel regions of altered DNA copy number in small cell lung tumors. *Genes Chromosomes Cancer* 13(3):175-185.

LeVois MF. 1997. Environmental tobacco smoke and coronary heart disease. *Circulation* 96(6):2086-2089.

Lewin F, Norell SE, Johansson H, et al. 1998. Smoking tobacco, oral snuff, and alcohol in the etiology of squamous cell carcinoma of the head and neck: a population-based case-referent study in Sweden. *Cancer* 82(7):1367-1375.

Lewis DF, Brantom PG, Ioannides C, Walker R, Parke DV. 1997. Nitrosamine carcinogenesis: rodent assays, quantitative structure-activity relationships, and human risk assessment. *Drug Metab Rev* 29(4):1055-1078.

Liao ML, Wang JH, Wang HM, Ou AQ, Wang XJ, You WQ. 1996. A study of the association between squamous cell carcinoma and adenocarcinoma in the lung, and history of menstruation in Shanghai women, China. *Lung Cancer* 14 Suppl 1:S215-S221.

Li D, Wang M, Cheng L, Spitz MR, Hittelman WN, Wei Q. 1996. In vitro induction of benzo(a)pyrene diol epoxide-DNA adducts in peripheral lymphocytes as a susceptibility marker for human lung cancer. *Cancer Res* 56(16):3638-3641.

Liber HL, Danheiser SL, Thilly WG. 1985. Mutation in single-cell systems induced by low-level mutagen exposure. *Basic Life Sci* 33:169-204.

Liloglou T, Scholes AG, Spandidos DA, Vaughan ED, Jones AS, Field JK. 1997. p53 mutations in squamous cell carcinoma of the head and neck predominate in a subgroup of former and present smokers with a low frequency of genetic instability. *Cancer Res* 57(18):4070-4074.

Lioy PJ, Greenberg A. 1990. Factors associated with human exposures to polycyclic aromatic hydrocarbons. *Toxicol Ind Health* 6(2):209-223.

Lippman SM, Peters EJ, Wargovich MJ, Stadnyk AN, Dixon DO, Dekmezian RH, Loewy JW, Morice RC, Cunningham JE, Hong WK. 1990. Bronchial micronuclei as a marker of an early stage of carcinogenesis in the human tracheobronchial epithelium. *Int J Cancer* 45:811-815.

Liu Q, Yan YX, McClure M, Nakagawa H, Fujimura F, Rustgi AK. 1995. MTS-1 (CDKN2) tumor suppressor gene deletions are a frequent event in esophagus squamous cancer and pancreatic adenocarcinoma cell lines. *Oncogene* 10(3):619-622.

Liu Y, Egyhazi S, Hansson J, Bhide SV, Kulkarni PS, Grafstrom RC. 1997. O6-methylguanine-DNA methyltransferase activity in human buccal mucosal tissue and cell cultures. Complex mixtures related to habitual use of tobacco and betel quid inhibit the activity in vitro. *Carcinogenesis* 18:1889-1895.

Liu Y, Sundqvist, K, Belinsky SA, Castonguay A, Tjalve H, Grafstrom RC. 1993. Metabolism and macromolecular interaction of the tobacco-specific carcinogen 4-(methylnitros-amino)-1-(3-pyridyl)-1-butanone in cultured explants and epithelial cells of human buccal mucosa. *Carcinogenesis* 14:2383-2388.

Liu Z. 1992. Smoking and lung cancer in China: combined analysis of eight case-control studies. *Int J Epidemiol* 21(2):197-201.

Lodovici M, Akpan V, Giovannini L, Migliani F, Dolara P. 1998. Benzo[a]pyrene diol-epoxide DNA adducts and levels of polycyclic aromatic hydrocarbons in autoptic samples from human lungs. *Chem Biol Interact* 116(3):199-212.

Loeb KR, Loeb LA. 2000. Significance of multiple mutations in cancer. *Carcinogenesis* 21(3):379-385.

Loft S, Poulsen HE. 1996. Cancer risk and oxidative DNA damage in man. *J Mol Med* 74(6):297-312.

Loft S, Vistisen K, Ewertz M, Tjonneland A, Overvad K, Poulsen HE. 1992. Oxidative DNA damage estimated by 8-hydroxydeoxyguanosine excretion in humans: influence of smoking, gender and body mass index. *Carcinogenesis* 13(12):2241-2247.

Lubin JH, Blot WJ. 1984. Assessment of lung cancer risk factors by histologic category. *J Natl Cancer Inst* 73(2):383-389.

Lubin JH, Blot WJ, Berrino F, et al. 1984a. Modifying risk of developing lung cancer by changing habits of cigarette smoking. *Br Med J (Clin Res Ed)* 288(6435):1953-1956.

Lubin JH, Blot WJ, Berrino F, et al. 1984b. Patterns of lung cancer risk according to type of cigarette smoked. *Int J Cancer* 33(5):569-576.

Lubin JH, Li JY, Xuan XZ, et al. 1992. Risk of lung cancer among cigarette and pipe smokers in southern China. *Int J Cancer* 51(3):390-395.

Lutz WK. 1999. Dose-response relationships in chemical carcinogenesis reflect differences in individual susceptibility.Consequences for cancer risk assessment, extrapolation, and prevention. *Hum Exp Toxicol* 18:707-712.

Lydiatt WM, Murty VV, Davidson BJ, et al. 1995. Homozygous deletions and loss of expression of the CDKN2 gene occur frequently in head and neck squamous cell carcinoma cell lines but infrequently in primary tumors. *Genes Chromosomes Cancer* 13(2):94-98.

Ma XL, Baraona E, Lasker JM, Lieber CS. 1991. Effects of ethanol consumption on bioactivation and hepatotoxicity of N-nitrosodimethylamine in rats. *Biochem Pharmacol* 42:585-591.

Macfarlane GJ, Zheng T, Marshall JR, et al. 1995. Alcohol, tobacco, diet and the risk of oral cancer: a pooled analysis of three case-control studies. *Eur J Cancer B Oral Oncol* 31B(3):181-187.

Maclure M, Katz RB, Bryant MS, Skipper PL, Tannenbaum SR. 1989. Elevated blood levels of carcinogens in passive smokers. *Am J Public Health* 79(10):1381-1384.

Maneckjee R, Minna JD. 1994. Opioids induce while nicotine suppresses apoptosis in human lung cancer cells. *Cell Growth Differ* 5(10):1033-1040

Mao L. 1996. Genetic alterations as clonal markers for bladder cancer detection in urine. *J Cell Biochem Suppl* 25:191-196.

Mao L, el-Naggar A. 1999. Molecular changes in the multistage pathogenesis of head and neck cancer. Srivastava S, Henson DE, Gazdar AF, eds. *Molecular Pathology of Early Cancer*. Amsterdam: IOS Press. Pp.189-205.

Mao L, Lee JS, Kurie JM, et al. 1997. Clonal genetic alterations in the lungs of current and former smokers. *J Natl Cancer Inst* 89(12):857-862.

Marnett LJ. 2000. Oxyradicals and DNA damage. *Carcinogenesis* 21(3):361-370.

Martin GC, Brown JP, Eifler CW, Houston GD. 1999. Oral leukoplakia status six weeks after cessation of smokeless tobacco use. *J Am Dent Assoc* 130(7):945-954.

Martin JC, Martin DD, Radow B, Day HE. 1979. Life span and pathology in offspring following nicotine and methamphetamine exposure. *Exp Aging Res* 5(6):509-522.

Martone T, Airoldi L, Magagnotti C, et al. 1998. 4-Aminobiphenyl-DNA adducts and p53 mutations in bladder cancer. *Int J Cancer* 75(4):512-516.

Mashberg A, Boffetta P, Winkelman R, Garfinkel L. 1993. Tobacco smoking, alcohol drinking, and cancer of the oral cavity and oropharynx among U.S. veterans. *Cancer* 72(4):1369-1375.

Matsukura N, Willey J, Miyashita M, et al. 1991. Detection of direct mutagenicity of cigarette smoke condensate in mammalian cells. *Carcinogenesis* 12(4):685-689.

Matthias C, Bockmuhl U, Jahnke V, Harries LW, Wolf CR, Jones PW, Alldersea J, Worrall SF, Hand P, Fryer AA, Strange RC. 1998. The glutathione S-transferase GSTP1 polymorphism: effects on susceptibility to oral/pharyngeal and laryngeal carcinomas. *Pharmacogenetics* 8:1-6.

Mayne ST, Buenconsejo J, Janerich DT. 1999a. Familial cancer history and lung cancer risk in United States nonsmoking men and women. *Cancer Epidemiol Biomarkers Prev* 8:1065-1069.

Mayne ST, Buenconsejo J, Janerich DT. 1999b. Previous lung disease and risk of lung cancer among men and women nonsmokers. *Am J Epidemiol* 149:13-20.

McLemore TL, Adelberg S, Liu MC, et al. 1990. Expression of CYP1A1 gene in patients with lung cancer: evidence for cigarette smoke-induced gene expression in normal lung tissue and for altered gene regulation in primary pulmonary carcinomas. *J Natl Cancer Inst* 82(16):1333-1339.

McWhorter WP, Schatzkin AG, Horm JW, Brown CC. 1989. Contribution of socioeconomic status to black/white differences in cancer incidence. *Cancer* 63:982-987.

Melnick RL, Huff J. 1992. 1,3-Butadiene: toxicity and carcinogenicity in laboratory animals and in humans. *Rev Environ Contam Toxicol* 124:111-144.

Mendilaharsu M, De Stefani E, Deneo-Pellegrini H, Carzoglio JC, Ronco A. 1998. Consumption of tea and coffee and the risk of lung cancer in cigarette-smoking men: a case-control study in Uruguay. *Lung Cancer* 19(2):101-107.

Mills NE, Fishman CL, Scholes J, Anderson SE, Rom WN, Jacobson DR. 1995. Detection of K-ras oncogene mutations in bronchoalveolar lavage fluid for lung cancer diagnosis. *J Natl Cancer Inst* 87:1056-1060.

Minamoto T, Mai M, Ronai Z. 1999. Environmental factors as regulators and effectors of multistep carcinogenesis. *Carcinogenesis* 20(4):519-527.

Mizuno S, Akiba S. 1989. Smoking and lung cancer mortality in Japanese men: estimates for dose and duration of cigarette smoking based on the Japan vital statistics data. *Jpn J Cancer Res* 80(8):727-731.

Mizusaki S, Okamoto H, Akiyama A, Fukuhara Y. 1977. Relation between chemical constituents of tobacco and mutagenic activity of cigarette smoke condensate. *Mutat Res* 48(3-4):319-325.

Mollerup S, Ryberg D, Hewer A, Phillips DH, Haugen A. 1999. Sex differences in lung CYP1A1 expression and DNA adduct levels among lung cancer patients. *Cancer Res* 59(14):3317-3320.

Mommsen S, Aagaard J. 1986. Susceptibility in urinary bladder cancer: acetyltransferase phenotypes and related risk factors. *Cancer Lett* 32(2):199-205.

Monro AM. 1994. Risk from low-dose exposures. *Science* 266:1141.

Moolgavkar SH, Dewanji A, Luebeck G. 1989. Cigarette smoking and lung cancer: reanalysis of the British doctors' data. *J Natl Cancer Inst* 81(6):415-420.

Moolgavkar SH, Luebeck EG, Krewski D, Zielinski JM. 1993. Radon, cigarette smoke, and lung cancer: a re-analysis of the Colorado Plateau uranium miners' data. *Epidemiology* 4(3):204-217.

Morita S, Yano M, Tsujinaka T, Akiyama Y, Taniguchi M, Kaneko K, Miki H, Fujii T, Yoshino K, Kusuoka H, Monden M. 1999. Genetic polymorphisms of drug-metabolizing enzymes and susceptibility to head-and-neck squamous-cell carcinoma. *Int J Cancer* 80:685-688.

Morrow JD, Frei B, Longmire AW, et al. 1995. Increase in circulating products of lipid peroxidation (F2-isoprostanes) in smokers. Smoking as a cause of oxidative damage. *N Engl J Med* 332(18):1198-1203.

Murdoch DJ, Krewski D, Wargo J. 1992. Cancer risk assessment with intermittent exposure. *Risk Anal* 12:569-577.

Muscat JE, Richie JP Jr, Thompson S, Wynder EL. 1996. Gender differences in smoking and risk for oral cancer. *Cancer Res* 56(22):5192-5197.

Muscat JE, Wynder EL. 1995. Lung cancer pathology in smokers, ex-smokers and never smokers. *Cancer Lett* 88(1):1-5.

Muscat JE, Wynder EL. 1998. A case/control study of risk factors for major salivary gland cancer. *Otolaryngol Head Neck Surg* 118(2):195-198.

Mustonen R, Schoket B, Hemminki K. 1993. Smoking-related DNA adducts: 32P-postlabeling analysis of 7-methylguanine in human bronchial and lymphocyte DNA. *Carcinogenesis* 14(1):151-154.

Nair UJ, Nair J, Mathew B, Bartsch H. 1999. Glutathione S-transferase M1 and T1 null genotypes as risk factors for oral leukoplakia in ethnic Indian betel quid/tobacco chewers. *Carcinogenesis* 20(5):743-748.

Nair J, Ohshima H, Nair UJ, Bartsch H. 1996. Endogenous formation of nitrosamines and oxidative DNA-damaging agents in tobacco users. *Crit Rev Toxicol* 26(2):149-161.

Nakachi K, Imai K, Hayashi S, Watanabe J, Kawajiri K. 1991. Genetic susceptibility to squamous cell carcinoma of the lung in relation to cigarette smoking dose. *Cancer Res* 51(19):5177-5180.

Nakajima T, Elovaara E, Anttila S, et al. 1995. Expression and polymorphism of glutathione S-transferase in human lungs: risk factors in smoking-related lung cancer. *Carcinogenesis* 16(4):707-711.

Nakayama J, Yuspa SH, Poirier MC. 1984. Benzo(a)pyrene-DNA adduct formation and removal in mouse epidermis in vivo and in vitro: relationship of DNA binding to initiation of skin carcinogenesis. *Cancer Res* 44:4087-4095.

Nandakumar A, Thimmasetty KT, Sreeramareddy NM, Venugopal TC, Rajanna, Vinutha AT, Srinivas, Bhargava MK. 1990. A population-based case-control investigation on cancers of the oral cavity in Bangalore, India. *Br J Cancer* 62(5):847-851.

NCI (National Cancer Institute). 1996. *The FTC Cigarette Test Method for Determining Tar, Nicotine, and Carbon Monoxide Yields of US Cigarettes: Report of the NCI Expert Committee. Smoking and Tobacco Control Monograph 7.* Bethesda, MD: NCI, NIH.

Nesnow S, Beck S, Rosenblum S, et al. 1994. N-nitrosodiethylamine and 4-(methylnitros-amino)-1-(3-pyridyl)-1-butanone induced morphological transformation of C3H/10T1/2CL8 cells expressing human cytochrome P450 2A6. *Mutat Res* 324(3):93-102.

Neugut AI, Murray T, Santos J, et al. 1994. Increased risk of lung cancer after breast cancer radiation therapy in cigarette smokers. *Cancer* 73(6):1615-1620.

Nilsson RA 1998. Qualitative and quantitative risk assessment of snuff dipping. *Regul Toxicol Pharmacol* 28(1):1-16.

Nordlund LA, Carstensen JM, Pershagen G. 1997. Cancer incidence in female smokers: a 26-year follow-up. *Int J Cancer* 73(5):625-628.

Nyberg F, Agudo A, Boffetta P, Fortes C, Gonzalez CA, Pershagen GA. 1998. European validation study of smoking and environmental tobacco smoke exposure in nonsmoking lung cancer cases and controls. *Cancer Causes Control* 9:173-182.

Nyberg F, Isaksson I, Harris JR, Pershagen G. 1997. Misclassification of smoking status and lung cancer risk from environmental tobacco smoke in never-smokers. *Epidemiology* 8(3):304-309.

Obe G, Vogt HJ, Madle S, Fahning A, Heller WD. 1982. Double-blind study on the effect of cigarette smoking on the chromosomes of human peripheral blood lymphocytes in vivo. *Mutat Res* 92:309-319.

O'Hara P, Portser SA, Anderson BP. 1989. The influence of menstrual cycle changes on the tobacco withdrawal syndrome in women. *Addict Behav* 14:595-600.

Olshan AF, Weissler MC, Pei H, Conway K. 1997. p53 mutations in head and neck cancer: new data and evaluation of mutational spectra. *Cancer Epidemiol Biomarkers Prev* 6(7):499-504.

Okkels H, Sigsgaard T, Wolf H, Autrup H. 1996. Glutathione S-transferase mu as a risk factor in bladder tumours. *Pharmacogen* 6:251-256.

Okkels H, Sigsgaard T, Wolf H, Autrup H. 1997. Arylamine N-acetyltransferase 1 (NAT1) and 2 (NAT2) polymorphisms in susceptibility to bladder cancer: the influence of smoking. *Cancer Epidemiol Biomarkers Prev* 6:225-231.

Omenn GS, Goodman GE, Thornquist MD, et al. 1996. Effects of a combination of beta carotene and vitamin A on lung cancer and cardiovascular disease. *N Engl J Med* 334(18):1150-1155.

Orani GP, Anderson JW, Sant'Ambrogio G, Sant'Ambrogio FB. 1991. Upper airway cooling and l-menthol reduce ventilation in the guinea pig. *J Appl Physiol* 70(5):2080-2086.

Oren M. 1999. Regulation of the p53 tumor suppressor protein. *J Biol Chem* 274(51):36031-36034.

Orlow I, LaRue H, Osman I, et al. 1999. Deletions of the INK4A gene in superficial bladder tumors. Association with recurrence. *Am J Pathol* 155(1):105-113.

Osann KE, Anton-Culver H, Kurosaki T, Taylor T. 1993. Sex differences in lung-cancer risk associated with cigarette smoking. *Int J Cancer* 54(1):44-48.

Osann KE, Lowery JT, Schell MJ. 2000. Small cell lung cancer in women: risk associated with smoking, prior respiratory disease, and occupation. *Lung Cancer* 28:1-10.

Otteneder M, Lutz WK. 1999. Correlation of DNA adduct levels with tumor incidence: carcinogenic potency of DNA adducts. *Mutat Res* 424(1-2):237-247.

Owen PE, Glaister JR, Gaunt IF, Pullinger DH. 1987. Inhalation toxicity studies with 1,3-butadiene. 3. Two year toxicity/carcinogenicity study in rats. *Am Ind Hyg Assoc J* 48(5):407-413.

Parazzini F, La Vecchia C, Negri E, Moroni S, Chatenoud L. 1995. Smoking and risk of endometrial cancer: results from an Italian case-control study. *Gynecol Oncol* 56(2):195-199.

Park IW, Wistuba II, Maitra A, et al. 1999. Multiple clonal abnormalities in the bronchial epithelium of patients with lung cancer. *J Natl Cancer Inst* 91(21):1863-1868.

Park JW, Cundy KC, Ames BN. 1989. Detection of DNA adducts by high-performance liquid chromatography with electrochemical detection. *Carcinogenesis* 10(5):827-832.

Parkin DM, Pisani P, Ferlay J. 1999. Global cancer statistics. *CA Cancer J Clin* 49(1):33-64, 1.

Parsons WD, Carmella SG, Akerkar S, Bonilla LE, Hecht SS. 1998. A metabolite of the tobacco-specific lung carcinogen 4-(methylnitrosamino)-1-(3-pyridyl)-1-butanone in the urine of hospital workers exposed to environmental tobacco smoke. *Cancer Epidemiol Biomarkers Prev* 7(3):257-260.

Pastorelli R, Guanci M, Cerri A, Negri E, LaVecchia C, Fumagalli F, Mexxetti M, Cappelli R, Panigalli T, Fanellli R, Airoldi L. 1998. Impact of inherited polymorphisms in glutathione S-transferase M1, microsomal epoxide hydrolase, cytochrome P450 enzymes on DNA, and blood protein adducts of benzo(a)pyrene-diolepoxide. *Cancer Epidemiol Biomarkers Prev* 7(8):703-709.

Pathak DR, Samet JM, Humble CG, Skipper BJ. 1986. Determinants of lung cancer risk in cigarette smokers in New Mexico. *J Natl Cancer Inst* 76(4):597-604.

Patten CJ, Smith TJ, Friesen MJ, Tynes RE, Yang CS, Murphy SE. 1997. Evidence for cytochrome P450 2A6 and 3A4 as major catalysts for N'-nitrosonornicotine alpha-hydroxylation by human liver microsomes. *Carcinogenesis* 18:1623-1630.

Pelkonen O, Vahakangas K, Nebert DW. 1980. Binding of polycyclic aromatic hydrocarbons to DNA: comparison with mutagenesis and tumorigenesis. *J Toxicol Environ Health* 6:1009-1020.

Peluso M, Airoldi L, Armelle M, Martone T, Coda R, Malaveille C, Giacomelli G, Terrone C, Casetta G, Vineis P. 1998. White blood cell DNA adducts, smoking, and NAT2 and GSTM1 genotypes in bladder cancer: a case-control study. *Cancer Epidemiol Biomarkers Prev* 7(4):341-346.

Perera FP. 1987. Molecular cancer epidemiology: a new tool in cancer prevention. *J Nat Cancer Inst* 78:887-898.

Perera FP. 1997. Environment and cancer: who are susceptible? *Science* 278(5340):1068-1073.

Perez-Stable EJ, Herrera B, Jacob P 3rd, Benowitz NL. 1998. Nicotine metabolism and intake in black and white smokers. *JAMA* 280(2):152-156.

Petersen I, Bujard M, Petersen S, et al. 1997. Patterns of chromosomal imbalances in adenocarcinoma and squamous cell carcinoma of the lung. *Cancer Res* 57(12):2331-2335.

Peto R, Darby S, Deo H, Silcocks P, Whitley E, Doll R. 2000. Smoking, smoking cessation, and lung cancer in the UK since 1950: combination of national statistics with two case-control studies. *BMJ* 321(7257):323-329.

Petruzzelli S, De Flora S, Bagnasco M, et al. 1989. Carcinogen metabolism studies in human bronchial and lung parenchymal tissues. *Am Rev Respir Dis* 140(2):417-422.

Pezzotto SM, Mahuad R, Bay ML, Morini JC, Poletto L. 1993. Variation in smoking-related lung cancer risk factors by cell type among men in Argentina: a case-control study. *Cancer Causes Control* 4:231-237.

Phillips, DH. 1996. DNA adducts in human tissues: biomarkers of exposure to carcinogens in tobacco smoke. *Environ Health Perspect* 104 Suppl 3:453-458.

Phillips DH, Hewer A. 1993. DNA adducts in human urinary bladder and other tissues. *Environ Health Perspect* 99:45-49.

Phillips DH, Hewer A, Martin CN, Garner RC, King MM. 1988. Correlation of DNA adduct levels in human lung with cigarette smoking. *Nature* 336(6201):790-792.

Picard E, Seguin C, Monhoven N, et al. 1999. Expression of retinoid receptor genes and proteins in non-small-cell lung cancer. *J Natl Cancer Inst* 91(12):1059-1066.

Pirkle JL, Flegal KM, Bernert JT, Brody DJ, Etzel RA, Maurer KR. 1996. Exposure of the US population to environmental tobacco smoke: the Third National Health and Nutrition Examination Survey, 1988 to 1991. *JAMA* 275(16):1233-1240.

Pledger WJ, Stiles CD, Antoniades HN, Scher CD. 1977. Induction of DNA synthesis in BALB/c 3T3 cells by serum components: reevaluation of the commitment process. *Proc Natl Acad Sci U S A* 74(10):4481-4485.

Pohlabeln H, Jockel KH, Muller KM. 1997. The relation between various histological types of lung cancer and the number of years since cessation of smoking. *Lung Cancer* 18(3):223-229.

Poirier MC, Beland FA. 1992. DNA adduct measurements and tumor incidence during chronic carcinogen exposure in animal models: implications for DNA adduct-based human cancer risk assessment. *Chem Res Toxicol* 5(6):749-755.

Poli P, Buschini A, Spaggiari A, Rizzoli V, Carlo-Stella C, Rossi C. 1999. DNA damage by tobacco smoke and some antiblastic drugs evaluated using the Comet assay. *Toxicol Lett* 108(2-3):267-276.

Portier CJ, Lucier GW, Edler L. 1994. Risk from low-dose exposures. *Science* 266(5188):1141-1142.

Poulsen HE, Prieme H, Loft S. 1998. Role of oxidative DNA damage in cancer initiation and promotion. *Eur J Cancer Prev* 7(1):9-16.

Pratesi G, Cervi S, Balsari A, Bondiolotti G, Vicentini LM. 1996. Effect of serotonin and nicotine on the growth of a human small cell lung cancer xenograft. *Anticancer Res* 16(6B):3615-3619.

Prescott E, Osler M, Andersen PK, et al. 1998. Mortality in women and men in relation to smoking. *Int J Epidemiol* 27(1):27-32.

Prieme H, Loft S, Klarlund M, Gronbaek K, Tonnesen P, Poulsen HE. 1998. Effect of smoking cessation on oxidative DNA modification estimated by 8-oxo-7,8-dihydro-2'-deoxyguanosine excretion. *Carcinogenesis* 19(2):347-351.

Pryor WA, Stone K, Zang LY, Bermudez E. 1998. Fractionation of aqueous cigarette tar extracts: fractions that contain the tar radical cause DNA damage. *Chem Res Toxicol* 11(5):441-448.

Puntoni R, Toninelli F, Zhankui L, Bonassi S. 1995. Mathematical modelling in risk/exposure assessment of tobacco related lung cancer. *Carcinogenesis* 16(7):1465-1471.

Purchase IF, Auton TR. 1995. Thresholds in chemical carcinogenesis. *Regul Toxicol Pharmacol* 22:199-205.

Qiao YL, Taylor PR, Yao SX, Schatzkin A, Mao BL, Lubin J, Rao JY, McAdams M, Xuan, XZ, Li JY. 1989. Relation of radon exposure and tobacco use to lung cancer among tin miners in Yunnan Province, China. *Am J Ind Med* 16:511-521.

Rao DN, Ganesh B, Rao RS, Desai PB. 1994. Risk assessment of tobacco, alcohol and diet in oral cancer—a case-control study. *Int J Cancer* 58(4):469-473.

Rao J, Hemstreet G III, Hurst R. 1999. Molecular pathology and biomarkers of bladder cancer. Srivastava S, Henson D, Gazdar A, eds. *Molecular Pathology of Early Cancer*. Amsterdam: IOS Press. Pp.53-78.

Rebbeck TR. 1997. Molecular epidemiology of the human glutathione S-transferase genotypes GSTM1 and GSTT1 in cancer susceptibility. *Cancer Epidemiol Biomarkers Prev* 6(9):733-743.

Repace JL, Lowrey AH. 1990. Risk assessment methodologies for passive smoking-induced lung cancer. *Risk Anal* 10(1):27-37.

Reynolds SH, Stowers SJ, Patterson RM, Maronpot RR, Aaronson SA, Anderson MW. 1987. Activated oncogenes in B6C3F1 mouse liver tumors: implications for risk assessment. *Science* 237:1309-1316.

Richie JP Jr, Carmella SG, Muscat JE, Scott DG, Akerkar SA, Hecht SS. 1997. Differences in the urinary metabolites of the tobacco-specific lung carcinogen 4-(methylnitrosamino)-1-(3-pyridyl)-1-butanone in black and white smokers. *Cancer Epidemiol Biomarkers Prev* 6(10):783-790.

Rimington J. 1981. The effect of filters on the incidence of lung cancer in cigarette smokers. *Environ Res* 24(1):162-166.

Risch A, Wallace DM, Bathers S, Sim E. 1995. Slow N-acetylation genotype is a susceptibility factor in occupational and smoking related bladder cancer. *Hum Mol Genet* 4(2):231-236.

Risch HA, Howe GR, Jain M, Burch JD, Holowaty EJ, Miller AB. 1993. Are female smokers at higher risk for lung cancer than male smokers? A case-control analysis by histologic type. *Am J Epidemiol* 138(5):281-293.

Risch HA, Howe GR, Jain M, Burch JD, Holowaty EJ, Miller AB. 1994. Lung cancer risk for female smokers. *Science* 263(5151):1206-1208.

Rivenson A, Djordjevic MV, Amin S, Hoffmann R. 1989. A study of tobacco carcinogenesis XLIV. Bioassay in A/J mice of some N- nitrosamines. *Cancer Lett* 47:111-114.

Robertson IGC, Guthenberg C, Mannervik B, Jernstrom B. 1988. The glutathione conjugation of benzo[a]pyrene diol-epoxide by human glutathione transferases. Cooke M, Dennis AJ, eds. *Polynuclear Aromatic Hydrocarbons: A Decade of Progress.* Columbus, OH: Batelle Press. Pp.799-808.

Rodenhuis S, van de Wetering ML, Mooi WJ, Evers SG, van Zandwijk N, Bos JL. 1987. Mutational activation of the K-ras oncogene. A possible pathogenetic factor in adenocarcinoma of the lung. *N Engl J Med* 317(15):929-935.

Rojas M, Alexandrov K, Cascorbi I, Brockmoller J, Likhachev A, Pozharisski K, Bouvier G, Auburtin G, Mayer L, Kopp-Schneider A, Roots I, Bartsch H. 1998. High benzo[a]pyrene diol-epoxide DNA adduct levels in lung and blood cells from individuals with combined CYP1A1 MspI/Msp-GSTM1*0/*0 genotypes. *Pharmacogenetics* 8(2):109-118.

Rojas M, Camus AM, Alexandrov K, Husgafvel-Pursiainen K, Anttila S, Vainio H, Bartsch H. 1992. Stereoselective metabolism of (-)-benzo[a]pyrene-7,8-diol by human lung microsomes and peripheral blood lymphocytes: effect of smoking. *Carcinogenesis* 13:929-933.

Rom WN, Hay JG, Lee TC, Jiang Y, Tchou-Wong KM. 2000. Molecular and genetic aspects of lung cancer. *Am J Respir Crit Care Med* 161(4 Pt 1):1355-1367.

Romert L, Dock L, Jenssen D, Jernstrom B. 1989. Effects of glutathione transferase activity on benzo[a]pyrene 7,8-dihydrodiol metabolism and mutagenesis studied in a mammalian cell co-cultivation assay. *Carcinogenesis* 10(9):1701-1707.

Romkes M, Chern HD, Lesnick TG, et al. 1996. Association of low CYP3A activity with p53 mutation and CYP2D6 activity with Rb mutation in human bladder cancer. *Carcinogenesis* 17(5):1057-1062.

Ronai ZA, Gradia S, Peterson LA, Hecht SS. 1993. G to A transitions and G to T transversions in codon 12 of the Ki-ras oncogene isolated from mouse lung tumors induced by 4-(methylnitrosamino)-1-(3-pyridyl)-1-butanone (NNK) and related DNA methylating and pyridyloxobutylating agents. *Carcinogenesis* 14(11):2419-2422.

Rosenkranz HS, Klopman G. 1993. Structural relationships between mutagenicity, maximum tolerated dose, and carcinogenicity in rodents. *Environ Mol Mutagen* 21:193-206.

Ross J, Nelson G, Kligerman A, et al. 1990. Formation and persistence of novel benzo(a)pyrene adducts in rat lung, liver, and peripheral blood lymphocyte DNA. *Cancer Res* 50(16):5088-5094.

Ross RK, Jones PA, Yu MC. 1996. Bladder cancer epidemiology and pathogenesis. *Semin Oncol* 23(5):536-545.

Rosvold EA, McGlynn KA, Lustbader ED, Buetow KH. 1995. Identification of an NAD(P)H:quinone oxidoreductase polymorphism and its association with lung cancer and smoking. *Pharmacogenetics* 5(4):199-206.

Royce JM, Hymowitz N, Corbett K, Hartwell TD, Orlandi MA. 1993. Smoking cessation factors among African Americans and whites. COMMIT Research Group. *Am J Public Health* 83(2):220-226.

Ryberg D, Hewer A, Phillips DH, Haugen A. 1994a. Different susceptibility to smoking-induced DNA damage among male and female lung cancer patients. *Cancer Res* 54(22):5801-5803.

Ryberg D, Kure E, Lystad S, et al. 1994b. p53 mutations in lung tumors: relationship to putative susceptibility markers for cancer. *Cancer Res* 54(6):1551-1555.

Ryberg D, Skaug V, Hewer A, Phillips DH, Harries LW, Wolf CR, Ogreid D, Ulvik A, Vu P, Haugen A. 1997. Genotypes of glutathione transferase M1 and P1 and their significance for lung DNA adduct levels and cancer risk. *Carcinogenesis* 18(7):1285-1289.

Rylander R, Axelsson G, Andersson T, Liljequist T, Bergman B. 1996. Lung cancer, smoking and diet among Swedish men. *Lung Cancer* 14(Suppl 1):S75-83.

Sanderson RJ, de Boer MF, Damhuis RA, Meeuwis CA, Knegt PP. 1997. The influence of alcohol and smoking on the incidence of oral and oropharyngeal cancer in women. *Clin Otolaryngol* 22(5):444-448.

Sankaranarayanan R, Duffy SW, Padmakumary G, Day NE, Krishan Nair M. 1990. Risk factors for cancer of the buccal and labial mucosa in Kerala, southern India. *J Epidemiol Community Health* 44(4):286-292.

Sant'Ambrogio FB, Anderson JW, Sant'Ambrogio G. 1991. Effect of l-menthol on laryngeal receptors. *J Appl Physiol* 70(2):788-793.

Saranath D, Tandle AT, Teni TR, Dedhia PM, Borges AM, Parikh D, Sanghavi V, Mehta AR. 1999. p53 inactivation in chewing tobacco-induced oral cancers and leukoplakias from India. *Oral Oncol.* 35(3):242-250.

Satcher D. 1999. Cigars and public health. *N Engl J Med* 340(23):1829-1831.

Scanlon EF, Suh O, Murthy SM, Mettlin C, Reid SE, Cummings KM. 1995. Influence of smoking on the development of lung metastases from breast cancer. *Cancer* 75(11):2693-2699.

Schantz SP, Zhang ZF, Spitz MS, Sun M, Hsu TC. 1997. Genetic susceptibility to head and neck cancer: interaction between nutrition and mutagen sensitivity. *Laryngoscope* 107(6):765-781.

Scherer G, Conze C, Tricker AR, Adlkofer F. 1992. Uptake of tobacco smoke constituents on exposure to environmental tobacco smoke (ETS). *Clin Investig* 70(3-4):352-367.

Schildt EB, Eriksson M, Hardell L, Magnuson A. 1998. Oral snuff, smoking habits and alcohol consumption in relation to oral cancer in a Swedish case-control study. *Int J Cancer* 77(3):341-346.

Schlecht NF, Franco EL, Pintos J, et al. 1999. Interaction between tobacco and alcohol consumption and the risk of cancers of the upper aero-digestive tract in Brazil. *Am J Epidemiol* 150(11):1129-1137.

Schmeltz I, Schlotzhauer WS. 1968. Benzo(a)pyrene, phenols and other products from pyrolysis of the cigarette additive, (d,1)-menthol. *Nature* 219(5152):370-371.

Schmezer P, Eckert C, Liegibel UM, Zelezny O, Klein RG. 1998. Mutagenic activity of carcinogens detected in transgenic rodent mutagenicity assays at dose levels used in chronic rodent cancer bioassays. *Mutat Res* 405(2):193-198.

Schmid W. 1975. The micronucleus test. *Mutat Res* 31(1):9-15.

Schoket B, Phillips DH, Kostic S, Vincze I. 1998. Smoking-associated bulky DNA adducts in bronchial tissue related to CYP1A1 MspI and GSTM1 genotypes in lung patients. *Carcinogenesis* 19(5):841-846.

Schreiber G, Fong KM, Peterson B, Johnson BE, O'Briant KC, Bepler G. 1997. Smoking, gender, and survival association with allele loss for the LOH11B lung cancer region on chromosome 11. *Cancer Epidemiol Biomarkers Prev* 6(5):315-319.

Schuller HM, McGavin MD, Orloff M, Riechert A, Porter B. 1995. Simultaneous exposure to nicotine and hyperoxia causes tumors in hamsters. *Lab Invest* 73(3):448-456.

Schwartz AG, Swanson GM. 1997. Lung carcinoma in African Americans and whites. A population-based study in metropolitan Detroit, Michigan. *Cancer* 79(1):45-52.

Scott FM, Modali R, Lehman TA, et al. 1997. High frequency of K-ras codon 12 mutations in bronchoalveolar lavage fluid of patients at high risk for second primary lung cancer. *Clin Cancer Res* 3(3):479-482.

Sekido Y, Fong KM, Minna JD. 1998. Progress in understanding the molecular pathogenesis of human lung cancer. *Biochim Biophys Acta* 1378(1):F21-59.

Sekine I, Nagai K, Tsugane S, et al. 1999. Association between smoking and tumor progression in Japanese women with adenocarcinoma of the lung. *Jpn J Cancer Res* 90(2):129-135.

Sellers TA, Bailey-Wilson JE, Potter JD, Rich SS, Rothschild H, Elston RC. 1992a. Effect of cohort differences in smoking prevalence on models of lung cancer susceptibility. *Genet Epidemiol* 9(4):261-271.

Sellers TA, Potter JD, Bailey-Wilson JE, Rich SS, Rothschild H, Elston RC. 1992b. Lung cancer detection and prevention: evidence for an interaction between smoking and genetic predisposition. *Cancer Res* 52:2694s-2697s.

Seto H, Ohkubo T, Kanoh T, Koike M, Nakamura K, Kawahara Y. 1993. Determination of polycyclic aromatic hydrocarbons in the lung. *Arch Environ Contam Toxicol* 24(4):498-503.

Shanks TG, Burns DM. 1998. Disease consequence of cigar smoking. U.S. Department of Health and Human Services. *Cigars: Health Effects and Trends. Smoking and Tobacco Control Monograph 9*. Bethesda, MD: NIH. Pp 105-158.

Shapiro JA, Jacobs EJ, Thun MJ. 2000. Cigar smoking in men and risk of death from tobacco-related cancers. *J Natl Cancer Inst* 92(4):333-337.

Shaten BJ, Kuller LH, Kjelsberg MO, Stamler J, Ockene JK, Cutler JA, Cohen JD. 1997. Lung cancer mortality after 16 years in MRFIT participants in intervention and usual-care groups. Multiple Risk Factor Intervention Trial. *Ann Epidemiol* 7:125-136.

Shields PG. 2000. Epidemiology of tobacco carcinogenesis. *Current Oncology Reports* 2(3):257-262.

Shields PG, Harris CC. 2000. Cancer risk and low-penetrance susceptibility genes in gene-environment interactions. *J Clin Oncol* 18(11):2309-2315.

Shimada T, Hayes CL, Yamazaki H, et al. 1996a. Activation of chemically diverse procarcinogens by human cytochrome P-450 1B1. *Cancer Res* 56(13):2979-2984.

Shimada T, Yamazaki H, Mimura M, Wakamiya N, Ueng YF, Guengerich FP, Inui Y. 1996b. Characterization of microsomal cytochrome P450 enzymes involved in the oxidation of xenobiotic chemicals in human fetal liver and adult lungs. *Drug Metab Dispos* 24:515-522.

Shopland DR, Eyre HJ, Pechacek TF. 1991. Smoking-attributable cancer mortality in 1991: is lung cancer now the leading cause of death among smokers in the United States? *J Natl Cancer Inst* 83(16):1142-1148.

Shriver SP, Bourdeau HA, Gubish CT, et al. 2000. Sex-specific expression of gastrin-releasing peptide receptor: relationship to smoking history and risk of lung cancer. *J Natl Cancer Inst* 92(1):24-33.

Sidney S, Tekawa IS, Friedman GD. 1993. A prospective study of cigarette tar yield and lung cancer. *Cancer Causes Control* 4(1):3-10.

Sidney S, Tekawa IS, Friedman GD, Sadler MC, Tashkin DP. 1995. Mentholated cigarette use and lung cancer. *Arch Intern Med* 155(7):727-732.

Sidransky D. 1997a. Molecular biology of head and neck tumors. DeVita VT, Hellman S Jr, Rosenberg S, eds. *Cancer: Principles and Practice of Oncology.* 5th ed. Philadelphia: Lippincott. Pp.735-740.

Sidransky D. 1997b. Nucleic acid-based methods for the detection of cancer. *Science* 278(5340):1054-1059.

Sidransky D, Frost P, Von Eschenbach A, Oyasu R, Preisinger AC, Vogelstein B. 1992. Clonal origin bladder cancer. *N Engl J Med* 326(11):737-740.

Sjalander A, Birgander R, Rannug A, Alexandrie AK, Tornling G, Beckman G. 1996. Association between the p21 codon 31 A1 (arg) allele and lung cancer. *Hum Hered* 46:221-225.

Slattery ML, Schumacher MC, West DW, Robison LM. 1988. Smoking and bladder cancer. The modifying effect of cigarettes on other factors. *Cancer* 61(2):402-408.

Slaughter TP, Southwick HW, Smejkel W. 1953. Field cancerization in oral stratefied epithelium. *Cancer* 6:963-968.

Slebos RJ, Hruban RH, Dalesio O, Mooi WJ, Offerhaus GJ, Rodenhuis S. 1991. Relationship between K-ras oncogene activation and smoking in adenocarcinoma of the human lung. *J Natl Cancer Inst* 83(14):1024-1027.

Slebos RJ, Rodenhuis S. 1992. The ras gene family in human non-small-cell lung cancer. *J Natl Cancer Inst Monogr* (13):23-29.

Smith LE, Denissenko MF, Bennett WP, et al. 2000. Targeting of lung cancer mutational hotspots by polycyclic aromatic hydrocarbons. *J Natl Cancer Inst* 92(10):803-811.

Smith TJ, Guo Z, Gonzalez FJ, Guengerich FP, Stoner GD, Yang CS. 1992. Metabolism of 4-(methylnitrosamino)-1-(3-pyridyl)-1-butanone in human lung and liver microsomes and cytochromes P-450 expressed in hepatoma cells. *Cancer Res* 52(7):1757-1763.

Smith TJ, Stoner GD, Yang CS. 1995. Activation of 4-(methylnitrosamino)-1-(3-pyridyl)-1-butanone (NNK) in human lung microsomes by cytochromes P450, lipoxygenase, and hydroperoxides. *Cancer Res* 55(23):5566-5573.

Sobue T, Suzuki T, Fujimoto I, Doi O, Tateishi R, Sato T. 1991. Prognostic factors for surgically treated lung adenocarcinoma patients, with special reference to smoking habit. *Jpn J Cancer Res* 82(1):33-39.

Sobue T, Yamaguchi N, Suzuki T, et al. 1993. Lung cancer incidence rate for male ex-smokers according to age at cessation of smoking. *Jpn J Cancer Res* 84(6):601-607.

Soloway MS, Perito PE. 1992. Superficial bladder cancer: diagnosis, surveillance and treatment. *J Cell Biochem* Suppl 16I:120-127.

Sorsa M, Falck K, Heinonen T, Vainio H, Norppa H, Rimpela M. 1984. Detection of exposure to mutagenic compounds in low-tar and medium-tar cigarette smokers. *Environ Res* 33(2):312-321.

Sozzi G, Huebner K, Croce CM. 1998a. FHIT in human cancer. *Adv Cancer Res* 74:141-166.

Sozzi G, Miozzo M, Pastorino U, et al. 1995. Genetic evidence for an independent origin of multiple preneoplastic and neoplastic lung lesions. *Cancer Res* 55(1):135-140.

Sozzi G, Pastorino U, Moiraghi L, et al. 1998b. Loss of FHIT function in lung cancer and preinvasive bronchial lesions. *Cancer Res* 58(22):5032-5037.

Speit G, Hartmann A. 1999. The comet assay (single-cell gel test). A sensitive genotoxicity test for the detection of DNA damage and repair. *Methods Mol Biol* 113:203-212.

Spitz MR, Fueger JJ, Halabi S, Schantz SP, Sample D, Hsu TC. 1993. Mutagen sensitivity in upper aerodigestive tract cancer: a case-control analysis. *Cancer Epidemiol Biomarkers Prev* 2(4):329-333.

Spitz MR, Hoque A, Trizna Z, et al. 1994. Mutagen sensitivity as a risk factor for second malignant tumors following malignancies of the upper aerodigestive tract. *J Natl Cancer Inst* 86(22):1681-1684.

Spitz MR, Hsu TC, Wu X, Fueger JJ, Amos CI, Roth JA. 1995. Mutagen sensitivity as a biological marker of lung cancer risk in African Americans. *Cancer Epidemiol Biomarkers Prev* 4(2):99-103.

Spruck CH 3d, Rideout WM 3d, Olumi AF, et al. 1993. Distinct pattern of p53 mutations in bladder cancer: relationship to tobacco usage. *Cancer Res* 53(5):1162-1166.

Stellman SD, Garfinkel L. 1986. Smoking habits and tar levels in a new American Cancer Society prospective study of 1.2 million men and women. *J Natl Cancer Inst* 76(6):1057-1063.

Stellman SD, Garfinkel L. 1989. Lung cancer risk is proportional to cigarette tar yield: evidence from a prospective study. *Prev Med* 18(4):518-525.

Stellman SD, Muscat JE, Hoffmann D, Wynder EL. 1997a. Impact of filter cigarette smoking on lung cancer histology. *Prev Med* 26(4):451-456.

Stellman SD, Muscat JE, Thompson S, Hoffmann D, Wynder EL. 1997b. Risk of squamous cell carcinoma and adenocarcinoma of the lung in relation to lifetime filter cigarette smoking. *Cancer* 80:382-388.

Sterling TD, Rosenbaum WL, Weinkam JJ. 1992. Analysis of the relationship between smokeless tobacco and cancer based on data from the National Mortality Followback Survey. *J Clin Epidemiol* 45(3):223-231.

Stern SJ, Degawa M, Martin MV, et al. 1993. Metabolic activation, DNA adducts, and H-ras mutations in human neoplastic and non-neoplastic laryngeal tissue. *J Cell Biochem Suppl* 17F:129-137.

Stevens RG, Moolgavkar SH. 1979. Estimation of relative risk from vital data: smoking and cancers of the lung and bladder. *J Natl Cancer Inst* 63(6):1351-1357.

Stockwell HG, Goldman AL, Lyman GH, et al. 1992. Environmental tobacco smoke and lung cancer risk in nonsmoking women. *J Natl Cancer Inst* 84(18):1417-1422.

Strong MS, Incze J, Vaughan CW. 1984. Field cancerization in the aerodigestive tract—its etiology, manifestation, and significance. *J Otolaryngol* 13(1):1-6.

Sturgis EM., Castillo EJ, Li L, Zheng R, Eicher SA, Clayman GL, Strom SS, Spitz MR, Wei Q. 1999. Polymorphisms of DNA repair gene XRCC1 in squamous cell carcinoma of the head and neck. *Carcinogenesis* 20:2125-2129.

Su JM, Hsu HK, Chang H, Lin SL, Chang HC, Huang MS, Tseng HH. 1996. Expression of estrogen and progesterone receptors in non-small-cell lung cancer: immunohisto-chemical study. *Anticancer Res* 16:3803-3806.

Sumida H, Watanabe H, Kugiyama K, Ohgushi M, Matsumura T, Yasue H. 1998. Does passive smoking impair endothelium-dependent coronary artery dilation in women? *J Am Coll Cardiol* 31(4):811-815.

Swenberg JA, Richardson FC, Boucheron JA, Deal FH, Belinsky SA, Charbonneau M, Short BG. 1987. High- to low-dose extrapolation: critical determinants involved in the dose response of carcinogenic substances. *Environ.Health Perspect* 76:57-63.

Szyfter K, Hemminki K, Szyfter W, Szmeja Z, Banaszewski J, Yang K. 1994. Aromatic DNA adducts in larynx biopsies and leukocytes. *Carcinogenesis* 15(10):2195-2199.

Takagi Y, Osada H, Kuroishi T, et al. 1998. p53 mutations in non-small-cell lung cancers occurring in individuals without a past history of active smoking. *Br J Cancer* 77(10):1568-1572.

Takezaki T, Hirose K, Inoue M, et al. 1996. Tobacco, alcohol and dietary factors associated with the risk of oral cancer among Japanese. *Jpn J Cancer Res* 87(6):555-562.

Talamini R, La Vecchia C, Levi F, Conti E, Favero A, Franceschi S. 1998. Cancer of the oral cavity and pharynx in nonsmokers who drink alcohol and in nondrinkers who smoke tobacco. *J Natl Cancer Inst* 90(24):1901-1903.

Talaska G, al-Juburi AZ, Kadlubar FF. 1991. Smoking related carcinogen-DNA adducts in biopsy samples of human urinary bladder: identification of N-(deoxyguanosin-8-yl)-4-aminobiphenyl as a major adduct. *Proc Natl Acad Sci U S A* 88(12):5350-5354.

Talaska G, Schamer M, Skipper P, et al. 1993. Carcinogen-DNA adducts in exfoliated urothelial cells: techniques for noninvasive human monitoring. *Environ Health Perspect* 99:289-291.

Tang DL, Rundle A, Warburton D, et al. 1998. Associations between both genetic and environmental biomarkers and lung cancer: evidence of a greater risk of lung cancer in women smokers. *Carcinogenesis* 19(11):1949-1953.

Tang D, Santella RM, Blackwood AM, et al. 1995a. A molecular epidemiological case-control study of lung cancer. *Cancer Epidemiol Biomarkers Prev* 4(4):341-346.

Tang JL, Morris JK, Wald NJ, Hole D, Shipley M, Tunstall-Pedoe H. 1995b. Mortality in relation to tar yield of cigarettes: a prospective study of four cohorts. *BMJ* 311(7019):1530-1533.

Taningher M, Saccomanno G, Santi L, Grilli S, Parodi S. 1990. Quantitative predictability of carcinogenicity of the covalent binding index of chemicals to DNA: comparison of the in vivo and in vitro assays. *Environ Health Perspect Mar* 84:183-192.

Taylor J, Umbach D, Stephens E, et al. 1995. The role of N-acetylation polymorphisms at NAT1 and NAT2 in smoing-associated bladder cancer. *Proc Am Assoc Cancer Res* 36:282.

Teeguarden JG, Dragan YP, Pitot HC. 1998. Implications of hormesis on the bioassay and hazard assessment of chemical carcinogens. *Hum Exp Toxicol* 17:254-258.

Tennant RW. 1993. Stratification of rodent carcinogenicity bioassay results to reflect relative human hazard. *Mutat Res* 286(1):111-118.

Tennant RW, Ashby J. 1991. Classification according to chemical structure, mutagenicity to Salmonella and level of carcinogenicity of a further 39 chemicals tested for carcinogenicity by the U.S. National Toxicology Program. *Mutat Res* 257(3):209-227.

Tennant RW, French JE, Spalding JW. 1995. Identifying chemical carcinogens and assessing potential risk in short-term bioassays using transgenic mouse models. *Environ Health Perspect* 103(10):942-950.

Tennant RW, Spalding J. 1996. Predictions for the outcome of rodent carcinogenicity bioassays: identification of trans-species carcinogens and noncarcinogens. *Environ Health Perspect* 104S(5):1095-1100.

Thilly, WG. 1985. Dead cells don't form mutant colonies: a serious source of bias in mutation assays. *Environ Mutagen* 7:255-258.

Thornton A, Lee P, Fry J. 1994. Differences between smokers, ex-smokers, passive smokers and non-smokers. *J Clin Epidemiol* 47(10):1143-1162.

Thun MJ, Day-Lally CA, Calle EE, Flanders WD, Heath CW Jr. 1995. Excess mortality among cigarette smokers: changes in a 20-year interval. *Am J Public Health* 85:1223-1230.

Thun MJ, Lally CA, Flannery JT, Calle EE, Flanders WD, Heath CW Jr. 1997. Cigarette smoking and changes in the histopathology of lung cancer. *J Natl Cancer Inst* 89(21):1580-1586.

Tiano HF, Hosokawa M, Chulada PC, et al. 1993. Retroviral mediated expression of human cytochrome P450 2A6 in C3H/10T1/2 cells confers transformability by 4-(methylnitrosamino)-1-(3-pyridyl)-1-butanone (NNK). *Carcinogenesis* 14(7):1421-1427.

Tiano HF, Wang RL, Hosokawa M, Crespi C, Tindall KR, Langenbach R. 1994. Human CYP2A6 activation of 4-(methylnitrosamino)-1-(3-pyridyl)-1-butanone (NNK): mutational specificity in the gpt gene of AS52 cells. *Carcinogenesis* 15(12):2859-2866.

Tokarskaya ZB, Okladnikova ND, Belyaeva ZD, Drozhko EG. 1995. The influence of radiation and nonradiation factors on the lung cancer incidence among the workers of the nuclear enterprise. Mayak. *Health Phys* 69:356-366.

Tokarskaya ZB, Okladnikova ND, Belyaeva ZD, Drozhko EG. 1997. Multifactorial analysis of lung cancer dose-response relationships for workers at the Mayak nuclear enterprise. *Health Phys* 73(6):899-905.

Tomar SL, Winn DM, Swango PA, Giovino GA, Kleinman DV. 1997. Oral mucosal smokeless tobacco lesions among adolescents in the United States. *J Dent Res* 76(6):1277-1286.

Tong L, Spitz MR, Fueger JJ, Amos CA. 1996. Lung carcinoma in former smokers. *Cancer* 78(5):1004-1010.

Travis WD, Travis LB, Devesa SS. 1995. Lung cancer. *Cancer* 75(1 Suppl):191-202.

Tricker AR, Ditrich C, Preussmann R. 1991. N-nitroso compounds in cigarette tobacco and their occurrence in mainstream tobacco smoke. *Carcinogenesis* 12(2):257-261.

Trivedi AH, Dave BJ, Adhvaryu SC. 1993. Genotoxic effects of nicotine in combination with arecoline on CHO cells. *Cancer Lett* 74(1-2):105-110.

Trivedi AH, Dave BJ, Adhvaryu SG. 1990. Assessment of genotoxicity of nicotine employing in vitro mammalian test system. *Cancer Lett* 54(1-2):89-94.

Trizna Z, Clayman GL, Spitz MR, Briggs KL, Goepfert H. 1995. Glutathione s-transferase genotypes as risk factors for head and neck cancer. *Am J Surg* 170:499-501.

Tsuda H, Hirohashi S, Shimosato Y, Ino Y, Yoshida T, Terada M. 1989. Low incidence of point mutation of c-Ki-ras and N-ras oncogenes in human hepatocellular carcinoma. *Jpn J Cancer Res* 80(3):196-199.

Tverdal A, Thelle D, Stensvold I, Leren P, Bjartveit K. 1993. Mortality in relation to smoking history: 13 years' follow-up of 68,000 Norwegian men and women 35-49 years. *J Clin Epidemiol* 46:475-487.

Tweedie RL, Mengersen KL. 1992. Lung cancer and passive smoking: reconciling the biochemical and epidemiological approaches. *Br J Cancer* 66(4):700-705.

U.S. DHHS (U.S. Department of Health and Human Services). 1998. *Tobacco Use Among U.S. Racial/Ethnic Minority Groups-African Americans, American Indians and Alaska Natives, Asian Americans and Pacific Islanders, and Hispanics. A Report of the Surgeon General.* Washington, DC: U.S. DHHS, Centers for Disease Control and Prevention.

U.S. PHS (Public Health Service). US Department of Health, Education, and Welfare. 1964. *Smoking and Health: Report of the Advisory Committee to the Surgeon General of the Public Health Service.* PHS Publication No. 1103. Princeton, NJ: Van Nostrand Compnay, Inc.

Valkonen M, Kuusi T. 1998. Passive smoking induces atherogenic changes in low-density lipoprotein. *Circulation* 97(20):2012-2016.

van Delft JH, Baan RA, Roza L. 1998. Biological effect markers for exposure to carcinogenic compound and their relevance for risk assessment. *Crit Rev Toxicol* 28(5):477-510.

van Schooten FJ, Hillebrand MJ, van Leeuwen FE, Lutgerink JT, van Zandwijk N, Jansen HM, Kriek E. 1990. Polycyclic aromatic hydrocarbon-DNA adducts in lung tissue from lung cancer patients. *Carcinogenesis* 11(9):1677-1681.

Vander Martin R, Cummings SR, Coates TJ. 1990. Ethnicity and smoking: differences in white, black, Hispanic, and Asian medical patients who smoke. *Am J Prev Med* 6(4):194-199.

Vineis P, Bartsch H, Caporaso N, et al. 1994. Genetically based N-acetyltransferase metabolic polymorphism and low-level environmental exposure to carcinogens. *Nature* 369(6476):154-156.

Vineis P, Esteve J, Terracini B. 1984. Bladder cancer and smoking in males: types of cigarettes, age at start, effect of stopping and interaction with occupation. *Int J Cancer* 34(2):165-170.

Vineis P, Frea B, Uberti E, Ghisetti V, Terracini B. 1983. Bladder cancer and cigarette smoking in males: a case-control study. *Tumori* 69:17-22.

Vineis P, Martone T. 1996. Molecular epidemiology of bladder cancer. *Ann Ist Super Sanita* 32(1):21-27.

Vineis P, Pirastu R. 1997. Aromatic amines and cancer. *Cancer Causes Control* 8(3):346-355.

Virmani AK, Fong KM, Kodagoda D, et al. 1998. Allelotyping demonstrates common and distinct patterns of chromosomal loss in human lung cancer types. *Genes Chromosomes Cancer* 21(4):308-319.

Voorrips LE, Goldbohm RA, Verhoeven DT, et al. 2000. Vegetable and fruit consumption and lung cancer risk in the Netherlands Cohort Study on diet and cancer. *Cancer Causes Control* 11(2):101-115.

Vutuc C. 1984. Quantitative aspects of passive smoking and lung cancer. *Prev Med* 13:698-704.

Vutuc C, Kunze M. 1983. Tar yields of cigarettes and male lung cancer risk. *J Natl Cancer Inst* 71(3):435-437.

Wagenknecht LE, Cutter GR, Haley NJ, et al. 1990. Racial differences in serum cotinine levels among smokers in the Coronary Artery Risk Development in (Young) Adults Study. *Am J Public Health* 80(9):1053-1056.

Wald NJ, Watt HC. 1997. Prospective study of effect of switching from cigarettes to pipes or cigars on mortality from three smoking related diseases. *BMJ* 314(7098):1860-1863.

Waldman JM, Lioy PJ, Greenberg A, Butler JP. 1991. Analysis of human exposure to benzo(a)pyrene via inhalation and food ingestion in the Total Human Environmental Exposure Study (THEES). *J Expo Anal Environ Epidemiol* 1(2):193-225.

Wang LE, Bondy ML, de Andrade M, et al. 2000. Gender difference in smoking effect on chromosome sensitivity to gamma radiation in a healthy population. *Radiat Res* 154(1):20-27.

Wang L, Zhu D, Zhang C, Mao X, Wang G, Mitra S, Li BF, Wang X, Wu M. 1997. Mutations of O6-methylguanine-DNA methyltransferase gene in esophageal cancer tissues from Northern China. *Int J Cancer* 71(5):719-723.

Wasnik KS, Ughade SN, Zodpey SP, Ingole DL. 1998. Tobacco consumption practices and risk of oro-pharyngeal cancer: a case-control study in Central India. *Southeast Asian J Trop Med Public Health* 29(4):827-834.

Wei Q, Gu J, Cheng L, et al. 1996. Benzo(a)pyrene diol epoxide-induced chromosomal aberrations and risk of lung cancer. *Cancer Res* 56(17):3975-3979.

Weiss W, Altan S, Rosenzweig M, Weiss WA. 1977. Lung cancer type in relation to cigarette dosage. *Cancer* 39:2568-2572.

Wiencke JK, Kelsey KT, Varkonyi A, et al. 1995. Correlation of DNA adducts in blood mononuclear cells with tobacco carcinogen-induced damage in human lung. *Cancer Res* 55(21):4910-4914.

Wiencke JK, Spitz MR, McMillan A, Kelsey KT. 1997. Lung cancer in Mexican-Americans and African-Americans is associated with the wild-type genotype of the NAD(P)H: quinone oxidoreductase polymorphism. *Cancer Epidemiol Biomarkers Prev* 6(2):87-92.

Wilcox HB, Schoenberg JB, Mason TJ, Bill JS, Stemhagen A. 1988. Smoking and lung cancer: risk as a function of cigarette tar content. *Prev Med* 17(3):263-272.

Windmill KF, McKinnon RA, Zhu X, Gaedigk A, Grant DM, McManus ME. 1997. The role of xenobiotic metabolizing enzymes in arylamine toxicity and carcinogenesis: functional and localization studies. *Mutat Res* 376(1-2):153-160.

Wingo PA, Ries LA, Giovino GA, et al. 1999. Annual report to the nation on the status of cancer, 1973-1996, with a special section on lung cancer and tobacco smoking. *J Natl Cancer Inst* 91(8):675-690.

Winn DM. 1997. Epidemiology of cancer and other systemic effects associated with the use of smokeless tobacco. *Adv Dent Res* 11(3):313-321.

Winn DM, Blot WJ, Shy CM, Pickle LW, Toledo A, Fraumeni JF Jr. 1981. Snuff dipping and oral cancer among women in the southern United States. *N Engl J Med* 304(13):745-749.

Winter E, Yamamoto F, Almoguera C, Perucho M. 1985. A method to detect and characterize point mutations in transcribed genes: amplification and overexpression of the mutant c-Ki-ras allele in human tumor cells. *Proc Natl Acad Sci U S A* 82(22):7575-7579.

Wistuba II, Berry J, Behrens C, et al. 2000. Molecular changes in the bronchial epithelium of patients with small cell lung cancer. *Clin Cancer Res* 6(7):2604-2610.

Wistuba II, Lam S, Behrens C, et al. 1997. Molecular damage in the bronchial epithelium of current and former smokers. *J Natl Cancer Inst* 89(18):1366-1373.

Witschi H, Espiritu I, Maronpot RR, Pinkerton KE, Jones AD. 1997a. The carcinogenic potential of the gas phase of environmental tobacco smoke. *Carcinogenesis* 18:2035-2042.

Witschi H, Espiritu I, Peake JL, Wu K, Maronpot RR, Pinkerton KE. 1997b. The carcinogenicity of environmental tobacco smoke. *Carcinogenesis* 18(3):575-586.

Wright SC, Zhong J, Zheng H, Larrick JW. 1993. Nicotine inhibition of apoptosis suggests a role in tumor promotion. *FASEB J* 7(11):1045-1051.

Wynder EL, Graham EA. 1950. Tobacco Smoking as a possible etiologic factor in bronchiogenic carcinoma. *JAMA* 143:329-336.

Wynder EL, Stellman SD. 1979. Impact of long-term filter cigarette usage on lung and larynx cancer risk: a case-control study. *J Natl Cancer Inst* 62(3):471-477.

Wynder E, Bross I, Feldman R. 1957. A study of the etiological factors in cancer of the mouth. *Cancer* 10:1300-1323.

Xavier F, Henn LD, Oliveira M, Orlandine L. 1996. Smoking and its relation to the histological type, survival, and prognosis among patients with primary lung cancer. *Rev Paul Med* 114:1298-1302.

Xinarianos G, Scott FM, Liloglou T, et al. 1999. Telomerase activity in non-small cell lung carcinomas correlates with smoking status. *Int J Oncol* 15(5):961-965.

Xu J, Gimenez-Conti IB, Cunningham JE, Collet AM, Luna MA, Lanfranchi HE, Spitz MR, Conti CJ. 1998. Alterations of p53, cyclin D1, Rb, and H-ras in human oral carcinomas related to tobacco use. *Cancer* 83(2):204-212.

Xu X, Stower MJ, Reid IN, Garner RC, Burns PA. 1997. A hot spot for p53 mutation in transitional cell carcinoma of the bladder: clues to the etiology of bladder cancer. *Cancer Epidemiol Biomarkers Prev* 6(8):611-616.

Xue KX, Wang S, Ma GJ, et al. 1992. Micronucleus formation in peripheral-blood lymphocytes from smokers and the influence of alcohol- and tea-drinking habits. *Int J Cancer* 50(5):702-705.

Yakubovskaya MS, Spiegelman V, Luo FC, et al. 1995. High frequency of K-ras mutations in normal appearing lung tissues and sputum of patients with lung cancer. *Int J Cancer* 63(6):810-814.

Yang CS, Yoo JS, Ishizaki H, Hong JY. 1990. Cytochrome P450IIE1: roles in nitrosamine metabolism and mechanisms of regulation. *Drug Metab Rev* 22(2-3):147-159.

Yim SH, Hee SS. 1995. Genotoxity of nicotine and cotinine in the bacterial luminescence test. *Mutat Res* 335(3):275-283.

Yong LC, Brown CC, Schatzkin A, et al. 1997. Intake of vitamins E, C, and A and risk of lung cancer. The NHANES I epidemiologic followup study. First National Health and Nutrition Examination Survey. *Am J Epidemiol* 146(3):231-243.

Yu MC, Ross RK, Chan KK, et al. 1995. Glutathione S-transferase M1 genotype affects aminobiphenyl-hemoglobin adduct levels in white, black and Asian smokers and non-smokers. *Cancer Epidemiol Biomarkers Prev* 4(8):861-864.

Zang EA, Wynder EL. 1992. Cumulative tar exposure. A new index for estimating lung cancer risk among cigarette smokers. *Cancer* 70(1):69-76.

Zang EA, Wynder EL. 1996. Differences in lung cancer risk between men and women: examination of the evidence. *J Natl Cancer Inst* 88(3-4):183-192.

Zaridze DG, Safaev RD, Belitsky GA, Brunnemann KD, Hoffmann D. 1991. Carcinogenic substances in Soviet tobacco products. *IARC Sci Publ* (105):485-488.

Zeiger E. 1987. Carcinogenicity of mutagens: predictive capability of the Salmonella mutagenesis assay for rodent carcinogenicity. *Cancer Res* 47(5):1287-1296.

Zeller WJ, Berger MR. 1989. Nicotine and estrogen metabolism—possible implications of smoking for growth and outcome of treatment of hormone-dependent cancer? Discussion of experimental results. *J Cancer Res Clin Oncol* 115(6):601-603.

Zhang YJ, Chen CJ, Lee CS, et al. 1991. Aflatoxin B1-DNA adducts and hepatitis B virus antigens in hepatocellular carcinoma and non-tumorous liver tissue. *Carcinogenesis* 12(12):2247-2252.

Zhang ZF, Sarkis AS, Cordon-Cardo C, et al. 1994. Tobacco smoking, occupation, and p53 nuclear overexpression in early stage bladder cancer. *Cancer Epidemiol Biomarkers Prev* 3(1):19-24.

Zhang ZF, Shu XM, Cordon-Cardo C, et al. 1997. Cigarette smoking and chromosome 9 alterations in bladder cancer. *Cancer Epidemiol Biomarkers Prev* 6(5):321-326.

13

Cardiovascular Disease

The impact of cigarette smoking on the U.S. national burden of cardiovascular disease (CVD) has been well documented. Each year, about 150,000 cardiovascular deaths are attributable to cigarette smoking, and of these, about 30,000 are attributable to environmental cigarette smoke exposure among nonsmokers (McGinnis and Foege, 1999, Taylor et al., 1992). The cardiovascular burden of smoking is amplified substantially because smoking significantly increases the risk for many types of cardiovascular morbidity: myocardial infarction (MI), sudden cardiac death, stroke, peripheral vascular disease, and abdominal aortic aneurysms (Green et al., 1993). Importantly, the risk of cardiac ischemic events is substantially and relatively rapidly reversible on cessation of smoking (U.S. DHHS, 1983).

In 1990-1994, an average of 430,700 Americans died each year of smoking-related illness. The largest portion of these deaths were cardiovascular-related illnesses. Approximately one in five deaths from cardiovascular diseases is attributable to smoking. According to the World Health Organization (WHO), 1 year after quitting, the risk of coronary heart disease (CHD) decreases by 50%, and within 15 years, the relative risk of dying from CHD for an ex-smoker approaches that of a long-time nonsmoker.

Despite this well-recognized association between smoking, and CVD, and sudden cardiac death, and the increasing amount of pathogenic information available, more needs to be known about the mechanisms of smoking-induced CVD injury. This is of particular importance because one constituent of tobacco smoke—nicotine—is currently used as short-term therapy for smoking cessation, and as discussed elsewhere in this volume, nicotine is being considered for extended use as an aid to smoking reduction. This necessitates a detailed consideration of the cardiovascular pharmacology of nicotine. The results of the Lung Health Study (Anthonisen et al., 1994) is notable in the consideration of the cardiovascular effects of long-term nicotine replacement therapy (NRT) since it was found that long-term use of nicotine gum for the purpose of smoking cessation did not result in an increased incidence of cardiovascular complications. Furthermore, it is pertinent to strategies being deployed by tobacco product manufacturers to eliminate selectively discrete constituents of tobacco and cigarettes and then market them as "safer." This implies a knowledge of the relative contributions of a myriad of tobacco constituents to CVD, which simply does not exist. This chapter summarizes the current scientific basis of our understanding of smoking and CVD, suggests modern methodologies that might usefully be applied to enhance understanding in this area, and highlights areas for further research.

CORONARY HEART DISEASE

The incidence of coronary artery disease (CAD), including sudden cardiac death, is more than doubled in cigarette smokers as a group and is increased fourfold in heavy smokers. There is a dose-response relationship between cigarette smoking and CAD, such that the risk increases with the number of cigarettes smoked daily, the extent of inhalation, and the number of years of smoking. Cigarettes that nominally deliver less tar or nicotine have not been shown to confer any protection from ischemic heart disease.

The clearest understanding of smoking-induced ischemic heart disease emerges from integration of data from epidemiological and pathophysiologic investigations. In considering the participation of smoking in the etiology of CHD, it is useful to separate the effects of smoking on the development of atherosclerotic stenosis of the coronary arteries from its effects on the process that converts coronary atherosclerosis to acute coronary events.

Acute coronary events represent an abrupt transition from stable chronic CAD to one of the major consequences of ischemia: unstable angina, myocardial infarction, and sudden cardiac death. Abundant evidence supports the concept that the rupture of a lipid-rich atherosclerotic plaque with attendant thrombus formation is the initiating event in the development of the vast majority of these ischemic syndromes (Davies and Thomas, 1984; Oates, 1989).

Coronary Atherosclerosis

Age is a powerful predictor of coronary atherosclerosis, and angiographic studies indicate that the relative risk for coronary stenosis associated with having previously smoked is greatest in the youngest age groups, suggesting an acceleration of the process by smoking. Some of the best evidence is provided by a prospective autopsy study performed in Hawaii (Reed et al., 1987) and from the study of autopsies on young men who died violently (PDAY Research Group, 1990). These studies indicate that there is a strong dose-related association of smoking with atherosclerosis of the abdominal aorta, as well as an association with atherosclerosis of the coronary arteries that is significant but less robust. From these two autopsy studies, as well as from an overview of all relevant studies, one gains the impression that the significant increase in fibrous atherosclerotic plaques and atheroma of the coronary arteries is not of sufficient magnitude to account fully for the greater increase in acute coronary events that are linked to cigarette smoking.

The finding that the increase in involvement of the coronary arteries with atherosclerosis is of small magnitude is in agreement with the prospective epidemiological evidence that stable angina pectoris is increased modestly, if at all, in cigarette smokers over age 40. In men less than 40 years of age, an increase in stable angina has been detected (twofold), but at 40 years and older the risk is increased only slightly, even after adjusting the incidence of stable angina for the loss of persons at risk owing to acute coronary events (Dawber, 1980). However, excess smoking-attributable mortality rates due to heart disease continue to increase with age and smoking duration, suggesting that smoking continues to be an important independent risk factor in the elderly (Burns, 2000).

Acute Coronary Events

Prospective epidemiological studies have consistently demonstrated a substantial increase in acute coronary ischemic events in individuals who smoke (U.S. DHHS, 1983). The pernicious effect of cigarette smoking on MI and sudden cardiac death is seen at least to age 70, but the increase in risk for a given individual appears to be greatest during middle age. The largest body of prospectively acquired North American data indicates that middle-aged men who smoke have a tenfold greater risk of sudden cardiac death and a 3.6-fold increased risk of MI (Kannel et al., 1984). The risk for sudden cardiac death is disproportionately greater than that for MI, a finding that is replicated to varying degrees in other studies. Of all the coronary risk factors, cigarette smoking is the strongest predictor of sudden cardiac death. The risk for both of these acute coronary events (MI and sudden death) clearly exceeds that for stable angina. Although the incidence of coronary deaths in women during middle age is lower than that in men, cigarette smoking accounts for about half of these deaths. Pipe and cigar smoking are also associated with increased rates of heart attack rates (Nyboe et al., 1991) and coronary mortality, as well as stroke occurrence.

In most studies that have explored the relationship of cigarette smoking rates to CVD outcomes, a dose-response has been observed. Cumulative exposures in epidemiological studies are not easy to determine because of interindividual short- and long-term variations in smoking patterns and choices of products. Also, individual inhalation depth and retention may vary in order to titrate the delivery of certain levels of nicotine (Hee et al., 1995). However, population studies employing even a relatively simple measure of exposure, such as cigarettes per day, usually reveal increasing adverse CVD outcomes with increasing numbers of cigarettes consumed. An example is the Cancer Prevention Study (CPS), shown in Table 13-1. Here, although some cells have small numbers of participants, there is a general trend of increased risk of CAD death with increasing baseline cigarette usage, most prominent up to 20 cigarettes per day.

Cessation of Smoking

Smoking cessation in individuals without known coronary heart disease makes an important contribution to reducing the risk of MI (Cook et al., 1986; Rosenberg et al., 1985, 1990; U.S. DHHS, 1983). The excess risk of MI falls by about 50% within the first two years after cessation of smoking, consistent with a substantial component of the risk of acute ischemic events being reversible. An example of the rate of decline in risk of fatal CAD, total CAD, and nonfatal MI is shown in Table 13-2 from the CPS (Stellman and Garfinkel, 1986). For those who have ceased smoking for 10-14 years, their general CVD risk declines to that of "never smokers." A similar or even more accelerated risk reduction can be seen for stroke in women, as shown in Figure 13-1 (Kawachi et al., 1993). Even among individuals over 60 years of age, a decrease in risk for ischemic heart disease postcessation has been found, though smaller than for younger individuals (Burns, 2000).

For smokers who already have CAD, cessation is also very effective in reducing the incidence of further acute coronary events. Survivors of MI have a greater risk of reinfarction, and survivors of sudden cardiac death have a greater risk of sudden death if they continue to smoke. For individuals who have angina pectoris or a positive exercise test or who have had coronary artery bypass graft surgery, continuing the smoking

TABLE 13-1 Age-Specific and Age-Adjusted Death Rates From Coronary Heart Disease, by Number of Cigarettes Currently Smoked

Cigarettes per Day

Age	Never Smoked		1-9		10-19		20		21-39		40		41+	
	M	F	M	F	M	F	M	F	M	F	M	F	M	F
30-34	—	—	—	—	—	—	—	—	()	—	()	—	()	—
35-39	(0.9)	()	—	—	—	—	—	—	—	()	—	()	—	—
40-44	[1.3]	0.6	()	—	()	—	(10.4)	(1.4)	[5.7]	()	12.4	—	(13.1)	()
45-49	2.7	0.4	17.7	[2.1]	14.1	3.3	14.1	2.8	9.9	2.1	19.5	4.6	18.1	—
50-54	5.6	0.8	14.7	3.4	25.1	4.2	21.4	4.0	21.1	6.8	21.9	4.6	20.9	(6.4)
55-59	11.9	2.4	28.0	2.4	31.0	5.7	35.9	9.7	28.4	8.1	31.2	11.3	41.3	14.6
60-64	22.9	6.2	49.4	11.1	57.0	13.5	64.0	19.0	46.1	19.8	59.0	17.3	28.9	()
65-69	40.5	12.6	65.2	22.6	76.9	30.4	82.5	35.4	82.1	29.2	65.6	29.1	78.5	[32.9]
70-74	68.5	25.4	114.1	35.6	110.4	52.6	132.9	49.2	101.0	51.4	114.4	54.8	39.2	—
75-79	123.1	53.0	128.9	72.9	148.7	76.9	176.6	94.4	222.5	103.2	177.9	92.9	[110.0]	()
80-84	189.5	97.5	248.5	127.2	279.6	114.2	309.7	149.1	201.9	81.4	315.1	(103.7)	—	—
85+	329.7	265.5	363.3	167.1	308.6	479.4	430.0	333.9	852.3	()	()	()	—	—
Age adjusted[a]	24.1	11.5	37.0	13.1	38.2	20.3	44.7	20.2	46.9	16.2	37.6	15.9	30.0	13.3

NOTE: —=0 deaths; ( )=1 death; (rate)=2 deaths; [rate]=3 deaths.

[a] Age-adjusted death rates per 10,000 person-years standardized to 1980 U.S. population.

SOURCE: Thun et al., 1997. Reprinted with permission from author.

TABLE 13-2 Time Since Quitting and Age-Adjusted and Multivariate RRs of Fatal Coronary Heart Disease and Nonfatal Myocardial Infarction, Compared with Current Smokers.[a]

Event	Never Smoker	Current Smoker	Years Since Quitting				
			<2	2-4	5-9	10-14	≥15
Fatal coronary heart disease							
cases	49	123	7	9	14	4	13
RR[b]	0.24	1.00	0.53	0.68	0.68	0.26	0.31
	(0.18-0.33)		(0.25-1.12)	(0.35-1.34)	(0.39-1.19)	(0.10-0.65)	(0.18-0.53)
RR[c]	0.23	1.00	1.47	0.58	0.72	0.28	0.32
	(0.17-0.33)		(0.42-5.20)	(0.23-1.44)	(0.36-1.42)	(0.09-0.87)	(0.16-0.66)
Nonfatal myocardial infarction							
cases	166	418	36	22	26	13	41
RR[b]	0.26	1.00	0.85	0.51	0.40	0.26	0.29
	(0.22-0.30)		(0.60-1.19)	(0.34-0.78)	(0.27-0.59)	(0.15-0.43)	(0.21-0.39)
RR[c]	0.24	1.00	0.81	0.43	0.38	0.26	0.27
	(0.20-0.28)		(0.51-1.29)	(0.25-0.74)	(0.23-0.62)	(0.13-0.49)	(0.18-0.41)
Total coronary heart disease							
cases	215	541	43	31	40	17	54
RR[b]	0.25	1.00	0.77	0.55	0.47	0.26	0.29
	(0.22-0.30)		(0.57-1.05)	(0.39-0.79)	(0.34-0.64)	(0.17-0.40)	(0.23-0.38)
RR[c]	0.24	1.00	0.75	0.46	0.44	0.26	0.28
	(0.20-0.28)		(0.49-1.15)	(0.29-0.74)	(0.30-0.66)	(0.14-0.45)	(0.20-0.40)

NOTE: RR=relative risk.

[a]Missing for 29 cases, including 6 fatal coronary heart disease and 23 nonfatal myocardial infarction.
[b]Age-adjusted RR.
[c]Adjusted for age in 5-year intervals, during follow-up period (1976-1978, 1978-1980, 1980-1982, 1982-1984, 1984-1986, or 1986-1988), history of hypertension, diabetes, high cholesterol levels, body mass index, past use of oral contraceptives, menopausal status, postmenopausal estrogen therapy, parental history of myocardial infarction before age 60, and daily number of cigarettes consumed.

SOURCE: Kawachi et al., 1994. Copyright (1994), American Medical Association.

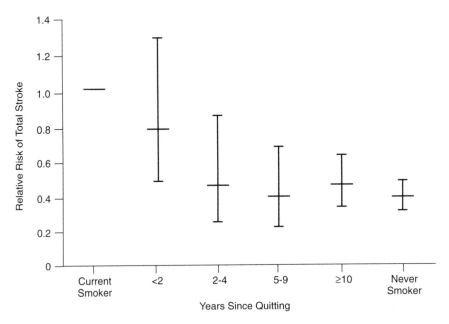

FIGURE 13-1 Risk of total stroke by time since quitting.

NOTE: Age-adjusted relative risk of total stroke in relation to time since stopping smoking. Current smoker was the reference category. Error bars represent 95% confident intervals.
SOURCE: Kawachi et al., 1993. Copyright (1993), American Medical Association.

habit confers a worse prognosis. Cessation of smoking after coronary angioplasty or vascular surgery reduces the rate of restenosis by one-third. Thus, there is a major incentive for smoking cessation in this group of patients. When efforts to cease smoking are effective, they probably confer a greater benefit than any pharmacological or surgical intervention aimed at coronary disease.

Overall, these observations are encouraging for the motivation to quit. The decline of risk upon smoking cessation can be variable in different risk groups and according to behaviors during cessation, potentially affording insight into interindividual differences in mechanisms relevant to healing smoking-induced cardiovascular injury. An initial interpretation of these data suggested that acute factors, such as hemostatic activation, dominated over more gradual processes, such as occlusive atherosclerotic disease contributed to the smoking-induced CVD burden. It is now appreciated that anatomic atherosclerosis may also restructure toward more normal vessels at rates comparable to the clinical and epidemiological findings, bringing vascular factors relevant to plaque stability into consideration. In summary, the kinetics of declining CVD risk after smoking cessation do not serve to discriminate easily between hemostatic and other vascular factors as mediators of smoking-induced disease.

However, the pattern of decline in CVD risk after quitting suggests that CVD may offer a particular advantage in assessing tobacco-related harm reduction over a reasonable time. That is, patterns of change in morbidity and mortality rates after changes from a conventional to a risk reduction tobacco product may help reveal whether lower toxic exposures are actually being achieved and whether they lead to altered CVD outcomes, without waiting the decades necessary to evaluate the outcomes of smoking these products beginning from initiation of the smoking habit. The issues of compensation, nicotine dose-response relationships, and aspects of nicotine-related cardiovascular risk are discussed in Chapter 9.

Smoking and Other Cardiac Risk Factors

Cigarette smoking, hypercholesterolemia, and hypertension have been the most extensively studied risk factors for CAD. The risk of CAD imposed by cigarette smoking is magnified by the presence of several other factors that cause coronary heart disease. Cigarette smoking alone imposes a risk for CHD that is independent of other risk factors. However, smoking in conjunction with another risk factor increases the actual rate of coronary heart disease events to a greater extent than smoking alone, and these risk factors may work additively or geometrically. In one large study (The Pooling Project Research Group, 1978), smoking increased the ten-year rate of a first CAD event (MI or sudden cardiac death) by 31 per 1,000 persons when neither hypercholesterolemia nor hypertension was present. In conjunction with either hypercholesterolemia or hypertension, cigarette smoking increased the rate by 49 per 1,000 persons, and when both hypercholesterolemia and hypertension were present, the superimposition of smoking increased the rate by 97 per 1,000 persons. In women, there is a tenfold increase in the risk of MI among oral contraceptive users who smoke. Therefore, hypercholesterolemia, hypertension, and oral contraceptive use provide an incentive to not smoke or to cease smoking that exceeds the abundant benefit of avoiding this addiction in the rest of the population. Further, control of elevated cholesterol and high blood pressure reduces the risk of CAD (Brown, 2000; Mormando, 2000; The sixth report, 1997), as does smoking cessation.

The pathophysiological impact of various environmental exposures and risk factor states may be modified by host genetic factors. For example, a certain polymorphism in coagulation Factor V (Q506) confers resistance to interaction with the endogenous antithrombotic protein activated protein C. Following presentation with an acute coronary syndrome, smokers with this polymorphism have an increased risk of MI and death compared to smokers with a normal genotype, extending to at least two years after the initial event (Holm et al., 1999). Another example is the interaction between smoking and the guanosine-adenosine at the G455A polymorphism in the b-fibrinogen gene. This polymorphism is associated with higher levels of circulating fibrinogen, which in turn have been associated with increased risk of CAD (Humphries et al., 1999). This association is more pronounced in smokers (Green et al., 1993; Humphries et al., 1999). Similarly, polymorphisms in the coagulation factor VII gene have been associated with reductions in risk of CAD (Donati et al., 2000). Cigarette smoking interacts to a variable degree with genes related to other aspects of cardiovascular risk, including antioxidant enzymes such as paraoxonase (Sen-Banerjee et al., 2000), apoliprotein B (Glisic et al., 1995), DNA repair genes (Abdel-Rahman and El-Zein, 2000), and proteins that regulate the availability of cardiotoxic autacoids such as the serotonin transporter (Arinami et al., 1999). Finally, the cardiovascular hazards of tobacco may be modulated by the presence of functionally important polymorphisms in enzymes that detoxify harmful constituents of tobacco smoke (Li et al., 2000). Clearly, emerging information about genetic polymorphisms that are of functional significance and relevance to the cardiovascular system will afford an increasing opportunity to understand and predict the effect of smoking on CVD risk at the individual level.

Environmental Tobacco Smoke

There is considerable evidence that exposure to environmental tobacco smoke (ETS; passive smoking) has an adverse effect on cardiovascular health (Kawachi et al., 1997a). Pooled results of epidemiological studies indicates a 20% excess coronary heart disease death rate among nonsmoking spouses of smokers (Steenland et al., 1996). As many as 40,000 deaths from MI each year may be the result of passive smoking. The mechanisms linking ETS exposure and cardiovascular disease are probably similar to those for active smoking.

Pathophysiology

A review of the mechanisms linking cigarette smoking and coronary heart disease must consider the fact that whereas smoking accelerates coronary atherosclerosis, it has an even greater effect on the process that abruptly converts atherosclerosis to the acute ischemic events of MI, unstable angina, and sudden cardiac death. The effect of chronic smoking on the initiation of acute ischemic events appears to be largely reversible, given the substantial and rapid reduction in the incidence of acute ischemic events after smoking cessation. Although there is likely overlap between the mechanisms by which smoking accelerates atherosclerosis and promotes acute ischemic events, it is useful to consider these adverse effects of smoking separately.

Acceleration of Atherosclerosis

As for other risk factors, vascular endothelial dysfunction likely plays a central role in the promotion of atherosclerosis in chronic smokers (FitzGerald et al., 1988; Heitzer et al., 1996a). Extensive endothelial abnormalities are present in the umbilical arteries of infants born to smoking mothers, and lesions of endothelial cells, subendothelial damage, and platelet adhesion have been described in vessels of experimental animals exposed to cigarette smoke. Increased in vivo prostacyclin biosynthesis and functional abnormalities of vascular endothelium also result from cigarette smoking. Impaired endothelium-dependent vasodilatation in both forearm and coronary vascular beds has been demonstrated even in young smokers (Campisi et al., 1998; Heitzer et al., 1996b; Zeiher et al., 1991). These functional abnormalities may be due to smoking-induced inhibition of nitric oxide (NO) release (Campisi et al., 1999; Kugiyama et al., 1996) or to acceleration of NO breakdown. Structural endothelial damage may result either from a direct toxic effect of nicotine or other components of cigarette smoke on endothelial cells or from smoking-induced oxidative stress (Heitzer et al., 1996a, b). Smokers have reduced levels of antioxidant vitamins and increased levels of oxidized low-density lipoprotein (LDL), a potent inhibitor of endothelial function (Heitzer et al., 1996b; Morrow et al., 1995). Smoking-induced acute and chronic systemic hemodynamic changes may also contribute to vascular endothelial dysfunction. For example, some but not all studies suggest that smoking a single cigarette causes an acute rise in systemic blood pressure and heart rate, and chronic smoking results in a persistent elevation in daytime blood pressure. The effect of chronic smoking on blood pressure is more complex as outlined later.

Smoking is associated with lipid abnormalities that may contribute to the development of atherosclerosis (Duthie et al., 1993; Craig et al., 1989). In addition to increased levels of oxidized low-density lipoprotein, smoking produces an increase in very low density lipoprotein (VLDL) and triglycerides and a reduction in high-density lipoprotein (HDL) levels.

Smoking has been shown to increase monocyte adhesion to endothelial cells, an important early event in atherosclerosis (Adams et al., 1997; Weber et al., 1996). Additional factors that may contribute to the development of coronary and other atherosclerotic events in smokers include smoking-induced platelet activation, increased fibrinogen levels, and increased blood viscosity.

Promotion of Acute Coronary Events

In addition to the effect of smoking on the development of atherosclerosis, there is a dramatically increased risk of acute ischemic events, including MI and sudden death. Acute systemic and coronary hemodynamic effects of smoking are likely to play an important role in the development of these ischemic events. After smoking a single cigarette, systemic arterial pressure, heart rate, and myocardial contractility increase, resulting in a rise in myocardial oxygen demand (Cryer et al., 1976; Nicod et al., 1984). Simultaneously, in patients with atherosclerosis, smoking causes acute vasoconstriction of both conduit and resistance coronary vessels, with a decrease in coronary blood flow (Czernin et al., 1995; Nicod et al., 1984; Quillen et al., 1993). Even in the absence of a hemodynamically significant stenosis, coronary flow may fall by more than 20% despite a significant increase in myocardial oxygen demand. In some individuals, smoking causes intense focal vasoconstriction or spasm that can lead to myocardial ischemia (Moauad et al., 1986). These acute hemodynamic effects of smoking in the coronary bed are most likely adrenergically mediated, since they can be prevented by α-adrenergic blockade (Winniford et al., 1986). Adrenergic stimulation has been shown to cause exaggerated coronary vasoconstriction in the setting of endothelial dysfunction (Vita et al., 1992; Zeiher et al., 1989). Plasma norepinephrine and epinephrine levels rise acutely after smoking (Cryer et al., 1976); this catecholamine release may lower arrhythmia threshold and increase the risk of sudden death.

In addition to lowering anginal threshold, repeated episodes of coronary vasoconstriction and elevations of systemic arterial pressure may increase hemodynamic stresses at the site of an atherosclerotic plaque. An increase in shear stress at the site of a vulnerable plaque is considered a potentially important cause of plaque rupture.

Smoking has also been shown to exacerbate the cardiovascular effects of cocaine (Moliterno et al., 1994). Both the epicardial coronary constriction and the increase in myocardial oxygen demand caused by cocaine are potentiated by concomitant smoking, perhaps increasing the risk of MI and sudden death attributed to cocaine use.

Inhaled carbon monoxide (CO) in cigarette smoke binds to hemoglobin, reducing oxygen delivery to myocardial cells. The carbon monoxide levels found in smokers have been shown to lower anginal threshold (Allred et al., 1989; Aronow, 1981) and increase ventricular fibrillation threshold (Aronow et al., 1979). Carbon monoxide may also have a deleterious effect on vascular endothelial cells (Thom et al., 1997).

Coronary thrombosis plays a major role in acute coronary events, and smoking is associated with several changes in the hemostatic system that lead to a hypercoagulable state. These hemostatic changes include an increase in fibrinogen (De Maat et al., 1996), red cell mass, and blood viscosity; platelet activation with thromboxane A_2 release (Benowitz et al., 1993); and impaired release of tissue plasminogen activator from endothelial cells (Newby et al., 1999).

Most research on the effects of smoking on the cardiovascular system has focused specifically on nicotine and carbon monoxide, but there are more than 4,000 other components in cigarette smoke, some of which may also contribute to smoking-induced vascular disease (Penn and Snyder 1996).

Coronary Vasospasm

A much less common cause of ischemia than plaque rupture, primary coronary vasospasm usually manifests as variant angina but may also cause MI and sudden cardiac death. Coronary vasospasm is strongly associated with cigarette smoking (Okasha et al., 2000).

EXTRACARDIAC VASCULAR DISEASE

Smoking markedly accelerates atherosclerosis in the abdominal aorta, and occlusive disease of its branches is increased as well (Reed et al., 1987; PDAY, 1990). Aortic aneurysm, peripheral vascular disease, and renal artery stenosis are increased in smokers. Renovascular hypertension should be strongly suspected in cigarette smokers who have severe hypertension. Cigarette smoking is an independent risk factor in the development of atherosclerosis in the internal pudendal and penile arteries of young men with impotence.

Two types of stroke are increased in cigarette smokers (U.S. DHHS, 1989). Smoking is a risk factor for cerebral infarction, and the magnitude of risk increases with greater smoking exposure. This occurs in conjunction with an increase in atherosclerosis of the carotid arteries in smokers. Subarachnoid hemorrhage is markedly increased in cigarette smokers, and in women who take oral contraceptives, this risk if further enhanced.

OTHER VARIABLES INFLUENCED BY SMOKING

Blood Pressure

The effect of cigarette smoking on chronic blood pressure levels in uncertain, in part because of concomitant effects of lower body mass index (BMI) and more frequent alcohol consumption among smokers. Some studies have reported lower ambulatory blood pressure levels (Mikkelsen, et al., 1997; Okasha et al., 2000), while others have reported increases (Minami et al., 1999; Mundal et al., 1997). Findings may depend in part on when in the day pressure recordings are made and whether smoking is assessed by self-report or by biochemical verification (Istvan et al., 1999). Smoking in healthy volunteers has been reported to alter the change in blood pressure response to changes in body position (Nardo et al., 1999) and to cause abnormal baroreceptor function (Gerhardt et al., 1999). Of note, maternal smoking has been associated with increased blood pressure levels in the children of these pregnancies (Blake et al., 2000).

Insulin Resistance

The insulin resistance syndrome, as well as insulin resistance per se, has been associated with increased risk of atherosclerotic vascular disease, independent of most other major risk factors and is a risk factor for non-insulin-dependent diabetes mellitus (NIDDM) (Mikhailidis et al., 1998). Increased insulin resistance has been associated with increased intimal-medial thickening in the presence of vasospastic angina (Shinozaki et al., 1997). The syndrome occurs in healthy young men and is sixfold more common in cigarette smokers (Tahtinen et al., 1998). Insulin resistance is also increased among smokers with existing NIDDM (Targher et al., 1997). Smoking cessation improves insulin sensitivity among healthy middle-aged men (Eliasson et al., 1997).

Other Lipid or Lipoprotein Measures

Lipoprotein (a) (Lp(a)) cholesterol levels are positively and independently correlated with the risk of coronary events in some but not all prospective studies (Seman et al., 1999). Cigarette smoking does not appear to alter blood Lp(a) levels (Chien et al., 1999; Kamboh et al., 2000). However, smoking appears to alter the ratio of LDL subfractions in the direction associated with increased CVD risk (Griffin et al., 1994).

Inflammation

There is substantial evidence that inflammation plays a role in the pathogenesis of atherosclerotic lesions in many parts of the arterial tree and that elevated blood levels of cytokines such as interleukin-6 (IL-6) and C-reactive protein (CRP) are associated with increased risk of coronary events (Ridker et al., 2000; Koenig et al., 1999). Smoking is associated with increased levels of CRP, suggesting that inflammation is one mechanism by which smoking promotes the pathogenesis of these lesions. Smoking has also been shown to increase local tissue levels of some cytokines in organs other than the vascular tree, such as tumor necrosis factor-alpha (TNF-α) in gingival cervicular fluid, but not IL-6 (Bostrom et al., 1999) or IL-8 and ICAM-1 in small respiratory airways (Takizawa et al., 2000).

Homocysteine

Blood homocysteine levels have been related to an increased risk of coronary heart disease and other atherosclerotic diseases in several studies (Saw, 1999). Cigarette smoking has been shown to be associated with higher blood homocysteine levels, both in the basal state and after a methionine load (Reis et al., 2000; Nygard et al., 1998). In a study in which the dose-response was explored, there was a clear positive correlation between the number of cigarettes smoked and blood homocysteine levels. This raises the possibility that smoking cessation may have its effects in part through this mechanism and that homocysteine level might be developed as a biomarker of smoking exposure.

SURROGATE MARKERS

The identification and validation of quantitative indices that reflect biochemical mechanisms of cardiovascular injury caused by cigarette smoking would be useful. Such surrogate indices might permit the study of both progression and regression of smoking-related cardiovascular disease and clarify the factors that contribute to interindividual differences in susceptibility to such complications. They might also be useful correlates of the relative bioavailability of individual constituents of cigarette smoke and aid in the evaluation of harm reduction strategies, including potentially "safer" smoking devices and nicotine replacement therapies. They might also prove of value in assessment of the cardiovascular risk of secondhand smoke. Presently, little information is available relating to the mechanisms of cigarette-induced cardiovascular damage, but the emergence of promising technologies relating to potentially important mechanisms, including platelet activation and oxidative stress, illustrates the potential of the approach.

Cigarette smoking causes platelet activation in vivo, as reflected by an accelerated platelet turnover time (Fuster et al., 1981). However, the development of noninvasive approaches to assessment of platelet activation facilitates repeated investigation of the effects of smoking on platelet function. Platelets are extremely susceptible to platelet activation ex vivo during blood sampling, and this has constrained the usefulness of circulating biomarkers of platelet activation, such as circulating platelet aggregates and β-thromboglobulin or, indeed, of more contemporary approaches such as P selectin expression on platelets ex vivo. Studies of the platelet aggregation response ex vivo have also proven uninformative. Platelets activated by cigarette smoke in vivo might be less available for harvesting for such ex vivo studies. Indeed, platelet responses to adenosine 5´-diphosphate (ADP) were reportedly diminished ex vivo in smokers in one large-scale epidemiological study (Meade et al., 1985). By contrast, urinary excretion of the 2,3-dinor and 11-hydro metabolites of thromboxane have proven to be noninvasive indices of platelet activation in vivo (FitzGerald et al., 1983). Thus, phasic increases in excretion of these metabolites occur during the ischemic episodes of unstable angina and in patients undergoing therapeutic thrombolysis (Fitzgerald et al., 1988; Lewis et al., 1983), two situations in which the functional importance of thromboxane dependent platelet activation is implied by the therapeutic efficacy of aspirin (Intravenous streptokinase, 1987; Lewis et al., 1983).

Excretion of thromboxane metabolites is increased in apparently healthy individuals who smoke and decreases on short-term cessation (Nowak et al., 1987). This observation has been confirmed, both in a large population study and in a case-control study of monozygotic twins discordant for the smoking habit (Lassila et al., 1988; Wennmalm et al., 1991). Interestingly, excretion of the major metabolite of prostacyclin, 2,3-dinor-6-keto-$PGF_{1\alpha}$ (prostaglandin $F_{1\alpha}$) (Brash et al., 1983), is also elevated in chronic cigarette smokers. Prostacyclin, like thromboxane, is a product of cyclooxygenase (COX) catalyzed metabolism of arachidonic acid. While prostacyclin is the major product of COX in endothelial cells, thromboxane is the principal product of the same enzyme in platelets. Platelets express exclusively COX-1, and the suppression of thromboxane biosynthesis by low-dose aspirin (Nowak et al., 1987) implies that this is the major source of its formation in smokers (Lassila et al., 1988). However, COX-2 in the vessel wall or in infiltrating cells, such as monocyte or macrophages, represents a second theoretical source of thromboxane formation. Indeed, COX-2 seems the likely source of prostacyclin formation (Wennmalm et al., 1991; Brash et al., 1983; McAdam et al., 1999), although the effects of selective inhibitors of COX-2 on the formation of either eicosanoid in cigarette smokers remains to be reported. Although cigarette smoke reduces the capacity of endothelial cells to form prostacyclin in vitro (Reinders et al., 1986), the capacity of cells to form this and other eicosanoids greatly exceeds actual biosynthetic rates in vivo (FitzGerald et al., 1983). The increment in prostacyclin biosynthesis coincident with that of thromboxane is consistent with the capacity of thromboxane analogues to evoke its formation by endothelial cells in vitro (Nicholson et al., 1984) and its role as a homeostatic response to accelerated platelet vessel wall interactions in vivo (FitzGerald et al., 1984). Deletion of the prostacyclin receptor increases the response to thrombotic stimuli (Murata et al., 1997). While an abundant literature supports the use of these indices of platelet-vascular interactions, they could be usefully related to other, more functional surrogates of the cardiovascular response to smoking, such as impaired endothelial function (Raitakari et al., 2000) and noninvasive assessment of atherosclerosis progression, using ultrasonography or electron beam computerized tomography (EBCT) (O'Malley et al., 2000; Raggi et al., 2000). Such surrogates could represent an intermediate link between the noninvasively acquired biochemical indices and true functional outcomes of smoking, such as myocardial infarction and sudden death.

Another mechanism that has been widely implicated in cigarette-induced tissue injury is oxidative stress. Free-radical-catalyzed damage has been suggested to be of relevance to both atherogenesis and tumor progression and, as such, may well be relevant to the harm caused by cigarette smoking. Indeed, cigarette smoke contains abundant radicals, and smoking is known to cause activation of platelets and neutrophils, which may in turn increase radical generation (Pryor, 1997; Shinagawa et al., 1998; Yamaguchi et al., 2000). Until recently, the study of oxidative damage in vivo has been constrained by the specificity of available biomarkers. However, the discovery of isoprostanes (iPs), free-radical-catalyzed products of arachidonic acid (Lawson et al., 1999; Roberts and Morrow, 2000), has transformed this situation. These products are formed initially in situ in the phospholipid domain of cell membranes, where they may be immunolocalized or quantitated directly (Roberts and Morrow, 2000). Following phospholipase-dependent cleavage, iPs circulate and are excreted in urine. Urinary iPs reflect the oxidant stress attendant on inflammation (McAdam et al., 2000) and are increased in a dose-dependent manner in cigarette smokers (Reilly et al., 1996). Some evidence exists suggesting their elevation by secondhand smoke and in the children of smokers (Sinzinger et al., 2000). Elevated urinary iPs in smokers appear to fall rapidly on quitting (McAdam et al., 2000; Morrow et al., 1995; Reilly et al., 1996; Sinzinger et al., 2000). It is possible that iPs may be of functional relevance to smoking-induced cardiovascular injury besides serving as a marker of the process. Thus, individual iPs may act as incidental ligands both at membrane receptors for prostanoids (Kunapuli et al., 1998; Morrow et al., 1995) and at peroxisomal proliferator activated receptors (PPARs) (Audoly et al., 2000). Thus, certain iPs activate platelets via thromboxane receptors in smokers (McNamara et al., 1999) and may, in turn, lead to generation of thromboxane itself, creating a vicious cycle (Davi et al., 1999). In this setting, effective antioxidants might suppress both iP generation and thromboxane biosynthesis. Cigarette smokers are susceptible to depletion of endogenous vitamin C, and supplementation of smokers with vitamin C depresses elevated iPs (Valkonen and Kuusi, 2000). By contrast, their endogenous levels of vitamin E are replete, and exogenous vitamin E does not suppress elevated iPs in smokers (Reilly et al., 1996). Other noninvasive indices of lipid peroxidation, such as 4-hydroxynonenal (Benedetti et al., 1980), and of radical-induced protein (Pennathur et al., 1999) and DNA (Marnett, 2000) damage have begun to emerge in clinical investigation. No information is available on relative effects of these distinct targets for radical-induced damage or their relevance to smoking-induced cardiovascular injury.

In summary, the examples of urinary thromboxane, prostacyclin, and isoprostanes illustrate the potential value of developing and validating quantitative, noninvasive biochemical indices of the mechanisms of smoking-induced cardiovascular damage and tissue injury in general. Initially, the methodology involved (e.g., mass spectrometry) is often expensive and labor intensive. However, once the proof of principle for such approaches has been established, effort can be rationally devoted to the development of simpler, more widely applicable, but validated methodologies, such as enzyme immunoassays. The power of this approach, which can also be applied to animal models (Pratico et al., 1998; Heinecke, 1999), will be greatly enhanced by its integration with surrogate markers of functional impairment and, ultimately, with actual clinical events in the assessment of the cardiovascular risks of smoking.

Potential surrogate markers to be used in animal models developed to detect potentially adverse effects of smoking and of the use of potential reduced-exposure products (PREPs) on arteries throughout the body and in the lungs and heart include measurements of platelet activation, blood pressure and heart rate, and oxidative stress. Other surrogate markers that may be utilized in clinical studies include measurements of oxidative stress, measurements of platelet activation, elements of the coagulation system, lipids and lipoproteins, pulmonary function studies, studies of endothelial function, myocardial stress perfusion studies, and periodic evaluation by ultrasound of the carotid, abdominal aortic, and lower-extremity arteries.

CONCLUSIONS

Dose-Response Relationship

Highly informative information on the existence of a dose-response relationship between cigarette exposure and cardiovascular risk comes from many studies such as the CPS-2 (Thun et al., 1997) and Harvard Nurses' Health Studies (Kawachi et al., 1997b). In both instances, there is a relationship between the number of cigarettes smoked and the incidence of cardiovascular events. This is illustrated for the incidence of myocardial infarction (Table 13-3) and stroke (Table 13-4). In both cases, the most striking difference is between nonsmokers and individuals who smoke the least number of cigarettes recorded. The relationship becomes somewhat less pronounced as the number of cigarettes smoked per day increases. Interestingly, ex-smokers tend to occupy a space intermediate between nonsmokers and those with the lowest daily smoking frequency.

This dose-response curve prompts several considerations. First, there is no persuasive evidence of a threshold below which a cardiovascular risk does not exist. This observation affirms the primary objective of encouraging smokers to quit completely. Second, the shallow dose-response relationship, with the impression of a plateau, accords with similar observations relating the number of cigarettes smoked and measurements of systemic bioavailability, such as nicotine and cotinine. The same is true of the relationship with CO (Gori and Lynch, 1985). Although this may represent saturation kinetics of nicotine or CO, the most likely explanation is compensation for lower numbers of cigarettes smoked. Thus, the smoker titrates nicotine delivery toward a range of convergence that is reflected by the measurement of nicotine delivery, which in turn, may reflect dose-dependent convergence of the delivery of additional toxic, but unmeasured, constituents of cigarette smoke. The relative contribution of distinct constituents of cigarette smoke to smoking-related cardiovascular morbidity and mortality is unknown. In this regard, the available information is incomplete regarding any differences between the dose-response relationship of filter or low-tar versus high-tar or unfiltered cigarettes.

Feasibility of Harm Reduction in Therapy

There are no data that are directly informative on the issue of harm reduction. Thus, although a dose-response relationship exists for cardiovascular and cerebrovascular events, we do not know if reducing the number of cigarettes smoked results in a quantitative reduction in the risk of these events. However, the intuitively appealing prospect that this is indeed the case is supported by evidence from individuals who quit smoking. Thus, quitting results in a time-dependent reduction in the incidence of myocardial infarction (Table 13-2) and stroke. The latter is most nicely illustrated by data relating to subarachnoid hemorrhage, which was significantly elevated in women smokers in the Nurses' Health Study (Figure 13-1). In addition to evidence from such unequivocal clinical events,

TABLE 13-3 Number of Cigarettes Smoked Daily and Age-Adjusted and Multivariate RRs of Fatal Coronary Heart Disease and Nonfatal Myocardial Infarction, Compared with Never Smokers[a]

Event	Never Smoker	Former Smoker	Cigarettes Smoked per Day Among Current Smokers					
			1-4	8-14	15-24	25-34	35-44	≥45
Fatal coronary heart disease cases	49	53	4	18	53	28	14	4
RR[b]	1.00	1.63	1.87	2.78	4.29	5.36	5.56	10.00
		(1.11-2.40)	(0.69-5.09)	(1.66-4.67)	(3.00-6.15)	(3.53-8.14)	(3.26-9.50)	(4.35-22.97)
RR[c]	1.00	1.62	←――2.85――→		4.85	6.96	←――7.84――→	
		(1.09-2.40)	(1.53-5,32)		(3.01-7.81)	(3.90-12.43)	(3.71-16.57)	
Nonfatal myocardial infarction cases	166	161	15	56	189	95	54	7
RR[b]	1.00	1.47	1.97	2.46	4.21	4.87	5.58	4.64
		(1.19-1.83)	(1.17-3.30)	(1.83-3.29)	(3.48-5.11)	(3.87-6.13)	(4.24-7.35)	(2.34-9.21)
RR[c]	1.00	1.44	←――2.45――→		4.77	5.21	←――5.32――→	
		(1.16-1.79)	(1.69-3.56)		(3.64—6.26)	(3.73-7.28)	(3.61-7.86)	
Total coronary heart disease cases	215	214	19	74	242	123	68	11
RR[b]	1.00	1.51	1.94	2.53	4.22	4.97	5.57	5.74
		(1.25-1.82)	(1.23-3.08)	(1.96-3.26)	(3.56-5.00)	(4.06-6.08)	(4.36-7.11)	(3.36-9.81)
RR[c]	1.00	1.48	←――2.53――→		4.79	5.49	←――5.49――→	
		(1.22-1.79)	(1.84-3.50)		(3.78-6.08)	(4.10-7.35)	(3.87-7.77)	

NOTE: RR=relative risk.

[a]Daily number smoked was missing for four cases, including two cases of fatal coronary heart disease and two cases of nonfatal myocardial infarction.

[b]Age-adjusted RR.

[c]Adjusted for age in five-year intervals, during follow-up period (1976-1978, 1978-1980, 1980-1982, 1982-1984, 1984-1986, or 1986-1988), history of hypertension, diabetes, high cholesterol levels, body mass index, past use of oral contraceptives, menopausal status, postmenopausal estrogen therapy, and age at starting smoking.

SOURCE: Kawachi et al., 1994. Copyright (1994), American Medical Association.

TABLE 13-4 Age-Adjusted RRs of Stroke (Fatal and Nonfatal Combined), by Daily Number of Cigarettes Consumed Among Current Smokers[a]

Event	Never Smoker	Former Smoker	Current Smoker	Cigarettes Smoked per Day Among Current Smokers			
				1-14	15-24	25-34	≥35
Total stroke cases							
RR[b]	126	114	208	40	92	38	34
	1.00	1.34	2.58	1.79	2.84	2.70	4.23
		(1.04-1.73)	(2.08-3.19)	(1.26-2.54)	(2.19-3.67)	(1.91-3.84)	(2.99-6.00)
RR[c]	1.00	1.35	2.73	2.02	3.34	3.08	4.48
		(0.98-1.85)	(2.18-3.41)	(1.29-3.14)	(2.38-4.70)	(1.94-4.87)	(2.78-7.23)
Subarachnoid hemorrhage cases							
RR[b]	19	25	64	13	21	17	11
	1.00	2.01	4.96	3.68	4.05	7.31	8.28
		(1.12-3.61)	(3.13-7.87)	(1.91-7.11)	(2.30-7.14)	(4.15-12.85)	(4.45-15.42)
RR[c]	1.00	2.26	4.85	4.28	4.02	7.95	10.22
		(1.16-4.42)	(2.90-8.11)	(1.88-9.77)	(1.90-8.54)	(3.50-18.07)	(4.03-25.94)
Ischemic stroke cases							
RR[b]	85	70	120	23	58	19	18
	1.00	1.20	2.25	1.54	2.69	2.06	3.43
		(0.88-1.65)	(1.72-2.95)	(0.98-2.44)	(1.95-3.72)	(1.27-3.36)	(2.13-5.51)
RR[c]	1.00	1.27	2.53	1.83	3.57	2.73	3.97
		(0.85-1.89)	(1.91-3.35)	(1.04-3.23)	(2.36-5.42)	(1.49-5.03)	(2.09-7.53)
Cerebral hemorrhage cases							
RR[b]	19	16	18	4	10	<—— 4[d] ——>	
	1.00	1.27	1.46	1.18	2.01	1.18	
		(0.66-2.44)	(0.77-2.78)	(0.40-3.46)	(0.94-4.28)	(0.41-3.46)	
RR[c]	1.00	1.24	1.24	1.68	2.53	1.41	
		(0.64-2.42)	(0.64-2.42)	(0.34-5.28)	(0.71-6.05)	(0.39-5.05)	

NOTE: RR=relative risk.

[a]Number unknown for four cases, including two subarachnoid hemorrhage and two ischemic stroke.
[b]Age-adjusted RR.
[c]Adjusted for age in five-year intervals, during follow-up period (1976-1978, 1978-1980, 1980-1982, 1982-1984, 1984-1986, or 1986-1988), history of hypertension, diabetes, high cholesterol levels, body mass index, past use of oral contraceptives, postmenopausal estrogen therapy, and age at starting smoking.
[d]These two categories were combined due to small numbers.

SOURCE: Kawachi et al., 1993. Copyright (1993), American Medical Association.

there is evidence that the increase in biomarkers of oxidant stress, platelet activation, and inflammation (Benowitz et al., 1993), all of potential mechanistic relevance to tobacco-related cardiovascular injury, rapidly falls toward the normal range on quitting cigarettes. The offset kinetics of more functional surrogates, such as endothelial dysfunction, remain to be determined in smokers. In summary, the data from quitters encourage the prospect that a graded reduction in cardiovascular risk and in biomarkers of this risk may accompany a reduction in the number of cigarettes smoked in pursuit of a harm reduction strategy. It is possible, indeed likely, that individuals and perhaps populations, differ in their susceptibility to tobacco-induced cardiovascular risk and, indeed, in their potential benefit from a harm reduction strategy. Data from quitting studies indicate a considerable variance in the rate of offset of risk, which declines with time. No data are available to address such issues across ethnic groups or gender. Acquisition of such information and research on the environmental and genetic factors that condition interindividual variability in exposure-risk relationships are necessary.

Utility in a Preclinical Setting

Studies in cell culture and model systems can afford much needed information on tobacco-related cardiovascular risk. These might include a profiling of gene expression and translation in cardiovascular tissues in response to cigarette smoke, constituents of smoke, and potential risk reduction substituents. These might identify proteins of potential functional relevance to the transduction of cardiovascular risk. Such studies might be coupled with gene inactivation and overexpression studies to address the role of these proteins in vivo. Similarly, studies of exposure to cigarette smoke or to discrete constituents of smoke might be deployed to investigate effects on atherosclerosis progression, susceptibility to vascular injury, thrombotic stimuli, graft rejection, cardiovascular development, or endothelial dysfunction in model systems such as mice. Studies of cardiovascular genomics and ultimately proteomics can also be extended to model systems to investigate gene expression and translation in response to exposure to tobacco-related products in vivo. These observations may, in turn, be related to the pattern of gene expression and translation in cardiovascular tissues obtained from cigarette smokers.

Biomarkers of Tobacco-Related Disease

The predominant mechanisms by which cigarette smoking induces cardiovascular injury is unknown. However, small studies in smokers of potentially relevant biomarkers of platelet and vascular activation, lipid peroxidation, and inflammation afford evidence of a dose-response relationship and a decline on quitting. There is even evidence of a signal in individuals exposed to ETS in the case of some of these markers. More mechanism-based clinical studies are required to confirm and expand these findings. Where possible, these should be related to surrogate measurements of cardiovascular function, such as hemodynamics, flow-mediated endothelial function and estimates of plaque progression by ultrasound or EBCT. Furthermore, biomarker studies can usefully be integrated into many studies in model systems (see above) as well as studies of clinical outcome (discussed below) to afford their ultimate validation.

Clinical Assessment of Tobacco-Related Disease

The time course of offset of myocardial infarction and stroke in people who stop smoking suggests that cardiovascular disease represents a tractable scenario in which one might evaluate harm reduction strategies. Clearly, assessment of the impact of such events can occur in a fairly reasonable time compared to that which is possible for cancer or lung disease.

RESEARCH AGENDA

1. Determine whether physiological and biochemical measures that are altered by tobacco products may be used as indicators of disease risk and as a means to distinguish various forms of tobacco use in a broad range of exposures (doses), including:
   - activation of platelet function;
   - endothelial function;
   - endothelial thickening and plaque formation;
   - blood pressure and heart rate alterations;
   - vascular aneurysm formation;
   - vascular thrombosis (arterial and venous);
   - cardiac electrophysiology, including ventricular and atrial ectopy, heart rate variability, and variations in conduction times, especially QT intervals; and
   - other risk factors for cardiovascular disease including lipid and carbohydrate abnormalities, homocysteine, folic acid and antioxidant vitamins, markers of inflammation, and immune function.

Based on current evidence, oxidative stress seems likely to play an important role in mediating the cardiovascular effects of smoking. Determining reliable biomarkers of oxidative stress and the interaction of unstable oxygen intermediates with lipids, proteins, and DNA should receive high priority. Since inflammation is a proximate cause of oxidative stress, further work on biomarkers of inflammation such as C-reactive protein, cytokines, and acute-phase proteins and the effects of tobacco exposure appears to be an important research direction. Similarly, further assessment of the effects of various tobacco products on biomarkers of hemostatic activation and platelet function should be fruitful in understanding and predicting atherogenesis.

2. Develop studies to determine whether new biomarkers, such as products of lipid peroxidation and corresponding adducts to DNA, and products of arachidonic acid metabolism can serve as intermediate indicators (risk factors) of cardiovascular disease risk related to different tobacco products, and tobacco-related PREPs in particular. Through well-designed clinical trials, determine the ability of antioxidant or related therapies to counter the tobacco-related perturbations due to oxidative stress that are likely to be caused by both conventional and new tobacco products.

3. Develop long-term observational studies in individuals exposed to conventional and modified tobacco products primarily *and* secondarily (i.e., passive smoking), better to identify actual risks for cardiovascular disease, including dose-response characteristics. Included should be evaluations of the variables described in items 1 and 2, subsequent cardiovascular disease development, and potential differences in different tobacco or smoking products.

4. Further apply modern in vitro and other model systems, and ex vivo in humans, to assist in elucidating the comparative toxicology or adverse effects of cigarette smoke and PREPs. Such studies might include the use of gene array and proteomic technologies to study differential gene expression and translation; the use of chronic exposures to investigate effects on atherosclerosis progression, response to thrombotic and arrhythmogenic stimuli, and cardiovascular disease development; and studies of the effects of cigarette smoke and its constituents on the function and integrity of cardiovascular cells in vitro, including platelets, leukocytes, endothelial cells, vascular smooth muscle cells, and myocytes.

5. Apply modern genomic and proteomic analyses to elucidate the array of genes of potential relevance to cardiovascular hazards differentially induced by exposure to conventional tobacco products and PREPs. Utilize such techniques as transgenic and knockout model systems to elucidate the relevance of single genes to vascular injury induced by cigarette smoke and its discrete constituents, including nicotine. Assess the possibility of allelic varia-

tion in such genes and its relevance to smoking-induced cardiovascular disease risk.

6. Continue to elucidate further aspects of the vascular biology of nicotine in view of its effects on flow-mediated vasodilation and the particular susceptibility of young smokers to prolonged exposure to nicotine products. Further careful evaluation of the dose-related effects of nicotine on platelet, leukocyte, hemostatic, and endothelial function and subsequent vascular injury, thrombogenesis, and atherogenesis is indicated.

REFERENCES

Abdel-Rahman SZ, El-Zein RA. 2000. The 399Gln polymorphism in the DNA repair gene XRCC1 modulates the genotoxic response induced in human lymphocytes by the tobacco-specific nitrosamine NNK. *Cancer Lett* 159(1):63-71.

Adams MR, Jessup W, Celermajer DS. 1997. Cigarette smoking is associated with increased human monocyte adhesion to endothelial cells: reversibility with oral L-arginine but not vitamin C. *J Am Coll Cardiol* 29 (491).

Allred EN, Bleecker ER, Chaitman BR, et al. 1989. Short-term effects of carbon monoxide exposure on the exercise performance of subjects with coronary artery disease. *N Engl J Med* 321:1426.

Anthonisen NR, Connett JE, Kiley JP, et al. 1994. Effects of smoking intervention and the use of an inhaled anticholinergic bronchodilator on the rate of decline of FEV1. The Lung Health Study. *JAMA* 272(19):1497-1505.

Arinami T, Ohtsuki T, Yamakawa-Kobayashi K, et al. 1999. A synergistic effect of serotonin transporter gene polymorphism and smoking in association with CHD. *Thromb Haemost* 81(6):853-856.

Aronow WS. 1981. Aggravation of angina pectoris by two percent carboxyhemoglobin. *Am Heart J* 101:154.

Aronow WS, Stemmer EA, Zweig S. 1979. Carbon monoxide and ventricular fibrillation threshold in normal dogs. *Arch Environ Health* 34:184.

Audoly LP, Rocca B, Fabre JE, et al. 2000. Cardiovascular responses to the isoprostanes iPF(2alpha)-III and iPE(2)-III are mediated via the thromboxane A(2) receptor in vivo. *Circulation* 101(24):2833-2840.

Benedetti A, Comporti M, Esterbauer H. 1980. Identification of 4-hydroxynonenal as a cytotoxic product originating from the peroxidation of liver microsomal lipids. *Biochim Biophys Acta* 620(2):281-296.

Benowitz NL, Fitzgerald GA, Wilson M, Zhang Q. 1993. Nicotine effects on eicosanoid formation and hemostatic function: comparison of transdermal nicotine and cigarette smoking. *J Am Coll Cardiol* 22(4):1159-1167.

Blake KV, Gurrin LC, Evans SF, et al. 2000. Maternal cigarette smoking during pregnancy, low birth weight and subsequent blood pressure in early childhood. *Early Hum Dev* 57(2):137-147.

Bostrom L, Linder LE, Bergstrom J. 1999. Smoking and cervicular fluid levels of IL-6 and TNF-alpha in periodontal disease. *J Clin Periodontol* 26(6):352-357.

Brash AR, Jackson EK, Saggese CA, Lawson JA, Oates JA, FitzGerald GA. 1983. Metabolic disposition of prostacyclin in humans. *J Pharmacol Exp Ther* 226(1):78-87.

Brown WV. 2000. Cholesterol lowering in atherosclerosis. *American Journal of Cardiology* 86(4B):29H-32H.

Burns DM. 2000. Cigarette smoking among the elderly: disease consequences and the benefits of cessation. *Am J Health Promot* 14(6):357-361.

Campisi R, Czernin J, Schoder H, et al. 1998. Effects of long-term smoking on myocardial blood flow, coronary vasomotion, and vasodilator capacity. *Circulation* 98(119).

Campisi R, Czernin J, Schoder H, et al. 1999. L-Arginine normalizes coronary vasomotion in long-term smokers. *Circulation* 99(491).

Chien KL, Lee YT, Sung FC, Su TC, Hsu HC, Lin RS. 1999. Lipoprotein (a) level in the population in Taiwan: relationship to sociodemographic and atherosclerotic risk factors. *Atherosclerosis* 143(2):267-273.

Cook DG, Shaper AG, Pollock SJ, et al. 1986. Giving up smoking and the risk of heart attacks. *Lancet* 2(1376).

Craig WY, Paolmaki GE, Haddow JE. 1989. Cigarette smoking and serum lipid and lipoprotein concentrations: an analysis of published data. *BMJ* 298(784).

Cryer PE, Haymond MW, Santiago JV, et al. 1976. Norepinephrine and epinephrine release and adrenergic mediation of smoking-associated hemodynamic and metabolic events. *New England Journal of Medicine* 295:573.

Czernin J, Sun K, Brunken R, et al. 1995. Effect of acute and long-term smoking on myocardial blood flow and flow reserve. *Circulation* 91:2891.

Davi G, Ciabattoni G, Consoli A, et al. 1999. In vivo formation of 8-iso-prostaglandin f2alpha and platelet activation in diabetes mellitus: effects of improved metabolic control and vitamin E supplementation. *Circulation* 99(2):224-229.

Davies MJ, Thomas A. 1984. Thrombosis and acute coronary artery lesions in sudden cardiac ischemic death. *New England Journal of Medicine* 310:1137.

Dawber TR. 1980. *The Framingham Study: The Epidemiology of Atherosclerotic Disease.* Chapter 8. Cambridge, MA: Harvard University Press.

DeMatt MP, Pietersma A, Kofflard M, et al. 1996. Association of plasma fibrinogen levels with coronary artery disease, smoking and inflammatory markers. *Atherosclerosis* 121:185.

Donati MB, Zito F, Castelnuovo AD, Iacoviello L. 2000. Genes, coagulation and cardiovascular risk. *J Hum Hypertens* 14(6):369-372.

Duthie GG, Arthus JR, Beattie JA, et al. 1993. Cigarette smoking antioxidants, lipid peroxidation, and coronary heart disease. *Ann N Y Acad Sci* 686(120).

Eliasson B, Attvall S, Taskinen MR, Smith U. 1997. Smoking cessation improves insulin sensitivity in healthy middle-aged men. *Eur J Clin Invest* 27(5):450-456.

Fitzgerald DJ, Catella F, Roy L, FitzGerald GA. 1988. Marked platelet activation in vivo after intravenous streptokinase in patients with acute myocardial infarction. *Circulation* 77(1):142-150.

FitzGerald GA, Oates JA, Novak J. 1988. Cigarette smoking and hemostatic function. *American Heart Journal* 115(267).

FitzGerald GA, Pedersen AK, Patrono C. 1983. Analysis of prostacyclin and thromboxane biosynthesis in cardiovascular disease. *Circulation* 67(6):1174-1177.

FitzGerald GA, Smith B, Pedersen AK, Brash AR. 1984. Increased prostacyclin biosynthesis in patients with severe atherosclerosis and platelet activation. *N Engl J Med* 310(17):1065-1068.

Fuster V, Chesebro JH, Frye RL, Elveback LR. 1981. Platelet survival and the development of coronary artery disease in the young adult: effects of cigarette smoking, strong family history and medical therapy. *Circulation* 63(3):546-551.

Gerhardt U, Vorneweg P, Riedasch M, Hohage H. 1999. Acute and persistent effects of smoking on the baroreceptor function. *J Auton Pharmacol* 19(2):105-108.

Glisic S, Savic I, Alavantic D. 1995. Apolipoprotein B gene DNA polymorphisms (EcoRI and MspI) and serum lipid levels in the Serbian healthy population: interaction of rare alleles and smoking and cholesterol levels. *Genet Epidemiol* 12(5):499-508.

Gori GB, Lynch CJ. 1985. Analytical cigarette yields as predictors of smoke bioavailability. *Regul Toxicol Pharmacol* 5(3):314-326.

Green F, Hamsten A, Blomback M, Humphries S. 1993. The role of beta-fibrinogen genotype in determining plasma fibrinogen levels in young survivors of myocardial infarction and healthy controls from Sweden. *Thromb Haemost* 70(6):915-920.

Griffin BA, Freeman DJ, Tait GW, et al. 1994. Role of plasma triglyceride in the regulation of plasma low density lipoprotein (LDL) subfractions: relative contribution of small, dense LDL to coronary heart disease risk. *Atherosclerosis* 106(2):241-253.

Hee J, Callais F, Momas I, Laurent AM, Min S, Molinier P, Chastagnier M, Claude JR, Festy B. 1995. Smokers' behaviour and exposure according to cigarette yield and smoking experience. *Pharmacology, Biochemistry and Behavior* 52(1):195-203.

Heinecke JW. 1999. Mass spectrometric quantification of amino acid oxidation products in proteins: insights into pathways that promote LDL oxidation in the human artery wall. *FASEB J* 13(10):1113-1120.

Heitzer T, Just H, Munzel T. 1996a. Antioxidant vitamin C improves endothelial dysfunction in chronic smokers. *Circulation* 94(6).

Heitzer T, Yia-Herttuala S, Luoma J, et al. 1996b. Cigarette smoking potentiates endothelial dysfunction of forearm resistance vessels in patients with hypercholesterolemia. Role of oxidized LDL. *Circulation* 93(1346).

Holm J, Hillarp A, Zoller B, Erhardt L, Berntorp E, Dahlback B. 1999. Factor V Q506 (resistance to activated protein C) and prognosis after acute coronary syndrome. *Thromb Haemost* 81(6):857-860.

Humphries SE, Henry JA, Montgomery HE. 1999. Gene-environment interaction in the determination of levels of haemostatic variables involved in thrombosis and fibrinolysis. *Blood Coagul Fibrinolysis* 10 (Suppl 1):S17-21.

Intravenous streptokinase given within 0-4 hours of onset of myocardial infarction reduced mortality in ISIS-2. 1987. *Lancet* 1(8531):502.

Istvan JA, Lee WW, Buist AS, Connett JE. 1999. Relation of salivary cotinine to blood pressure in middle-aged cigarette smokers. *Am Heart J* 137(5):928-931.

Kamboh MI, McGarvey ST, Aston CE, Ferrell RE, Bausserman L. 2000. Plasma lipoprotein(a) distribution and its correlates among Samoans. *Hum Biol* 72(2):321-336.

Kannel WB, McGee DL, Castelli WP. 1984. Latest perspectives on cigarette smoking and cardiovascular disease. The Framingham Study. *Journal of Cardiac Rehabilitation* 4:267.

Kawachi I, Colditz GA, Speizer FE, et al. 1997a. A prospective study of passive smoking and coronary heart disease. *Circulation* 95(2374).

Kawachi I, Colditz GA, Stampfer MJ, Willett WC, Manson JE, Rosner B, Speizer FE, Hennekens CH. 1994. Smoking cessation and time course of decreased risks of coronary heart disease in middle-aged women. *Arch Intern Med* 154(2):169-175.

Kawachi I, Colditz GA, Stampfer MJ, Willett WC, Manson JE, Rosner B, Speizer FE, Hennekens CH. 1993. Smoking cessation and decreased risk of stroke in women. *JAMA* 269(2):232-236.

Kawachi I, Colditz GA, Stampfer MJ, et al. 1997b. Smoking cessation and decreased risk of total mortality, stroke, and coronary heart disease incidence in women; a prospective cohort study. National Cancer Institute. *Changes in Cigarette-Related Disease Risks and Their Implication for Prevention and Control. NCI Smoking and Tobacco Control Monograph 8.* Washington, DC: NCI. Pp. 531-565.

Koenig W, Sund M, Frohlich M, et al. 1999. C-Reactive protein, a sensitive marker of inflammation, predicts future risk of coronary heart disease in initially healthy middle-aged men: results from the MONICA (Monitoring Trends and Determinants in Cardiovascular Disease) Augsburg Cohort Study, 1984 to 1992. *Circulation* 99(2):237-242.

Kugiyama K, Yasue H, Ohgushi M, et. al. 1996. Deficiency in nitric oxide bioactivity in epicardial coronary arteries of cigarette smokers. *J Am Coll Cardiol* 8(1161).

Kunapuli P, Lawson JA, Rokach JA, Meinkoth JL, FitzGerald GA. 1998. Prostaglandin F2alpha (PGF2alpha) and the isoprostane, 8, 12-iso-isoprostane F2alpha-III, induce cardiomyocyte hypertrophy. Differential activation of downstream signaling pathways. *J Biol Chem* 273(35):22442-22452.

Lassila R, Seyberth HW, Haapanen A, Schweer H, Koskenvuo M, Laustiola KE. 1988. Vasoactive and atherogenic effects of cigarette smoking: a study of monozygotic twins discordant for smoking. *BMJ* 297(6654):955-957.

Lawson JA, Rokach J, FitzGerald GA. 1999. Isoprostanes: formation, analysis and use as indices of lipid peroxidation in vivo. *J Biol Chem* 274(35):24441-24444.

Lewis HD, Davis JW, Archibald DG, et al. 1983. Protective effects of aspirin against acute myocardial infarction and death in men with unstable angina. Results of a Veterans Administration Cooperative Study. *N Engl J Med* 309(7):396-403.

Li R, Boerwinkle E, Olshan AF, et al. 2000. Glutathione S-transferase genotype as a susceptibility factor in smoking-related coronary heart disease. *Atherosclerosis* 149(2):451-462.

Marnett LJ. 2000. Oxyradicals and DNA damage. *Carcinogenesis* 21(3):361-370.

McAdam BF, Catella-Lawson F, Mardini IA, Kapoor S, Lawson JA, FitzGerald GA. 1999. Systemic biosynthesis of prostacyclin by cyclooxygenase (COX)-2: the human pharmacology of a selective inhibitor of COX-2. *Proc Natl Acad Sci U S A* 96(1):272-277.

McAdam BF, Mardini IA, Habib A, et al. 2000. Effect of regulated expression of human cyclooxygenase isoforms on eicosanoid and isoeicosanoid production in inflammation. *J Clin Invest* 105(10):1473-1482.

McNamara P, Lawson JA, Rokach J, FitzGerald GA. 1999. F2 and E2 isoprostanes are ligands for peroxisome prolifer-activated receptors. *FASEB J* 13:A1549.

McGinnis JM, Foege WH. 1999. Mortality and morbidity attributable to use of addictive substances in the United States. *Proceedings of the Association of American Physicians* 111(2):109-118.

Meade TW, Vickers MV, Thompson SG, Stirling Y, Haines AP, Miller GJ. 1985. Epidemiological characteristics of platelet aggregability. *Br Med J (Clin Res Ed)* 290(6466):428-432.

Mikhailidis DP, Papadakis JA, Ganotakis ES. 1998. Smoking, diabetes and hyperlipidaemia. *J R Soc Health* 118(2):91-93.

Mikkelsen KL, Winberg N, Hoegholm A, et al. 1997. Smoking related to 24-h ambulatory blood pressure and heart rate: a study in 352 normotensive Danish subjects. *Am J Hypertens* 10(5 Pt 1):483-491.

Minami J, Ishimitsu T, Matsuoka H. 1999. Effects of smoking cessation on blood pressure and heart rate variability in habitual smokers. *Hypertension* 33(1 Pt 2):586-590.

Moauad J, Fernandez F, Herbert JL, et al. 1986. Cigarette smoking during coronary angiography: diffuse or focal narrowing (spasm) of the coronary arteries in 13 patients with angina at rest and normal coronary angiograms. *Cathet Cardiovasc Diagn* 12:366.

Moliterno DJ, Willard JE, Lange RA, et al. 1994. Coronary-artery vasoconstriction induced by cocaine, cigarette smoking, or both. *New England Journal of Medicine* 330:454.

Mormando RM. 2000. Lipid levels. Applying the second National Cholesterol Educaton Program report to geriatric medicine. *Geriatrics* 55(8):48-53.

Morrow JD, Frei B, Longmire AW, et al. 1995. Increase in circulating products of lipid peroxidation (F2-isoprostanes) in smokers. Smoking as a cause of oxidative damage. *N Engl J Med* 332(18):1198-1203.

Mundal R, Kjeldsen SE, Sandvik L, Erikssen G, Thaulow E, Erikssen J. 1997. Predictors of 7-year changes in exercise blood pressure: effects of smoking, physical fitness and pulmonary function. *J Hypertens* 15(3):245-249.

Murata T, Ushikubi F, Matsuoka T, et al. 1997. Altered pain perception and inflammatory response in mice lacking prostacyclin receptor. *Nature* 388(6643):678-682.

Nardo CJ, Chambless LE, Light KC, et al. 1999. Descriptive epidemiology of blood pressure response to change in body position. The ARIC study. *Hypertension* 33(5):1123-1129.

Newby DE, Wright RA, Labinjoh C, et al. 1999. Endothelial dysfunction, impaired endogenous fibrinolysis, and cigarette smoking: a mechanism for arterial thrombosis and myocardial infarction. *Circulation* 99:1411.

Nicholson NS, Smith SL, Fuller GC. 1984. Effect of the stable endoperoxide analog U-46619 on prostacyclin production and cyclic AMP levels in bovine endothelial cells. *Thromb Res* 35(2):183-192.

Nicod P, Rehr R, Winniford MD, et al. 1984. Acute systemic and coronary hemodynamic and serologic responses to cigarette smoking in long-term smokers with atherosclerotic coronary artery disease. *J Am Coll Cardiol* 4:964.

Nowak J, Murray JJ, Oates JA, FitzGerald GA. 1987. Biochemical evidence of a chronic abnormality in platelet and vascular function in healthy individuals who smoke cigarettes. *Circulation* 76(1):6-14.

Nygard O, Refsum H, Ueland PM, Vollset SE. 1998. Major lifestyle determinants of plasma total homocysteine distribution: the Hordaland Homocysteine Study. *Am J Clin Nutr* 67(2):263-270.

Nyboe J, Jensen G, Appleyard M, Schnohr P. 1991. Smoking and the risk of first acute myocardial infarction. *American Heart Journal* 122:438-447.

O'Malley PG, Taylor AJ, Jackson JL, Doherty TM, Detrano RC. 2000. Prognostic value of coronary electron-beam computed tomography for coronary heart disease events in asymptomatic populations. *Am J Cardiol* 85(8):945-948.

Oates JA. 1989. Aspirin in the prevention of catastrophes of the coronary circulation. Samuelsson B, ed. *Advances in Prostaglandin, Thromboxane and Leukotriene Research.* New York: Raven. Pp. 57. Chapter 8.

Okasha M, McCarron P, McEwen J, Davey Smith G. 2000. Determinants of adolescent blood pressure: findings from the Glasgow University student cohort. *J Hum Hypertens* 14(2):117-124.

PDAY (Pathobiological Determinants of Atherosclerosis in Youth Research Group). 1990. The pathobiological determinants of atherosclerosis in Youth Research Group: relationship of atherosclerosis in young men to serum lipoprotein cholesterol concentrations and smoking. *JAMA* 265:3018.

Penn A, Snyder CA. 1996. 1,3 Butadiene, a vapor phase component of environmental tobacco smoke, accelerates arteriosclerotic plaque development. *Circulation* 93:552.

Pennathur S, Jackson-Lewis V, Przedborski S, Heinecke JW. 1999. Mass spectrometric quantification of 3-nitrotyrosine, ortho-tyrosine, and o,o'-dityrosine in brain tissue of 1-methyl-4-phenyl-1,2,3, 6-tetrahydropyridine-treated mice, a model of oxidative stress in Parkinson's disease. *J Biol Chem* 274(49):34621-34628.

Pratico D, Tangirala RK, Rader DJ, Rokach J, FitzGerald GA. 1998. Vitamin E suppresses isoprostane generation in vivo and reduces atherosclerosis in ApoE-deficient mice. *Nat Med* 4(10):1189-1192.

Pryor WA. 1997. Cigarette smoke radicals and the role of free radicals in chemical carcinogenicity. *Environ Health Perspect* 105 Suppl 4(1):875-882.

Quillen JE, Rossen JD, Oskarsson JJ, et al. 1993. Acute effect of cigarette smoking on the coronary circulation: constriction of epicardial and resistance vessels. *J Am Coll Cardiol* 22:642.

Raggi P, Callister TQ, Cooil B, et al. 2000. Identification of patients at increased risk of first unheralded acute myocardial infarction by electron-beam computed tomography. *Circulation* 101(8):850-855.

Raitakari OT, Adams MR, McCredie RJ, Griffiths KA, Stocker R, Celermajer DS. 2000. Oral vitamin C and endothelial function in smokers: short-term improvement, but no sustained beneficial effect. *J Am Coll Cardiol* 35(6):1616-1621.

Reed DM, MacLean CJ, Hayashi T. 1987. Predictors of atherosclerosis in the Honolulu Heart Program: I. Biologic, dietary and lifestyle characteristics. *American Journal of Epidemiology* 126:214.

Reilly M, Delanty N, Lawson JA, FitzGerald GA. 1996. Modulation of oxidant stress in vivo in chronic cigarette smokers. *Circulation* 94(1):19-25.

Reinders JH, Brinkman HJ, van Mourik JA, de Groot PG. 1986. Cigarette smoke impairs endothelial cell prostacyclin production. *Arteriosclerosis* 6(1):15-23.

Reis RP, Azinheira J, Reis HP, Pina JE, Correia JM, Luis AS. 2000. Influence of smoking on homocysteinemia at baseline and after methionine load. *Rev Port Cardiol* 19(4):471-474.

Ridker PM, Rifai N, Stampfer MJ, Hennekens CH. 2000. Plasma concentration of interleukin-6 and the risk of future myocardial infarction among apparently healthy men. *Circulation* 101(15):1767-1772.

Roberts LJ, Morrow JD. 2000. Measurement of F(2)-isoprostanes as an index of oxidative stress in vivo. *Free Radic Biol Med* 28(4):505-513.

Rosenberg L, Kaufman DW, Helrich SP, et. al. 1985. The risk of myocardial infarction after quitting smoking in men under 55 years of age. *New England Journal of Medicine* 313:1511.

Rosenberg L, Palmer JR, Shapiro S. 1990. Declining risk of myocardial infarction among women who stopped smoking. *New England Journal of Medicine* 322:213.

Saw SM. 1999. Homocysteine and atherosclerotic disease: the epidemiologic evidence. *Ann Acad Med Singapore* 28(4):565-568.

Seman LJ, DeLuca C, Jenner JL, et al. 1999. Lipoprotein(a)-cholesterol and coronary heart disease in the Framingham Heart Study. *Clin Chem* 45(7):1039-1046.

Sen-Banerjee S, Siles X, Campos H. 2000. Tobacco smoking modifies association between Gln-Arg192 polymorphism of human paraoxonase gene and risk of myocardial infarction. *Arterioscler Thromb Vasc Biol* 20(9):2120-2126.

Shinagawa K, Tokimoto T, Shirane K. 1998. Spin trapping of nitric oxide in aqueous solutions of cigarette smoke. *Biochem Biophys Res Commun* 253(1):99-103.

Shinozaki K, Hattori Y, Suzuki M, et al. 1997. Insulin resistance as an independent risk factor for carotid artery wall intima media thickening in vasospastic angina. *Arterioscler Thromb Vasc Biol* 17(11):3302-3310.

Sinzinger H, Granegger S, Oguogho A. 2000. Passive smoking in children is associated with increased isoprostane 8-EPI-PGF. *Atherosclerosis* 151:257.

Steenland K, Thun M, Lally C, Heath C. 1996. Environmental tobacco smoke and coronary heart disease in the American Cancer Society CPS-II cohort. *Circulation* 94(4):622-628.

Stellman SD, Garfinkel L. 1986. Smoking habits and tar levels in a new American Cancer Society prospective study of 1.2 million men and women. *Journal of the National Cancer Institute* 76(6):1057-1063.

Tahtinen TM, Vanhala MJ, Oikarinen JA, Keinanen-Kiukaanniemi SM. 1998. Effect of smoking on the prevalence of insulin resistance-associated cardiovascular risk factors among Finnish men in military service. *J Cardiovasc Risk* 5(5-6):319-323.

Takizawa H, Tanaka M, Takami K, et al. 2000. Increased expression of inflammatory mediators in small-airway epithelium from tobacco smokers. *Am J Physiol Lung Cell Mol Physiol* 278(5):L906-913.

Targher G, Alberiche M, Zenere MB, Bonadonna RC, Muggeo M, Bonora E. 1997. Cigarette smoking and insulin resistance in patients with noninsulin-dependent diabetes mellitus. *J Clin Endocrinol Metab* 82(11):3619-3624.

Taylor AE, Johnson DC, Kazemi H. 1992. Environmental tobacco smoke and cardiovascular disease. A position paper from the Council on Cardiopulmonary and critical Care, American Heart Association. *Circulation* 86(2):699-702.

The Pooling Project Research Group. 1978. Relationship of blood pressure, serum cholesterol, smoking habit, relative weight and ECG abnormalities to incidence of major coronary events: final report of the pooling project. *Journal of Chronic Disease* 31(201).

The sixth report of the Joint National Committee on prevention, detection, evaluation, and treatment of high blood pressure. 1997. *Archives of Internal Medicine* 157(21):2413-2446.

Thom SR, Xu YA, Ischiropoulos H. 1997. Vascular endothelial cells generate peroxynirite in response to carbon monoxide exposure. *Chem Res Toxicol* 10:1023.

Thun MJ, Myers DG, Day-Lally C, et al. 1997. Age and exposure-response relationships between cigarette smoking and premature death in cancer prevention study II. National Cancer Institute. *Changes in Cigarette-Related Disease Risks and Their Implication for Prevention and Control. NCI Smoking and Tobacco Control Monograph 8.* Washington, DC: NCI. Pp. 383-475.

U.S. DHHS (U.S. Department of Health and Human Services). 1989. *The Health Consequences of Smoking: 25 Years of Progress. A Report of the Surgeon General.* Chapter 8. Washington, DC: DHHS Publications.

U.S. DHHS (U.S. Department of Health and Human Services). 1983. *The Health Consequences of Smoking: Cardiovascular Disease.* Chapter 8. Washington, DC: DHHS Publications.

Valkonen MM, Kuusi T. 2000. Vitamin C prevents the acute atherogenic effects of passive smoking. *Free Radic Biol Med* 28(3):428-436.

Vita JA, Treasure CB, Yeung AC, et al. 1992. Patients with evidence of coronary endothelial dysfunction as assessed by acetylcholine infusion demonstrate marked increase in sensitivity to constrictor effects of catecholamines. *Circulation* 85:1390.

Weber C, Erl W, Weber K, et al. 1996. Increased adhesiveness of isolated monocytes to endothelium is prevented by vitamin C intake in smokers. *Circulation* 93(1488).

Wennmalm A, Benthin G, Granstrom EF, Persson L, Petersson AS, Winell S. 1991. Relation between tobacco use and urinary excretion of thromboxane A2 and prostacyclin metabolites in young men. *Circulation* 83(5):1698-1704.

Winniford MD, Wheelan KR, Kremers MS, et al. 1986. Smoking-induced coronary vasoconstriction in patients with atherosclerotic coronary artery disease: evidence for adrenergically mediated alterations in coronary artery tone. *Circulation* 73:662.

Yamaguchi Y, Kagota S, Haginaka J, Kunitomo M. 2000. Peroxynitrite-generating species: good candidate oxidants in aqueous extracts of cigarette smoke. *Jpn J Pharmacol* 82(1):78-81.

Zeiher AM, Drexler H, Wollschlaeger H, et al. 1989. Coronary vasomotion in response to sympathetic stimulation in humans: importance of the functional integrity of the endothelium. *J Am Coll Cardiol* 14:1181.

Zeiher AM, Drexler H, Wollschlager H, et al. 1991. Endothelial dysfunction of the coronary microvasculature is associated with coronary blood flow regulation in patients with early atherosclerosis. *Circulation* 84(1984).

14

Nonneoplastic Respiratory Diseases

In persons who smoke, the cells that line the bronchi and alveoli come into direct contact with high concentrations of tobacco toxicants. Not surprisingly, respiratory diseases such as chronic obstructive pulmonary disease (COPD) are major health problems in smokers (Murin and Silvestri, 2000). Tobacco-related respiratory diseases predominately affect male smokers, but the prevalence of COPD in women is rising rapidly and it appears to follow the prevalence of smoking by 20-30 years (Tanoue, 2000). Children exposed to environmental tobacco smoke (ETS) are also affected (Joad, 2000). Because low levels of tobacco toxicants from ETS come in direct contact with the lung, it is necessary to consider the health effects of both mainstream and secondary smoke.

In evaluating harm reduction strategies for tobacco-related lung diseases, three major nonneoplastic respiratory diseases linked to cigarette smoking must be considered: COPD, asthma, and respiratory infections. Numerous other respiratory diseases are strongly related to cigarette smoking as shown in Table 14-1 (Murin et al., 2000). The relative risks of mortality due to smoking-related nonneoplastic respiratory diseases are considerable, and approximately 91,000 Americans died annually of respiratory diseases attributed to smoking during 1990-1994 (Table 14-1 and 14-2) (Novotny and Giovino, 1998). Cigarette smoking is estimated to contribute to 80-90% of cases of COPD, and the amount and duration of cigarette smoking directly influence the progression of COPD. Asthma and respiratory infections, on the other hand, are not caused by tobacco smoke but are worsened by exposure to cigarette smoke. Of special im-

TABLE 14-1 Smoking-Affected Pulmonary Diseases

Disease Incidence or Severity Definitely Increased by Smoking
 Common cold
 Influenza
 Bacterial pneumonia
 Tuberculosis infection
 Invasive pneumococcal infection
 Pulmonary hemmorrhage
 Pulmonary metastatic disease
 Spontaneous pneumothorax
 Eosinophillic granuloma
 Respiratory bronchiolitis-associated interstitial lung disease
 Idiopathic pulmonary fibrosis
 Asbestosis
 Rheumatoid arthritis-associated interstitial lung disease
Disease Incidence or Severity Possibly Decreased by Smoking
 Sarcoidosis
 Hypersensitivity pneumonitis

NOTE: Modified with permission from Murin et al., 2000. Copyright (2000) by W.B. Saunders Company.

portance in considering harm reduction strategies is the contribution of ETS to asthma and respiratory infections in children. Abatement strategies for susceptible children exposed to environmental tobacco smoke may differ from those used to reduce harm in tobacco smokers.

An extensive knowledge base exists describing the contribution of tobacco smoke exposure to nonneoplastic respiratory disease, and key points are described briefly here. The major goal of this chapter is to summarize studies designed to test whether reducing exposure to tobacco toxicants improves health outcomes for respiratory diseases. As will be described below, there are considerable gaps in information about reducing harm and uncertainties about the quality of the existing knowledge base in this regard. Consequently, a research agenda is proposed to guide future studies aimed at reducing the harm from smoking in COPD.

BIOMARKERS OF RESPIRATORY DISEASES

There are currently no specific biomarkers of respiratory disease due to smoking tobacco products (see Chapter 11). The rare genetic deficiency of α_1-antitrypsin is a risk factor for disease, not a biomarker (see below). No unique molecular or genetic defect specific for tobacco-related respi-

TABLE 14-2 Relative Risk (RR) for Smoking-Attributable Mortality and Average Annual Smoking-Attributable Respiratory Disease Mortality (SAM) Among Current and Former Smokers, by Sex and Disease, United States, 1990-1994

Respiratory Disease	Men Current Smokers RR	Former Smokers RR	SAM	Women Current Smokers RR	Former Smokers RR	SAM	Total SAM
Pneumonia, influenza	2.0	1.6	11,267	2.2	1.4	8,060	19,327
Bronchitis, Emphysema	9.7	8.8	9,642	10.5	7.0	6,475	16,117
Chronic airway obstruction	9.7	8.8	32,132	10.5	7.0	21,893	54,025
Other respiratory diseases	2.0	1.6	776	2.2	1.4	721	1,497
Total			53,817			37,149	90,966

SOURCE: Reprinted with modifications and permission from Novotny and Giovino, 1998. Copyright (1998) by the American Public Health Association.

ratory disease has been identified. The processes involved, such as inflammation and increased levels of oxidants, are not unique to tobacco-related respiratory diseases. Identifying unique biomarkers is further confounded by the heterogeneous nature of these diseases, the complex mixture that makes up tobacco smoke, and the range of individual susceptibilities to the harmful effects of tobacco smoke among users (see Chapter 11). In COPD, for example, the majority of smokers develop abnormal lung function (Camilli et al., 1987), but only 15-20% will develop symptomatic COPD (Fletcher and Peto, 1977). There appears to be no specific clinical or physiological feature to predict which smokers exhibit a rapid decline in lung function (Habib et al., 1987). The most widely used markers of tobacco-related respiratory diseases in population studies are symptom questionnaires and pulmonary function testing. These have well-known limitations of specificity and sensitivity, particularly for detecting the early effects of tobacco smoke on lungs (U.S. DHHS, 1989). Subtle effects of tobacco smoke exposure on the lung can be detected by sampling fluid in the lower respiratory tract via a bronchoscope inserted into the airways, but the significance of these changes to clinically important pulmonary disease has not been established. Newer approaches such as sampling the subjects' urine (Pratico et al., 1998) or exhaled gas (Ichinose et al., 2000; Paolo et al., 2000) for metabolic products due to tissue injury have the advantage of noninvasive sampling but must be validated. Clearly, the greatest obstacle for developing a specific biomarker is the lack of fundamental information on mechanisms by which exposure to tobacco smoke causes specific respiratory diseases. Insight into understanding the molecular basis of diseases may come from unraveling the complex interactions between genetic makeup and the environment that could evolve from the Human Genome Project or similar molecular studies.

CHRONIC OBSTRUCTIVE PULMONARY DISEASE

Definition and Epidemiology

Chronic obstructive pulmonary disease is an all-inclusive term that encompasses chronic bronchitis (the presence of chronic productive cough) and emphysema (permanent enlargement of the distal airspaces) (Aubry et al., 2000). Usually, chronic bronchitis and emphysema occur in combination. The major clinical features of COPD are chronic cough, expectoration of sputum, breathlessness during exertion, and airflow obstruction on forced expiration that tends to worsen over time (Barnes, 2000; Piquette et al., 2000).

The natural history of COPD and its health care implications have been extensively reviewed (Hensley and Saunders, 1989). The course of COPD is characterized by loss of ventilatory function from the peak attained in early adulthood (Figure 14-1). Clinical symptoms of COPD develop after substantial loss of ventilatory function, typically in the fourth and fifth decades of life. In longitudinal studies, the average loss of function in nonsmokers begins at about age 30 as assessed by the relationship of change in forced expiratory volume at 1 second (FEV_1) with age. In nonsmokers, this amounts to about 20 ml per year. The rate of decline in FEV_1 is somewhat greater (40 ml per year) in smokers of 30 cigarettes per day compared to nonsmokers (Camilli et al., 1987; Fletcher and Peto, 1977). A subset of smokers, who will develop COPD, lose function more rapidly (60 ml per year; Habib et al., 1987). However, there is variation in these rates and the rate of decline increases with age (Camilli et al., 1987; Fletcher and Peto, 1977). Smokers who cease smoking do not regain lost function, but the rate of decline of FEV_1 slows to that of nonsmokers (Burrows, 1990). Loss of lung function appears to be a risk factor for mortality even among "never smokers" (Ashley et al., 1975; Beaty et al.,

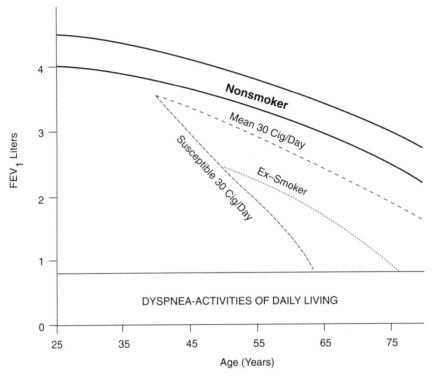

FIGURE 14-1 FEV_1

NOTE: Scheme showing loss of forced expiratory volume at 1 second (FEV_1) with age. Nonsmokers lose lung function with age, about 30 ml per year. Smokers of 30 cigarettes/day have a greater rate of loss (dashed line). A small proportion of smokers (10-15%) have a steeper rate of decline, about 60 ml per year (dot and dashed line). This susceptible group of smokers reaches an FEV_1 of 0.8 liter (the level at which shortness of breath occurs on activities of daily living) at approximately 60-70 years of age. If a susceptible smoker stops smoking at age 50, the rate of decline follows that of the nonsmoker (dotted line), showing that smoking cessation may prolong onset of symptoms.

SOURCE: Reprinted with permission from Piquette, et al., 2000. Copyright (2000) by W.B. Saunders Company.

1985). A loss of lung function may reflect other diseases such as heart disease that decrease life expectancy (Beaty et al., 1985). The annual number of deaths attributable to chronic airway obstruction, bronchitis, and emphysema in the United States during 1990-1994 was 70,000 (Table 14-2). The relative risk of death from COPD in smokers who have symptoms is reduced by cessation, but remains almost as high as that of current smokers. Many of those who eventually die of COPD suffer from prolonged disability due to dyspnea, cough, and sputum production.

Pathogenesis

The role of inflammation and exposure to toxins in cigarette smoke is central to the pathogenesis of COPD (Aubry et al., 2000). There is considerable evidence to support the theory that pulmonary emphysema is caused by excessive exposure to elastolytic enzymes in relation to inhibitors of these enzymes, the "protease-antiprotease" theory (Barnes, 2000; Piquette et al., 2000). Cigarette smoke promotes injury to the lungs by increasing the proteolytic burden and compromising the antiproteolytic defenses leading to a breakdown in lung structure. The major antiproteolytic protein in the lower respiratory tract is α_1-antitrypsin (α_1-AT), although other proteinase inhibitors play a lesser role (Senior and Shapiro, 1998). In chronic bronchitis, an inflammatory airway response caused by chronic exposure to airborne toxins (cigarette smoke, dust, air pollutants) is the central pathogenic mechanism (Barnes, 2000; Piquette et al., 2000). Inflammation leads to edema, cellular infiltration, fibrosis, smooth-muscle hypertrophy, and secretions that narrow the bronchioles.

Animal models of emphysema and chronic bronchitis have been used to study the pathophysiology of COPD (Drazen et al., 1999; Shapiro, 2000), including the contribution of tobacco smoke (Witschi et al., 1997) (see Chapter 10).

Several etiologic factors in COPD have been investigated with regard to cigarette smoking (Sethi and Rochester, 2000). One factor is airway hyperreactivity, measured by the provocation of airway constriction following inhalation of a bronchoconstricting drug or physical condition. Several longitudinal studies have shown that airway hyperreactivity is related to accelerated decline in lung function in cigarette smokers (Frew et al., 1992; O'Conner et al., 1995; Postma et al., 1986; Rijcken et al., 1995; Tashkin et al., 1996; Tracey et al., 1995). Hyperresponsiveness in the presence of blood eosinophilia increases the risk of developing respiratory symptoms (Jansen et al., 1999). If airway hyperreactivity plays a role in the progression of COPD, it is possible that regular use of bronchodilators may slow the rate of decline. However, this conclusion was not supported by the Lung Health Study (Anthonisen et al., 1994; see below).

Special tests of "small-airway" function were developed in the late 1960s and early 1970s to detect early changes in small airways (less than 2-mm diameter) that were not detected on standard tests of pulmonary function (U.S. DHHS, 1984). When applied to prospective studies of populations, tests of small-airway function were not predictive for susceptible subjects who would progress to clinically significant airflow obstruction (U.S. DHHS, 1984). In subjects who exhibit accelerated deterioration of lung function (greater than 60 ml per year), this type of physiological testing was not predictive of the rate of development of clinically significant airway obstruction, probably because of the heterogeneous nature of COPD (Habib et al., 1987).

Mucous hypersecretion and infections have been postulated to play roles in the accelerated decline of pulmonary function in COPD. In adults, prospective studies have failed to show an association between acute chest infections and the rate of decline of FEV_1 (Bates, 1973). Chronic mucous hypersecretion in earlier studies (Fletcher and Peto, 1977; Higgins et al., 1982; Kauffmann et al., 1989; Peto et al., 1983) was not shown to be related to a decline in FEV_1 in COPD. More recent studies in larger samples of the general population have shown associations between chronic mucous hypersecretion and a decline in FEV_1 (Lange at al., 1990; Sherman et al., 1992; Vestbo et al., 1996). In children, however, population studies suggest that childhood respiratory infection is a risk factor for the development of COPD in adults (Gold et al., 1989; Samet et al., 1983). The role of infections is unclear, but it is believed that bacterial colonization/ infection stimulates inflammatory responses that cause local damage and progression of disease. In addition, infections impair host defenses and tissue repair, leading to further infection and perpetuating tissue injury (Murphy and Sethi, 1992).

Considerable evidence from human and laboratory studies suggest that oxidant-antioxidant imbalance in favor of oxidants occurs in COPD (Figure 14-2). The evidence is based on a large number of studies showing an increased oxidant burden and markers of oxidative stress in the airspaces, breath, blood, and urine of smokers and patients with COPD (Koyama and Geddes, 1998; Macnee and Rahman, 1999). Oxidants in patients with COPD are derived from oxygen free radicals in both the gas and tar phases of cigarette smoke (Pryor and Stone, 1993; Zang et al., 1995). Inflammation itself induces oxidant stress in the lungs of smokers as suggested by studies showing greater production of oxygen free radicals by leukocytes in smokers compared to nonsmokers (Morrison et al., 1999) and increased iron content, which promotes formation of free radicals in alveolar macrophages of smokers (Mateos et al., 1998). Pathological examination of the alveolar regions of smokers' lungs has shown increases in the number and percentage of leukocytes compared to non-

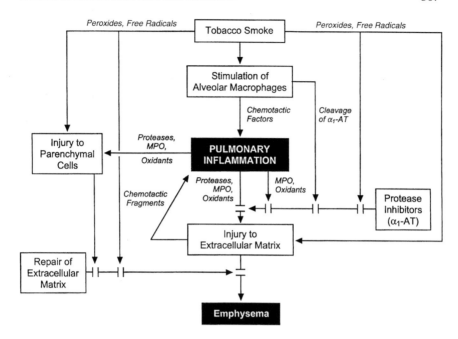

FIGURE 14-2 Pathogenesis of emphysema.

NOTE: Scheme of smoking-induced pulmonary emphysema. Smoking recruits inflammatory cells to the lower respiratory system via stimulation of alveolar macrophages. Inflammatory cells release enzymes (peroxidase, myeloperoxidase) and oxygen free radicals that degrade the extracellular matrix of the respiratory tissues and interfere with normal repair mechanisms. Tobacco smoke also contains oxygen free radicals that, together with products released from inflammatory cells, inactivate protease inhibitors such as α_1-AT.

SOURCE: Reprinted with permission from Senior and Shapiro, 1998. Copyright (1998) by McGraw-Hill Companies.

smokers (Hunninghake and Crystal, 1983). Recent studies have utilized analysis of exhaled gas in patients with COPD for gaseous products (i.e., reactive nitrogen species, ethane)(Ichinose et al., 2000; Paolo et al., 2000) and urinary products (isoprostane F_2-α-III) (Pratico et al., 1998) formed from oxidative stress. It is possible that measurement of products of oxidative stress in exhaled gas may be used as surrogate markers of inflammation in COPD. A possible mechanism whereby oxidants damage the lung is by inactivating the antielastase α_1-AT, thereby decreasing the capacity of α_1-antitrypsin to inhibit proteases (Carp et al., 1982; Johnson and Travis, 1979). This mechanism has been expanded to include proteinases other than neutrophil elastase and antiproteinases other than α_1-AT (Senior and Shapiro, 1998). Dietary deficiency of antioxidants has been proposed as a factor accounting for airflow limitation in elderly people (Dow et al., 1996), but dietary supplements of antioxidants have not been shown to modify clinical symptoms of COPD (Macnee and Rahman, 1999; Rautalahti et al., 1997; Traber et al., 2000).

Mainstream Smoke Exposure

The predominant risk factor for COPD is cigarette smoking, and it is estimated to account for 80-90% of the risk of developing COPD (U.S. DHHS, 1984). Cigarette smoking is associated with a lower FEV_1 in cross-sectional studies (Burrows et al., 1977a; Dockery et al., 1988; Knudson et al., 1976) and with an accelerated decline in FEV_1 in longitudinal studies (reviewed in Sherman, 1991). The lower FEV_1 in cross-sectional studies and accelerated rate of decline in FEV_1 exhibit a dose-response relationship. Both the duration of smoking and the amount smoked are significant predictors of lung function impairment (U.S. DHHS, 1989).

Individuals who smoke have age-adjusted death rates for COPD that are tenfold higher than those of never smokers (U.S. DHHS, 1989). Mortality and morbidity rates for COPD are higher in pipe and cigar smokers than nonsmokers, although the rates are lower than in cigarette smokers. Factors predictive of COPD mortality include age at starting, current smoking status, and total pack-years. For reasons that are not known, only about 15% of smokers develop clinically significant COPD (Fletcher and Peto, 1977).

Data suggest that 10-15% of COPD cases are attributable to causes other than smoking, including occupational exposures to coal dust (Marine et al., 1988), grain dust (Zejda et al., 1993), air pollution (Bates, 1973; Buist and Vollmer, 1994; Rokaw et al., 1980), childhood respiratory infections (Burrows et al., 1977b; Shaheen et al., 1995), and airway hyperresponsiveness (Buist and Vollmer, 1994). Genetic factors that are independent of personal smoking history or environmental exposures also contribute to COPD (Khoury et al., 1985). These include a_1-antitrypsin deficiency, which accounts for less than 1% of cases of COPD (Snider, 1989). Additional genetic factors other than α_1-AT deficiency that appear to play a role in susceptibility to COPD have not been identified (Khoury et al., 1985).

Prospective population studies have shown an additive effect of air pollution on the decline in lung function in smokers (Tashkin et al., 1994;

van der Lende et al., 1981). In a cohort study examining the additive effects of smoking and pollution on lung function decline, the mean adjusted decrement in FEV_1 attributable to smoking more than 24 cigarettes a day was 17.4 ml per year. An adjusted mean annual decrease of 8.9 ml per year was attributed to living in a moderately polluted environment compared to living in a clean environment (van der Lende et al., 1981). The conclusion of this study was that the decline in lung function attributable to smoking was twice as great when living in a polluted environment. In a prospective study of persons living in three communities with air pollution, a significant interaction between smoking more than 12 cigarettes a day and area of residence was found for mean decline of adjusted FEV_1 in males (Tashkin et al., 1994). These findings suggest an independent adverse effect of air pollution on decline of lung function in smokers.

Environmental Tobacco Smoke Exposure

Adults

Coultas (1998) has reviewed published reports of an association between ETS exposure and COPD. He classified three types of studies: two categories based on "indirect" measures of COPD (self-reports of symptoms of COPD; effects on lung function measurements) and a third category that used "direct" measures of COPD (mortality and hospitalizations). The studies focused on environmental tobacco smoke as a risk factor for developing COPD, not as factor that contributed to worsening of symptoms or lung function. In regard to self-reported outcomes, Coultas (1998) summarized the results of three studies (Dayal et al., 1994; Leuenberger et al., 1994; Robbins et al., 1993) published since the Environmental Protection Agency (EPA) report (EPA, 1993) that concluded environmental tobacco smoke "may increase the frequency of respiratory symptoms in adults." Results of one population-based survey of self-reported COPD suggest that 3-5% of nonsmokers may be affected (Whittemore et al., 1995). The three published reports combined asthma and COPD. These studies all report similar findings that passive smoking is associated with chronic respiratory symptoms found in adults with COPD.

The second type of study examined declines in lung function and passive smoking in the development of COPD. Epidemiological studies have investigated the association between environmental tobacco smoke, respiratory diseases, and reduction in pulmonary function tests (Sherman, 1991; Tredaniel et al., 1994). It is controversial whether ETS exposure is associated with COPD. Some studies have reported greater reduction in pulmonary function in nonsmokers married to smokers and exposed to environmental tobacco smoke in the workplace (Berglund et al., 1999; Hole et al., 1989; Leuenberger et al., 1994; Svendsen et al., 1987; White and Froeb, 1980). However, most of the studies used sensitive indicators of lung function, and the physiological significance of small changes in lung function to COPD is not established. In addition, the effect of bias and confounding factors were not taken into account. No relation between passive smoking and reduced lung function in adults was found in two large cross-sectional studies (Comstock et al., 1981; Schottenfeld, 1984) and one longitudinal study (Jones et al., 1983). Thus, whether environmental tobacco smoke causes COPD in adults remains uncertain.

Current evidence suggests that the development of COPD in adults results from impaired lung development and growth in childhood, premature onset of declines in lung function, and/or accelerated decline in lung function (Fletcher et al., 1976; Kerstjens et al., 1997; Samet and Lange, 1996). Although passive smoking is a biologically plausible risk factor, the impact of passive smoking during adulthood on the development of COPD remains controversial. In a review of this topic, Tredaniel and associates (1994) summarized the results of 18 relevant publications. Eight reports found no effect of ETS exposure on lung function, and ten demonstrated small decrements. The authors pointed out the methodological limitations of these studies and raised questions about the relevance of small declines in lung function to the development of COPD. Coultas (1998) reviewed the results of three additional studies published since the Tredaniel et al. (1994) review and concluded that the results were inconsistent (two showing no effect and one showing an effect). In the study showing an effect, a large sample of never-smoking adults in Switzerland (Leuenberger et al., 1994), the association between increased symptoms of chronic bronchitis and environmental tobacco smoke was dose-related. The adjusted odds ratio of chronic bronchitis symptoms increased by years of exposure to environmental tobacco smoke, number of smokers to whom the subject was exposed, and workplace exposures (Leuenberger et al., 1994). An additional report published after Coultas' review concluded that environmental tobacco smoke in adults is associated with small defects in lung function (Carey et al., 1999).

Children

Exposure to passive smoking in childhood has been associated with reduced rate of growth of the lung as determined by change of ventilatory function with age in children exposed to environmental smoke compared to unexposed children (Berkey et al., 1986; Tager et al., 1979, 1983, 1987) (see Chapter 15). However, one cohort study failed to find any effect of maternal smoking on lung growth (Lebowitz and Holberg, 1988), possibly due to differences in the amount of maternal smoking or differences in indoor environments (i.e., greater air mixing or exchange may decrease the concentration of environmental tobacco smoke). A recent study concluded that in utero exposure to maternal smoking is independently associated with decreased lung function in children of school age, especially for small-airway flows (Gilliland et al., 2000). Studies in neonates that excluded the effect of environmental tobacco smoke by measuring lung function after birth concluded that in utero exposure has an independent effect on reduced lung mechanics (Hanrahan et al., 1992; Stick et al., 1996). In a cross-sectional study by Milner et al. (1999), smoking was associated with a significant reduction in birthweight, but lung volume was not reduced when related to weight. Smoking was associated with a highly significant reduction in static compliance in boys and a small but significant reduction in respiratory conductance in girls. These authors concluded that smoking in pregnancy reduces static compliance in boys and conductance in girls although there was no evidence that maternal smoking adversely affected fetal development. Fetal lung hypoplasia has been produced in rats in a model of maternal cigarette smoke exposure (Collins et al., 1985). The weight of the evidence suggests that passive smoking contributes to COPD by causing a slight decrease in airway diameter that increases airway resistance.

Latency

There are limited data on the latency (Robbins et al., 1993) and dose-response (Dayal et al., 1994) between exposure to environmental tobacco smoke and the development of respiratory symptoms. Additional studies are needed, particularly of latency and dose-response relationships, to establish a causal relationship between passive smoking and COPD.

Review of Harm Reduction Strategies Applied to COPD

Impact of Smoking Reduction

It is plausible that reducing the number of cigarettes smoked per day would reduce disease risk. The presumption is that smoking causes morbidity and mortality in a dose-related manner. However, this presumption has never been formally tested (Hughes, 2000). One impediment to determining whether reducing smoking reduces risk is that a study would require a large number of subjects over a long time. Hughes (2000) estimated that 8,000 subjects would have to be followed for eight years. A large study would be required since there are no adequate surrogate markers of disease. Since a large study of this type would delay implementation of harm reduction measures, one school of thought is that the benefits from smoking reduction can be assumed and we need not wait eight years for the study to be completed (Hughes, 2000). Another school of thought argues that the decline in disease from smoking reduction may not be as great as presumed. This view is based on the following: (1) post hoc corrections for variables associated with smoking reduction may inflate mortality and morbidity among smokers, and (2) the expected reductions from "low-tar" cigarettes were not as great as expected (Hughes, 2000). It has been suggested that reducing smoking may undermine attempts at quitting, but based on limited evidence from three trials, Hughes (2000) concluded that there is no evidence to support this concern. It should be noted, however, that these results apply only to people who were self-motivated to reduce smoking, and the effects on quitting may not apply to the general population (Hughes, 2000).

There are data from retrospective cross-sectional, multicenter epidemiological studies suggesting that reducing smoking reduces the risk of COPD (Higgins et al., 1984). The risk for cigarette smokers of developing COPD was assessed in 958 men and 1,159 women ages 25 to 64 years over a 10-year period (1967-1969 and retested in 1978-1979). At entry, patients were excluded who already had borderline or definite COPD. Data for the change in daily cigarette consumption were available from only one center. The odds ratio (OR) for developing COPD within ten years predicted from the change in the number of cigarettes per day (number at follow-up minus number at entry into the study) was 1.7 for men and 2.4 for women. These results were interpreted as showing a benefit from stopping or reducing cigarette smoking. The limitations of this study are that it was not designed to show a change in mortality and morbidity for people who reduced smoking, that estimates are based only on survivors, and that it is a cross-sectional population study. Nevertheless, it does provide some evidence that reducing smoking might modify the risk of COPD.

The potential benefit to improving survival from COPD by reducing smoking can be analyzed from dose-response relationships between mortality and number of cigarettes smoked in a population. Such data are available from two large-scale prospective surveys of smoking and mortality among men and women in the United States sponsored by the American Cancer Society. The first survey of approximately one million people covered the period 1959 to 1972 and is referred to as Cancer Prevention Study I (Burns et al., 1997; CPS-I). The second survey was conducted from 1982 through 1986 on approximately 1.2 million people and is referred to as Cancer Prevention Study II (CPS-II; Thun et al., 1997).

Methodological issues pertaining to calculations of risk-attributable mortality in these studies are discussed in the 1989 Surgeon General's report on reducing the health consequences of smoking (U.S. DHHS, 1989). Data are available from CPS-I that illustrate the relationship between cigarettes smoked per day and mortality rates from COPD among white male and female current smokers. Figure 14-3 shows data for two 15-year age groups (65-79 and 50-64 years). For both men and women age 65-79, the results show a monotonic increase in mortality rate from COPD as a function of number of cigarettes smoked per day, with higher rates in men than women. For men and women age 50-64, there is an apparent increase in mortality rate from COPD with number of cigarettes smoked per day, although the slopes are less steep than for the older group. These

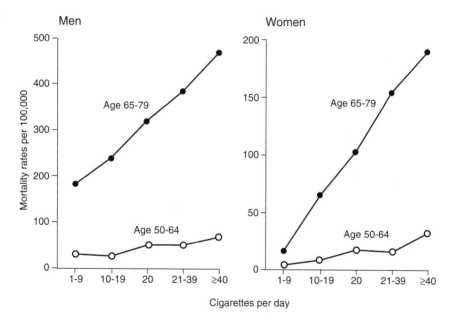

FIGURE 14-3 Mortality rates.

NOTE: Graphs showing relationships between mortality rates from COPD among current white smokers by level of cigarette consumption. Plotted are standardized mortality rates per 100,000 adjusted within 15-year age-specific rates to 1980 U.S. standard population for men and women.

SOURCE: Data for figure gathered from Burns et al., 1997.

findings strongly suggest that increased cigarette smoking causes greater mortality due to COPD in current smokers.

One cannot infer from these results that reducing smoking will decrease mortality from COPD. However, the apparent by linear dose-response relationships suggest that decreasing the amount of cigarettes smoked may lead to fewer deaths from COPD. This may be particularly important to heavy smokers (more than 40 cigarettes per day). Caution is needed, however, in interpreting these results too simply. For example, it is unlikely that someone who smoked 40-plus cigarettes per day during adulthood will experience a substantial reduction in the risk of dying from COPD (e.g., comparable to that of a lifelong 20-cigarette per day smoker) if he or she were able to cut down on cigarette consumption. Only prospective studies of mortality in smokers who reduce their exposure to cigarettes will provide definitive answers to the question of how much mortality rates are decreased by reducing smoking.

Interventional Studies

A landmark study in smoking cessation in people at risk for developing symptomatic COPD is the Lung Health Study (Anthonisen et al., 1994). The study population consisted of 5,887 subjects from North America, 35 to 60 years old, who had presymptomatic COPD (no symptoms but abnormal lung function tests) and included a substantial fraction (37%) of women. Participants were moderate to heavy smokers (mean consumption, 31 cigarettes per day) who were motivated to quit smoking. They were randomized into three groups: (1) smoking intervention, (2) smoking intervention plus a brononchodilator, and (3) usual care. Smoking intervention was intensive and involved 12 group meetings in ten weeks. Behavior modification was stressed, and nicotine replacement therapy (NRT) with nicotine gum was used aggressively. Those who ceased smoking entered a maintenance program. Relapses were treated individually. The main outcome was the rate of change in FEV_1 measured annually over five years. The cessation rate was among the highest reported in any trial. In the intervention group, 35% were abstinent one year and 22% were abstinent for five years (compared to 5% in the non-intervention group). The most important finding of this intent-to-treat analysis was that smoking cessation reduced the rate of decline of FEV_1. In a separate analysis of the study population, symptoms of cough and sputum production, and dyspnea improved after smoking cessation (Kanner et al., 1999). The administration of a bronchodilator (ipratropium bromide) resulted in a small improvement in FEV_1 at the onset of use. However, this effect reversed when the drug was discontinued at the end of the study. The major conclusion of the Lung Health Study was that an aggressive intervention program of smoking cessation reduces the age-related rate of decline in FEV_1 in middle-aged smokers with mild airway obstruction in the first year. After that time, however, the rate of lung function decline was not different in the two groups. These results provide a measure of the impact of aggressive smoking cessation compared with the usual care in reducing the harm of cigarette smoking.

A particularly important audience at which to target harm reduction is young smokers who want to quit. To determine how rapidly lung function returns to normal after smoking is stopped in this group of subjects, Ingram and O'Cain (1971) used sensitive tests of lung function (frequency dependence of compliance) in six asymptomatic young smokers. Within eight weeks of smoking cessation, lung function had returned to normal in all subjects, suggesting that structural changes in the peripheral airways are reversible in these subjects after cessation of smoking.

There is a dearth of information on whether reducing cigarette smoking results in improvement in lung function. Two prospective studies in the 1970s and one in 1989 (Buist et al., 1976; Lange et al., 1989; McCarthy et al., 1976) examined the effect of reduced smoking over a one year period using volunteers with small-airway disease and either normal or mild reductions in FEV_1. Subjects who were able to reduce smoking consumption by 25% or more were considered to have reduced smoking. Tests of small-airway disease improved after smoking reduction, and in one study, FEV_1 showed slight improvement at one year (McCarthy et al., 1976). An observational study by Lange and colleagues in (1989) examined the effect of smoking habits on change in FEV_1 over a five-year period in more than 7,000 Danish men and women. Among these were a sample of 189 subjects who were grouped as reducing smoking from more than 15 cigarettes per day to less than 15 cigarettes per day between intake and final examination. Regression analysis of change in age-adjusted FEV_1 over five years showed that smoking reduction resulted in a beneficial effect in younger subjects (less than 55 years old) but not in older subjects (more than 55 years old). In the younger group, the annual decline in FEV_1 unadjusted for age in women who were able to reduce smoking was 14 ml per year versus 30 ml per year in women who continued to smoke more than 15 cigarettes a day. For younger men who were able to cut back on smoking, the reduction was 17 ml per year compared to 42 ml per year in those who continued to smoke more than 15 cigarettes a day. Limitations of this study were that it was an observational study, smoking habits were based on self-reported data, and there were wide interindividual differences in the rate of decline of FEV_1. Further limitations of these studies are that they were not designed primarily to test whether reducing smoking results in benefit as measured by less decline in lung function, but rather were designed to study the effect of cessation. Although they suggest that reducing smoking (at least in younger subjects) may slow the rate of decline in function, more rigorously designed prospective studies are needed to determine whether reducing smoking results in less harm to the lungs.

Decreasing Smoking Slows Decline in Lung Function

Numerous cross-sectional studies show that the level of smoking is a strong determinate of decline in lung function (U.S. DHHS, 1984, 1989). Multiple regression and analysis has been applied to these cross-sectional studies to determine the quantitative relationships between the rate of decline in FEV_1 and pack-years of cigarettes smoked (Burrows et al., 1977a; Dockery et al., 1988). In one study of over 8,000 men and women from U.S. cities (Dockery et al., 1988), it was estimated that the average male smoker lost 7.4 ml of FEV_1 for every pack-year of cigarettes smoked and that women lost 4.4 ml of FEV_1. FEV_1 values in the population were skewed toward lower values for the smokers compared to the nonsmokers. This suggests that the loss in FEV_1 was relatively larger in smokers than nonsmokers. It is uncertain whether these data could be used to predict a slower rate of decline in lung function if a smoker reduced the number of cigarettes smoked over a prolonged period of time. There are several uncertainties about using studies of longitudinal change in lung function to predict slowing of lung function after reducing smoking. For example, the observations were made over a limited span of the natural history of COPD and it is unclear whether the measured changes in reduced FEV_1 correlated with progression of disease. Function of pack-years smoked remains constant throughout the natural history of the disease. It is possible, for instance, that a more rapid decline in lung function takes place after an event such as an infection or that the rate of decline accelerates with age. Individual factors that contribute to the rate of decline of lung function are poorly understood and also influence progression of disease. Because of uncertainties about the natural history of COPD, it is not possible to use longitudinal studies of decline in lung function as a function of pack-years smoked to estimate the reduction in harm in terms of decline in lung function from reducing cigarette smoking.

Effect of Short-Term Smoking Reduction on Lung Inflammation

The beneficial effect of short-term smoking reduction has been studied in current smokers using a technique that enables inflammation in the lower respiratory tract to be examined directly (ECLIPSE, 2000; Rennard,

2000; Rennard et al., 1990). In current cigarette smokers, three approaches to reduce smoking consumption were used: administration of nicotine medication, switching to a low-tar and nicotine cigarette, and using Eclipse™ (R.J. Reynolds Tobacco Company), a product that results in less combustion of tobacco and potentially generates relatively less tar than nicotine. All studies used a similar approach. Heavy smokers (at least two packs per day) who had no clinical or physiological evidence of COPD were enrolled. During the reduced smoking period, cigarette smoke exposure was monitored by exhaled carbon monoxide (CO) concentration, serum cotinine level, and self-monitoring the number of cigarettes smoked. Lung inflammation was assessed before and two to three months after reduced smoking using bronchoalveolar lavage. Inflammatory cells and protective antiproteinase molecules were measured in the fluid sample, and results were compared to those from nonsmokers. The degree of inflammation of the airways was scored by directly visualizing the degree of erythema and edema though the bronchoscope.

Reduced smoking by approximately 50% (measured by self-reporting and exhaled CO) was achieved by the use of nicotine medication (Rennard et al., 1990). The amount of visible inflammation decreased but remained greater than that of nonsmokers. In lavage fluid, the number of inflammatory cells (particularly the pigment-laden macrophages and neutrophils) and the amount of elastase complexed to its a_1-antiproteinase inhibitor decreased after smoking reduction. These results show that reducing smoking from two packs to one pack per day produces a significant decrease in lung inflammation within two to three months. Whether the reduced inflammation results in less risk of developing COPD remains to be determined.

A second study examined the effect of switching to low-tar and nicotine cigarettes (ECLIPSE, 2000; Rennard, 2000). This study differed from the nicotine replacement study in several ways: the subjects had moderate to heavy cigarette smoking (approximately 30 cigarettes per day) at enrollment, subjects chewed either nicotine gum (2 or 4 mg) or a placebo gum, and the duration of the study was six months. The groups that used NRT reduced their cigarette consumption (measured by self-reporting and exhaled CO). However, the group that used the low-tar and nicotine cigarettes did not reduce smoking because they changed their smoking strategy to compensate. The results showed no reduction in the inflammation score in the group that used low-tar and nicotine cigarettes, which was attributed to altered smoking strategy.

The study using heated tobacco (Eclipse™) was similar in design to the reports cited above and involved 12 smokers studied after a three-month intervention. Exhaled CO was used to monitor cigarette use. Although one-third of the study group smoked conventional cigarettes, the number of conventional cigarettes used was small (97% of all products consumed were Eclipse™). In Eclipse™ users, inflammation scores and number of recovered alveolar macrophages decreased after intervention as did the density of goblet cells (an index of the mucous secreting cells) on bronchial biopsies. (Results of neutrophil counts were confounded by outliers in the control group.) After three months of Eclipse™, inflammation scores and goblet cells density remained above levels found in non-smoking controls. The small number of patients and control subjects limits interpretation of this study.

In summary, asymptomatic smokers clearly have inflammation in their lower respiratory tract, and this inflammation is likely to be involved in the pathogenesis of COPD. The major conclusion of Rennard's (Rennard, 2000; ECLIPSE, 2000) studies is that smoking reduction has measurable effects on inflammation. It is plausible that reducing inflammation by reduced exposure to cigarettes will mitigate the development of COPD. However, the hypothesis that reduced smoking will lead to less severe COPD needs considerably more study using prospective intervention trials in persons susceptible to developing COPD.

ASTHMA

Definition and Epidemiology

Asthma is defined as a "chronic inflammatory disorder of the airways in which many cells and cellular elements play a role, in particular, mast cells, eosinophils, T lymphocytes, neutrophils, and epithelial cells. In susceptible individuals, this inflammation causes recurrent episodes of wheezing, breathlessness, chest tightness, and cough, particularly at night and in the early morning. These episodes are usually associated with widespread but variable airflow obstruction that is often reversible either spontaneously or with treatment. The inflammation also causes an associated increase in the existing airways hyperreactivity to a variety of stimuli" (NIH, 1997). There is no standardized method for measuring asthma, and the lack of standardization has hampered the investigation of asthma by making it difficult to compare results of different studies (Woolcock, 1994).

Asthma is an extremely common disorder in the United States, with approximately 10-15% of boys (Sly, 2000), 7-10% of girls (Sly, 2000), and 5% of adults having signs and symptoms consistent with asthma (Drazen, 2000). Asthma has its origins primarily in childhood, with 50% of all asthma diagnosed by three years of age and 80% by six years of age (Yunginger et al,, 1992), although asthma may develop at any time throughout life. The worldwide incidence of asthma has increased more than 30% since the 1970s (Weiss et al., 1993). A disproportional fraction of these cases have occurred among socioeconomically disadvantaged urban dwellers. The reason for the increased incidence of asthma is not known. In the United States, the overall age-adjusted prevalence of asthma increased 54% from 1980 to 1993-1994 (from 31 to 54 per 1,000 population) (NCHS, 1999).

Pathogenesis

The factors that influence the pathogenesis of asthma include atopy (production of abnormal amounts of immunoglobulin E [IgE]), early respiratory illness (before age two), and parental history of atopy (Woolcock, 1994). Allergens act as triggers of asthma attacks and provoke inflammation of the airways. Other triggers include exercise and air pollutants such as ozone and sulfur dioxide (Woolcock, 1994). Inflammation of the airways with lymphocytes, mast cells, and eosinophils and production of certain interleukins by these cells creates an environment that promotes the production of IgE. The links between infiltration of cells and the pathobiologic processes that account for episodic airway narrowing have not been clearly delineated. The constriction of airway smooth-muscle cells due to the local release of bioactive mediators or neurotransmitters is a widely accepted explanation for acute constriction of the airways, but thickening of airway epithelium and the presence of liquid in the lumen may also contribute. The consequences of airway obstruction are decreased flow rates throughout expiration. As the asthma attack resolves, airway narrowing reverses and flow rates return toward normal. These physiological and pathological changes account for intermittent symptoms of shortness of breath, coughing, and wheezing (Drazen, 2000).

Mainstream Smoke Exposure

Asthmatics have extreme sensitivity of the airways to chemical, physical, and pharmacological stimuli, which can cause pronounced airflow limitation (Drazen, 2000). Cigarette smoke is a complex mixture of irritant compounds. Smoking can potentially lead to amplification of the airway inflammation already present in asthmatics by a number of mechanisms including recruitment of inflammatory cells, enhancement of some cellular function, and release of proinflammatory mediators (Floreani and Rennard, 1999). For example, cigarette smoking stimulates production of cysteinyl leukotrienes (Fauler and Frolich, 1997), a class of mediators that is increased in asthmatics, causes rapid proliferation of cells in small airways (Sekhon et al., 1994). In addition to inflammation, cigarette smoking is associated with increased airway wall remodeling (U.S. DHHS, 1990), a change asociated with chronic asthma. Physiological studies have shown that direct cigarette smoking causes acute bronchoconstriction in subjects with asthma (Nadel and Comroe, 1961). Many patients with asthma report worsening respiratory symptoms upon active cigarette exposure. In a cross-sectional study of 225 adult asthmatics, current smokers had a higher asthma symptom scores than nonsmokers when adjusted for age, gender, and physician visits (Althuis et al., 1999). However, there are methodological issues that might obscure the associations between active smoking and asthma in cross-sectional studies. One form of selection bias, referred to as the healthy smoker effect (Weiss et al., 1999), refers to self-selection of subjects with less mild asthma as active smokers compared to those who remain smokers. The probability of an asthmatic becoming an active smoker may be less in individuals who have a greater airway hyperreactivity or who become more symptomatic when they smoke. Similarily, adults who had childhood asthma may be more likely to remain nonsmokers than those who did not have asthma in childhood. Thus, cross-sectional studies of smoking and asthma have limitations of interpretation. Nevertheless, observational studies show that asthmatics who are unable to cease smoking have more serious problems with control of asthma and have more frequent hospital admissions than nonsmoking asthmatics (Abramson et al., 1995; Floreani and Rennard, 1999).

Since the 1960s, it has been suggested that airways hyperresponsiveness and atopy are risk factors for the development of COPD, the so-called Dutch hypothesis (Orie et al., 1961). Data from several longitudinal studies now clearly show that airways hyperreactivity is a susceptibility factor for accelerated decline of lung function in active cigarette smokers compared to smokers without airways hyperreactivity (Frew et al., 1992; O'Connor et al., 1995; Rijcken et al., 1995; Tashkin et al., 1996; Tracey et al., 1995). Data from the Lung Health Study, a five-year prospective study of subjects with mild COPD, showed that responsiveness to methocholine at the initiation of the study was a strong predictor of change in FEV_1 in continuing smokers even after controlling for baseline lung function, number of cigarettes smoked, and demographic variables (Tashkin et al., 1996). Moreover, airway reactivity appeared to be a more important determinant of subsequent lung function decline then baseline lung function. In addition, the magnitude of the decline in FEV_1 increased with greater levels of airways hyperreactivity. The design of the study did not permit the distinction of whether the risk factor for progression of COPD was airway hyperreactivity itself or an underlying mechanism of reactivity such as inflammation. The finding that the effect of reactivity on lung function decline was greatest in continuing smokers compared to subjects who quit smoking suggests that the effect is at least partially due to progressive airway damage induced by smoking. Since subjects who reduced smoking were not analyzed, there is insufficient information to determine whether decreasing the number of cigarettes smoked leads to less airways hyperreactivity.

Environmental Tobacco Smoke Exposure

Adults

The three major reviews of the health effects of passive smoking published between 1986 and 1992 cited no literature directly examining the association between passive smoking and asthma in adults (NRC, 1986; U.S. DHHS, 1986). Coultas (1998) reviewed studies on this topic from 1992 to 1998 and classified them as etiological studies (the association between passive smoking and the diagnosis of asthma) and morbidity studies (the role of passive smoking in causing symptoms or worsening lung function in patients with asthma). Among the papers reviewed was that of Hu and coworkers (1997) who observed in young adults that physician-diagnosed asthma was associated with parental smoking. The odds ratio was 1.6 (95% confidence interval [CI]=1.1, 2.3) for maternal smoking and 1.3 (95% CI=0.9, 1.8) for paternal smoking. Their investigators also found a dose-response relationship with the amount smoked and the number of parents smoking. Greer and associates (1993) studied a large, nonsmoking population over a ten-year period to examine the association between workplace exposure to environmental tobacco smoke and the development of asthma. After controlling for confounding factors, workplace ETS exposure was a significant risk for asthma (relative risk [RR]=1.5, 95% CI=1.2, 1.8). In another study of more than 4000 never-smoking adults, Leuenberger and associates (1994) found exposure to environmental tobacco smoke at home or work was associated with physician-diagnosed asthma (OR=1.4; 95% CI=1.0, 1.9).

Several studies have examined the role of environmental tobacco smoke as a factor for asthma in adults. There are limitations of present epidemiological studies including potential biases in both subject selection and misclassification of exposure and differing designs. The effects of environmental tobacco smoke on pulmonary function in otherwise healthy adults are likely to be small and unlikely to result in clinically significant chronic respiratory disease (Zeise et al., 1997). Although the results tend to indicate potential effects of environmental tobacco smoke as a cause of asthma in adults, the present epidemiological data are limited and have to be interpreted cautiously.

The results of published reports on the effects of ETS exposure on respiratory symptoms and lung function in adults have been summarized by Tredaniel and associates (1994) and Coultas (1998). Six observational studies in patients and six experimental studies of patients with asthma are reviewed. In general, these studies report that environmental tobacco smoke worsens respiratory symptoms and lung function among adult asthmatics. There are no published intervention trials in adults examining whether decreasing exposure reduces asthma symptoms or impairment of lung function.

A recent review evaluated the quality of the studies of the relation of environmental tobacco smoke as an exacerbating factor for asthma in adults (Weiss et al., 1999). Although exposure of adult asthmatics to environmental tobacco smoke was associated with increased symptoms of asthma, these studies are subject to recall and information bias, and no firm conclusions can be made. On the other hand, a number of controlled exposure studies have examined the effect of exposure to environmental smoke in healthy adult volunteers and subjects with asthma (Weiss et al., 1999). In general, the experimental studies show that brief exposure to ETS produces symptoms such as eye and nasopharyngeal irritation with less consistent responses in terms of lung function. Airway hyperreactivity after environmental smoke exposure was assessed by bronchoprovocation tests using methacholine. The majority of these studies failed to document an effect on ventilatory function or measures of airway responsiveness. In general, studies evaluating acute exposure of adult asthmatics to environmental tobacco smoke have generated conflicting data, yet there is evidence that some asthmatics and groups of asthmatics do respond to ETS levels that do not elicit responses in healthy volunteers. However, because of the limited number of studies and potential problems in their design, definitive conclusions cannot be drawn at this time.

Children

The relationship between environmental tobacco smoke and asthma, lung function, and respiratory symptoms in childhood has been reviewed (Cook et al., 1998; Cook and Strachan, 1997; NRC, 1986; Strachan and Cook, 1998; U.S. DHHS, 1986). Most of the studies support the notion that small but significant differences can be detected in respiratory symptom prevalence and pulmonary function studies comparing children exposed to environmental tobacco smoke and children who are unexposed. The incidence of wheezing illnesses in children whose parents smoke is greatest in early life, whereas the incidence of asthma during school years in less strongly affected by parental smoking. This may be due to a stronger influence of viral-associated wheezing in early childhood or less exposure to maternal smoke among older children.

The results of effects on spirometric indices of passive smoking have been reviewed (Cook et al., 1998; Cook and Strachan, 1999). These analyses concluded that passive smoking is associated with a small but significant reduction in FEV_1 and other spirometric indices in school-age children without asthma. For example, the large Six Cities Study showed a very small but significant effect of maternal smoking on lung growth (minus 3.8 ml per year for FEV_1) (Tager et al., 1983). Such subtle reductions are unlikely to impact the rates of development of chronic obstruction in airflow unless data emerge demonstrating that children exposed in early life have more rapid decline of lung function. Unfortunately, none of the longitudinal studies have looked at changes in lung function in relation to change in maternal smoking behavior. Therefore, it is unknown whether their effects are reversible. Similarly, a study in infants demonstrated that a family history of parental smoking contributes to airway hyperresponsiveness at an early age (Young et al., 1991). Whether these normal, healthy infants with airway hyperresponsiveness develop asthma as adults is not known.

An important issue in evaluating potential strategies for reducing harm from environmental cigarette smoke among asthmatic children in smoking households is to assess the relationship between biomarkers of cigarette smoke exposure and severity of asthma. If a dose-effect relationship exits, these data might indicate the approximate extent of reduction in exposure that may be expected to reduce the severity of asthma in children residing with smokers. Several studies have examined the dose-response relationship in asthmatic children using cross-sectional or case-control designs (Table 14-3). Asthma was assessed by self-reporting, physical examination, or lung function studies, and cotinine was measured in urine or saliva. Two general types of analyses were done: (1) associations of mean levels of cotinine in asthmatic or control groups (Group A) and (2) relationships between cotinine levels and asthma severity (Group B). Two studies (Ehrlich et al., 1992; Willers et al., 1991) found higher levels of cotinine in urine of asthmatics than controls, and one study found that cotinine levels in saliva were no different in asthmatics and control subjects (Chang et al., 2000). Four studies showed a relationship between urinary cotinine levels and the severity of asthma (Chilmonczyk et al., 1993; Ehrlich et al., 1996; Oddoze et al., 1999; Rylander et al., 1995) (see Table 14-3).

There are methodological problems with these studies, however, since most have not controlled for confounding factors such as respiratory illnesses or exposure to house mite antigen. It is possible that children of parents who smoke are more likely to become allergic or to have respiratory infections. Moreover, the effects are applicable only to some age groups or some symptoms of asthma, leading to inconsistencies in the

TABLE 14-3 Relationships Between Cotinine Levels in Asthmatic Children Exposed to Environmental Tobacco Smoke

Author and Year	Study Type and Sample Size	Asthma Population	Cotinine	Asthma	Finding
Group A—Association of high cotinine levels with asthma					
Willers et al., 1991	Case-control 49/77	3-15 yr. old, clinic	Urinary	Self-reporting symptoms	Cotinine levels greater in cases than controls (p<0.005)
Ehrlich et al., 1992	Case-control 72/121	3-14 yr. old, ER and clinic	Urinary, grouped <30 ng/ml or >30 ng/ml	Clinical exam	Cases had higher cotinine levels than control (OR=1.9, 95% CI 1.0-3.3)
Chang et al., 2000	Case-control 165/106	2 mo.-16 yr. old, ER	Salivary	Clinical exam	No difference in cotinine levels between case and control groups
Group B—Dose-effect relationship between cotinine and asthma severity					
Chilmonczyk et al., 1993	Case-control 199/199	8 mo.-13 yr. old, clinic	Urinary, grouped, <10, 10-39, >39 ng/ml	Asthma exacerbations; lung function	Monotonic relationship for ↑ exacerbations and ↓ function
Rylander et al., 1995	Case-control 199/351	4 mo.-4 yr. old, hospital	Urinary	Clinical exam	Risk of wheezing increased with cotinine concentration
Ehrlich et al., 1996	Case-control 368/294	7-9 yr. old; heavy household smokers	Urinary	Self-reported symptoms	Positive exposure-response relationship for asthma and cotinine by quartile (p=0.02)
Oddoze et al., 1991	Cross sectional 90	4-14 yr. old, clinic	Urinary	Bronchial responsiveness	Bronchial responsiveness correlated with cotinine level (p<0.03)

NOTE: ER=emergency room; OR=odds ratio; CI=confidence interval reported results. Finally, the clinical significance of the small differences in lung function or symptoms is unknown. Therefore, there is uncertainty about whether a dose-response relationship exists between cigarette smoke exposure as measured by cotinine levels and severity of asthma, and this subject requires more study before harm reduction studies can be planned.

Review of Harm Reduction Strategies Applied to Asthma

Several studies have examined the effect of reducing tobacco smoke exposure on wheezing symptoms. In one study (O'Connell and Logan, 1974), parents of a subpopulation of children in whom ETS exposure was considered a "significant factor" were given antismoking advice and followed for 6-24 months. Symptoms improved in 90% of children whose parents stopped smoking and in 27% of children who remained exposed to passive smoking. These results are difficult to interpret because this is an uncontrolled trial. In a second study (Murray and Morrison, 1989), severity of asthma was scored in a group of children with the diagnosis of asthma whose mothers smoked. When the group was scored three years later, there was a highly significant decline in lung function in children of mothers who continued to smoke but an improvement in lung function of children whose mothers no longer smoked. This reduction was attributed to an alteration of maternal smoking habits, but this explanation is based on anecdotal evidence. A randomized, controlled trial to investigate whether brief intervention by advising parents about the impact of passive smoking would reduce salivary cotinine levels in asthmatic children was found to be ineffective (Irvine et al., 1999). A study of the relationship between modifications of parental smoking behavior and nicotine exposure found that smoking outside the home was associated with lower urinary cotinine levels only when the parent was the only smoker in the house (Winkelstein et al., 1997). These intervention studies lack assessment of confounders and require a more rigorous design. Intervention studies, utilizing dose-response relationships between asthma severity and levels of environmental tobacco smoke, represent a future strategy for evaluating harm reduction from ETS in asthmatic children.

RESPIRATORY INFECTIONS

Definition and Epidemiology

Viral rhinitis, tracheobronchitis, and bacterial bronchitis are acute infections of the epithelium of the upper respiratory tract. According to the 1997 National Center for Health Statistics survey (NCHS, 1999), upper respiratory infections are the most common medical reason for school or work absenteeism in the United States. These infections can occur in healthy people during epidemics. Immunocompromised patients are at special risk of developing pneumonia and tuberculosis. More serious infections of the lower respiratory tract such as pneumonia are less common in normal hosts. In elderly hospitalized individuals, nosocomial pneumonia is a major health problem and risk to life. Tuberculosis is a disease of the lower respiratory tract present predominately in socially and economically disadvantaged populations and in immunosuppressed patients in the United States, but it is a major health problem worldwide.

Respiratory tract infections and pneumonia are particularly important causes of death in the United States. In 1997, the combined cause-of-death category "pneumonia and influenza" ranked sixth among the leading causes of death (CDC, 1999). The age-adjusted mortality rate for pneumonia and influenza was 12.9 deaths per 100,000 population. During the interval from 1977 to 1997, the death rate from pneumonia and influenza increased 15.2% (CDC, 1999). The number of deaths in the United States in 1990-1994 due to respiratory infections that were attributed to cigarette smoking was estimated to be 8,000 per year (Table 14-2). Pneumonia and influenza are particularly important in the elderly, and the death rate reaches 20% in community-acquired pneumonia (Feldman, 1999; Fine et al., 1996). During 1997, in children aged 1-4 years, the combined category pneumonia-influenza was the sixth leading cause of death, and it was the seventh leading cause of death in children 5-14 years of age (NCHS, 1999).

Pathogenesis

The respiratory tract in the normal host is effective in containing microbes present in the environment. The airways below the level of the major bronchi are mostly sterile, and special situations predispose to respiratory tract infections. These include the introduction of a new species into the environment, such as an antigenic shift in influenza virus. A second situation involves an overwhelming inoculum of organisms, such as exposure to contaminated ventilatory equipment. A third situation results from impairment of part of the host respiratory defense apparatus or systemic immune response. The major components of the respiratory host defenses are the mechanical barriers of the upper respiratory tract, locally secreted immunoglobulins, lymphoid structures of the bronchi, phagocytic macrophages, and a number of immune and nonimmune substances that line the alveolar surfaces (Reynolds, 1997). Components of the host defense system are specially designed to deal with large particulates (3-10 μm) in the upper respiratory tract and small particles (0.5-<3 mm) that may reach alveoli. If inflammation occurs in the alveoli, various systemic immune components can enter the airspaces by leakage through the alveolar surface. If local host defenses are insufficient, the host can recruit additional phagocytic cells such as polymorphonuclear leukocytes (PMNs) and create a local inflammatory response. The creation of an inflammatory response considerably increases the repertoire of systemic immune responses that can be recruited to contain the spread of microbes (Delves and Roitt, 2000). If the infection is successfully contained, inflammation is terminated and lung tissue is restored to its normal state. If it is not contained, inflammation can spread locally, leading to pneumonia. If severe, the local responses may produce permanent damage such as scarring to lung structure.

The effects of exposure to cigarette smoke on immune and inflammatory responses have been studied extensively in humans (Holt, 1987; Johnson et al., 1990). Alteration of immune and inflammatory processes by cigarette smoke are also relevant to asthma and COPD. Both acute exposure and chronic exposure to cigarette smoke alter the responsiveness of the immune system. The magnitude and direction of the alteration depend on the quantity and duration of exposure and the particular immune functions being studied. In general, low concentrations and short exposures enhance immune response, whereas high concentrations and long-term exposures suppress the responsiveness of the immune system. Many of the effects appear reversible within several weeks or months following cessation of exposure even after long periods (years) of exposure. The pulmonary tissues directly in contact with the smoke and the associated regional lymphatics are more affected than the peripheral immune system. Smokers have serum immunoglobulin levels that are 10-20% lower than those of nonsmokers (Holt, 1987; Mili et al., 1991). In addition to impairment of the peripheral immune system, smoking impairs mucociliary clearance, enhances bacterial adherence to respiratory epithelium, and disrupts respiratory epithelium (Dye and Adler, 1994; Fainstein and Musher, 1979; Green and Carolin, 1967; Raman et al., 1983). These altered functions may explain the higher rates of nasopharyngeal colonization with meningococcus among active and passive smokers compared to nonsmokers (Stuart et al., 1989).

Mainstream Smoke Exposure

Cigarette smoking is associated with increased susceptibility to certain types of respiratory infections (Aronson et al., 1982; Finklea et al., 1969; Fischer et al., 1997; Haynes et al., 1966; Imrey et al., 1996; Kark et al., 1982; Nuorti et al., 2000; Parnell et al., 1966; Peters and Ferris, 1967; Stanwell-Smith et al., 1994; Straus et al., 1996). The dose-response relationship between smoking and mortality from influenza and pneumonia has been studied in several cohort studies. In the American Cancer Society CPS-I, the standardized mortality ratio for deaths due to influenza and pneumonia for ever smokers compared to never smokers was 1.9-1.7 for males and 1.3 for females (Hammond, 1965). In the British Physician Study (Doll and Peto, 1976), there was a slight increase in risk for current smokers, and a dose-response relationship was noted by the amount smoked. Similar results were found in the U.S. Veterans Study (Rogot and Murray, 1980) and in a study of Swedish men (Carstensen et al., 1987).

Cigarette smoking is an independent risk factor for invasive pneumococcal disease among immunocompetent, nonelderly adults (Nuorti et al., 2000). A dose-response relation was found for invasive pneumococcal disease and number of cigarettes smoked daily, pack-years of smoking, and time since quitting. Thus, cigarette smoking increases the risk of certain respiratory diseases and predisposes to increased invasiveness of pneumococcal disease.

Smoking has been studied as a possible risk factor for tuberculosis. Although research has been sparse, it has been postulated that smokers are at increased risk of developing active pulmonary tuberculosis. A case-control study by Alcaide and associates (1996) found smoking to be linked to the development of pulmonary tuberculosis in infected subjects who were in close contact with a case of active pulmonary tuberculosis. The risk of active tuberculosis was increased by exposure to environmental tobacco smoke. The significance of the association remained after controlling for social and economic status, age, and gender, but it became not significant in those exposed only to passive smoking. A dose-response effect was also described in this study, although findings have been inconsistent in other studies. Children exposed to passive smoking were found to have increased risk of developing active pulmonary tuberculosis after having been infected (Altet et al., 1996). This association persisted after controlling for age and social and economic status. Although there are few studies, smoking has been linked to positive skin test conversions (Anderson et al., 1997). The mechanism postulated is that smoking decreases immune defenses and increases susceptibility to pulmonary tuberculosis (Plit et al., 1998).

Secondary Smoke Exposure

Compelling evidence exists for a causal relationship between parental smoking and lower respiratory infections (bronchiolitis, pneumonia) during childhood. Strachan and Cook (1998) have reviewed 40 published series on this topic, which included papers in previous reviews (DiFranza and Lew, 1996; EPA, 1993; U.S. DHHS, 1986). The studies sought to ascertain the relationship between lower respiratory illnesses and parental smoking in children less than three years old.

In one group, the occurrence of lower respiratory infections was based on data from a community or ambulatory care setting and included longitudinal studies, controlled trials, case-control studies, and retrospective prevalence surveys. The illnesses included unspecified lower respiratory illness, bronchitis, bronchiolitis, and pneumonia.

The most widely derived measures of effect related to either parent smoking (compared with neither parent) and the effect of maternal smoking (compared with father only or neither parent). Information was sought to determine whether there was a dose-response relationship by relating illness to the amount smoked by either parent. A second series of studies analyzed the association between hospital admissions for lower respiratory tract infections and parental smoking. Diagnoses included undifferentiated chest illness, bronchitis and/or pneumonia, wheezing illness, and bronchiolitis with or without confirmation of respiratory syncytial virus.

All 14 studies in this group found an increased risk associated with parental smoking. The results of the community and hospital studies were broadly consistent, with only one study reporting a reduced risk among children of smokers. The pooled odds ratios were 1.6 (95% CI=1.4, 1.9) for smoking by either parent and 1.7 (95% CI=1.6, 1.9) for maternal smoking. The associations for parental smoking were robust to adjustments for confounding factors. In general, the evidence was more convincing for an association with smoking by the mother only than by the father only, probably related to a greater degree of exposure to the mother. Twelve cohort studies presented evidence for dose-response within smoking families, either to the number of smokers or to the amount smoked in the household. A statistically significant dose-response relationship was found in 10 of the 13 cohort studies (Cook and Strachan, 1997; Cook et al., 1998). In two case-control studies (Reese et al., 1992; Rylander et al., 1995), urinary cotinine levels were measured, but in neither study was it possible to determine if the cotinine levels were related to the risk of respiratory illnesses. The effect of paternal smoking on respiratory illnesses was most marked in the first year of life (Colley et al., 1974; Fergusson et al., 1981; Fergusson and Horwood, 1985), but the effects of maternal smoking on hospital admissions for respiratory infections were similar at all ages up to five years (Taylor and Wadsworth, 1987). There appears to be no increased risk of respiratory infections in susceptible subgroups (i.e., history of parental allergy, prematurity, low birthweight) (Burr et al., 1989; Chen, 1994; Ehrlich et al., 1996; Lucas et al., 1990).

Although the studies are generally consistent with an association between parental smoking and lower respiratory tract infections in children, the summary odds ratios derived from the pooled studies should be interpreted with caution. Difficulties with the pooled analysis include the process by which the variables were selected, the different types of data collection and analytical approaches applied to each study, and the inconsistent methods used to account for confounding effects. In addition, the precise nature of the specific diagnosis of lower respiratory tract infections and overlap with childhood asthma are important barriers to the analysis. The quality and sizes of the populations in these studies varied considerably, possibly introducing certain biases. Despite these limitations, it is reasonable to conclude that there is a causal relationship between parental smoking and acute lower respiratory illnesses at least during the first two years of life.

Latency

The effect of active smoking and smoking cessation on age-adjusted mortality from infectious diseases was examined in the American Cancer Society CPS-II study. Male former smoker of less than 15 cigarettes per day have mortality ratios after ten years approaching unity, whereas former smokers of more than 21 cigarettes per day have mortality ratios near unity at 15 years. Female former smokers of any amount have mortality ratios from influenza and pneumonia approaching unity within 3-5 years. These results suggest that impaired host defenses are reversible after smoking cessation.

Review of Harm Reduction Strategies Applied to Respiratory Infections

One controlled intervention study has examined the effect of acute lower respiratory tract infection after an intervention designed to modify postnatal exposure to tobacco smoke (Greenberg et al., 1994). There was a reduced prevalence of lower respiratory symptoms among the intervention group of infants of smoking mothers whose education level was high school or less versus the control group (Greenberg et al., 1994). However, there was uncertainty about the effectiveness of the intervention since the mean cotinine levels did not differ between study groups despite a reduction in smoking levels in the intervention group.

SUMMARY AND CONCLUSIONS

In evaluating the use of potential reduced-exposure products (PREPs) in tobacco-related lung disease, three major nonneoplastic respiratory diseases linked to cigarette smoking should be considered: COPD, asthma, and respiratory infections. Respiratory diseases are major tobacco-related illnesses, and there is a clear need to mitigate the harmful effects of exposure to both mainstream and secondary tobacco smoke. It is plausible that decreasing smoking will reduce the severity of chronic lung diseases and the incidence of respiratory infections, but there is no adequate scientific evidence to support this conclusion because the effects of reduced smoking on harm reduction have not been extensively studied.

Important considerations are determining dose-effect relationships and use of respiratory disease biomarkers. Rational design of studies of harm reduction would rely on dose-effect data, but such data for respiratory diseases are limited and of uncertain quality. High-quality dose-effect data are required for adequate study design, and such data should be generated. Rational study design would also incorporate biomarkers of disease, and the testing of current and new biomarkers might be done concurrently in the populations studied for dose-effects. The Cancer Prevention Studies I and II, large-scale prospective studies, however, do suggest a linear dose-effect relationship between number of cigarettes smoked per day and mortality rates from COPD, indicating that decreasing the amount of cigarettes smoked may lead to fewer deaths from COPD.

There are currently no specific biomarkers of respiratory disease due to smoking tobacco products. No unique molecular or genetic defect specific for tobacco-related respiratory disease has been identified. The processes involved, such as inflammation and increased levels of oxidants, are not unique to tobacco-related respiratory diseases. Identifying unique biomarkers is further confounded by the heterogeneous nature of these diseases, the complex mixture of tobacco smoke, and the range of individual susceptibilities to the harmful effects of tobacco smoke among users. The most widely used markers of tobacco-related respiratory diseases in population studies are symptom questionnaires and pulmonary function testing. These have well-known limitations of specificity and sensitivity, particularly for detecting the early effects of tobacco smoke on the lungs (U.S. DHHS, 1989). Subtle effects of tobacco smoke exposure on the lung can be detected by sampling fluid in the lower respiratory tract via a bronchoscope inserted into the airways, but the significance of these changes to clinically important pulmonary disease has not been established. Newer approaches such as sampling the subjects' urine (Pratico et al., 1998) or exhaled gas (Ichinose et al., 2000; Paolo et al., 2000) for metabolic products due to tissue injury have the advantage of noninvasive sampling but must be validated. Clearly, the greatest obstacle for developing a specific biomarker is the lack of fundamental information on mechanisms by which exposure to tobacco smoke causes specific respiratory diseases.

The availability of dose-effect data and validated biomarkers may improve the quality of, and provide greater confidence in, the results of contemplated intervention studies. However, the time frame for generating dose-effect data and testing biomarkers is uncertain, and it is unclear whether inclusion of dose-effect considerations and biomarkers will improve the quality of clinical trials of harm reduction in respiratory diseases.

An alternative is to proceed with interventional trials based on current knowledge if there are uncertainties about the added value of dose-effect data or untested biomarkers to study design. As an example, an intervention study of the effect of smoking reduction on COPD could be considered that is similar in design to the Lung Health Study, a large prospective trial of the effects of smoking cessation on rate of decline of FEV_1 in middle-aged smokers with mild COPD (Anthonisen et al., 1994). Another approach is to conduct a trial using a low-tar/moderate-nicotine product from a noncommercial source to avoid product endorsement issues. (A more detailed research agenda can be found in the next section.)

Design of population studies for harm reduction of major respiratory diseases is challenging because of uncertainties about effectiveness and long-term compliance with harm reduction interventions. Reducing the burden of tobacco-related respiratory diseases through harm reduction strategies should be a major priority of the nation's public health.

RESEARCH AGENDA

This section outlines a suggested research agenda for studying harm reduction due to cigarette smoking in respiratory diseases (i.e., COPD, asthma, respiratory infections).

Several specific suggestions for research design arise from this review. An interventional study of the effect of smoking reduction on COPD could be considered that is similar in design to the Lung Health Study, a large prospective trial of the effects of smoking cessation on rate of decline of FEV_1 in middle-aged smokers with mild COPD (Anthonisen et al., 1994). In a proposed smoking reduction trial, it might be possible to include an intervention group of smokers who are able to reduce their smoking spontaneously and maintain significant reductions for a long period using behavior intervention and/or nicotine replacement products. The goal would be to decrease the number of cigarettes per day to eight to ten, a point below which it is very difficult to reduce the number of cigarettes smoked (Hughes, 2000). The primary end point could be the change in FEV_1 over five years, and secondary end points could be exhaled carbon monoxide concentration, serum cotinine level, survival, and comorbid smoking-related disorders. If successful, this study should be able to determine whether reduced smoking as measured by change in FEV_1 mitigates harm to the lungs. Potential weaknesses are the self-selection of subjects, the large number of subjects required, ethical issues regarding the nonintervention group, and a probable large dropout rate in the intervention group.

Another approach is to conduct a trial using a low-tar/moderate-nicotine product from a noncommercial source to avoid product endorsement issues. This study might be conducted in two phases. Phase I would be a controlled-exposure study in human smokers to compare the effects of the low-tar/moderate-nicotine product versus a reference cigarette on inflammatory changes in the lower respiratory tract, similar to the observations of Rennard et al. (Rennard, 2000). The objective of Phase I would be to determine if inflammation is reduced by exposure to the low-tar/moderate-nicotine product. If so, a Phase II intervention trial would compare the low-tar/moderate nicotine product to reference cigarettes in cohorts of current smokers. The objectives and design of the Phase II trial would be similar to that described above for intervention using behavioral modification or NRT. Potential advantages of this trial are that the low-tar/moderate-nicotine product, if it reduces harm, could be used as a reference product for future regulation of marketed products. Uncertainties related to such a study include inference from results of short-term controlled exposures to longer-term studies, controlling use of the intervention product, and the large effort that would be needed to complete the study.

REFERENCES

Abramson M, Kutin J, et al. 1995. Morbidity, medication and trigger factors in a community sample of adults with asthma. *Med J Aust* 162:78-81.

Alcaide J, Altet M, et al. 1996. Cigarette smoking as a risk factor for tuberculosis in young adults: a case-control study. *Tubercle and Lung Disease* 77:112-116.

Altet M, Alcaide J, et al. 1996. Passive smoking and risk of pulmonary tuberculosis in children immediately following infection. A case-control study. *Tubercle and Lung Disease* 77:537-544.

Althuis MD, Sexton M, Prybylski D. 1999. Cigarette smoking and asthma symptom severity among adult asthmatics. *J Asthma* 36(3):257-264.

Anderson R, Sy F, et al. 1997. Cigarette smoking and tuberculin skin test conversion among incarcerated adults. *Amer J Prev Med* 13:175-181.

Anthonisen NR, Connett JE, Kiley JP, et al. 1994. Effects of smoking intervention and the use of an inhaled anticholinergic bronchodilator on the rate of decline of FEV1. The Lung Health Study. *JAMA* 272(19):1497-1505.

Aronson MD, Weiss ST, Ben RL, Komaroff AL. 1982. Association between cigarette smoking and acute respiratory tract illness in young adults. *JAMA* 248(2):181-183.

Ashley F, Kannel WB, Sorlie PD, Masson R. 1975. Pulmonary function: relation to aging, cigarette habit, and mortality. The Framingham Study. *Ann Intern Med* 82:739-745.

Aubry MC, Wright JL, Myers JL. 2000. The pathology of smoking-related lung diseases. *Clin Chest Med* 21(1):11-35.

Barnes PJ. 2000. Chronic obstructive pulmonary disease. *N Engl J Med* 343(4):269-280.

Bates DV. 1973. The fate of the chronic bronchitic: a report of the ten-year follow-up in the Canadian Department of Veteran's Affairs coordinated study of chronic bronchitis. The J. Burns Amberson Lecture of the American Thoracic Society. *Am Rev Respir Dis* 108(5):1043-1065.

Beaty TH, Newill CA, Cohen BH, Tockman MS, Bryant SH, Spurgeon HA. 1985. Effects of pulmonary function on mortality. *J Chron Dis* 38:703-710.

Berglund DJ, Abbey DE, Lebowitz MD, Knutsen SF, McDonnell WF. 1999. Respiratory symptoms and pulmonary function in an elderly nonsmoking population. *Chest* 115(1):49-59.

Berkey CS, Ware JH, Dockery DW, Ferris BG Jr, Speizer FE. 1986. Indoor air pollution and pulmonary function growth in preadolescent children. *Am J Epidemiol* 123(2):250-60.

Buist AS, Sexton GJ, Nagy JM, Ross BB. 1976. The effect of smoking cessation and modification on lung function. *Am Rev Respir Dis* 114(1):115-122.

Buist A, Vollmer W. 1994. Smoking and other risk factors. Murray J, Nadel J, eds. *Textbook of Respiratory Medicine*. Philadelphia, PA: W.B. Saunders Company. Pp. 1259-1287.

Burns DM, Shanks TG, Choi W, Thun MJ, Heath CW, Garfinkel L. 1997. The American Cancer Society Cancer Prevention Study: 12-year follow-up of one million men and women. *Changes in Cigarette-Related Diseases Risks and Their Implication for Prevention and Control*. Smoking and Tobacco Control Monograph 8. Washington, DC: National Cancer Institute. Pp. 113-304.

Burr ML, Miskelly FG, Butland BK, Merrett TG, Vaughan-Williams E. 1989. Environmental factors and symptoms in infants at high risk of allergy. *J Epidemiol Community Health* 43(2):125-132.

Burrows B. 1990. Airways obstructive diseases: pathogenetic mechanisms and natural histories of the disorders. *Med Clin North Am* 74(3):547-559.

Burrows B, Knudson RJ, Cline MG, Lebowitz MD. 1977a. Quantitative relationships between cigarette smoking and ventilatory function. *Am Rev Respir Dis* 115(2):195-205.

Burrows B, Knudson RJ, Lebowitz MD. 1977b. The relationship of childhood respiratory illness to adult obstructive airway disease. *Am Rev Respir Dis* 115(5):751-760.

Camilli AE, Burrows B, Knudson RJ, Lyle Sk, Lebowitz MD. 1987. Longitudinal changes in forced expiratory volume in one second in adults. Effects of smoking and smoking cessation. *Am Rev Respir Dis* 135(4):794-799.

Carey IM, Cook DG, Strachan DP. 1999. The effects of environmental tobacco smoke exposure on lung function in a longitudinal study of British adults. *Epidemiology* 10(3):319-326.

Carp H, Miller F, Hoidal JR, Janoff A. 1982. Potential mechanism of emphysema: alpha 1-proteinase inhibitor recovered from lungs of cigarette smokers contains oxidized methionine and has decreased elastase inhibitory capacity. *Proc Natl Acad Sci U S A* 79(6):2041-2045.

Carstensen JM, Pershagen G, Eklund G. 1987. Mortality in relation to cigarette and pipe smoking: 16 years' observation of 25,000 Swedish men. *J Epidemiol Community Health* 41(2):166-172.

CDC (Centers for Disease Control and Prevention). 1999. Mortality Patterns-United States, 1997. *Morbidity and Mortality Weekly Reports* 48:664-668.

Chang M, Hogan A, et al. 2000. Salivary cotinine levels in children presenting with wheezing in an emergency department. *Ped Pulm* 29:257-263.

Chen Y. 1994. Environmental tobacco smoke, low birth weight, and hospitalization for respiratory disease. *Am J Respir Crit Care Med* 150(1):54-58.

Chilmonczyk B, Salmun L, et al. 1993. Association between exposure to environmental tobacco smoke and exacerbations of asthma in children. *N Engl J Med* 328:1665-1669.

Colley JR, Holland WW, Corkhill RT. 1974. Influence of passive smoking and parental phlegm on pneumonia and bronchitis in early childhood. *Lancet* 2(7888):1031-1034.

Collins M, Moessinger A, et al. 1985. Fetal lung hypoplasia associated with maternal smoking: a morphometric analysis. *Ped Research* 19:408-412.

Comstock GW, Meyer MB, Helsing KJ, Tockman MS. 1981. Respiratory effects on household exposures to tobacco smoke and gas cooking. *Am Rev Respir Dis* 124(2):143-148.

Cook DG, Strachan DP. 1997. Health effects of passive smoking. Parental smoking and prevalence of respiratory symptoms and asthma in school age children. *Thorax* 52(12):1081-1094.

Cook DG, Strachan DP. 1999. Health effects of passive smoking. Summary of effects of parental smoking on the respiratory health of children and implications for research. *Thorax* 54(4):357-366.

Cook DG, Strachan DP, Carey IM. 1998. Health effects of passive smoking. Parental smoking and spirometric indices in children. *Thorax* 53(10):884-893.

Coultas DB. 1998. Health effects of passive smoking. Passive smoking and risk of adult asthma and COPD: an update. *Thorax* 53(5):381-387.

Dayal HH, Khuder S, Sharrar R, Trieff N. 1994. Passive smoking in obstructive respiratory disease in an industrialized urban population. *Environ Res* 65(2):161-171.

Delves P, Roitt I. 2000. The immune system. *NEJM* 343:37-49.

DiFranza JR, Lew RA. 1996. Morbidity and mortality in children associated with the use of tobacco products by other people. *Pediatrics* 97(4):560-568.

Dockery DW, Speizer FE, Ferris BG Jr, Ware JH, Louis TA, Spiro A 3d. 1988. Cumulative and reversible effects of lifetime smoking on simple tests of lung function in adults. *Am Rev Respir Dis* 137(2):286-292.

Doll R, Peto R. 1976. Mortality in relation to smoking: 20 years' observations on male British doctors. *Br Med J* 2(6051):1525-1536.

Dow L, Tracey M, Villar A, et al. 1996. Does dietary intake of vitamins C and E influence lung function in older people? *Am J Respir Crit Care Med* 154(5):1401-1404.

Drazen J. 2000. Asthma. Goldman L, Bennett J, eds. *Cecil Textbook of Medicine.* Philadelphia, PA: W.B. Saunders Company. Pp. 387-393.

Drazen JM, Takebayashi T, Long NC, De Sanctis GT, Shore SA. 1999. Animal models of asthma and chronic bronchitis. *Clin Exp Allergy* 9 (Suppl 2):37-47.

Dye JA, Adler KB. 1994. Effects of cigarette smoke on epithelial cells of the respiratory tract. *Thorax* 49(8):825-834.

ECLIPSE Expert Panel. 2000. A safer cigarette? A comparative study. A consensus report. *Inhalation Toxicology* 12 (Suppl.5): 1-48.

Ehrlich RI, Du Toit D, Jordaan E, et al. 1996. Risk factors for childhood asthma and wheezing. Importance of maternal and household smoking. *Am J Respir Crit Care Med* 154(3 Pt 1):681-688.

Ehrlich R, Kattan M, Godbold J, et al. 1992. Childhood asthma and passive smoking. Urinary cotinine as a biomarker of exposure. *Am Rev Respir Dis* 145(3):594-599.

EPA (Environmental Protection Agency). 1993. *Respiratory Health Effects of Passive Smoking: Lung Cancer and Other Disorders.* NIH Publication No. 93-3605. Washington, DC: U.S. Environmental Protection Agency.

Fainstein V, Musher DM. 1979. Bacterial adherence to pharyngeal cells in smokers, nonsmokers, and chronic bronchitics. *Infect Immun* 26(1):178-182.

Fauler J, Frolich JC. 1997. Cigarette smoking stimulates cysteinyl leukotriene production in man. *Eur J Clin Invest* 27(1):43-47.

Feldman C. 1999. Pneumonia in the elderly. *Clin Chest Med* 20(3):563-573.

Fergusson DM, Horwood LJ. 1985. Parental smoking and respiratory illness during early childhood: a six-year longitudinal study. *Pediatr Pulmonol* 1(2):99-106.

Fergusson DM, Horwood LJ, Shannon FT, Taylor B. 1981. Parental smoking and lower respiratory illness in the first three years of life. *J Epidemiol Community Health* 35(3):180-184.

Fine MJ, Smith MA, Carson CA, et al. 1996. Prognosis and outcomes of patients with community-acquired pneumonia. A meta-analysis. *JAMA* 275(2):134-141.

Finklea JF, Sandifer SH, Smith DD. 1969. Cigarette smoking and epidemic influenza. *Am J Epidemiol* 90(5):390-399.

Fischer M, Hedberg K, Cardosi P, et al. 1997. Tobacco smoke as a risk factor for meningococcal disease. *Pediatr Infect Dis J* 16(10):979-983.

Fletcher C, Peto R. 1977. The natural history of chronic airflow obstruction. *Br Med J* 1(6077):1645-1648.

Fletcher C, Peto R, et al. 1976. *The Natural History of Bronchitis and Emphysema*. Oxford, England: Oxford University Press.

Floreani AA, Rennard SI. 1999. The role of cigarette smoke in the pathogenesis of asthma and as a trigger for acute symptoms. *Curr Opin Pulm Med* 5(1):38-46.

Frew AJ, Kennedy SM, Chan-Yeung M. 1992. Methacholine responsiveness, smoking, and atopy as risk factors for accelerated FEV1 decline in male working populations. *Am Rev Respir Dis* 146(4):878-883.

Gilliland FD, Berhane K, McConnell R, et al. 2000. Maternal smoking during pregnancy, environmental tobacco smoke exposure and childhood lung function. *Thorax* 55(4):271-276.

Gold DR, Tager IB, Weiss ST, Tosteson TD, Speizer FE. 1989. Acute lower respiratory illness in childhood as a predictor of lung function and chronic respiratory symptoms. *Am Rev Respir Dis* 140(4):877-884.

Green GM, Carolin D. 1967. The depressant effect of cigarette smoke on the in vitro antibacterial activity of alveolar macrophages. *N Engl J Med* 276(8):421-427.

Greenberg RA, Strecher VJ, Bauman KE, et al. 1994. Evaluation of a home-based intervention program to reduce infant passive smoking and lower respiratory illness. *J Behav Med* 17(3):273-290.

Greer JR, Abbey DE, Burchette RJ. 1993. Asthma related to occupational and ambient air pollutants in nonsmokers. *J Occup Med* 35(9):909-915.

Habib MP, Klink ME, Knudson DE, Bloom JW, Kaltenborn WT, Knudson RJ. 1987. Physiologic characteristics of subjects exhibiting accelerated deterioration of ventilatory function. *Am Rev respir Dis* 136:638-645.

Hammond E. 1965. Evidence on the effects of giving up cigarette smoking. *Am J Public Health* 55:682-691.

Hanrahan JP, Tager IB, Segal MR, Tosteson TD, Castile RG, Van Vunakis H, Weiss ST, Speizer FE. 1992. The effect of maternal smoking during pregnancy on early infant lung function. *Am Rev Respir Dis* 145(5):1129-1135.

Haynes WF Jr, Krstulovic VJ, Bell AL Jr. 1966. Smoking habit and incidence of respiratory tract infections in a group of adolescent males. *Am Rev Respir Dis* 93(5):730-735.

Hensley M, Saunders N. 1989. Clinical epidemiology of chronic obstructive lung disease. *Lung Biology in Health and Disease*. New York: Marcel Dekker, Inc.

Higgins MW, Keller JB, Becker M, Howatt W, Landis JR, Rotman H, et al. 1982. An index of risk for obstructive airways disease. *Am Rev Respir Dis* 125:144-151.

Higgins MW, Keller JB, Landis JR, et al. 1984. Risk of chronic obstructive pulmonary disease. Collaborative assessment of the validity of the Tecumseh index of risk. *Am Rev Respir Dis* 130(3):380-385.

Hole DJ, Gillis CR, Chopra C, Hawthorne VM. 1989. Passive smoking and cardiorespiratory health in a general population in the west of Scotland. *BMJ* 299(6696):423-427.

Holt PG. 1987. Immune and inflammatory function in cigarette smokers. *Thorax* 42(4):241-249.

Hu FB, Persky V, Flay BR, Richardson J. 1997. An epidemiological study of asthma prevalence and related factors among young adults. *J Asthma* 34(1):67-76.

Hughes J. 2000. Reduced smoking: an introduction and review of the evidence. *Addiction* 95(Suppl 1):S3-7.

Hunninghake GW, Crystal RG. 1983. Cigarette smoking and lung destruction. Accumulation of neutrophils in the lungs of cigarette smokers. *Am Rev Respir Dis* 128(5):833-838.

Ichinose M, Sugiura H, Yamagata S, Koarai A, Shirato K. 2000. Increase in reactive nitrogen species production in chronic obstructive pulmonary disease airways. *Am J Respir Crit Care Med* 162(2 Pt 1):701-706.

Imrey PB, Jackson LA, Ludwinski PH, et al. 1996. Outbreak of serogroup C meningococcal disease associated with campus bar patronage. *Am J Epidemiol* 143(6):624-630.

Ingram RJ, O'Cain C. 1971. Frequency dependence of compliance in apparently healthy smokers versus non-smokers. *Bull European de Physiopathologie Resp* 7:195-212.

Irvine L, Crombie I, et al. 1999. Advising parents of asthmatic children on passive smoking: randomized control trial. *BMJ* 318:1456-1459.

Jansen DF, Schouten JP, Vonk JM, Rijcken B, Timens W, Kraan J, Weiss S, Postma DS. 1999. Smoking and airway hyperresponsiveness especially in the presence of blood eosinophilia increase the risk to develop respiratory symptoms: a 25-year follow-up study in the general adult population. *Am J Respir Crit Care Med* 160(1):259-264.

Joad JP. 2000. Smoking and pediatric respiratory health. *Clin Chest Med* 21(1):37-46.

Johnson D, Travis J. 1979. The oxidative inactivation of human alpha-1-proteinase inhibitor. Further evidence for methionine at the reactive center. *J Biol Chem* 254(10):4022-4026.

Johnson JD, Houchens DP, Kluwe WM, Craig DK, Fisher GL. 1990. Effects of mainstream and environmental tobacco smoke on the immune system in animals and humans: a review. *Crit Rev Toxicol* 20(5):369-395.

Jones JR, Higgins IT, Higgins MW, Keller JB. 1983. Effects of cooking fuels on lung function in nonsmoking women. *Arch Environ Health* 38(4):219-222.

Kanner RE, Connett JE, Williams DE, Buist AS. 1999. Effects of randomized assignment to a smoking cessation intervention and changes in smoking habits on respiratory symptoms in smokers with early chronic obstructive pulmonary disease: the Lung Health Study. *Am J Med* 106(4):410-416.

Kark JD, Lebiush M, Rannon L. 1982. Cigarette smoking as a risk factor for epidemic a(h1n1) influenza in young men. *N Engl J Med* 307(17):1042-1046.

Kauffman F, Dockery D, et al. 1989. Respiratory symptoms and lung function in relation to passive smoking: a comparative study of American and French women. *Int J Epidemiol* 18:334-344.

Kerstjens HA, Rijcken B, Schouten JP, Postma DS. 1997. Decline of FEV1 by age and smoking status: facts, figures, and fallacies. *Thorax* 52(9):820-827.

Khoury MJ, Beaty TH, Tockman MS, Self SG, Cohen BH. 1985. Familial aggregation in chronic obstructive pulmonary disease: use of the loglinear model to analyze intermediate environmental and genetic risk factors. *Genet Epidemiol* 2(2):155-166.

Knudson RJ, Burrows B, Lebowitz MD. 1976. The maximal expiratory flow-volume curve: its use in the detection of ventilatory abnormalities in a population study. *Am Rev Respir Dis* 114(5):871-879.

Koyama H, Geddes DM. 1998. Genes, oxidative stress, and the risk of chronic obstructive pulmonary disease. *Thorax* 53(Suppl 2):S10-14.

Lange P, Groth S, Nyboe J, et al. 1989. Effects of smoking and changes in smoking habits on the decline of FEV1. *Eur Respir J* 2(9):811-816.

Lange P, Nyboe J, Appleyard M, Jensen G, Schnohr P. 1990. Relation of ventilatory impairment and of chronic mucus hypersecretion to mortality from obstructive lung disease and from all causes. *Thorax* 45:579-585.

Lebowitz MD, Holberg CJ. 1988. Effects of parental smoking and other risk factors on the development of pulmonary function in children and adolescents. Analysis of two longitudinal population studies. *Am J Epidemiol* 128(3):589-597.

Leuenberger P, Schwartz J, Ackermann-Liebrich U, et al. 1994. Passive smoking exposure in adults and chronic respiratory symptoms (SAPALDIA Study). Swiss Study on Air Pollution and Lung Diseases in Adults, SAPALDIA Team. *Am J Respir Crit Care Med* 150(5 Pt 1):1222-1228.

Lucas A, Brooke OG, Cole TJ, Morley R, Bamford MF. 1990. Food and drug reactions, wheezing, and eczema in preterm infants. *Arch Dis Child* 65(4):411-415.

Macnee W, Rahman I. 1999. Oxidants and antioxidants as therapeutic targets in chronic obstructive pulmonary disease. *Am J Respir Crit Care Med* 160(5 Pt 2):S58-65.

Marine WM, Gurr D, Jacobsen M. 1988. Clinically important respiratory effects of dust exposure and smoking in British coal miners. *Am Rev Respir Dis* 137(1):106-112.

Mateos F, Brock JH, Perez-Arellano JL. 1998. Iron metabolism in the lower respiratory tract. *Thorax* 53(7):594-600.

McCarthy DS, Craig DB, Cherniack RM. 1976. Effect of modification of the smoking habit on lung function. *Am Rev Respir Dis* 114(1):103-113.

Mili F, Flanders WD, Boring JR, Annest JL, Destefano F. 1991. The associations of race, cigarette smoking, and smoking cessation to measures of the immune system in middle-aged men. *Clin Immunol Immunopathol* 59(2):187-200.

Milner AD, Marsh MJ, Ingram DM, Fox GF, Susiva C. 1999. Effects of smoking in pregnancy on neonatal lung function. *Arch Dis Child Fetal Neonatal Ed* 80(1):F8-14.

Morrison D, Rahman I, Lannan S, MacNee W. 1999. Epithelial permeability, inflammation, and oxidant stress in the air spaces of smokers. *Am J Respir Crit Care Med* 159(2):473-479.

Murin S, Bilello KS, Matthay R. 2000. Other smoking-affected pulmonary diseases. *Clin Chest Med* 21(1):121-137.

Murin S, Silvestri G, eds. 2000. Smoking and pulmonary and cardiovascular diseases. *Clinics in Chest Medicine*. Philadelphia, PA: W.B. Saunders Co.

Murphy TF, Sethi S. 1992. Bacterial infection in chronic obstructive pulmonary disease. *Am Rev Respir Dis* 146(4):1067-1083.

Murray AB, Morrison BJ. 1989. Passive smoking by asthmatics: its greater effect on boys than on girls and on older than on younger children. *Pediatrics* 84(3):451-459.

Nadel J, Comroe J. 1961. Acute effects of inhalation of cigarette smoke on airway conductance. *J Applied Physiol* 16: 713-716.

NCHS (National Center for Health Statistics). 1999. *Health, United States 1999 with Health and Aging Chartbook*. Hyattsville, MD: NCHS.

NIH (National Institutes of Health). 1997. *Guidelines for the Diagnosis and Management of Asthma*. Bethesda, MD: NIH, National Heart, Lung, and Blood Institute.

Novotny TE, Giovino GE. 1998. Tobacco Use. Brownson RE, Renington PL, Davis JR, eds. *Chronic Disease Epidemiology and Control*. 2nd ed. Washington, DC: American Public Health Association Press. Pp.117-148.

NRC (National Research Council). 1986. *Environmental Tobacco Smoke: Measuring Exposures and Assessing Health Effects*. Washington, DC: National Academy Press.

Nuorti JP, Butler JC, Farley MM, et al. 2000. Cigarette smoking and invasive pneumococcal disease. Active Bacterial Core Surveillance Team. *N Engl J Med* 342(10):681-689.

O'Connell EJ, Logan GB. 1974. Parental smoking in childhood asthma. *Ann Allergy* 32(3):142-145.

O'Conner G, Sparrow D, et. al. 1995. A prospective longitudinal study of methacholine airway responsiveness as a predictor of pulmonary function decline: Normative Aging Study. *American Journal of Respiratory and Critical Care Medicine* 152:87-92.

Oddoze C, Dunus J, et al. 1999. Urinary cotinine and exposure to parental smoking in a population of children with asthma. *Clinical Chemistry* 45:505-509.

Orie NGM, Sluiter HJ, De Vries K, Tammeling GJ, Witkop J. 1961. The host factor in bronchitis. Orie NGM, Sluiter HJ, ed. *Bronchitis.* Assen, Holland: Royal Van Gorcum. Pp. 43-59.

Paolo P, Kharitonov S, et al. 2000. Exhaled ethane, a marker of lipid peroxidation, is elevated in chronic obstructive pulmonary disease. *Am J Resp Crit Care Med* 162:369-373.

Parnell JL, Anderson DO, Kinnis C. 1966. Cigarette smoking and respiratory infections in a class of student nurses. *N Engl J Med* 274(18):979-984.

Peters JM, Ferris BG Jr. 1967. Smoking and morbidity in a college-age group. *Am Rev Respir Dis* 95(5):783-789.

Peto R, Speizer FE, Cochrane AL, et al. 1983. The relevance in adults of air-flow obstruction, but not of mucus hypersecretion, to mortality from chronic lung disease. Results from 20 years of prospective observation. *Am Rev Respir Dis* 128(3):491-500.

Piquette C, Rennard S, et. al. 2000. Chronic bronchitis and emphysema. Murray J, Nadel J, Mason R, Boushey HJ. *Textbook of Respiratory Medicine.* Philadelphia, PA: W.B. Saunders Company. Pp. 1187-1245.

Plit M, Theron A, et al. 1998. Influence of antimicrobial chemotherapy and smoking status on the plasma concentrations of vitamin C, vitamin E, beta-carotene, acute phase reactants, iron and lipid peroxides in patients with pulmonary tuberculosis. *Inter J Tuber Lung Dis* 2(7):590-596.

Postma DS, de Vries K, Koeter GH, Sluiter HJ. 1986. Independent influence of reversibility of air-flow obstruction and nonspecific hyperreactivity on the long-term course of lung function in chronic air-flow obstruction. *Am Rev Respir Dis* 134(2):276-280.

Practico D, Basili S, Vieri M, Cordova C, Violi F, Fitzgerald GA. 1998. Chronic obstructive pulmonary disease is associated wiht an increase in urinary levels of isoprostane F2alpha-III, an index of oxidant stress. *Am J Respir Crit Care Med* 158(6):1709-1714.

Pryor WA, Stone K. 1993. Oxidants in cigarette smoke. Radicals, hydrogen peroxide, peroxynitrate, and peroxynitrite. *Ann N Y Acad Sci* 686:12-27, Discussion 27-28.

Raman AS, Swinburne AJ, Fedullo AJ. 1983. Pneumococcal adherence to the buccal epithelial cells of cigarette smokers. *Chest* 83(1):23-27.

Rautalahti M, Virtamo J, Haukka J, et al. 1997. The effect of alpha-tocopherol and beta-carotene supplementation on COPD symptoms. *Am J Respir Crit Care Med* 156(5):1447-1452.

Reese AC, James IR, Landau LI, Lesouef PN. 1992. Relationship between urinary cotinine level and diagnosis in children admitted to hospital. *Am Rev Respir Dis* 146(1):66-70.

Rennard SI. 2000 Apr 25. Presentation to the IOM Committee to Assess the Science Base for Tobacco Harm Reduction. Washington, DC.

Rennard SI, Daughton D, Fujita J, et al. 1990. Short-term smoking reduction is associated with reduction in measures of lower respiratory tract inflammation in heavy smokers. *Eur Respir J* 3(7):752-759.

Reynolds H. 1997. Integrated host defense against infections. Crystal R, West J, Barnes P, Weibel E, eds. *The Lung: Scientific Foundations.* Vol. 2. Philadelphia, PA: Lippincott-Raven Publishers. Pp. 2353-2365.

Rijcken B, Schouten JP, Xu X, Rosner B, Weiss ST. 1995. Airway hyperresponsiveness to histamine associated with accelerated decline in FEV1. *Am J Respir Crit Care Med* 151(5):1377-1382.

Robbins AS, Abbey DE, Lebowitz MD. 1993. Passive smoking and chronic respiratory disease symptoms in non-smoking adults. *Int J Epidemiol* 22(5):809-817.

Rogot E, Murray JL. 1980. Smoking and causes of death among U.S. veterans: 16 years of observation. *Public Health Rep* 95(3):213-222.

Rokaw SN, Detels R, Coulson AH, Sayre JW, Tashkin DP, Allwright SS, Massey FJ. 1980. The UCLA population studies of chronic obstructive respiratory disease. Comparison of pulmonary function in three communities exposed to photochemical oxidants, multiple primary pollutants, or minimal pollutants. *Chest* 78(2):252-262.

Rylander E, Pershagen G, Eriksson M, Bermann G. 1995. Parental smoking, urinary cotinine, and wheezing bronchitis in children. *Epidemiology* 6(3):289-293.

Samet JM, Lange P. 1996. Longitudinal studies of active and passive smoking. *Am J Respir Crit Care Med* 154(6 Pt 2):S257-265.

Samet JM, Tager IB, Speizer FE. 1983. The relationship between respiratory illness in childhood and chronic air-flow obstruction in adulthood. *Am Rev Respir Dis* 127(4):508-523.

Schottenfeld D. 1984. Epidemiology of cancer of the esophagus. *Semin Oncol* 11(2):92-100.

Sekhon HS, Wright JL, Churg A. 1994. Cigarette smoke causes rapid cell proliferation in small airways and associated pulmonary arteries. *Am J Physiol* 267(5 Pt 1):L557-563.

Senior RM, Shapiro SD. 1998. Chronic obstructive lung disease: epidemiology, pathophysiology, and pathogenesis. Fishman A, Elias J, Fishman J, et al., eds. *Fishman's Pulmonary Diseases and Disorder*. 3rd ed. New York: McGraw-Hill. Pp. 680.

Sethi JM, Rochester CL. 2000. Smoking and chronic obstructive pulmonary disease. *Clin Chest Med* 21(1):67-86.

Shaheen SO, Barker DJ, Holgate ST. 1995. Do lower respiratory tract infections in early childhood cause chronic obstructive pulmonary disease? *Am J Respir Crit Care Med* 151(5):1649-51, discussion 1651-1652.

Shapiro SD. 2000. Animal models for COPD. *Chest* 117(5 Suppl 1):223S-227S.

Sherman CB. 1991. Health effects of cigarette smoking. *Clin Chest Med* 12(4):643-658.

Sherman CB, Xu X, Speizer FE, Ferris Jr BG, Weiss ST, Dockery DW. 1992. Longitudinal lung function decline in subjects with respiratory symptoms. *Am Rev Respir Dis* 146:855-859.

Sly R. 2000. Asthma. Behrman R, Kliegman R, Jenson H, eds. *Nelson Textbook of Pediatrics*. Philadelphia: WB Saunders Company.

Snider GL. 1989. Pulmonary disease in alpha-1-antitrypsin deficiency. *Ann Intern Med* 111(12):957-959.

Stanwell-Smith RE, Stuart JM, Hughes AO, Robinson P, Griffin MB, Cartwright K. 1994. Smoking, the environment and meningococcal disease: a case control study. *Epidemiol Infect* 112(2):315-328.

Stick SM, Burton PR, Gurrin L, Sly PD, LeSouef PN. 1996. Effects of maternal smoking during pregnancy and a family history of asthma on respiratory function in newborn infants. *Lancet* 348(9034):1060-1064.

Strachan DP, Cook DG. 1998. Health effects of passive smoking. Parental smoking and childhood asthma: longitudinal and case-control studies. *Thorax* 53(3):204-212.

Straus WL, Plouffe JF, File TM Jr, et al. 1996. Risk factors for domestic acquisition of legionnaires disease. Ohio Legionnaires Disease Group. *Arch Intern Med* 156(15):1685-1692.

Stuart J, Cartwright K, et al. 1989. Effect of smoking on meningococcal cariage. *Lancet* 2:723-725.

Svendsen KH, Kuller LH, Martin MJ, Ockene JK. 1987. Effects of passive smoking in the Multiple Risk Factor Intervention Trial. *Am J Epidemiol* 126(5):783-795.

Tager IB, Segal MR, Munoz A, Weiss ST, Speizer FE. 1987. The effect of maternal cigarette smoking on the pulmonary function of children and adolescents. Analyses of data from two populations. *Am Rev Respir Dis* 136(6):1366-1370.

Tager IB, Weiss ST, Munoz A, Rosner B, Speizer FE. 1983. Longitudinal study of the effects of maternal smoking on pulmonary function in children. *N Engl J Med* 309(12):699-703.

Tager IB, Weiss ST, Rosner B, Speizer FE. 1979. Effect of parental cigarette smoking on the pulmonary function of children. *Am J Epidemiol* 110(1):15-26.

Tanoue LT. 2000. Cigarette smoking and women's respiratory health. *Clin Chest Med* 21(1):47-65.

Tashkin DP, Altose MD, Connett JE, Kanner RE, Lee WW, Wise RA. 1996. Methacholine reactivity predicts changes in lung function over time in smokers with early chronic obstructive pulmonary disease. The Lung Health Study Research Group. *Am J Respir Crit Care Med* 153(6 Pt 1):1802-1811.

Tashkin DP, Detels R, Simmons M, Liu H, Coulson AH, Sayre J, Rokaw S. 1994. The UCLA population studies of chronic obstructive respiratory disease: XI. Impact of air pollution and smoking on annual change in forced expiratory volume in one second. *Am J Respir Crit Care Med* 149(5):1209-1217.

Taylor B, Wadsworth J. 1987. Maternal smoking during pregnancy and lower respiratory tract illness in early life. *Arch Dis Child* 62(8):786-791.

Thun MJ, Myers DG, Day-Lally C, et al. 1997. Age and exposure-response relationships between cigarette smoking and premature death in cancer prevention study II. National Cancer Institute. *Changes in Cigarette-Related Disease Risks and Their Implication for Prevention and Control. NCI Smoking and Tobacco Control Monograph 8.* Washington, DC: NCI. Pp. 383-475.

Traber MG, van der Vliet A, Reznick AZ, Cross CE. 2000. Tobacco-related diseases. Is there a role for antioxidant micronutrient supplementation? *Clin Chest Med* 21(1):173-187.

Tracey M, Villar A, Dow L, Coggon D, Lampe FC, Holgate ST. 1995. The influence of increased bronchial responsiveness, atopy, and serum IgE on decline in FEV1. A longitudinal study in the elderly. *Am J Respir Crit Care Med* 151(3 Pt 1):656-662.

Tredaniel J, Boffetta P, Saracci R, Hirsch A. 1994. Exposure to environmental tobacco smoke and risk of lung cancer: the epidemiological evidence. *Eur Respir J* 7(10):1877-1888.

U.S. DHHS (U.S. Department of Health and Human Services). 1984. *The Health Consequences of Smoking: Chronic Obstructive Lung Disease. A Report of the Surgeon General.* Washington, DC: U.S. Department of Health and Human Services, Public Health Service, Office of Smoking and Health.

U.S. DHHS. 1986. *The Health Consequences of Involuntary Smoking: A Report of the Surgeon General.* Atlanta, GA: U.S. Department of Health and Human Services, Centers for Disease Control and Prevention.

U.S. DHHS. 1989. *Reducing the Health Consequences of Smoking; A Report of the Surgeon General.* Washington, DC: U.S. Department of Health and Human Services, Centers for Disease Control and Prevention.

U.S. DHHS. 1990. *The Health Benefits of Smoking Cessation; A Report of the Surgeon General.* Washington, DC: U.S. Department of Health and Human Services, Centers for Disease Control and Prevention.

van der Lende R, Kok T, et al. 1981. Decreases in VC and FEV1 with time: indicators for effects of smoking and air pollution. *Bulletin Europeen de Physiopathologie Respiratoire* 17:775-792.

Vestbo J, Prescott E, Lange P. 1996. Association of chronic mucus hypersecretion with FEV1 decline and chronic abstructive pulmonary disease morbidity. Copenhagen City Heart Study Group. *Am J Respir Crit Care Med* 153(5):1530-1535.

Weiss K, Gergen P, et al. 1993. Breathing better or worse? The changing epidemiology of asthma morbidity and mortality. *Ann Rev Public Health* 14:491-513.

Weiss ST, Utell MJ, Samet JM. 1999. Environmental tobacco smoke exposure and asthma in adults. *Environ Health Perspect* 107(suppl 6):891-895.

White JR, Froeb HF. 1980. Small-airways dysfunction in nonsmokers chronically exposed to tobacco smoke. *N Engl J Med* 302(13):720-723.

Whittemore AS, Perlin SA, DiCiccio Y. 1995. Chronic obstructive pulmonary disease in lifelong nonsmokers: results from NHANES. *Am J Public Health* 85(5):702-706.

Willers S, Svenonius E, Skarping G. 1991. Passive smoking and childhood asthma. Urinary cotinine levels in children with asthma and in referents. *Allergy* 46(5):330-334.

Winkelstein M, Tarzian A, et al. 1997. Paternal smoking behavior and passive smoke exposure in children with asthma. *Ann Allergy Asthma Immunol* 78:419-423.

Witschi H, Joad JP, Pinkerton KE. 1997. The toxicology of environmental tobacco smoke. *Annu Rev Pharmacol Toxicol* 37:29-52.

Woolcock A. 1994. Asthma. Murray J, Nadel J, eds. *Textbook of Respiratory Medicine.* Philadelphia, PA: W.B. Saunders and Company. Pp. 1288-1330.

Young S, Le Souef P, et al. 1991. The influence of a family history of asthma and parental smoking on airway responsiveness in early infancy. *N Engl J Med* 324:1168-1173.

Yunginger JW, Reed CE, O'Connell EJ, Melton LJ 3rd, O'Fallon WM, Silverstein MD. 1992. A community-based study of the epidemiology of asthma. Incidence rates, 1964-1983. *Am Rev Respir Dis* 146(4):888-894.

Zang LY, Stone K, Pryor WA. 1995. Detection of free radicals in aqueous extracts of cigarette tar by electron spin resonance. *Free Radic Biol Med* 19(2):161-167.

Zeise L, Dunn A, Haroun L, Ting D, Windham G, Golub M, Waller K, Lipsett M, Shusterman D, Mann J, Wu A. 1997. *Health Effects of Exposure to Environmental Tobacco Smoke. Office of Health and Environmental Health Hazard Assessment.* California: Environmental Protection Agency.

Zejda J, McDuffie H, et al. 1993. Respiratory effects of exposure to grain dust. *Seminars in Respiratory Medicine* 14:20-30.

15

Reproductive and Developmental Effects

Smoking among women of reproductive age is a critical risk factor for reproductive health problems including fetal and infant mortality and impaired fetal development. Cigarette smoking has numerous well-documented adverse effects on pregnancy and fetal health, including low birthweight, preterm delivery, perinatal morbidity, placental complications, and increased risk of sudden infant death syndrome (USDHHS, 1988, 1990). The harmful effects of cigarette smoke exposure during pregnancy have been well known for decades; nevertheless, a significant fraction of pregnant women continue to smoke and smoking continues to account for an estimated 10% of all fetal mortality (Kleinman et al., 1988). The percentage of women who smoke during pregnancy declined from 13.6% in 1996 to 12.9% in 1998 with rates being highest for non-Hispanic whites, American Indian, and Hawaiian women and for women of lower socioeconomic and educational levels (CDC, 1998, 1999, 2000). The number of cigarettes smoked per day has also steadily declined over the last decade with about a third of maternal smokers reporting smoking at least a half a pack per day in 1996 compared to more than 40% in 1990 (CDC, 1998). Among pregnant teenagers, however, the smoking rate increased from 18.8% in 1997 to 19.2% in 1998 (CDC, 2000).

Trends in maternal smoking behavior, based on data from the Behavioral Risk Factor Surveillance System, showed a decline in overall smoking initiation among women aged 18-44 to a reported rate of 38.2% in 1996, with no difference between pregnant and nonpregnant women. Reported rates of quitting in the same population have shown little change among pregnant and nonpregnant women, 25.2% and 14.4%, respectively in 1996 (Ebrahim et al., 2000). Even after learning that they are pregnant, 54% of women continue to smoke (Ebrahim et al., 2000). Additionally, maternal smokers who have experienced previous preterm delivery or small-for-gestational age infants do not show greater quit rates than smokers who have had uncomplicated deliveries (Cnattigius et al., 1999). According to the National Health Interview Survey (1992-1993), among female smokers in general, 72.5% reported that they wanted to quit smoking with 34% attempting to quit each year and 2.5% being successful.

The contribution of environmental tobacco smoke (ETS), including paternal smoking, to adverse reproductive health outcomes is uncertain, but ETS exposure is widespread among women of reproductive age. Data from the third National Health and Nutrition Examination Survey report a 32.9% prevalence of ETS exposure at home or at work among non-tobacco-using females age 17 and over.

FERTILITY IMPAIRMENT

Smoking has been associated with increased time to conception, decreased pregnancy rate in assisted reproduction, increased risk of ectopic pregnancy, and menstrual changes including early menopause. The risk of being unable to conceive within a year of trying is increased two- to threefold among smokers (Werler, 1997). Consistent with many previous studies, a recent cohort study of current and past smokers during assisted reproduction cycles suggested a dose-related decrease in ovarian function and a 50% reduction in pregnancy rates of current smokers (Van Voorhis et al., 1996). Also, a positive relationship (odds ratio, OR≅1.3-2.2) between cigarette smoke exposure at conception and during pregnancy and the risk of subsequent ectopic pregnancy has been documented, with mixed results regarding dose-response (Coste et al., 1991; Handler et al., 1989; Saraiya et al., 1998; Stergachis et al., 1991). Animal studies have suggested altered gonadotropin release, decrease in luteinizing hormone (LH) surge, inhibition of prolactin release, altered tubal motility, and impairment of blastocyst formation and implantation as possible mechanisms of fertility impairment among smokers (reviewed in Hughes and Brennan, 1996). Additional studies in rats have shown follicle destruction and oocyte depletion when exposed to benzo[a]pyrene (BaP), a tobacco smoke toxin (Cooper et al., 1999). Furthermore, an evaluation of the Women's Health Study found that current and former smokers, after adjusting for age, race, education, marital status, number of sexual partners, frequency of intercourse, history of gonorrhea, and current method of contraception, had a significantly increased risk of pelvic inflammatory disease, possibly related to impairment of immunity and altered tubal factors (Marchbanks et al., 1990; Scholes et al., 1992). Pelvic inflammatory disease is an independent risk factor for ectopic pregnancy and infertility.

The association of altered male fertility with cigarette smoking is less consistent. Studies have shown mixed results with respect to sperm and semen quality and have not supported detrimental effects on male fertility (reviewed in Hughes and Brennan, 1996). Several animal and human studies have established an increased risk of vascular erectile dysfunction associated with cigarette smoking (Juenemann et al., 1987; Shabsigh et al., 1991; U.S. DHHS, 1990), including a large survey of Vietnam-era veterans, age 31 to 49, which found a 50% increase in the risk of impotence in smokers after controlling for all other major risk factors for impotence (Mannino et al., 1994).

SPONTANEOUS ABORTIONS

Numerous studies suggest a positive, dose-related association between smoking and spontaneous abortions, with the risk reported to be increased 20-80% (Kline et al., 1977, 1983; Ness et al., 1999). A prospective study by Ness et al. (1999) of adolescent girls and women that presented to an emergency room found a strong correlation between the risk of spontaneous abortion and the presence of cotinine (threshold concentration=500 ng/ml) in the urine (OR=1.8). The association is found primarily in the second trimester and for chromosomally normal spontaneous abortions (Kline et al., 1995) thought to be associated with intrauterine growth retardation. Postulated causal mechanisms include fetal hypoxia mediated by carbon monoxide (CO) and placental and uterine vascular insufficiency or teratogenic effects mediated by nicotine (Kline et al., 1995; Ness et al., 1999). The effect is modified by alcohol consumption, caffeine use, and history of previous spontaneous abortions (Ness et al., 1999; Windham et al., 1992). Although recent large prospective study found no consistent evidence of an association between environmental tobacco smoke and spontaneous abortion (Windham et al., 1999), a few earlier studies have described such a relationship (Ahlborg and Bodin, 1991; Chatenoud et al., 1998; Windham et al., 1992).

PLACENTAL COMPLICATIONS

Findings have been consistent regarding the positive association of smoking with placenta previa (obstruction of the internal cervical os by the placenta), with a relative risk (RR)=1.3-2.6 and placental abruption (premature separation of the placenta from the uterus), RR=1.4-1.6 (reviewed in Andres and Day, 2000; Castles et al., 1999). These pregnancy complications cause at least one-fifth of all prenatal deaths (Ananth et al.,

1996). Also, among smokers the perinatal death rate after placental abruption is two to three times higher than among nonsmokers (Werler, 1997). There has been consistent evidence that smoking is an independent risk factor for placenta previa and placental abruption after control for potential confounders including maternal age and parity, hypertension, preeclampsia, and alcohol use.

Studies have suggested a dose-dependent association between reported numbers of cigarettes smoked and placental complications, but the data have been inconclusive (Ananth et al., 1996; Handler et al., 1994; reviewed in Andres, 1996). Although the exact mechanism by which maternal smoking causes placental complications is unknown, the placentas of smokers exhibit anatomical and histological changes that suggest hypoxia and underperfusion (Voigt et al., 1990). The placentas of smokers have been found to be larger and heavier than those of nonsmokers (Christianson, 1979; Pfarrer et al., 1999). Pfarrer et al. (1999) found increased angiogenesis within the placental villi of smokers, thought to be a response to hypoxic stress caused by the components of cigarette smoke. Pfarrer and colleagues speculated that this adaptive response contributes to the large placentas of maternal smokers. Histologic studies have found necrotic and hemorrhagic changes of the decidua basalis, including calcification, hypertrophy, and thickening of the basement membrane (reviewed in Voigt et al., 1990). The hypoxic and ischemic changes in placental tissues are possibly due to damage of the endothelial cells of placental vessels resulting in decreased tissue perfusion and to the increase in carboxyhemoglobin resulting in decreased oxygen delivery (Ananth et al., 1996). It has also been postulated that smokers are more prone to placental inflammation and infection secondary to a smoking-related decreased immune response.

PRETERM DELIVERY

Preterm delivery is defined as delivery before 37 weeks gestation and is an important cause of perinatal mortality. Studies have suggested an increase in the risk of preterm delivery among maternal smokers but have been inconsistent regarding the magnitude of risk (20% to >100% increased risk). Evidence of a dose-response association has been demonstrated (Cnattingius, et al., 1999, reviewed in Shah and Bracken, 2000; Shiono et al., 1986). Risk of infant mortality was more pronounced for very preterm births for preterm delivery (≤32 weeks gestation) (Kyrklund-Blomberg and Cnattingius, 1998; Windham et al., 2000) and with spontaneous versus induced preterm births after controlling for complications of pregnancy such as placental abruption, placenta previa, premature rupture of membranes, and hypertension (Kyrklund-Blomberg and Cnattingius, 1998).

The literature has supported the reduction of risk of preterm delivery associated with smoking cessation. A large population-based cohort study of Swedish women by Cnattingius et al. (1999) found that among non-smokers with a term first delivery, those who initiated smoking after the first pregnancy had a greater risk of subsequent preterm pregnancies compared to women who did not smoke during pregnancy. Maternal smokers with an initial term delivery reduced their risk of subsequent preterm deliveries by stopping smoking prior to the second pregnancy. An earlier prospective interventional trial by Li (1993) found no improvement of gestational age with smoking reduction but did show a significant improvement after smoking cessation.

The mechanism of smoke-attributed preterm delivery is not certain but may be related to intrauterine infections secondary to decreased immunity, structural abnormalities especially the loss of integrity of type III collagen, or the increase in production of prostaglandin (PGE_2), causing myometrial muscle contractions (Cnattigius et al., 1999).

LOW BIRTHWEIGHT

The detrimental effect of smoking on birthweight has been extensively studied and well documented. This effect has been labeled "fetal tobacco syndrome," which is described as maternal smoking of five or more cigarettes per day during pregnancy, no evidence of maternal hypertension during pregnancy, symmetrical growth retardation of newborn at term, and no other explicit cause of intrauterine growth retardation (Benowitz, 1991; Nieburg et al., 1985). Twelve percent of infants born to all mothers who are smokers weighed less than 2,500 grams (low birthweight) in 1998, and eleven percent of infants born to mothers who report smoking as few as one to five cigarettes per day have low birthweight (CDC, 2000). Among pregnant smokers the risk of low birthweight babies is doubled compared to nonsmokers, and about 20% of all low birthweight babies are attributable to smoking (U.S. DHHS, 1983). Infants born to mothers who smoke during pregnancy are on average 200 grams lighter and 1.4 cm shorter than infants of nonsmokers (Wang et al., 1997). The effect of smoking is particularly prominent if exposure occurs after the first trimester. The association persists even after controlling for confounding factors such as maternal age, maternal nutrition, socioeconomic level, education, maternal weight gain, and alcohol consumption (reviewed in ACOG, 1997). Long-term effects of maternal smoking on growth have not been described, but small-for-gestational-age infants have a higher incidence of certain illnesses and disabilities into childhood and adulthood, such as cardiovascular disease, non-insulin-dependent diabetes mellitus, and hypertension (reviewed in Barker, 1997, 1999; Foresen et al., 2000).

Environmental tobacco smoke has also been suggested to have a significant association with intrauterine growth retardation. A recent Swedish study found an OR of 2.4 for low birthweight infants among non-smoking mothers exposed to ETS and an OR of 3.6 for smoking mothers exposed to ETS (Dejin-Karlsson et al., 1998). Paternal smoking has also been shown to have an adverse effect on infant birthweight. Martinez et al. (1994) report an 88-gram average reduction in birthweight of infants whose fathers smoked more than 20 cigarettes per day.

The dose-response evidence has been studied, extensively, and the findings suggest reversibility of risk with smoking reduction (Ahlsten et al., 1993; Hebel et al., 1988; Li et al., 1993; Sexton and Hebel, 1984). Li and his colleagues (1993) in a prospective intervention trial found that reduction as well as cessation of smoking (defined by serum cotinine levels) resulted in significantly increased birthweights compared to women who continued to smoke. Secker-Walker et al. (1998) reported that women who stop smoking before 20 weeks' gestation obtain maximum benefits of smoking reduction. Others have found that cessation even late in pregnancy led to normal-weight infants (Ahlsten et al., 1993; Hebel et al., 1988). Attempts have been made to quantify the relationship of cigarette consumption and fetal growth outcomes. One such study, which evaluated taking the average of serial cotinine measurements over the entire pregnancy, reported that for every 1,000-ng/ml increase in urine cotinine concentration, there was an associated 59 ± 9 gram reduction in birthweight, an $\cong 0.25$-cm reduction in length, and an $\cong 0.12$-cm reduction in head circumference (Wang et al., 1997). Maternal serum and urine cotinine levels have been reliable markers of maternal nicotine intake and cigarette use and useful tools in predicting infant birthweight (Haddow et al., 1987; Li et al., 1993). Cotinine levels in infant cord blood are highly correlated with maternal serum and urine cotinine concentrations, $r=.91$ and $r=.72$, respectively (Wang et al., 1997). Among African Americans, a few studies have noted differences in the relationship between cigarette consumption and cotinine concentrations. Specifically, these studies have shown higher cotinine levels for all cigarette dose levels among African Americans (English et al., 1994). This group also reported no significant difference in birthweight reduction per 1 ng/ml maternal cotinine among black infants compared to white infants, and Li et al. (1993) found that among mothers with high baseline cotinine levels (>200 ng/ml), the birthweights of black infants were less sensitive to smoking reduction than those of white infants.

The pathogenesis of low birthweight secondary to smoking is not known but is generally thought to be multifactorial. The probable mechanisms involved include CO formation of carboxyhemoglobin in maternal and fetal circulation, causing decrease in oxygen delivery and resulting in tissue hypoxia and increased viscosity that may affect placental perfusion (Benowitz et al., 2000). Cyanide in tobacco smoke can decrease stores of vitamin B_{12}, a cofactor for fetal growth (Ness et al., 1999). Additionally, smoking-related maternal and fetal nutritional deficits caused by cigarette smoking have been postulated as mechanisms for fetal growth retardation. Reduction in uteroplacental blood flow due to the vasoconstrictive catecholamine-mediated effects of nicotine has been suggested as a possible mechanism of intrauterine growth retardation and has been studied in animal models (Bassi et al., 1984). These findings have not been fully supported by measurements of placental blood flow in humans (Benowitz et al., 2000).

SUDDEN INFANT DEATH SYNDROME (SIDS)

It is estimated that more than 50% of the risk of SIDS may be attributed to exposure to cigarette smoke (Dwyer et al., 1999). A positive association between maternal smoking and SIDS has been consistently supported in the literature (reviewed in Golding, 1997; Leach et al., 1999). Generally, the risk of SIDS increases two- to fourfold among infants of mothers who smoke during pregnancy, and the risks increase even further when combined with postnatal exposure to tobacco smoke (ACOG, 1997). However, it has been difficult to separate the effects of postnatal smoke exposure from the effects of prenatal tobacco smoke exposure (Golding, 1997; Spiers, 1999). A recent prospective study found an OR of 2-3 for the association between prenatal maternal smoking and risk of SIDS, and an OR of roughly 3 for postnatal maternal smoking; however, no significant effect was seen for smoke exposure from other household members or maternal smoking restricted to rooms without the infant (Dwyer et al., 1999). The mechanism of effect is not fully known. It has been suggested, based on animal models, that fetal nicotine exposure results in loss of the normal response of the adrenomedullary system to hypoxia (Benowitz, 1998; Slotkin et al., 1995, 1997). Furthermore, the increased susceptibility of infants of smoking mothers to respiratory infections may play a role.

CONGENITAL MALFORMATIONS

Oral clefts is the most extensively studied malformation thought to be associated with maternal smoking. Research has failed to show a consistent association, although a moderate association with cleft lip ± cleft palate (CLP) has been described (Lieff et al., 1999). A recent study found a strong positive association between cigarette smoking and CLP, with a dose-response effect. This study found an overall 55% increase in the risk of infant CLP among all pregnant smokers, with the risk almost 80% higher for women who smoked more than a pack a day compared to nonsmoking mothers (Chung, 2000). Results have also been inconsistent regarding the interaction of maternal smoking with the rare transforming growth factor-α (TGF-α) allele and association with CLP (Christensen et al., 1999).

COGNITIVE AND BEHAVIORAL DEFICITS IN CHILDHOOD

Tobacco use during pregnancy has been linked to neurological damage that may be expressed during childhood as intellectual deficits, behavioral problems, and poor school achievement. Studies have found significant differences in IQ scores and increased likelihood of mental retardation in children of mothers who smoked during pregnancy compared to children of nonsmokers (Drews et al., 1996; Olds et al., 1994). The differences are decreased but generally persist after control for environmental and parental factors that may influence intelligence. Attention deficit hyperactivity disorder (ADHD) has been associated with smoking during pregnancy. An OR of 2.7 has been found for the risk of children of mothers who smoked during pregnancy after controlling for socioeconomic status, parental ADHD, and parental IQ. Furthermore, conduct disorder and disruptive behavior have been found to have a weak dose-related association with maternal cigarette smoking after control for confounding social factors (Fergusson et al., 1993).

Postulated mechanisms of cognitive impairment involve effects of cigarette smoking on the fetus including chronic hypoxia, decreased nutrition, and direct toxicity to cortical tissue by toxins such as CO, nicotine, and lead in cigarette smoke. Small-for-gestational-age infants, a significant adverse effect of maternal cigarette smoking as discussed previously, have been independently linked in several studies to decreased cognitive abilities (measured by IQ scores) into childhood and adolescence, compared to children of normal birthweight (McCarton et al., 1996; Seidman et al., 1992; Sommerfelt et al., 2000). Animal models have linked fetal nicotine exposure to upregulation of nicotinic receptors causing hyperactivity in infant mice (Milberger et al., 1996). Also, fetal nicotine exposure has potentially detrimental effects on fetal brain development through premature stimulation of nicotinic cholinergic receptors in the fetal brain causing disruption of the development of cholinergic neurons, which affects numerous other neurotransmitters and hormones, including catecholamines, serotonin, and dopamine (Benowitz et al., 2000).

FETAL LUNG DEVELOPMENT

Maternal smoking during pregnancy has been demonstrated to be associated with abnormal effects on fetal lung function and development, possibly contributing to increased incidence of early childhood respiratory diseases, including bronchitis, pneumonia, and asthma (Joad, 2000; Morgan and Martinez, 1998). Based on a review by Joad (2000), in utero cigarette smoke exposure more than doubles the risk for disease of airway hyperresponsiveness in childhood, including wheezing illnesses and asthma. In a large cross-sectional study, Cunningham and colleagues (1994) found significantly lower measures of flow in pulmonary function tests among 8-12 year olds who were exposed to maternal smoking compared to children of nonsmokers. After controlling for prenatal exposure, no significant effect of current ETS exposure was found. This finding was further supported in a study by Tager et al. (1995) that found significantly lower expiratory flow measurements of infants up to 18 months of age among mothers who smoked during pregnancy. Lodrup Carlsen and colleagues (1997) found supporting evidence that flow parameters were diminished when measured within days of birth after in utero smoke exposure. This study also found an effect on volume measurements and a dose-response relationship to number of cigarettes smoked per day. A study of infants who were seven weeks premature (Hoo et al., 1998) suggests that the detrimental effects on respiratory function occur before the last trimester and further supports the predominant effects of in utero smoke exposure compared to postnatal ETS exposure.

Suggested mechanisms of airway dysfunction caused by cigarette smoke exposure during pregnancy include decreased airway compliance, poor bronchial tree development, and emphysema-like changes of the alveoli (Cunningham et al., 1994; Lodrup Carlsen et al., 1997). Several animal studies have shown impaired fetal lung growth, including decreased lung interstitium and elastic tissue (Maritz et al., 1993; Moessinger, 1989). It has been suggested that nicotine interferes with the synthesis of elastic tissue needed for stability of the alveoli and that other smoke constituents may promote protease function (Maritz et al., 1993). See Chapter 14 for further discussion of this topic.

SMOKELESS TOBACCO

Reproductive effects of smokeless tobacco have been less extensively studied than those of smoking. Many of the human data come from countries in which consumption of tobacco in other forms is more widespread among women. Animal studies have shown significant weight reduction and perinatal mortality among rat fetuses of mothers exposed to smokeless tobacco, administered through gastric intubation of smokeless tobacco aqueous extract (Krishna, 1978; Paulson et al., 1994). Researchers in India have found a significant decrease in birthweight of infants born to mothers who use smokeless tobacco (Deshmukh et al., 1998; Krishnamurthy and Joshi, 1993; Verma et al., 1983). Changes in placental anatomy were also noted by Agrawal and colleagues (1983), who discovered the placenta of tobacco-chewing mothers to be on average 65.9 grams heavier than that of non-using mothers. It has been suggested that nicotine-induced vasoconstriction and decreased perfusion may be the primary mechanism leading to decreased fetal growth in mothers who use smokeless tobacco (Verma et al., 1983).

CONCLUSIONS

Exposure to cigarette smoking is a major cause of fetal and infant morbidity and mortality. This is particularly true for the association with low birthweight and it consequences, as well as for preterm delivery and SIDs. For several important adverse reproductive effects of maternal smoking, a decrease in smoking has been found to decrease or be associated with a decrease in risks to the fetus and infant. The greatest benefit, of course, comes from smoking cessation. However, many women continue to smoke during pregnancy, despite knowledge of the harmful effects of smoking and personal experience with adverse fetal and infant conditions. Moreover, as current rates of smoking among adolescent women slowly rise, these adverse effects associated with tobacco smoke exposure while pregnant may worsen.

On average, infants exposed to maternal smoking in utero are 200 grams lighter and 1.4 cm shorter than those who are unexposed. A strong dose-response has been supported in numerous studies and a decrease in dose (number of cigarettes) in controlled studies has led to increased birthweights in a predictable pattern. What is known about the mechanism of effect of cigarette smoke on the fetus suggests a multifactorial etiology, with CO considered to play a major role in growth retardation through increased tissue hypoxia. Nicotine has also been thought to play a role through increasing vasoconstriction and decreasing perfusion through the placenta.

Although nicotine replacement products and bupropion SR are currently not approved by the Food and Drug Administration for use by pregnant women, the Agency for Healthcare Research and Quality's (AHCRQ) Clinical Practice Guidelines for Treating Tobacco Use and Dependence (Fiore et al., 2000) recommend that "Pharmacotherapy should be considered when a pregnant woman is otherwise unable to quit, and when the likelihood of quitting, with its potential benefits, outweighs the risks of the pharmacotherapy and potential continued smoking". It is generally thought that nicotine replacement therapy can reasonably be used with pregnant patients if prior behavioral modifications have failed and the patient continues to smoke at least 10-15 cigarettes per day (ACOG, 1997). There are no data regarding the efficacy of potential reduced-exposure products (PREPs) during pregnancy, but there is the presumption that the tobacco-related PREPs are likely to have toxic effects at some level and that, until further evidence is produced, existing guidelines concerning pharmacological PREPs still pertain.

RECOMMENDATIONS

Surveillance of Tobacco Use Patterns Among Pregnant Women

Central to understanding exposure to tobacco products is continuous population information on patterns of tobacco use among pregnant women. This may not be attainable by general population survey methods, due to inadequate sample sizes and insufficient representation of various geographic or demographic groups or of the earliest stages of pregnancy. Thus, surveys should be devoted specifically to pregnant women in all stages of gestation, irrespective of receipt of medical care. Survey content should include other known or putative causes of adverse maternal or fetal outcomes, as well as detailed product types and usage patterns as delineated in Chapter 6, in recommendations for general population surveillance.

Biochemical and toxicological exposure measures should be a routine part of surveillance for exposure to conventional products as well as PREPs. These will be necessary to conduct more precise, coordinated toxicological studies and also to assess actual exposure rates more accurately. For example, dose may be measured by maternal serum and urine cotinine levels, which have shown reliable correlations with maternal and consequently fetal tobacco smoke exposure. Self-reported smoking data can be unreliable, since pregnant women who have been advised to quit tend to under report tobacco use because of the stigma attached to smoking (Kendrick et al., 1995). Also, self-reports do not adequately account for differences in depth and frequency of puffs among smokers.

Assessment of Fetal and Maternal Outcomes
Associated with New Tobacco Product Exposure

To practically assess the health effects of PREPs, reliable measures of health outcomes that can be utilized in a relatively short time are desired. Among the reproductive outcomes of maternal smoking, intrauterine growth retardation resulting in low birthweight babies has been extensively studied, and a large body of evidence has supported a causal link with cigarette smoke exposure. The committee recommends, based on currently available scientific knowledge, that fetal birthweight and intrauterine growth retardation be used as the outcome measure in evaluating the harm reduction potential of the use of PREPs. Study designs might include repeated cohort or case-control studies of pregnant women with an appropriate distribution of exposures to both PREPs and conventional products, and suitable contrast groups. Concomitant, coordinated toxicological studies should be performed to provide biological correlations with clinical outcomes. Such outcomes as fetal birthweight and incidence of other reproductive and developmental health outcomes (e.g., fertility outcomes, placental complications, gestational age at birth, incidence of SIDS, spontaneous abortion, etc.) should be considered primary objects of study.

Findings in pregnant women exposed to PREPs may have value beyond maternal/fetal outcomes. The nature of adverse effects from PREP exposure will likely be determined much sooner in pregnant women (several months) than findings on chronic disease outcomes such as various cancers and cardiovascular disease in nonpregnant tobacco users. Should adverse findings become apparent, there may substantial implications for risk of chronic illnesses among nonpregnant adults, and coordinated pathogenic studies might allow conclusions on new tobacco product outcomes in advance of studies exploring longer "incubation periods."

Studies on the Component Exposures of PREPs

The committee recommends that further basic research be undertaken to elucidate the components of cigarette smoke that are primarily responsible for adverse health outcomes. In order to evaluate the safety of many PREPs, it is important to understand the effect of smoke components, especially nicotine and CO, on the pathogenesis of intrauterine growth retardation, spontaneous abortions and other health outcomes. In addition, better understanding of the risks of bupropion SR use by pregnant women (i.e., seizure risk) and the teratogenic effects of nicotine on the central nervous system, including dose-response and periods of vulnerability during gestation, is needed for adequate risk-benefit analysis of the harm reduction potential of these products.

REFERENCES

ACOG (American College of Obstetricians and Gynecologists). 1997. Educational Bulletin: Smoking and women's health. *International Journal of Gynecology & Obstetrics* 60:71-82.

Agrawal P, Chansoriya M, Kaul KK. 1983. Effect of tobacco chewing by mothers on placental morphology. *Indian Pediatrics* 20(8):561-565.

Ahlborg G Jr, Bodin L. 1991. Tobacco smoke exposure and pregnancy outcome among working women. A prospective study at prenatal care centers in Orebro County, Sweden. *Am J Epidemiol* 133:338-347.

Ahlsten G, Cnattingius S, Lindmark G. 1993. Cessation of smoking during pregnancy improves fetal growth and reduces infant morbidity in the neonatal period. A population-based prospective study. *Acta Paediatr* 82(2):177-181.

Ananth CV, Savitz DA, Luther ER. 1996. Maternal cigarette smoking as a risk factor for placental abruption, placenta previa, and uterine bleeding in pregnancy. *Am J Epidemiol* 144(9):881-889.

Andres RL. 1996. The association of cigarette smoking with placenta previa and abruptio placentae. *Semin Perinatol* 20(2):154-159.

Andres RL, Day MC. 2000. Perinatal complications associated with maternal tobacco use. *Semin Neonatol* 5(3):231-241.

Barker D. 1997. Intrauterine programming of coronary heart disease and stroke. *Acta Paediatr Suppl* 423:178-182.

Barker D. 1999. Fetal origins of cardiovascular disease. *Ann Med* 31(SUPP 1):3-6.

Bassi JA, Rosso P, Moessinger AC, Blanc WA, James LS. 1984. Fetal growth retardation due to maternal tobacco smoke exposure in the rat. *Pediatr Res* 18(2):127-130.

Benowitz NL. 1991. Nicotine replacement therapy during pregnancy. *JAMA* 266(22):3174-3177.

Benowitz NL. 1998. *Nicotine Safety and Toxicity.* PAF2. New York: Oxford University Press.

Benowitz NL, Dempsey DA, Goldenberg RL, Hughes JR, Dolan-Mullen P, Ogburn PL, Oncken C, Orleans CT, Slotkin TA, Whiteside HP, Yaffe S. 2000. The use of pharmacotherapies for smoking cessation during pregnancy. *Tob Control* 9 Suppl 3(2):III91-4.

Castles A, Adams EK, Melvin CL, Kelsch C, Boulton ML. 1999. Effects of smoking during pregnancy. Five meta-analyses. *Am J Prev Med* 16(3):208-215.

CDC (Centers for Disease Control and Prevention). 1998. Smoking during pregnancy, 1990-96. *National Vital Statistics Reports* 47 (10):1-12.

CDC (Centers for Disease Control and Prevention). 1999. Prevalence of selected maternal and infant characteristics, pregnancy risk assessment monitoring system (PRAMS), 1997. *MMWR* 48(SS05):1-37.

CDC (Centers for Disease Control and Prevention). 2000. Tobacco use during pregnancy. *National Vital Statistics Report* 48(3):10-11.

Chatenoud L, Parazzini F, di Cintio E, Zanconato G, Benzi G, Bortolus R, La Vecchia C. 1998. Paternal and maternal smoking habits before conception and during the first trimester: relation to spontaneous abortion. *Ann Epidemiol* 8(8):520-526.

Christensen K, Olsen J, Norgaard-Pedersen B, Basso O, Stovring H, Milhollin-Johnson L, Murray JC. 1999. Oral clefts, transforming growth factor alpha gene variants, and maternal smoking: a population-based case-control study in Denmark, 1991-1994. *Am J Epidemiol* 149(3):248-255.

Christianson RE. 1979. Gross differences observed in the placentas of smokers and non-smokers. *Am J Epidemiol* 110(2):178-187.

Chung KC, Kowalski CP, Kim HM, Buchman SR. 2000. Maternal cigarette smoking during pregnancy and the risk of having a child with cleft lip/palate. *Plast Reconstr Surg* 105(2):485-491.

Cnattingius S, Granath F, Petersson G, Harlow BL. 1999. The influence of gestational age and smoking habits on the risk of subsequent preterm deliveries. *N Engl J Med* 341(13):943-948.

Cooper GS, Sandler DP, Bohlig M. 1999. Active and passive smoking and the occurrence of natural menopause. *Epidemiology* 10(6):771-773.

Coste J, Job-Spira N, Fernandez H. 1991. Increased risk of ectopic pregnancy with maternal cigarette smoking. *Am J Public Health* 81(2):199-201.

Cunningham J, Dockery DW, Speizer FE. 1994. Maternal smoking during pregnancy as a predictor of lung function in children. *Am J Epidemiol* 139(12):1139-1152.

Dejin-Karlsson E, Hanson BS, Ostergren PO, Sjoberg NO, Marsal K. 1998. Does passive smoking in early pregnancy increase the risk of small-for-gestational-age infants? *Am J Public Health* 88(10):1523-1527.

Deshmukh JS, Motghare DD, Zodpey SP, Wadva SK. 1998. Low birth weight and associated maternal factors in an urban area. *Indian Pediatrics* 35(1):33-36.

Drews CD, Murphy CC, Yeargin-allsopp M, Decoufle P. 1996. The relationship between idiopathic mental retardation and maternal smoking during pregnancy. *Pediatrics* 97(4):547-553.

Dwyer T, Ponsonby AL, Couper D. 1999. Tobacco smoke exposure at one month of age and subsequent risk of SIDS—a prospective study. *Am J Epidemiol* 149(7):593-602.

Ebrahim SH, Floyd RL, Merritt RK 2nd, Decoufle P, Holtzman D. 2000. Trends in pregnancy-related smoking rates in the United States, 1987-1996. *JAMA* 283(3):361-366.

English PB, Eskenazi B, Christianson RE. 1994. Black-white differences in serum cotinine levels among pregnant women and subsequent effects on infant birthweight. *Am J Public Health* 84(9):1439-1443.

Fergusson DM, Horwood LJ, Lynskey MT. 1993. Maternal smoking before and after pregnancy: effects on behavioral outcomes in middle childhood. *Pediatrics* 92(6):815-822.

Fiore M, Bailey W, Cohen S, Dorfman S, Goldstein M, Gritz E, Heyman R, Jaen C, Kottke T, Lando H, Mecklenburg R, Mullen P, Nett L, Robinson L, Stitzer M, Tommasello A, Villejo L, Wewers M. (2000). *Treating Tobacco Use and Dependence. Clinical Practice Guideline*. Rockville, MD: U.S. Department of Health and Human Services, Public Health Service.

Foresen T, Eriksson J, Tuomilheto J, Reunanen A, Osmond C, Barker D. 2000. The fetal and childhood growth of persons who develop type 2 diabetes. *Ann Intern Med* 133(3):176-182.

Golding J. 1997. Sudden infant death syndrome and parental smoking-a literature review. *Paediatric and Perinatal Epidemiology* 11:67-77.

Haddow JE, Knight GJ, Palomaki GE, Kloza EM, Wald NJ. 1987. Cigarette consumption and serum cotinine in relation to birthweight. *Br J Obstet Gynaecol* 94(7):678-681.

Handler AS, Mason ED, Rosenberg DL, Davis FG. 1994. The relationship between exposure during pregnancy to cigarette smoking and cocaine use and placenta previa. *Am J Obstet Gynecol* 170(3):884-889.

Handler A, Davis F, Ferre C, Yeko T. 1989. The relationship of smoking and ectopic pregnancy. *Amer J Pub Health* 79(9):1239-1242.

Hebel JR, Fox NL, Sexton M. 1988. Dose-response of birth weight to various measures of maternal smoking during pregnancy. *J Clin Epidemiol* 41(5):483-489.

Hoo AF, Henschen M, Dezateux C, Costeloe K, Stocks J. 1998. Respiratory function among preterm infants whose mothers smoked during pregnancy. *Am J Respir Crit Care Med* 158(3):700-705.

Hughes EG, Brennan BG. 1996. Does cigarette smoking impair natural or assisted fecundity? *Fertil Steril* 66(5):679-689.

Joad JP. 2000. Smoking and pediatric respiratory health. *Clinics in Chest Medicine* 21(1):37-46.

Jueneman K, Lue TF, Luo J, Benowitz NL, Abozeid M, Tanagho EA. 1987. The effect of cigarette smoking on penile erection. *J Urology* 138:438-441.

Kendrick JS, Zahniser SC, Miller N, Salas N, Stine J, Gargiullo PM, Floyd RL, Spierto FW, Sexton M, Metzger RW, et al. 1995. Integrating smoking cessation into routine public prenatal care: the Smoking Cessation in Pregnancy project. Am J Public Health 85(2):217-222.

Kleinman JC, Pierre MB, Madans JH, Land GH, Schramm WF. 1988. The effects of maternal smoking on fetal and infant mortality. *Am J Epidemiol* 127(2):274-282.

Kline J, Levin B, Kinney A, Stein Z, Susser M, Warburton D. 1995. Cigarette smoking and spontaneous abortion of known karyotype. Precise data but uncertain inferences. *Am J Epidemiol* 141(5):417-427.

Kline J, Levin B, Shrout P, Stein Z, Susser M, Warburton D. 1983. Maternal smoking and trisomy among spontaneously aborted conceptions. *Am J Hum Genet* 35:421-431.

Kline J, Stein ZA, Susser M, Warburton D. 1977. Smoking: a risk factor for spontaneous abortion. *NEJM* 297(15):793-796.

Krishna K. 1978. Tobacco chewing in pregnancy. *Br J Obstet Gynaecol* 85(10):726-728.

Krishnamurthy S, Joshi S. 1993. Gender differences and low birth weight with maternal smokeless tobacco use in pregnancy. *J Trop Pediatrics* 39(4):253-254.

Kyrklund-Blomberg NB, Cnattingius S. 1998. Preterm birth and maternal smoking: risks related to gestational age and onset of delivery. *Am J Obstet Gynecol* 179(4):1051-1055.

Leach CE, Blair PS, Fleming PJ, Smith IJ, Platt MW, Berry PJ, Golding J. 1999. Epidemiology of SIDS and explained sudden infant deaths. CESDI SUDI Research Group. *Pediatrics* 104(4):e43.

Li CQ, Windsor RA, Perkins L, Goldenberg RL, Lowe JB. 1993. The impact on infant birth weight and gestational age of cotinine-validated smoking reduction during pregnancy. *JAMA* 269(12):1519-1524.

Lieff S, Olshan AF, Werler M, Strauss RP, Smith J, Mitchell A. 1999. Maternal cigarette smoking during pregnancy and risk of oral clefts in newborns. *Am J Epidemiol* 150(7):683-694.

Lodrup Carlsen KC, Jaakola JJK, Nafstad P, Garlsen KH. 1997. In utero exposure to cigarette smoking influences lung function at birth. *Eur Respir J* 10(8):1774-1779.

Mannino DM, Klevins RM, Flanders WD. 1994. Cigarette smoking: an independent risk factor for impotence? *Amer J Epidemiol* 140(11):1003-1008.

Marchbanks PA, Lee NC, Peterson HB. 1990. Cigarette smoking as a risk factor for pelvic inflammatory disease. *Am J Obstet Gyncol* 162:639-644.

Maritz GS, Woolward KM, Du toit G. 1993. Maternal nicotine exposure during pregnancy and development of emphysema-like damage in the offspring. *S Afr Med J* 83(3):195-198.

Martinez FD, Wright AL, Taussig LM. 1994. The effect of paternal smoking on the birthweight of newborns whose mothers did not smoke. *Am J Public Health* 84 (9):1489-1491.

McCarton CM, Wallace IF, Divon M, Vaughan HG Jr. 1996. Cognitive and neurologic development of the premature, small for gestational age infant through age six: comparison by birth weight and gestational age. *Pediatrics* 98(6):1167-1178.

Milberger S, Biederman J, Faraone SV, Chen L, Jones J. 1996. Is maternal smoking during pregnancy a risk factor for attention deficit hyperactivity disorder in children? *Am J Psychiatry* 153(9):1138-1142.

Moessinger AC. 1989. Mothers who smoke and the lungs of their offspring. *Ann N Y Acad Sci* 562:101-104.

Morgan WJ, Martinez FD. 1998. Maternal smoking and infant lung function: further evidence for an in utero effect. *Am J Respir Crit Care Med* 158(3):689-690.

Ness RB, Grisso JA, Hirschinger N, Markovic N, Shaw LM, Day NL, Kline J. 1999. Cocaine and tobacco use and the risk of spontaneous abortion. *N Engl J Med* 340(5):333-339.

Nieburg P, Marks JS, Mclaren NM, Remington PL. 1985. The fetal tobacco syndrome. *JAMA* 253(20):2998-2999.

Olds DL, Henderson CR, Tatelbaum R. 1994. Intellectual impairment in children of women who smoke cigarettes during pregnancy. *Pediatrics* 93(2):221-227.

Paulson RB, Shanfeld J, Mullet D, Cole J, Paulson JO. 1994. Prenatal smokeless tobacco effects on the rat fetus. *J Craniofac Genet Dev Biol* 14(1):16-25.

Pfarrer C, Macara L, Leiser R, Kingdom J. 1999. Adaptive angiogenesis in placentas of heavy smokers. *Lancet* 354(9175):303.

Saraiya M, Berg CJ, Kendrick JS, Strauss LT, Atrash HK, Ahn YW. 1998. Cigarette smoking as a risk factor for ectopic pregnancy. *Am J Obstet Gynecol* 178(3):493-498.

Scholes D, Daling JR, Stergachis AS. 1992. Current cigarette smoking and risk of acute pelvic inflammatory disease. *Amer J Pub Health* 82(10):1352-1355.

Secker-Walker RH, Vacek PM, Flynn BS, Mead PB. 1998. Estimated gains in birth weight associated with reductions in smoking during pregnancy. *J Reprod Med* 43(11):967-74.

Seidman DS, Laor A, Gale R, Stevenson DK, Mashiach S, Danon YL. 1992. Birth weight and intellectual performance in late adolescence. *Obstet Gynecol* 79(4):543-546.

Sexton M, Hebel JR. 1984. A clinical trial of change in maternal smoking and its effect on birth weight. *JAMA* 251(7):911-915.

Shabsigh R, Fishman IJ, Schum C, Dunn JK. 1991. Cigarette smoking and other vascular risk factors in vasculogenic impotence. *Urology* 38(3):227-231.

Shah NR, Bracken MB. 2000. A systematic review and meta-analysis of prospective studies on the association between maternal cigarette smoking and preterm delivery. *Am J Obstet and Gynecol* 182(2):465-472.

Shiono PH, Klebanoff MA, Rhoads GG. 1986. Smoking and drinking during pregnancy. *JAMA* 255(1):82-84.

Slotkin TA, Lappi SE, McCook EC, Lorber BA, Seideler FJ. 1995. Loss of neonatal hypoxia tolerance after prenatal nicotine exposure: implications for sudden infant death syndrome. *Brain Research Bulletin* 38(1):69-75.

Slotkin TA, Saleh JL, McCook EC, Seidler FJ. 1997. Impaired cardiac function during postnatal hypoxia in rats exposed to nicotine prenatally: implications for perinatal morbidity and mortality, and for sudden infant death syndrome. *Teratology* 55:177-184.

Sommerfelt K, Andersson HW, Sonnander K, Ahlsten G, Ellersten B, Markestad T, Jacobsen G, Hoffman HJ, Bakketeig L. 2000. Cognitive development of term gestational age children at five years of age. *Arch Dis Child* 83:25-30.

Spiers PS. 1999. Disentangling the separate effects of prenatal and postnatal smoking on the risk of SIDS. *Am J Epidemiol* 149(7):603-606, discussion 607.

Stergachis A, Scholes D, Daling JR, Weiss NS, Chu J. 1991. Maternal smoking and risk of tubal pregnancy. *Amer J Epidemiol* 133(4):332-337.

Tager IB, Ngo L, Hanrahan JP. 1995. Maternal smoking during pregnancy. Effects on lung function during the first 18 months of life. *Am J Respir Crit Care Med* 152(3):977-983.

U.S. DHHS (U.S. Department of Health and Human Services). 1983. *The Health Consequences of Smoking; Cardiovascular Disease; A Report of the Surgeon General.* Atlanta, GA: U.S. Department of Health and Human Services, Centers for Disease Control and Prevention.

U.S. DHHS (U. S. Department of Health and Human Services). 1988. *The Health Consequences of Smoking: Nicotine Addiction; A Report of the Surgeon General.* Rockville, MD: U.S. Department of Health and Human Services, Centers for Disease Control and Prevention.

U.S. DHHS (U. S. Department of Health and Human Services). 1990. *The Health Benefits of Smoking Cessation; A Report of the Surgeon General.* Rockville, MD: U.S. Department of Health and Human Services, Centers for Disease Control and Prevention.

Van Voorhis BJ, Dawson JD, Stovall DW, Sparks AE, Syrop CH. 1996. The effects of smoking on ovarian function and fertility during assisted reproduction cycles. *Obstet Gynecol* 88(5):785-791.

Verma RC, Chansoriya M, Kaul KK. 1983. Effect of tobacco chewing by mothers on fetal outcome. *Indian Pediatrics* 20(2):105-111.

Voigt LF, Hollenbach KA, Krohn MA, Daling JR, Hickok DE. 1990. The relationship of abruptio placentae with maternal smoking and small for gestational age infants. *Obstet Gynecol* 75(5):771-774.

Wang X, Tager IB, Van Vunakis H, Speizer FE, Hanrahan JP. 1997. Maternal smoking during pregnancy, urine cotinine concentrations, and birth outcomes. A prospective cohort study. *Int J Epidemiol* 26(5):978-988.

Werler MM. 1997. Teratogen update: smoking and reproductive outcomes. *Teratology* 55(6):382-388.

Windham GC, Hopkins B, Fenster L, Swan SH. 2000. Prenatal active or passive tobacco smoke exposure and the risk of preterm delivery or low birth weight. *Epidemiology* 11(4):427-433.

Windham GC, Swan SH, Fenster L. 1992. Parental cigarette smoking and the risk of spontaneous abortion. *Am J Epidemiol* 135(12):1394-1403.

Windham GC, Von Behren J, Waller K, Fenster L. 1999. Exposure to environmental and mainstream tobacco smoke and risk of spontaneous abortion. *Am J Epidemiol* 149(3):243-247.

16

Other Health Effects

This chapter briefly reviews some of the health outcomes of tobacco consumption not covered in previous chapters. Although these outcomes by themselves are not as prevalent or threatening as cardiovascular disease, respiratory disease, or cancer, they do contribute significantly to the morbidity and reduced quality-of-life associated with smoking. In addition, cigarette smoking is associated with a small number of positive health outcomes in Parkinson's disease, ulcerative colitis, and preeclampsia, which are reviewed here. This is not intended to be an exhaustive review of smoking-attributable disease but does include peptic ulcer disease, wound healing, inflammatory bowel disease, rheumatoid arthritis, oral disease, dementia, osteoporosis, ocular disease, diabetes, dermatological disease, schizophrenia, and depression.

PEPTIC ULCERS

Cigarette smokers have been found to have an increased risk of peptic ulcer disease, increased rate of relapse after treatment, and increased risk of the complications associated with ulcer development (Kato et al., 1992; Smedley et al., 1988). A prospective study by Kato et al. (1992) is consistent with previous findings suggesting a positive association between cigarette smoking and gastric and duodenal ulcers, with relative risks (RR) of 3.4 and 3.0, respectively. This association was maintained after controlling for alcohol use. Ma et al. (1998) reviewed potential mechanisms of the effect of smoking on ulcer formation and healing. These mechanisms include increased levels of oxygen-derived free radicals, which are known to injure gastric mucosa; decreased mucosal blood flow mediated by decreased levels of nitric oxide (NO; a potent vasodilator); increased neutrophil infiltration; reduced epithelial cell proliferation and granulation tissue formation; and decreased prostaglandin synthesis. Ma and his colleagues (1999) demonstrated grossly the slowing of ulcer healing and an increase in ulcer size associated with smoking in rat models, which were attributed to reduced blood flow at the ulcer margins and decreased constitutive NO synthesis. Cigarette smoking has also been found to augment nonsteroidal anti-inflammatory drug (NSAID) and alcohol-attributable peptic ulcer disease (Guo et al., 1999; Ma et al., 1998).

Ulcer healing has been found to improve with cessation and reduction of smoking. Presenting results consistent with many other studies, Tatsuta et al. (1987) described the course of healing of patients who were advised to reduce their smoking (by at least 50%) or to stop. They found that after 12 weeks of treatment, 91.7% of the ulcers had healed in patients who reduced or stopped smoking, while only 25% had healed in patients who continued to smoke. Furthermore, about 23% of ulcers recurred in patients who stopped or reduced their smoking during the six-month follow-up, while 75% recurred in those who continued to smoke.

SURGICAL WOUND HEALING

Cigarette smoking and its adverse effects on wound healing have been well established in animal models and in surgical patients (Mosely et al., 1978). Surgical complications experienced by smokers include increased failures of amputations, skin flaps, and atrioventricular (AV) shunts (Kwiatkowski et al., 1996). Mechanisms suggested for the adverse effects on wound healing include the cutaneous vasoconstrictive properties of nicotine, poor oxygen delivery due to an increase in carboxyhemoglobin levels leading to cellular hypoxia and tissue ischemia, and possibly interference with cellular respiration and metabolism due to hydrogen cyanide (HCN) (Campanile et al., 1998; Jensen et al., 1991; Kwiatkowski et al., 1996; Silverstein, 1992). Further detrimental effects on wound healing attributed to smoking are decreased microcirculation due to increased platelet aggregation and microclot formation and reduction of epithelial cells, fibroblasts, and macrophages, which are important in scar formation (Campanile et al., 1998; Silverstein, 1992).

INFLAMMATORY BOWEL DISEASE

Smokers with Crohn's disease, especially women, have increased risk of developing severe disease and have a greater risk of requiring surgery and of having postsurgical complications (Thomas et al., 2000). Studies, reviewed by Rhodes and Thomas (1994), have attributed a three- to five-fold higher risk of developing Crohn's disease to smoking. Ex-smokers were found to have less risk, and nonsmokers or those who had quit for at least ten years were found to experience the least risk. Several studies have found a decrease in the need for surgery after smoking cessation and a decrease in recurrence after surgery (Cosnes et al., 1996; Yamamoto and Keighly, 2000).

In contrast, smoking has been shown to have a protective effect for ulcerative colitis (UC), with odds ratios (ORs) between 0.34 and 0.48, and smokers with UC exhibit a better clinical course (reviewed by Calkins et al., 1989). The literature indicates a negative association (pooled OR=0.41) of current smoking with the development of ulcerative colitis, a significant positive association among ex-smokers, and a positive or equivocal risk among nonsmokers (Thomas et al., 2000). Researchers have found a positive influence of smoking on the clinical course of UC, with a lower rate of recurrence found among smokers (Fraga et al., 1997). Studies have revealed an earlier age of onset of ulcerative colitis in lifetime nonsmokers, and among a sample of patients who quit smoking, almost 70% developed UC during the first year postcessation (Rhodes and Thomas, 1994; reviewed in Thomas et al., 1998). The pathological process is involved in the relationship between smoking and inflammatory bowel disease is unknown. It is speculated that nicotine affects cellular immunity, especially the levels of immunoglobulin A (IgA) that are important in the defense of mucosal tissue (Rhodes and Thomas, 1994). It is also postulated that nicotine is involved in the reduction of blood flow and ischemic changes that may contribute to the changes seen in Crohn's disease.

RHEUMATOID ARTHRITIS

Environmental risk factors are known to be important in the pathogenesis of rheumatoid arthritis (RA). The few studies that have evaluated smoking as a risk factor indicate a positive link between cigarette smoking and the development of rheumatoid arthritis (Uhlig et al., 1999). This relationship has been identified among men (Heliovaara et al., 1993), with a more modest association found among women smokers (Karlson et al., 1999). Smoking has also been more closely associated with seropositive RA compared to rheumatoid factor (RF) negative disease (Silman et al., 1996). Evidence of a dose-response relationship has not been consistently upheld. It is unclear whether smoking plays a causal role in the etiology or progression of rheumatoid arthritis or the formation of rheumatoid factor or whether it is linked to other unidentified factors that predispose patients to RA.

ORAL DISEASE

Research has consistently found that cigarette smoking is a major risk factor for periodontal disease (Tomar and Asma, 2000); studies report 30-75% of periodontitis linked to cigarette smoking. Smoking has been linked to increased frequency of diseased sites, reduced periodontal height, and a weak association with gingival bleeding (Bergstrom et al., 2000). A dose-related relationship has been described. In a review by Haber (1994), he noted that smokers had earlier onset of disease, resistance to treatment, and faster disease progression. Smoking cessation has been found to improve gingival health, and there is evidence of a decrease but not a complete reversal in the severity and prevalence of periodontitis among former smokers. Hypothesized mechanisms of action include smoking-induced impairment of immunity, reduced epithelial cell and fibroblast function involved in the healing process, decreased bone mineralization, and local nicotine-mediated gingival vasoconstriction leading to decreased blood flow and tissue oxygen delivery (Christen, 1992; Haber, 1994; Palmer et al., 1999).

Increased risk of oral lesions in immunocompromised smokers has been described. These lesions include hairy leukoplakia, oral candidiasis, and human papillomavirus (HPV) lesions; on the other hand, smoking has been consistently found to decrease the risk of aphthous ulcers (Palacio et al., 1997).

Smokeless tobacco use has been extensively linked to oral disease. The prevalence of leukoplakia, a white mucosal plaque that cannot be attributed to any other disease process, among smokeless tobacco users is very high (at least 50% by some reports), and it is generally accepted that smokeless tobacco use is an important cause (Spangler and Salisbury, 1995). The site of plaque formation in smokeless tobacco users corresponds to tobacco contact with the mucosa. The prevalence of tobacco-associated leukoplakia exhibits a dose-response relationship, and there has been evidence of reversal of these changes upon cessation of tobacco use (Grady et al., 1990). A review by Walsh and Epstein (2000) found the rate of malignant transformation of tobacco-attributable leukoplakia to range between 4 and 17.5%.

The dentitia and gingiva adjacent to the location of tobacco placement in the mouth are at a high risk of oral problems ranging from gingivitis and gingival recession to halitosis (Spangler and Salisbury, 1995). A study involving professional baseball players by Robertson et al. (1990) found that the only significant periodontal changes associated with smokeless tobacco use are gingival recession, loss of gingival attachment, and mucosal lesions at the area of tobacco placement. The evidence has been uncompelling regarding the risk of dental caries and gingivitis among smokeless tobacco users (Walsh and Epstein, 2000).

DEMENTIA

Many previous studies have suggested an inverse relationship between smoking and Alzheimer's dementia (Brenner et al., 1993; Forbes and Hayward, 1993; reviewed in Lee 1994). More recent studies, however, have suggested no relationship or possibly a positive association (Debanne et al., 2000; Doll et al., 2000; Merchant et al., 1999; Ott et al., 1998). In a large prospective study of British doctors, Doll et al. (2000) found no significant association of current or former smoking with the risk of Alzheimer's disease or other types of dementia including vascular dementia. In another large prospective study (the Rotterdam study) by Ott et al. (1998), it was found that smoking significantly increased the risk of vascular and Alzheimer's dementia and was linked to a younger age of onset. Smokers with the apolipoprotein-E4 (APOE4) allele have shown a reduction in dementia risk, consistent with many other studies (Carmelli et al., 1999; Merchant et al., 1999; Ott et al., 1998;). The association persisted when factors such as educational status, alcohol use, and ethnicity were controlled.

Nicotine, however, has been shown to have beneficial effects on cognition in human and animal studies. Evidence exists to suggest that nicotinic cholinergic receptors are decreased in Alzheimer's patients, and the receptors have been found to increase with chronic exposure to nicotine. Nicotine has also been found to decrease neuron loss in animal models of Alzheimer's disease (reviewed in Lee, 1994). Nicotine has been shown to acutely ameliorate recall, attentional, and visuospatial abilities (Leibovici et al., 1999; Newhouse et al., 1997) and to improve performance of memory-related tasks in animal studies (Buccafusco et al., 1999; Newhouse et al., 1997).

Attempting to reconcile the contradictory studies, investigators have postulated numerous reasons for the inconsistent findings. Investigators have raised the possibility that the protective effect of smoking in Alzheimer's disease (AD) may be mediated by the APOE4 allele (Ott et al., 1998; Van Duijn et al., 1995). Many of the earlier studies that pointed to a negative relationship between smoking and AD were case-control designs in which it is necessary to match cases and controls by age and other variables. In this design, there may be an increased effect of selective mortality among smokers, thus decreasing the proportion of smokers available (Debanne et al., 2000; Riggs, 1993). In addition to selection bias, other sources of bias that may have been a factor in the early studies include problems inherent in the design of retrospective studies such as recall bias and the use of surrogate informants for case patients (Doll et al., 2000) and, generally, poor control of confounding variables.

ORTHOPEDIC CONSEQUENCES

Cigarette smoking has been linked to adverse orthopedic consequences including osteoporosis, hip fracture, and delay in bone healing. A dose-response effect has been reported for the increased risk of hip fracture and decreased bone mineral density (BMD) in many but not all studies (Cornuz et al., 1999; Hoidrup et al., 1999; Hollenbach et al., 1993; la Vecchia et al., 1991;). Based on a meta-analysis by Law and Hackshaw (1997), one out of eight hip fractures is attributable to smoking. Smoking is associated with loss of bone mineral density, and this association increases with age especially in postmenopausal women. The study goes on to conclude that the risk of hip fracture is 17% greater in smokers at age 60, 41% greater at age 70, 71% greater at age 80, and 108% greater at age 90. Reversal of the risk for hip fractures has been described 10-20 years postcessation (Cornuz et al. 1999; la Vecchia et al., 1991). There have been few studies examining BMD and hip fracture risk among male smokers. An evaluation of the National Health and Nutrition Examination Survey (NHANES) I Epidemiologic Follow-up Study found a nonsignificant increased risk associated with smoking (Mussolino et al., 1998). A review by Kwiatkowski et al. (1996) stated that bone mineral content of smokers was 10-20% lower among men and 20-30% lower among women. Hypothesized mechanisms include impairment of osteoblastic function and consequently decreased bone formation, reduced calcium absorption, vasoconstrictive effects of nicotine that prevent revascularization of healing bone, and decreased body weight due to smoking, which is an independent risk factor for osteoporosis.

OCULAR DISEASE

Cigarette smoking has been associated with numerous diseases of the eye. Smoking can induce tobacco smoke-mediated vasoconstriction, atherosclerosis, hyperviscosity, hypercoagueability, and decreased oxygen availability in ocular tissues. There is a strong association with ischemic diseases including amaurosis fugax, retinal infarction, and anterior ischemic optic neuropathy (Solberg et al., 1998). Smoking has also been consistently associated with two of the most important causes of vision loss, cataracts and age-related macular degeneration. A dose-dependent relationship between cigarette smoking and risk and severity of cataracts has been established (Hankinson et al., 1992; West et al., 1995). A large prospective study by Delcourt et al. (2000) found an OR of 1.9 for nuclear-type cataracts in current smokers and a two- to fourfold increase in the rate of cataract surgery among current and former smokers. Most studies have found the increased risk limited to nuclear-type cataracts, although there has been evidence of an increased risk of posterior subscapular cataracts as well (Solberg et al., 1998). The risk of cataract formation among smokers appears to be related to lifetime cumulative cigarette dose, with less reduction in risk found among heavy smokers compared to moderate and light smokers after cessation (Christen et al., 2000; Hankinson et al., 1992). The mechanism of smoking-associated cataract formation is thought to be due to direct and indirect oxidative damage to the lens and exposure to heavy metals in tobacco smoke.

Age-related macular degeneration induced by free-radical formation and oxidative stress associated with cigarette smoking has also been well supported in the literature. Many studies have found a two- to threefold increase in risk of macular degeneration among smokers (reviewed in Solberg et al., 1998).

DERMATOLOGIC CONDITIONS

In many retrospective and prospective studies, cigarette smoking has been associated with premature wrinkling independent of age, sun exposure, pigmentation, and sex (Ernster et al., 1995; Smith and Fenske, 1996). Kadunce et al (1991) found a dose-response relationship associated with pack-years of exposure. Smokers with a 0.9 to 49 pack-year history were twice as likely to be wrinkled compared to nonsmokers. The risk increased to 4.7 among 50+ pack-year smokers. Tobacco smoke toxins are believed to have a detrimental effect on collagen and elastin integrity and on the microvasculature of skin, to increase oxidant stress free-radical activity, and to have possible antiestrogenic effects (reviewed in Frances, 1998; Kadunce et al., 1991; Leow and Maibach, 1998; Smith and Fenske, 1996).

Psoriasis also has been found to be associated with cigarette smoking, especially in women (Mills, 1993; Naldi et al., 1999; Poikolainen et al., 1994). A recent case-control study by Naldi et al. (1999) found an OR of 1.5-2.4 among smokers and exhibition of a dose-response relationship.

A few studies have also suggested a positive association of cigarette smoking with a variety of dermatological changes such as psoriasis palmoplantar pustulosis, atopic dermatitis, and yellowing of fingers and nails (reviewed in Smith and Fenske, 1996). These findings are not discussed in detail here.

DIABETES

Several studies over the last decade have indicated a positive dose-related association between cigarette smoking and risk of non-insulin dependent diabetes mellitus (NIDDM) (Nakanishi et al., 2000; Rimm et al., 1993, 1995; Simon et al., 1997; Uchimoto et al., 1999). In a Nurses' Health Study cohort, Rimm et al. (1993) found a positive, dose-related association between cigarette smoking and development of NIDDM that persisted after controlling for age, body mass index (BMI), family history, physical activity, alcohol, and so forth. Simon et al. (1997), in another large prospective study, found a significant link between smoking and increased waist-to-hip ratios and an odds ratio of 1.38 for the relationship of smoking more than ten cigarettes per day and the prevalence of reported diabetes mellitus. The odds ratio was modified when data were adjusted for waist-to-hip ratio, suggesting a relationship between smoking-associated body fat distribution and the occurrence of diabetes. Investigators have consistently suggested that smoking is an independent risk factor for increased insulin resistance, measured by insulin clamp test and oral glucose challenge, among both diabetics and nondiabetics (Facchini et al., 1992; Targher et al., 1997), and increased transient blood glucose levels during oral and intravenous glucose tolerance tests (Janzon et al., 1983). Muhlhauser (1994) in his review of smoking and diabetes discusses the epidemiological evidence suggesting increased risk of diabetic nephropathy and retinopathy and increased overall mortality among smokers.

RENAL DISEASE

Smoking has been linked to abnormal renal function in diabetics and more recently in nondiabetics (Cirillo et al., 1998; Halimi et al., 2000; Pinto-Sietsma et al., 2000; Ritz et al., 2000). Research has consistently established the association of cigarette smoking with a two- to threefold increased risk of microalbuminuria and proteinuria and an increased rate of progression to diabetic nephropathy and ultimately end-stage renal disease in type I and type II diabetics in a dose-responsive manner (reviewed in Orth, 2000; Ritz et al., 2000). Other studies have found an independent association of smoking with increased risk of developing renal failure in patients with autosomal dependent polycystic kidney disease, lupus nephritis, and glomeruloephritis (Ritz et al., 2000).

Recently, consistent with previous findings, in a cross-sectional study of nondiabetic patients by Pinto-Sietsma and colleagues (2000), current smokers were found to have a dose-responsive increased rate of high-normal albuminuria and microalbuminuria (markers of some degree of renal damage). Current smokers also had significantly abnormal glomerular filtration rates (increased or decreased) compared to nonsmokers. Former smokers had a risk for both events that fell between that of current smokers and nonsmokers, suggesting some degree of reversibility of the effects of smoking. Another large population-based, cross-sectional study (Halimi et al., 2000) of subjects without known renal disease showed that current and former smokers, even after adjusting for diagnoses of diabetes and hypertension, had greater risks for proteinuria by dipstick testing compared to nonsmokers (RR=2-3). Current male smokers in this study also had a higher creatinine clearance than nonsmokers, suggesting a degree of glomerular hyperfiltration.

Many mechanisms of smoking-attributable nephrotoxicity have been postulated including increased sympathetic nervous system activity, transient blood pressure elevation, endothelial cell damage and dysfunction of renal vasculature, direct toxic effects on tublar cells, and oxidative stress (Orth, 2000; Pinto-Sietsma et al., 2000).

SCHIZOPHRENIA

It has been reported that up to 80% of patients with schizophrenia smoke cigarettes (McCreadie and Kelly, 2000; Simpson et al., 1999). The rate of smoking among schizophrenics is also higher than rates among other mentally ill patients (Leonard et al., 2000; Tidey et al., 1999). Many different reasons for the high rate of smoking among schizophrenics have been postulated including "self-medication" of the symptoms of schizophrenia, attenuation of the adverse effects of antipsychotic medication, social factors of schizophrenic patients that predispose them to smoke, or genetic vulnerability for both conditions (reviewed in Levin and Rezvani, 2000; Tidey et al., 1999). Nicotine may attenuate the negative symptoms of schizophrenia by stimulating the release of dopamine and glutamate (Dursun and Kutcher, 1999). Nicotine has also been found to stabilize sensory deficits found in schizophrenia including smooth eye tracking movements and auditory gating, possibly through decreased expression of the α-7 nicotinic acetylcholine receptor (Durson and Kutcher, 1999; Leonard et al., 2000). Furthermore, cigarette smoking has been found to reduce monoamine oxidase activity, which is thought to increase vulnerability to the development of schizophrenia (Fowler et al., 1996; Simpson et al., 1999).

DEPRESSION

Depression has consistently been linked with smoking (Breslau and Johnson, 2000; Glassman, 1993). Studies have demonstrated that a history of major depression is associated with a greater prevalence of smoking and less success in smoking cessation (Anda et al., 1990; Balfour and Ridley, 2000; Kinnunen et al., 1996). Kinnunen and colleagues, looking at data from a randomized interventional trial, found that significantly fewer depressed smokers were able to remain abstinent three months after cessation compared to nondepressed smokers. The same study concluded that depressed patients were more responsive to the use of nicotine for cessation. Investigators have suggested that depressive symptoms may be a part of the withdrawal syndrome in those with a history of depression, presenting an obstacle to successful cessation (Glassman, 1993). It has also been proposed that depression increases the intensity of nicotine dependence (Breslau et al., 1998; Carton et al., 1994). A prospective study by Breslau and colleagues (1998) with a five-year follow-up found that baseline major depression tripled the risk for progression to daily smoking. Smokers with a history of depression or depressive symptoms are thought to "self-medicate" with nicotine and its antidepressant properties (Carton et al., 1994; Lerman et al., 1996). Other studies have hypothesized that smokers are predisposed to develop depression secondary to chronic nicotine central nervous system (CNS) effects. Investigators have suggested that certain genetic or environmental factors may predispose patients to depression and the tendency to smoke independently (reviewed in Breslau et al., 1998).

WEIGHT CHANGE

Weight loss is a commonly cited reason for smoking, especially among young females (reviewed in Perkins, 1993), and fear of weight gain has been an obstacle to successful cessation (Emont and Cummings, 1987; Froom et al., 1998). It has been reported that around 80% of smokers gain weight after cessation (Perkins, 1993). The average weight gain after smoking cessation is 3-4 kg. This weight gain has been found to peak within the first few weeks or months of cessation, and smokers often return to the weight range of nonsmokers. A recent large prospective study, however, found that significant weight gain continued five years after cessation (O'Hara et al., 1998). The amount of weight gain, however, is subject to many individual differences in baseline BMI, age, physical activity, genetic predisposition, and so forth (reviewed in Froom et al., 1998; O'Hara et al., 1998; Pomerleau et al., 2000; Williamson et al., 1991).

The physiologic effect of nicotine has been described as a combination of a reduction in caloric intake and an increase in caloric expenditure. Although the evidence has been inconsistent, short-term increases in intake postcessation occur, peaking within weeks or months (Froom et al., 1998), and decreases in intake upon relapse or initiation have been noted.

The research has supported acute changes in metabolic rate associated with changes in smoking, but chronic changes have not been as well supported. A further hypothesis for the weight change seen with change in smoking habits is the speculation that cigarette smoke or nicotine changes the weight set point of the smoker (reviewed in Perkins, 1993).

Investigators have found that weight changes upon smoking initiation and cessation are nicotine associated (Emont and Cummings, 1987; Grunberg et al., 1986). Consistent with many previous studies, Doherty et al. (1996) found that nicotine replacement during smoking cessation suppresses weight gain. In their randomized control study, they describe a linear relationship between nicotine dose and postcessation weight gain, with placebo users gaining the most weight. This relationship was maintained when smoking was biochemically validated by serum cotinine.

DRUG INTERACTIONS

Exposure to cigarette smoke affects the metabolism of many categories of drugs, which may make the action of a certain dose of a drug unpredictable. Certain constituents of tobacco smoke can affect drug action through pharmacokinetic (changes in absorption, distribution, metabolism, and elimination) and pharmacodynamic (changes in drug action or response) actions (Schein, 1995). Among the best-understood agents that affect enzymes of drug metabolism are polycyclic aromatic hydrocarbons (PAHs) and nicotine—though less defined—selectively induces various cytochrome P-450 (CYP) enzymes and uridine 5-diphosphate (UDP) glucuronosyltransferases, while carbon monoxide and heavy-metal constituents have been found to inhibit or decrease certain CYP enzymes (Zevin and Benowitz, 1999).

The effect of cigarette smoke exposure on drug metabolism can influence the action of many commonly used drugs, including certain antidepressants, antipsychotics, heart medications such as β-blockers and antiarrhythmics, anticoagulants, alcohol, caffeine, theophylline, and others (D'Arcy, 1984; Schein, 1995; Zevin and Benowitz, 1999). Many of the drug interactions are of unknown clinical significance, but others may necessitate the adjustment of medication dose among smokers.

Although not considered to be due to an interaction of cigarette smoke and oral contraceptive pills, it is important to note the significantly increased risk of cardiovascular and cerebrovascular events in female smokers using oral contraceptives.

FIRE SAFETY

Cigarettes are the leading cause of accidental fires in the United States, causing about 20-25% of all fire deaths and resulting in about 1,000 deaths and 2,500 injuries annually (Barillo et al., 2000; Brigham and McGuire, 1995). Often, fires are caused by the ignition of upholstered furniture and mattresses by lit cigarettes (Achauer and McGuire, 1989). Many fires also involve the use of alcohol by the smoker (Botkin, 1988). The fire hazard of cigarettes stems from the fact that a cigarette continues to burn until completely consumed.

There has been a push from the public health community to force cigarette manufacturers to modify cigarettes to reduce their combustibility (Chapman, 1999). In response to pressure from various organizations to decrease the danger of improperly discarded cigarettes, the Cigarette Safety Act of 1984 was signed. This act created the Technical Study Group in 1987 to study the technical and commercial feasibility of developing a fire-safe cigarette. This group was comprised of representatives of different medical organizations, federal agencies, the furniture industry, fire service, and four of the top tobacco companies (Barillo et al., 2000). The report of this group stated that it was technically and commercially feasible to develop fire-safe cigarettes and that a reduction of up to 90% in morbidity and mortality could be reached if such modifications were made (Achauer and McGuire, 1989). The main characteristics of cigarettes that were found to significantly decrease flammability included decreased packing density, smaller cigarette circumference, decreased paper porosity, and reduction of citrate—a paper-burning additive (Achauer and McGuire, 1989). The Fire Safe Act was passed in 1990 and mandated further study of the issue based on the recommendations of the Technical Study Group. The Technical Advisory Group was formed which developed cigarette fire-safety test methods. No federal legislation has been passed up to this time that has set a fire-safety standard for cigarettes. In June 2000, however, the state of New York passed a cigarette fire-safety bill that would require all cigarettes sold in New York to meet flammability standards by 2003 (Perez-Pena, 2000).

PARKINSON'S DISEASE

Parkinson's disease is a neurological disease affecting 1-2% of the population that is characterized by motor dysfunction and, in severe cases, cognitive dysfunction caused by loss of dopamine and other neurotransmitters through the destruction of neurons in the substantia nigra. Over the last decade there have been many epidemiological studies suggesting a biologically protective effect of cigarette smoking on the risk of Parkinson's disease. Numerous population based case-control and prospective studies have indicated an inverse dose-response relationship between smoking and Parkinson's disease (Checkoway and Nelson, 1999; Gorell et al., 1999; Grandinetti et al., 1994; Tzourio et al., 1997). In a large 26-year prospective follow-up of the Honolulu Heart Study, Grandinetti et al. (1994) found the relative risk of Parkinson's disease among smokers to be 0.39 with a dose-response relation to pack-years of smoking. Postulated mechanisms of the neuroprotective traits of smoking include direct neuronal effects of nicotine on dopamine release in the substantia nigra, reduction of the action of monoamine oxidase B, or reduction of free radical injury (Checkoway et al., 1998; Court et al., 1998; Grandinetti et al., 1994; Kelton et al., 2000). Critics of the protective association between smoking and Parkinson's disease have attributed the results to an artifact of selective mortality of smokers that would have acquired Parkinson's, diagnostic errors, cause-and-effect bias, or another unrecognized confounder (Morens et al., 1996; Riggs, 1996).

PREECLAMPSIA

Studies have described an inverse relationship between cigarette smoking and the risk of preeclampsia. Many studies have reported around a 30-70% decrease in risk for preeclampsia among smokers, using reported daily cigarette use and serum cotinine levels, when compared to nonsmokers. A finding of a dose-response effect has been less consistent (reviewed in Conde-Agudelo et al., 1999; Klonoff-Cohen and Savitz, 1993; Lain et al., 1999; Lindqvist and Marsal, 1999). A population-based study by Cnattingius and colleagues (1997) found a decreased incidence of preeclampsia among smokers but also found that smokers who did develop preeclampsia tended to have worse pregnancy outcomes. In a case-control study by Marcoux et al. (1989), the protective effect of smoking for preeclampsia was maintained only among pregnant smokers who continued to smoke throughout pregnancy or quit after 20 weeks' gestation. The postulated mechanisms for this protective effect of cigarette smoking includes the nicotine-mediated inhibition of fetal thromboxane A, stimulation of NO production, increased levels of thiocyanate, endothelial damage secondary to the oxidative stress effect of cigarette smoke, and altered immune responses (Cnattingius et al., 1997; Conde-Agudelo et al., 1999; reviewed in Lindqvist and Marsal, 1999).

SUMMARY

Several important diseases and conditions of adults, in addition to cardiovascular disease, chronic obstructive pulmonary disease, and various cancers, have been associated with tobacco use. Some of the associations are supported by substantial scientific evidence, and a causal linkage is likely. These illnesses must ultimately be subjected to the same evaluation with regard to changing risks and outcomes associated with potential reduced-exposure products (PREPs), because they are common and clinically important, even if not often as fatal as the diseases considered earlier in this report. Further, each of the conditions for which the association with tobacco use is substantial also offers the opportunity to address pathogenic mechanisms related to the varying constituents of PREPs, as well as the impact on disease incidence of coordinated behaviors and exposures such as alcohol use, various dietary elements, and certain medications.

RECOMMENDATIONS

Surveillance

The committee recommends that selected conditions, reviewed in this chapter, be part of a comprehensive, population-based surveillance program, as outlined in Chapter 6. This will allow determination of trends in occurrence for these tobacco-related conditions and assessment on a national basis of whether changes in tobacco product use have an effect on these important health problems. Based on surveillance findings, more specific population, clinical, and basic research studies can be directed and justified, in order to pursue causal mechanisms and suggest more effective interventions.

Applying Selected Conditions as Indicators of Clinical Harm Reduction

Some of the conditions reviewed in this chapter may be applied as indicators of the general biological effects of new tobacco products. For example, cigarette smoking has been consistently found to be an independent risk factor for an adverse clinical course of both peptic ulcer disease and wound healing. The effects of smoking on ulcer formation and healing have been clearly described clinically and in animal models. Peptic ulcers have been found to be larger, slower to heal, and more likely to recur among smokers and to exhibit clinically improved healing upon cessation. Surgical and traumatic wounds heal more slowly in cigarette smokers. The committee recommends that rigorous clinical studies be designed and executed to determine whether variations in ulcer and wound healing rates are related to different categories of tobacco products, including those with claims of harm reduction. This may offer the opportunity to define some clinical outcomes that have clinical relevance in their own right and to identify potential indicators of harm alteration that could be obtained much sooner after the introduction of PREPs than would be possible when evaluating heart disease and cancer.

Other candidate diseases for such evaluation might include periodontal disease and Crohn's disease. Here the outcomes to assess would be the effects of various conventional tobacco products and PREPs on the natural history of the condition, including intermittancy, progression or regression, and longitudinally collected biomarkers of disease severity. As noted above, PREPs that alter the history and outcomes of these conditions could be further evaluated for specific constituent exposures that are associated with this altered history. This may lead to a more refined understanding of pathogenic mechanisms as well.

There is also room for clinical and basic research on intermediate clinical outcomes. For example, as noted in this chapter, the risk of osteoporosis has also been strongly linked to cigarette smoking. In controlled observational studies, bone mineral density has been found to be significantly lower among cigarette smokers, which contributes to a higher risk of osteoporotic fractures among older populations. While the effects of smoking on fracture rates may take a few decades or longer to detect, it is possible that surveillance for bone mineral density among those using PREPs and conventional products may prove valuable in a shorter time period and thus serve to detect important outcomes over an interval in which tobacco policy and clinical preventive interventions may have their greatest effects.

REFERENCES

Achauer BM, McGuire A. 1989. Fire safe cigarettes: an update. *J Burn Care Rehabil* 10(2):173-174.

Anda RF, Williamson DF, Escobedo LG, Remington PL. 1990. Smoking and the risk of peptic ulcer disease among women in the United States. *Arch Intern Med* 150(7):1437-1441.

Balfour DJK, Ridley DL. 2000. The effects of nicotine on neural pathways implicated in depression: a factor in nicotine addiction? *Pharmacology Biochemistry and Behavior* 66(1):79-85.

Barillo DJ, Brigham PA, Kayden DA, Heck RT, McManus AT. 2000. The fire-safe cigarette: a burn prevention tool. *J Burn Care Rehabil* 21(2):162-164.

Bergstrom J, Eliasson S, Dock J. 2000. Exposure to tobacco smoking and periodontal health. *J Clin Periodontol* 27(1):61-68.

Botkin JR. 1988. The fire-safe cigarette. *JAMA* 260(2):226-229.

Brenner DE, Kukull WA, van Belle G, Bowen JD, McCormick WC, Teri L, Larson EB. 1993. Relationship between cigarette smoking and Alzheimer's disease in a population-based case-control study. *Neurology* 43(2):293-300.

Breslau N, Johnson EO. 2000. Predicting smoking cessation and major depression in nicotine-dependent smokers. *Am J Public Health* 90(7):1122-1127.

Breslau N, Peterson EL, Schultz LR, Chilcoat HD, Andreski P. 1998. Major depression and stages of smoking. *Arch Gen Psychiatry* 55:161-166.

Brigham PA, McGuire A. 1995. Progress towards a fire-safe cigarette. *J Public Health Policy* 16(4):433-439.

Buccafusco JJ, Jackson WJ, Jonnala RR, Terry AV Jr. 1999. Differential improvement in memory-related task performance with nicotine by aged male and female rhesus monkeys. *Behavioral Pharmacology* 10(6-7):681-690.

Calkins BM. 1989. A meta-analysis of the role of smoking in inflammatory bowel disease. *Dig Dis Sci* 34(12):1841-1854.

Campanile G, Hautmann G, Lotti T. 1998. Cigarette smoking, wound healing, and face-lift. *Clin Dermatol* 16(5):575-578.

Carmelli D, Swan GE, Reed T, et al. 1999. The effect of Apolipoprotein E epsilon4 in the relationships of smoking and drinking to cognitive function. *Neuroepidemiology* 18(3):125-133.

Carton S, Jouvent R, Widlocher D. 1994. Nicotine dependence and motives for smoking in depression. *J Substance Abuse* 6:67-76.

Chapman S. 1999. Where there's smoke, there's fire. *Tobacco Control* 8(1):12-13.

Checkoway H, Franklin GM, Costa-Mallen P, Smith-Weller T, Dilley J, Swanson PD, Costa LG. 1998. A genetic polymorphism of MAO-B modifies the association of cigarette smoking and Parkinson's disease. *Neurology* 50(5):1458-1461.

Checkoway H, Nelson LM. 1999. Epidemiologic approaches to the study of Parkinson's disease etiology. *Epidemiology* 10(3):327-336.

Christen AG. 1992. The impact of tobacco use and cessation on oral and dental diseases and conditions. *Amer J Medicine* 93(Suppl 1A):25S-31S.

Christen WG, Glynn RJ, Ajani UA, Schaumberg DA, Buring JE, Hennekens CH, Manson JE. 2000. Smoking cessation and risk of age-related cataract in men. *JAMA* 284(6):713-716.

Cirillo M, Senigalliesi L, Laurenzi M, Alfieri R, Stamler J, Panarelli W, De Santo NG. 1998. Microalbuminuria in nondiabetic adults: relation of blood pressure, body mass index, plasma cholesterol levels, and smoking: the Gubbio population study. *Arch Intern Med* 158(17):1933-1939.

Cnattingius S, Mills JL, Yuen J, Eriksson O, Ros HS. 1997. The paradoxical effect of smoking in preeclamptic pregnancies: smoking reduce the incidence but increases the rates of perinatal mortality, abruptio placentae, and intrauterine growth restriction. *Am J Obstet Gynecol* 177(1):156-161.

Conde-Agudelo A, Althabe F, Belizan JM, Kafury-Goeta AC. 1999. Cigarette smoking during pregnancy and risk of preeclampsia: a systematic review. *Am J Obstet Gynecol* 181(4):1026-1035.

Cornuz J, Feskanich D, Willett WC, Colditz GA. 1999. Smoking, smoking cessation, and risk of hip fracture in women. *Am J Med* 106(3):311-314.

Cosnes J, Carbonnel F, Beaugerie L, Le Quintrec Y, Gendre JP. 1996. Effects of cigarette smoking on the long-term course of Crohn's disease. *Gastroenterology* 110(2):424-431.

Court JA, Lloyd S, Thomas N, Piggott, MA, Marshall, EF, Morris, CM, Lamb H, Perry RH, Johnson, M, Perry EK. 1998. Dopamine and nicotinic receptor binding and the levels of dopamine and homovanillic acid in human brain related to tobacco use. *Neuroscience* 87(1):63-78.

D'Arcy PF. 1984. Tobacco smoking and drugs: a clinically important interaction? *Drug Intell Clin Pharm* 18(4):302-307.

Debanne SM, Rowland DY, Riedel TM, Cleves MA. 2000. Association of Alzheimer's disease and smoking: the case for sibling controls. *J Am Geriatr Soc* 48(7):800-806.

Delcourt C, Cristol J, Tessier F, Leger CL, Michel F, Papoz L. 2000. Risk factors for cortical, nuclear, and posterior subscapular cataracts. *Amer J Epidemiology* 151(5):497-504.

Doherty K, Militello FS, Kinnumen T, Garvey AJ. 1996. Nicotine gum dose and weight gain after smoking cessation. *J Consulting and Clinical Psychology* 64(4):799-807.

Doll R, Peto R, Boreham J, Sutherland I. 2000. Smoking and dementia in male British doctors: prospective study. *BMJ* 320(7242):1091-1102.

Durson SM, Kutcher S. 1999. Smoking, nicotine and psychiatric disorders: evidence for therapeutic role, controversies and implications for future research. *Medical Hypotheses* 52(2):101-109.

Emont SL, Cummings KM. 1987. Weight gain following smoking cessation: a possible role for nicotine replacement in weight management. *Addictive Behaviors* 12:151-155.

Ernster VL, Grady D, Miike R, Black D, Selby J, Kerlikowske K. 1995. Facial wrinkling in men and women, by smoking status. *Am J Public Health* 85(1):78-82.

Facchini FS, Hollenbeck CB, Jeppesen J, Chen YDI, Reaven GM. 1992. Insulin resistance and cigarette smoking. *Lancet* 339:1128-1130.

Forbes WF, Hayward LM. 1993. Concerning the link between tobacco smoking and impaired mental functioning. *Can J Public Health* 84(3):211-212.

Fowler JS, Volkow ND, Wang GJ, Pappas KJ, Logan J, MacGregor, R, Alexoff D, Shea G, Schlyer D, Wolf AP, Warner D, Zezulkova I, Cilento R. 1996. Inhibition of monoamine oxidase B in the brains of smokers. *Nature* 379:733-736.

Fraga XF, Vergara M, Medina C, Casellas F, Bermejo B, Malagelada JR. 1997. Effects of smoking on the presentation and clinical course of inflammatory bowel disease. *Eur J Gastroenterol Hepatol* 9(7):683-687.

Frances C. 1998. Smoker's wrinkles: epidemiological and pathogenic considerations. *Clin Dermatol* 16(5):565-570.

Froom P, Melamed S, Benbassat J. 1998. Smoking cessation and weight gain. *J Family Practice* 46(6):460-464.

Glassman AH. 1993. Cigarette smoking: implications for psychiatric illness. *Am J Psychiatry* 150:546-553.

Gorell JM, Rybicki BA, Johnson CC, Peterson EL. 1999. Smoking and Parkinson's disease: a dose-response relationship. *Neurology* 52(1):115-119.

Grady D, Greene J, Daniels TE, Ernster VL, Robertson PB, Hauck W, Greenspan D, Greenspan J, Silverman S. 1990. Oral mucosal lesions found in smokeless tobacco users. *J Am Dent Assoc* 121(1):117-123.

Grandinetti A, Morens DM, Reed D, MacEachern D. 1994. Prospective study of cigarette smoking and the risk of developing idiopathic Parkinson's disease. *Am J Epidemiology* 139(12):1129-1138.

Grunberg NE, Bowen DJ, Winders SE. 1986. Effects of nicotine on body weight and food consumption in female rats. *Psychopharmacology* 90:101-105.

Guo X, Mei QB, Cho CH. 1999. Potentiating effect of passive cigarette smoking on gastrointestinal damage induced by indomethacin in rats. *Dig Dis Sci* 44(5):896-902.

Haber J. 1994. Smoking is a major risk factor for periodontitis. *Curr Opin Periodontol*:12-8.

Halimi JM, Giraudeau B, Vol S, Caces E, Nivet H, Lebranchu Y, Tichet J. 2000. Effects of current smoking and smoking discontinuation on renal function and proteinuria in the general population. *Kidney Int* 58(3):1285-1292.

Hankinson SE, Willett WC, Colditz GA, Seddon JM, Rosner B, Speizer FE, Stampfer MJ. 1992. A prospective study of cigarette smoking and risk of cataract surgery in women. *JAMA* 268(8):994-998.

Heliovaara M, Aho K, Aromaa A, Knekt P, Reunanen A. 1993. Smoking and risk of rheumatoid arthritis. *J Rheumatol* 20(11):1830-1835.

Hoidrup S, Gronbaek M, Pedersen AT, Lauritzen JB, Gottschau A, Schroll M. 1999. Hormone replacement therapy and hip fracture risk: effect modification by tobacco smoking, alcohol intake, physical activity, and body mass index. *Am J Epidemiol* 150(10):1085-1093.

Hollenbach KA, Barrett-Connor E, Edelstein SL, Holbrook T. 1993. Cigarette smoking and bone mineral density in older men and women. *Am J Public Health* 83(9):1265-1270.

Janzon L, Berntorp K, Hanson M, Lindell SE, Trell E. 1983. Glucose tolerance and smoking: a population study of oral and intravenous glucose tolerance tests in middle-aged men. *Diabetologia* 25(2):86-88.

Jensen JA, Goodson WH, Hopf HW, Hunt TK. 1991. Cigarette smoking decreases tissue oxygen. *Arch Surg* 126:1131-1134.

Kadunce DP, Burr R, Gress R, et al. 1991. Cigarette smoking: risk factor for premature facial wrinkling. *Annals of Internal Medicine* 114:840-844.

Karlson EW, Lee IM, Cook NR, Manson JE, Buring JE, Hennekens CH. 1999. A retrospective cohort study of cigarette smoking and risk of rheumatoid arthritis in female health professionals. *Arthritis Rheum* 42(5):910-917.

Kato I, Nomura AM, Stemmermann GN, Chyou PH. 1992. A prospective study of gastric and duodenal ulcer and its relation to smoking, alcohol, and diet. *Am J Epidemiol* 135(5):521-530.

Kelton MC, Kahn HJ, Conrath CL, Newhouse PA. 2000. The effects of nicotine on Parkinson's disease. *Brain Cogn* 43(1-3):274-282.

Kinnunen T, Doherty K, Militello FS, Garvey AJ. 1996. Depression and smoking cessation: Characteristics of depressed smokers and effects of nicotine replacement. *J Consulting and Clinical Psychology* 64(4):791-798.

Klonoff-Cohen HES, Savitz D. 1993. Cigarette smoking and preeclampsia. *Obstet Gynecol* 81(4):541-544.

Kwiatkowski TC, Hanley Jr EN, Ramp WK. 1996. Cigarette smoking and its orthopedic consequences. *Amer J Orthopedics* 25(9):590-597.

la Vecchia C, Negri E, Levi F, Baron JA. 1991. Cigarette smoking, body mass and other risk factors for fractures of the hip in women. *Int J Epidemiol* 20(3):671-677.

Lain KY, Powers RW, Krohn MA, Ness RB, Crombleholme WR, Roberts JM. 1999. Urinary cotinine concentration confirms the reduced risk of preeclampsia with tobacco exposure. *Am J Obstet Gynecol* 181(5 pt. 1):1192-1196.

Law MR, Hackshaw AK. 1997. A meta-analysis of cigarette smoking, bone mineral density and risk of hip fracture: recognition of a major effect. *BMJ* 315(7112):841-846.

Lee PN. 1994. Smoking and Alzheimer's disease: a review of the epidemiological evidence. *Neuroepidemiology* 13(4):131-144.

Leibovici D, Ritchie K, Ledesert B, Touchon J. 1999. The effects of wine and tobacco consumption on cognitive performance in the elderly: a longitudinal study of relative risk. *Int J Epidemiol* 28(1):77-81.

Leonard S, Breese C, Adams C, Benhammou K, Gault J, Stevens K, Lee M, Adler L, Olincy A, Ross R, Freedman R. 2000. Smoking and schizophrenia: abnormal nicotinic receptor expression. *European J Pharmacology* 393:237-242.

Leow YH, Maibach HI. 1998. Cigarette smoking, cutaneous vasculature, and tissue oxygen. *Clin Dermatol* 16(5):579-584.

Lerman C, Audrain J, Orleans CT, Boyd R, Gold K, Main D, Caporaso N. 1996. Investigation of mechanisms linking depressed mood to nicotine dependence. *Addictive Behaviors* 21(1):9-19.

Levin ED, Rezvani AH. 2000. Development of nicotinic drug therapy for cognitive disorders. *European J Psychopharmacology* 393:141-146.

Lindqvist PG, Marsal K. 1999. Moderate smoking during pregnancy is associated with a reduced risk of preeclampsia. *Acta Obstet Gynecol Scand* 78:693-697.

Ma L, Chow JYC, Cho CH. 1998. Effects of cigarette smoking on gastric ulcer formation and healing: possible mechanisms of action. *J Clin Gastroenterol* 27(Suppl 1):S80-S86.

Ma L, Chow JY, Cho CH. 1999. Cigarette smoking delays ulcer healing: role of constitutive nitric oxide synthase in rat stomach. *Am J Physiol* 276(1 Pt 1):G238-48.

Marcoux S, Brisson J, Fabia J. 1989. The effect of cigarette smoking on the rise of preeclampsia and gestational hypertension. *Am J Epidemiol* 130(5):950-957.

McCreadie RG, Kelly C. 2000. Patients with schizophrenia who smoke: private disaster, public resource. *Br J Psychiatry* 176:109.

Merchant C, Tang MX, Albert S, Manly J, Stern Y, Mayeux R. 1999. The influence of smoking on the risk of Alzheimer's disease. *Neurology* 52(7):1408-1412.

Mills CM. 1993. Smoking and skin disease. *Int J Dermatol* 32(12):864-865.

Morens DM, Grandinetti A, Davis JW, Ross GW, White LR, Reed D. 1996. Evidence against the operation of selective mortality in explaining the association between cigarette smoking and reduced occurrence of idiopathic Parkinson disease. *Am J Epidemiology* 144(4):400-404.

Mosely LH, Finseth F, Goody M. 1978. Nicotine and its effect on wound healing. *Plastic and Reconstructive Surgery* 61(4):570-574.

Muhlhauser I. 1994. Cigarette smoking and diabetes: an update. *Diabet Med* 11(4):336-343.

Mussolino ME, Looker AC, Madans JH, Langlois JA, Orwoll ES. 1998. Risk factors for hip fracture in white men: the NHANES I Epidemiologic Follow-up Study. *J Bone Miner Res* 13(6):918-924.

Nakanishi N, Nakamura K, Matsuo Y, Susuki K, Tatara K. 2000. Cigarette smoking and risk for impaired fasting glucose and Type 2 diabetes in middle-aged Japanese men. *Ann Intern Med* 133(3):183-191.

Naldi L, Peli L, Parazzini F. 1999. Association of early-stage psoriasis with smoking and male alcohol consumption: evidence from an Italian case-control study. *Arch Dermatol* 135(12):1479-1484.

Newhouse PA, Potter A, Levin ED. 1997. Nicotinic system involvement in Alzheimer's and Parkinson's diseases. Implications for therapeutics. *Drugs Aging* 11(3):206-228.

O'Hara P, Connett JE, Lee WW, Nides M, Murray R, Wise R. 1998. Early and late weight gain following smoking cessation in the Lung Health Study. *Am J Epidemiol* 148(9):821-830.

Orth SR. 2000. Smoking-a renal risk factor. *Nephron* 86(1):12-26.

Ott A, Slooter AJ, Hofman A, van Harskamp F, Witteman JC, van Broeckhaven C, van Duijn Cm, Breteler MM. 1998. Smoking and risk of dementia and Alzheimer's disease in a population-based cohort study: the Rotterdam Study. *Lancet* 351(9119):1840-1843.

Palacio H, Hilton JF, Canchola AJ, Greenspan D. 1997. Effect of cigarette smoking on HIV-related oral lesions. *J Acquir Immune Defic Syndr Hum Retrovirol* 14(4):338-342.

Palmer RM, Scott DA, Meekin TN, Poston RN, Odell EW, Wilson RF. 1999. Potential mechanisms of susceptibility to periodontitis in tobacco smokers. *J Periodontal Res* 34(7):363-369.

Perez-Pena R. 2000. June 15. New York pact on fire-safety for cigarettes. [Online]. Available: www.nytimes.com.

Perkins KA. 1993. Weight gain following smoking cessation. *J Consulting and Clinical Psychology* 61(5):768-777.

Pinto-Sietsma SJ, Mulder J, Jansses WM, Hillege HL, de Zeeuw D, de Jong PE. 2000. Smoking is related to albuminuria and abnormal renal function in nondiabetic persons. *Ann Intern Med* 133(8):585-591.

Poikolainen K, Reunala T, Karvonen J. 1994. Smoking, alcohol and life events related to psoriasis among women. *Br J Dermatol* 130(4):473-477.

Pomerleau CS, Pomerleau OF, Namenek RJ, Mehringer AM. 2000. Short-term weight gain in abstaining women smokers. *J Substance Abuse Treatment* 18:339-342.

Rhodes J, Thomas GA. 1994. Smoking: good or bad for inflammatory bowel disease? *Gastroenterology* 106(3):807-810.

Riggs JE. 1993. Smoking and Alzheimer's disease: protective effect or differential survival bias? *Lancet* 342:793-794.

Riggs JE. 1996. The "protective" influence of cigarette smoking on Alzheimer's and Parkinson's diseases. *Neurol Clin* 14(2):353-358.

Rimm EB, Chan J, Stampfer MJ, Colditz GA, Willett WC. 1995. Prospective study of cigarette smoking, alcohol use, and the risk of diabetes in men. *BMJ* 310(6979):555-559.

Rimm EB, Manson JE, Stampfer MJ, Colditz GA, Willett WC, Rosner B, Hennekens CH, Speizer FE. 1993. Cigarette smoking and the risk of diabetes in women. *Am J Public Health* 83(2):211-214.

Ritz E, Ogata H, Orth SR. 2000. Smoking: a factor promoting onset and progression of diabetic nephropathy. *Diabetes Metab* 26(supp. 4):54-63.

Robertson PB, Walsh M, Greene J, Ernster V, Grady D, Hauck W. 1990. Periodontal effects associated with the use of smokeless tobacco. *J Periodontol* 61(7):438-443.

Schein JR. 1995. Cigarette smoking and clinically significant drug interactions. *Ann Pharmacother* 29(11):1139-1148.

Silman AJ, Newman J, MacGregor AJ. 1996. Cigarette smoking increases the risk of rheumatoid arthritis. Results from a nationwide study of disease-discordant twins. *Arthritis Rheum* 39(5):732-735.

Silverstein P. 1992. Smoking and wound healing. *Amer J Medicine* 93(suppl 1A):22S-24S.

Simon JA, Seeley DG, Lipschutz RC, Vittinghoff E, Browner WF. 1997. The relation of smoking to waist-to-hip ratio and diabetes mellitus among elderly women. *Prev Med* 26(5 pt. 1):639-644.

Simpson GM, Shih JC, Chen K., Flowers C, Kumazawa T, Spring B. 1999. Schizophrenia, monoamine oxidase activity, and cigarette smoking. *Neuropsychopharmacology* 20(4):392-394.

Smedley F, Hickish T, Taube M, Yale C, Leach R, Wastell C. 1988. Perforated duodenal ulcer and cigarette smoking. *J R Soc Med* 81(2):92-94.

Smith JB, Fenske NA. 1996. Cutaneous manifestation and consequences of smoking. *J Am Acad Dermatol* 34:717-732.

Solberg Y, Rosner M, Belkin M. 1998. The association between cigarette smoking and ocular diseases. *Surv Ophthalmol* 42(6):535-547.

Spangler JG, Salisbury PL. 1995. Smokeless tobacco: epidemiology, health effects and cessation strategies. *Am Fam Physician* 52(5):1421-1430, 1433-1434.

Targher G, Alberiche M, Zenere MB, Bonadonna RC, Muggeo M, Bonora E. 1997. Cigarette smoking and insulin resistance in patients with noninsulin-dependent diabetes mellitus. *J Clin Endocrinol Metab* 82(11):3619-3624.

Tatsuta M, Iishi H, Okuda S. 1987. Effects of cigarette smoking on the location, healing and recurrence of gastric ulcers. *Hepatogastroenterology* 34(5):223-228.

Thomas GA, Rhodes J, Green JT. 1998. Inflammatory bowel disease and smoking—a review. *Am J Gastroenterol* 93(2):144-149.

Thomas GA, Rhodes J, Green JT, Richardson C. 2000. Role of smoking in inflammatory bowel disease: implications for therapy. *Postgrad Med J* 76(895):273-279.

Tidey JW, O'Neill SC, Higgens ST. 1999. Effects of abstinence on cigarette smoking among outpatients with schizophrenia. *Experimental and Clinical Psychopharmacology* 7(4):347-353.

Tomar SL, Asma S. 2000. Smoking-attributable periodontitis in the United States: findings from NHANES III. National Health and Nutrition Examination Survey. *J Periodontol* 71(5):743-751.

Tzourio C, Rocca WA, Breteler MM, Baldereschi M, Dartigues JF, Lopez-Pousa S, Manubens-Bertran JM, Alperovitch A. 1997. Smoking and Parkinson's disease: an age-dependent risk factor? *Neurology* 49:1267-1272.

Uchimoto S, Tsumura K, Hayashi T, Suematso C, Endo G, Fujii S, Okada K. 1999. Impact of cigarette smoking on the incidence of Type 2 diabetes mellitus in middle-aged Japanese men: the Osaka Health Survey. *Diabet Med* 16(11):951-955.

Uhlig T, Hagen KB, Kvien TK. 1999. Current tobacco smoking, formal education, and the risk of rheumatoid arthritis. *J Rheumatol* 26(1):47-54.

Van Duijn CM, Havekes LM, Van Broeckhoven C, de Knijff P, Hofman A. 1995. Apolipoprotein E genotype and association between smoking and early onset Alzheimer's disease. *BMJ* 310:627-631.

Walsh PM, Epstein JB. 2000. The oral effects of smokeless tobacco. *J Can Dent Assoc* 66(1):22-25.

West S, Munoz B, Schein OD, Vitale S, Macguire M, Taylor HR, Bressler NM. 1995. Cigarette smoking and risk for progression of nuclear opacities. *Arch Ophthalmol* 113(11):1377-1380.

Williamson DF, Madans J, Anda RF, Kleinman JC, Giovino GA, Byers T. 1991. Smoking cessation and severity of weight gain in a national cohort. *N Engl J Med* 324(11):739-745.

Yamamoto T, Keighley MR. 2000. Smoking and disease recurrence after operation for Crohn's disease. *Br J Surg* 87(4):398-404.

Zevin S, Benowitz NL. 1999. Drug interactions with tobacco smoking. An update. *Clin Pharmacokinet* 36(6):425-438.

Appendixes

A

Presentations and Submissions

The committee would like to thank the following people who shared their expertise.

PRESENTATIONS

December 15, 1999, Washington, D.C.

Presentation of the Charge to the Institute of Medicine Committee
    Mitch Zeller, Food and Drug Administration (current affiliation, American Legacy Foundation
    Bern Schwetz, Food and Drug Administration

March 2, 2000, Washington, D.C.

Harm Reduction and Lessons Learned: Historical Perspective and Policy Advice
    John Slade, University of Medicine and Dentistry at New Jersey School of Public Health
Harm Reduction and Lessons Learned: Low Tar and Nicotine Story
    Don Shopland, National Cancer Institute
Tobacco Toxicology Review: Emphasis on Carcinogenesis
    Steve Hecht, University of Minnesota
Exposure Assays
    David Ashley, Centers for Disease Control and Prevention
Nicotine Pharmacology and Toxicology
    Neal Benowitz, University of California, San Francisco

April 25, 2000, Washington, D.C.

Risk Perception and Decision Making
   Paul Slovic, Decision Research, University of Oregon
Epidemiology of Low-Tar Cigarettes
   David Burns, University of California, San Diego
Clinical Trial Design Considerations
   Jack Henningfield, Pinney Associates, Johns Hopkins School of
      Medicine
   Janine Pillitteri, Pinney Associates
   John Hughes, University of Vermont
Clinical Indicators of Chronic Obstructive Pulmonary Disease
   Steve Rennard, University of Nebraska
Physical and Chemical Characteristics of Tobacco Smoke
   Robert Phalen, University of California, Irvine
State of Massachusetts Regulation of Tobacco Products
   Gregory Connolly, Massachusetts Department of Public Health
   Michael Borgerding, R.J. Reynolds Tobacco Company

CONTRIBUTIONS

Thoughtful contributions were also provided by the following:

Industry Representatives

Brown and Williamson Tobacco Company
   Scott Appleton
   Sharon Boyse
   Rufus Honeycutt
   Tilford Riehl
Lorillard Tobacco Company
   Christopher Coggins and colleagues
   Chuck Gaworski
   J. Dan Heck
Philip Morris Tobacco Company
   Richard A. Carchman
   Bruce D. Davies
   Hans-Juergen Haussmann
   Robin D. Kinser
   Donald E. Leyden
   George J. Patskan
   Richard P. Solana
   Anthony R. Ticker

R.J. Reynolds Tobacco Company
    Michael Borgerding
    J. Donald deBethizy
    David Doolittle
    David Iauco
    Arnold T. Mosberg
    Seth Moskowitz
    John H. Robinson
    Robert L. Suber
    James E. Swauger
Star Scientific, Inc.
    Scott D. Ballin
    Jerome H. Jaffe
    Paul L. Perito
GlaxoWellcome Pharmacutical
    Brenda Jamerson
Pharmacia Corporation
    Mikael Franzon
    Vibeke Kronborg
Smith Kline Beecham

Individuals

Gregory Dalack, VA Ann Arbor Healthcare System
Karl Fagerstrom, Fagerstrom Consulting
Nigel Gray, European Institute of Oncology
Jack Henningfield, Pinney Associates
Thomas Houston, American Medical Association
John Hughes, University of Vermont
Anne Kirchner, Food and Drug Administration
Robert Nilsson, National Chemicals Inspectorate, Sweden
Brad Radu, University of Alabama
Elsy-Britt Schildt, University Hospital, Sweden
Saul Shiffman, University of Pittsburgh and Pinney Associates
John Slade, University of Medicine and Dentistry at New Jersey School
    of Public Health
Steven Stellman, American Health Foundation
David Sweanor, Smoking and Health Action Foundation, Canada
Puzant Torigian
Dennis Ole Wik
Judy Wilkenfeld, Campaign for Tobacco Free Kids
Richard Windsor, University of Alabama

B

Committee Biographical Sketches

Stuart Bondurant, M.D. *(chair)*, is professor of medicine and dean emeritus at the University of North Carolina School of Medicine. Previously, Dr. Bondurant was director of the Center for Urban Epidemiologic Studies in New York City, chairman of the Department of Medicine at Albany Medical Center Hospital, and dean of Albany Medical College. He has been an active member of professional organizations including president of the Association of American Physicians and American College of Physicians, member of the Committee on Smoking and Health for the American Medical Association, and chairman of the Association of American Medical Colleges. He has served on numerous scientific advisory committees including the Directors Advisory Committee of the National Institutes of Health (NIH), the Special Medical Advisory Group of the Veterans Administration, and the Scientific Advisory Board of the U.S. Air Force. As a member of the Institute of Medicine (IOM), Dr. Bondurant has served as acting president as well as vice-chair of the IOM Council. He was chair of the IOM Committee on the Safety of Silicone Breast Implants, co-chair of the Committee on Public Health, and a member of the Committee on Science, Engineering and Public Policy. Dr. Bondurant's research interests have been in the area of cardiovascular and pulmonary medicine as well as issues concerning medical education and public health.

J. Richard Crout, M.D., is president of Crout Consulting, a company that provides regulatory and drug development advice to pharmaceutical and biotechnology companies. Dr. Crout formerly served as the director of the Bureau of Drugs at the Food and Drug Administration (FDA) (1973-1982) and as vice president of medical and scientific affairs at Boehringer Mannheim Pharmaceuticals Corporation (1984-1993). Dr. Crout was an IOM scholar-in-residence in 1994 and has served on various Institute of Medicine committees including the Committee to Study Medication Development and Research at the National Institute on Drug Abuse (NIDA) and as adviser to the Project on the Medical Use of Marijuana. He has published extensively in the area of clinical pharmacology and drug regulatory policy. Dr. Crout's awards include the Distinguished Service Medal of the U.S. Public Health Service and the Oscar B. Hunter Award in Therapeutics from the American Society for Clinical Pharmacology and Therapeutics.

Garret FitzGerald, M.D., is chair of the Department of Pharmacology, professor of medicine and pharmacology, and director of the Center for Experimental Therapeutics at the University of Pennsylvania. Before coming to the University of Pennsylvania, Dr. FitzGerald was professor and chairman of medicine and experimental therapeutics at University College in Dublin, Ireland, and professor of medicine and pharmacology and director of the Division of Clinical Pharmacology at Vanderbilt University School of Medicine. He has received many honors and appointments for his work, including the Robinette Foundation Professor of Cardiovascular Medicine and appointment as chair of the Arterosclerosis, Thrombosis, Vascular Biology Council of the American Heart Association. Dr. FitzGerald is on the editorial board of *Circulation* and was formerly on the board of the *Journal of Biological Chemistry*. His principle area of research involves isoeicosanoids as indices of oxidant stress, pharmacology of cyclooxygenase-2 and eicosanoid receptors and pathways, and development of antithrombotic and anti-inflammatory medications.

Adi Gazdar, M.D., is professor of pathology and deputy director of the Hamon Center for Therapeutic Oncology Research at University of Texas Southwestern Medical Center, Dallas. Prior to coming to Dallas, Dr. Gazdar worked at the National Cancer Institute (NCI) where he served as head of the Tumor Cell Biology Section and deputy head of the NCI-Navy Medical Oncology Branch. His research focuses on the molecular and genetic analysis of cancer, especially cancers of the lung, breast, and cervix, in order to further cancer diagnosis, prognosis, and treatment. Dr. Gazdar manages an extensive cancer tissue specimen data bank and a collection of in vitro developed tumor cell lines. He is also developing molecular and cellular techniques for early cancer detection.

Gary Giovino, M.S., Ph.D., is senior research scientist in the Department of Cancer Prevention, Epidemiology, and Biostatistics at Roswell Park Cancer Institute, Buffalo, New York. He was previously chief of the Epidemiology Branch (1991-1998) at the Office on Smoking and Health of the Centers for Disease Control and Prevention (CDC). Dr. Giovino serves on multiple professional organizations that are concerned with tobacco issues, including Adviser to the American Cancer Society on Epidemiology and Surveillance Research and expert consultant on tobacco issues for the National Household Survey on Drug Abuse, of the Substance Abuse and Mental Health Services Administration. He is a senior editor of *Tobacco Control: An International Journal* and currently serves as a core group member in the Research Network on the Etiology of Nicotine Dependence and as director of the State Tobacco Research Component of Project ImpacTeen, both funded by the Robert Wood Johnson Foundation. Dr. Giovino has earned numerous awards for his tobacco-related work and has published extensively in the areas of patterns, determinants, consequences, and control of tobacco use.

Dorothy Hatsukami, Ph.D., is professor in the Department of Psychiatry, adjunct professor in the Department of Psychology, and adjunct professor in the Division of Epidemiology at the University of Minnesota. She is director of the Tobacco Research Use Center at the University of Minnesota. Dr. Hatsukami is former president of the Society for Research on Nicotine and Tobacco and president-elect of the College on Problems of Drug Dependence. She serves on numerous scientific advisory boards and committees on substance abuse and addiction. Dr. Hatsukami recently served as an adviser for the IOM Project on Medical Use of Marijuana: Assessment of the Science Base. Her research interests include nicotine addiction and treatment, smoking behavior, smokeless tobacco use, and gender and age differences in tobacco use. She has received much recognition for her work, including being a co-recipient of the Ove Ferno Award for contributions in the area of tobacco dependence.

Rogene Henderson, Ph.D., is deputy director of the National Environmental Respiratory Center and senior biochemist and toxicologist at the Lovelace Respiratory Research Institute. Dr. Henderson also holds appointments as clinical professor of pharmacy at the University of New Mexico, Albuquerque, and as adjunct professor in the Department of Veterinary Microbiology, Pathology, and Public Health in the School of Veterinary Medicine, Purdue University. She has chaired the National Research Council (NRC) Subcommittee on Toxicological Hazard and Risk Assessment, Subcommittee on Pulmonary Toxicology, and standing Committee on Toxicology for six years, and she is currently a member of the Board on Environmental Studies and Toxicology. She has served as a member of the advisory council of the National Institute of Environmental Health Sciences and the U.S. Environmental Protection Agency's (EPA) Science Advisory Board, Environmental Health Committee. Dr. Henderson has received numerous appointments on scientific advisory committees, including her current appointment as a member of the Health Effects Institute Research Committee. Dr. Henderson has done extensive research in the areas of lung biochemistry, the pharmacokinetics of inhaled toxins and their metabolites, and biological markers of exposure.

Peter Reuter, Ph.D., is professor of public policy and of criminology at the University of Maryland, where he also directs the public policy Ph.D. program. In July 1999, he became editor of the *Journal of Policy Analysis and Management*. Dr. Reuter previously held the position of codirector of the Drug Policy Research Center and senior economist at the RAND Corporation. His research has included an examination of the organization of illegal markets and alternative approaches to controlling drug problems. Dr. Reuter's policy analysis of harm reduction approaches to illegal drug use has produced numerous articles on the subject and a forthcoming book: *Drug War Heresies: Learning from Other Vices, Times and Places* (Cambridge University Press with Robert MacCoun). He is a member of the NRC Committee on Law and Justice and served on the IOM Committee on the Federal Regulation of Methadone.

David J. Riley, M.D., is professor in the Division of Pulmonary and Critical Care Medicine at the University of Medicine and Dentistry of New Jersey (UMDNJ)-Robert Wood Johnson Medical School. Dr. Riley holds additional academic appointments as adjunct professor of physiology and biophysics, member of the Joint Graduate Program in Toxicology, and affiliate member of Environmental and Occupational Health Sciences Institute at the University of Medicine and Dentistry of New Jersey (UMDNJ)-Robert Wood Johnson Medical School. He has received numerous awards for his work in the field of pulmonology including the Sir William Osler Humanitarian Award from the New Jersey Thoracic Society and a Pulmonary Academic Award from the National Heart, Lung, and Blood Institute (NHLBI) of NIH. Dr. Riley has served on many scientific advisory committees, including several for the NIH, the Review Committee of the Health Effects Institute, chairman of the Pulmonary Disease Study Section of the Tobacco-Related Research Program, University of California at San Francisco (1993-1994), and a variety of committees of the American Thoracic Society. Dr. Riley holds NIH-supported grants for his research on vascular and interstitial diseases of the lung.

Peter Shields, M.D., is professor of medicine and oncology at Georgetown University Medical Center, where he is director of the Division of Cancer Genetics and Epidemiology in the Department of Oncology. Dr. Shields is associate director for population sciences at the Lombardi Cancer Center at Georgetown University. Prior to his position at Georgetown, Dr. Shields was a tenured investigator and chief of the Molecular Epidemiology Section of the Laboratory of Human Carcinogenesis in the Division of Basic Sciences at the NCI. Dr. Shields serves on the editorial board or as reviewer for a number of scientific journals including *Journal of the National Cancer Institute, Journal of the American Medical Association, Carcinogenesis, Cancer Epidemiology Biomarkers and Prevention*, and others. He has been a member of various professional and scientific committees, including the Tobacco Research Implementation Group for NCI (1998). Dr. Shields was the first elected chairperson of the Molecular Epidemiology Group of the American Association of Cancer Research. His research activities involve identifying and using biomarkers of cancer risk and gene-environment interactions to enhance risk assessment. More specifically, he is examining the relationship between tobacco and alcohol use as environmental elements in predicting cancer risk in susceptible subgroups of the population.

Robert Wallace, M.D., M.Sc. *(vice-chair)*, is professor of epidemiology and internal medicine at the University of Iowa Colleges of Public Health and Medicine. He was formerly head of the Department of Preventive Medicine at the University of Iowa College of Medicine and director of the University of Iowa Cancer Center. He was a member of the U.S. Preventive Services Task Force, Second Panel. Dr. Wallace's research interests include cancer epidemiology and prevention; the causes and prevention of chronic, disabling diseases among older persons; women's health issues; and risk factors for cardiovascular disease. Dr. Wallace received the Irene Ensminger Stecher Professorship in April 1999 for cancer-related research. He is currently a member of the IOM Board on Health Promotion and Disease Prevention. He recently served as a member of the Committee on Leading Health Indicators for Healthy People 2010 and the NRC Panel on a Research Agenda and New Data for an Aging World.

James T. Willerson, M.D., is President of the University of Texas Health Sciences Center at Houston. He is the Edward Randall III Professor and chairman of the Department of Internal Medicine at the University of Texas Medical School. Dr. Willerson serves as chief of medical services at Memorial Hermann Hospital and Lyndon B. Johnson General Hospital. He is also medical director, chief of cardiology, director of cardiology research, and codirector of the Cullen Cardiovascular Research

Laboratories at the Texas Heart Institute. He is a member of numerous professional and scientific organizations including the Council of Clinical Cardiology of the American Heart Association, a fellow of the American College of Cardiology, editor-in-chief of *Circulation* (1993-2004), and a member of the IOM. Dr. Willerson was a member of the IOM Committee to Provide Review of the Fialuridine Clinical Trials. He is the author of multiple textbooks and hundreds of articles on a wide variety of cardiovascular issues including gene therapy, basic mechanisms involved in the development of acute coronary syndromes, and detection of vulnerable atherosclerotic plaques.

Liaisons

Kenneth E. Warner, Ph.D., is the Richard D. Remington Collegiate Professor of Public Health in the Department of Health Management and Policy at the University of Michigan School of Public Health. He is also director of the university's Tobacco Research Network and associate director of the university's Robert Wood Johnson Foundation Clinical Scholars Program. Dr. Warner's research has focused on economic and policy aspects of tobacco and health. Dr. Warner served as senior scientific editor of the twenty-fifth anniversary Surgeon General's report on smoking and health. He is on the editorial board of numerous journals and chairs the board of *Tobacco Control*. He serves on the Board of Directors of the American Legacy Foundation and on the Council and Membership Committee of the IOM. Among his numerous professional awards, Dr. Warner has been honored twice by Delta Omega, the national public health honorary society, for "outstanding achievement in public health." He received the Surgeon General's Medallion from Dr. C. Everett Koop in 1989. Dr. Warner was named to the first class of fellows of the Association for Health Services Research in 1996; was elected to the Institute of Medicine in 1996; and received the Excellence in Research Award from the University of Michigan School of Public Health in 1997.

Richard Bonnie, LL.B., is the John S. Battle Professor of Law at the University of Virginia School of Law and director of the Institute of Law, Psychiatry and Public Policy at the University of Virginia. Dr. Bonnie's professional expertise is in law and public policy pertaining to public health and mental health, including prevention and treatment of addiction and substance abuse. He has written extensively about regulation of alcohol, tobacco, and controlled substances and has served as a member of the National Advisory Council on Drug Abuse and on the Board of Directors of the College on Problems of Drug Dependence. Also, Professor Bonnie has written about psychiatry and the law and the rights of mental health patients, and serves as an adviser to the American Psychiatric Association's Council on Psychiatry and Law. He is a member of the Institute of Medicine and serves on the Board on Neuroscience and Behavioral Health. Professor Bonnie has served as chair of the IOM Committee on Opportunities in Drug Abuse Research and Committee on Injury Prevention and Control, and as a member of the Committee on Prevention of Nicotine Dependence in Children and Adolescents. He is currently a member of the NRC Committee on Data and Research for Policy on Illegal Drugs.

Institute of Medicine Staff

Kathleen R. Stratton, Ph.D. (senior program officer), received a B.A. in natural sciences from Johns Hopkins University. She received her Ph.D. in neurotoxicology from the University of Maryland at Baltimore. After completing postdoctoral fellowships in neuropharmacology at the University of Maryland School of Medicine and in neurophysiology of second-messenger systems at the Johns Hopkins University School of Medicine, she joined the IOM staff in 1990. Dr. Stratton has worked in the IOM Divisions of Health Sciences Policy, Biobehavioral Medicine and Mental Disorders, and Health Promotion and Disease Prevention, as well as the NRC Board on Environmental Studies and Toxicology. She has worked on projects in environmental risk assessment, neurotoxicology, the organization of research and service programs in the Public Health Service, vaccine safety, fetal alcohol syndrome, and vaccine development. She directed the Division of Health Promotion and Disease Prevention from 1997 until 1999.

Padma Shetty, M.D. (program officer), finished her B.A. in Psychology at Northwestern University and her M.D. at the Ohio State University College of Medicine in 1998. After completing an internship in Internal Medicine, she joined the IOM staff for the Committee to Assess the Science Base of Tobacco Harm Reduction in 1999.

Ann W. St. Claire is senior project assistant for the IOM Committee to Assess the Science Base for Tobacco Harm Reduction. Ms. St. Claire joined the IOM Board on Health Promotion and Disease Prevention in 1999 and has worked on the IOM study, *Reducing the Burden of Injury* (1999). Ms. St. Claire graduated from Mount Holyoke College in 1999 with a degree in politics and a certificate in the culture, health and science program.

C

Time Line of Tobacco Events

Science	Products	Policies or Regulation
	1700s • Cigarettes are first made from scraps of cigars	
1761 • Dr. John Hill reports in *Cautions Against the Immoderate Use of Snuff* on two case histories and observed that ("snuff is able to produce swellings and excrescences") in the nose, and he believed these to be cancerous		
1807 • Cerioli isolates the "essential oil" or "essence of tobacco"		
1828 • Posselt and Reimann isolate nicotine from tobacco		
1843 • Melsens describes nicotine's chemical empirical formula		

Science	Products	Policies or Regulation

Science

1893
• Pictet and Crepieux synthesize nicotine

1900
• Brosch appears to be the first investigator involved in experimental tobacco carcinogenesis. He applies tobacco ("juices") to guinea pigs, observing epithelial proliferation

1928
• Lombard and Doerring find an association between heavy smoking and buccal cancer

1938
• Pearl, a statistician and biometrician at Johns Hopkins, publishes the first statistical analysis comparing the health of smokers and nonsmokers and finds that individuals who smoked could expect shorter lives

1950
• Epidemiological studies report that lung cancer is particularly prevalent among cigarette smokers

Products

1854-1856
• Cigarette popularity grows between 1854 and 1856 during the Crimean War

1881
• James Bonsack of Virginia patents the first cigarette-making machine.

1913
• Camel brand is produced by R.J. Reynolds in 1913

1950
• Filters introduced

Science	Products	Policies or Regulation
1954	**1954**	**1954**
• Doll and Hill publish The Mortality of Doctors and Their Smoking Habits in the *BMJ*	• Winston by RJR features a filter (cellulose acetate) and reconstituted sheet tobacco • Tobacco Industry Research Committee is formed	• Industry faces first liability lawsuit by lung cancer victim claiming negligence and breach of warranty. Suit is dropped 13 years later
	1956	
	• RJR introduces Salem as its first filter-tipped menthol cigarette	
	Early 1960s	
	• More porous cigarette paper is introduced	
		1960
		• Ban on advertising tar and nicotine levels as less harmful 1960-1966
1962		
• Royal College of Physicians reports that lung cancer is prevalent among smokers in *Smoking and Health*		
1964		**1964**
• Surgeon General's report concludes that smoking causes cancer and other serious diseases		• Public Health Service establishes the National Clearinghouse for Smoking and Health (NCSH), later to become the Office on Smoking and Health
		1965
		• Federal Cigarette Labeling and Advertising Act is passed requiring warning labels on all cigarette packs but not on advertisements. The act also requires the Secretary of Health, Education, and Welfare to issue annual reports to congress on the health consequences of smoking

Science	Products	Policies or Regulation

<table>
<tr><td></td><td></td><td>• Federal excise tax on smokeless tobacco products is repealed.</td></tr>
</table>

Science

1966
• The Cambridge Filter method is ultimately adopted by the Federal Trade Commission

1967
• Surgeon General's report concludes, "Cigarette smoking is the most important of the causes of chronic non-neoplastic bronchiopulmonary diseases in the United States." The report also identifies measures of morbidity associated with smoking

Products

Late 1960s
• Expanded or "puffed" tobacco appears in cigarettes

1968
• National Cancer Institute begins work on a safer cigarette by establishing the Less Hazardous Cigarette Working Group

1969
• Surgeon General's report makes solid conclusions regarding the relationship between maternal smoking and infant low birthweight. It also defines evidence of increased incidence of prematurity, spontaneous abortion, stillbirth, and neonatal death
• Use of chlorinated pesticides begins to be faded out

Science	Products	Policies or Regulation
	Early 1970s • Ventilated filter tips are used and modified to be longer with increased efficiency • Introduction of puffed, expanded, and freeze-dried tobaccos	
1971 • Surgeon General's report finds smoking associated with cancers of the oral cavity and esophagus		**1971** • Broadcast ads for cigarettes banned
1972 • Surgeon General's report studies immunological effects of tobacco and tobacco smoke, and identifies carbon monoxide, nicotine, and tar as smoke constituents the most likely to produce health hazards from smoking		**1972** • Officials declare that airlines must create nonsmoking sections
1973 • Surgeon General's report presents evidence on the health effects of smoking pipes, cigars, and "little cigars"		
1975 • Surgeon General's report issues further evidence regarding health effects from involuntary (passive) smoking, especially the relationship between parental smoking and rates of bronchitis and pneumonia in children's first year of life	**1975** • RJR introduces NOW with lower-tar and nicotine while preserving tobacco taste	
1977–1978 • Surgeon General's report focuses on health effects of smoking on women, noting in particular the effects of oral contraceptives and smoking on the cardiovascular system		

Science	Products	Policies or Regulation

1979
- Surgeon General's report addresses the role of adult and youth education in preventing smoking habits. Report also reviews health effects of smokeless tobacco
- Bandury conference and report

1980
- Surgeon General's report projects that lung cancer in women will surpass breast cancer as the leading cause of cancer mortality in women. Report also notes prevalence of smoking by adolescent females

1981
- Surgeon General's report examines the health consequences of lower-tar and nicotine cigarettes. Concludes that lower-yield cigarettes decrease the risk of lung cancer, but have little effect on rates of cardiovascular disease, chronic obstructive pulmonary disease, and fetal damage. The report also reviews risks related to various additives and their combustion. The Surgeon General reinforces the fact that there is no safe cigarette

1981
- Insurers offer discounts on life insurance premiums to nonsmokers

Science	Products	Policies or Regulation

1982
- Surgeon General's report releases epidemiological evidence from a study of nonsmoking wives and their smoking husbands, finding that the risk of lung cancer in wives was not causal, but a possible serious public health problem. Report notes possible low-cost smoking cessation interventions

1983
- Surgeon General's report evaluates health consequences of smoking for cardiovascular disease, declaring cigarette smoking as one of the three primary causes of coronary heart disease

1984
- Surgeon General's report examines health effects of smoking on chronic obstructive lung disease (COLD). Smoking accounts for 80-90% of COLD deaths in the United States

1984
- Nicotine-based chewing gum approved as an aid to quitting

1984
- Warnings strengthened on cigarette packages and ads
- San Francisco requires business to accommodate nonsmokers

1985
- Surgeon General's report focuses on smoking and hazardous substances in the workplace. Smoking alone is found to be a greater risk than the average workplace environment. Workplace exposure to asbestos and other such substances is found to compound health risks

Science	Products	Policies or Regulation

1986
- Surgeon General's report states that "Involuntary smoking is a cause of disease, including lung cancer, in healthy nonsmokers." Report further notes the health of children of smokers and nonsmokers, as well as the exposure to smoke of passengers in smoking and nonsmoking sections of airplanes
- A special report of the advisory committee appointed by the Surgeon General examines the health effects of smokeless tobacco, concluding that it too leads to nicotine addiction and can cause cancer

1987
- Tobacco Institute Testing Laboratory (TITL) assumes cigarette-testing responsibilities from the Federal Trade Commission Test Center using its approved methodology

1988
- Surgeon General Koop's report states that nicotine (cigarettes and other forms of tobacco) are addicting

1988
- Government bans smoking on short domestic airline flights

1989
- Surgeon General's report reports that cigarette smoking is a major cause of cerebrovascular disease (stroke). Report also addresses the future of nicotine addiction in light of new nicotine delivery systems test marketed in 1988

Late 1980s
- RJR Premier is developed and introduced to the public

Science	Products	Policies or Regulation
1990 • Surgeon General's report identifies the health benefits of smoking cessation: "Smoking cessation has major and immediate health benefits for men and women of all ages." Report examines life expectancy, smoking-related diseases, and reproductive health issues of smokers and former smokers		**1990** • Smoking banned on interstate buses and domestic airline flights of six hours or less
	1992 • Nicotine patches introduced • 97.5% of cigarettes in the U.S. have filters	
		1993 • Vermont bans smoking in all indoor public places
1994 • Surgeon General's report looks at "preventing tobacco use among young people." Report examines and discusses age at first initiation, issues or problems encountered with youth cessation, tobacco as a "gate-way drug," effect of advertising, and school-based tobacco use prevention programs	**1994** • Major U.S. cigarette companies release a list of 599 additives used in the manufacture of cigarettes	**1994** • Executives of seven largest U.S. tobacco companies swear in congressional testimony that nicotine is not an addictive and deny manipulating nicotine levels in cigarettes • Amtrak bans smoking on short-and medium-distance trips • Brown and Williamson documents provide evidence that tobacco executives discovered smoking risks before the surgeon general made declaration • Mississippi files first of 24 state lawsuits seeking to recoup millions from tobacco companies for smokers' Medicaid bills

Science	Products	Policies or Regulation
	1996	**1996**
	• RJR Eclipse test marketed in United States, Germany, and Sweden	• Liggett Group settles claims with five state attorneys-general and promises to help them against other companies
		1997
		• Federal judges rules that government can regulate tobacco as a drug, but industry is allowed to continue advertising
		• Landmark settlement, subject to congressional approval, calls for restrictions on cigarettes and on tobacco maker's liability in lawsuits. Industry is required to spend $368 billion over 25 years, run antismoking campaigns, issue bold health warnings on packs, decrease advertising, and pay fines if youth smoking doesn't drop significantly
		• Mississippi is first state to settle, agreeing to $3.6 billion deal with tobacco companies
		• Florida settles at $11.3 billion

Science	Products	Policies or Regulation
1998	**1998**	**1998**
• Surgeon General's report examines tobacco use among U.S. racial and ethnic minority groups: African Americans, American Indians and Alaska Natives, Asian Americans and Pacific Islanders, and Hispanics. Concludes that cigarette smoking is one of the major health hazards among different racial and ethnic groups. Tobacco use and patterns of use vary among these groups as well	• Phillip Morris Accord presented at a poster presentation at the Society of Toxicology in Seattle	• Texas settles with industry at $15.3 billion over 25 years • Tobacco executives testify before Congress that nicotine is addictive under current definitions of the word and smoking may cause cancer • Minnesota and Blue Cross/Blue Shield of Minnesota settle at $6.6 billion with the tobacco industry • Senate vetoes a proposal of $1.50 tax increase per pack on cigarettes • McCain Universal Tobacco Settlement Bill dies in congressional filibuster. Bill addressed tobacco product regulation of ingredients, sales, and advertising. It also addresses education and nicotine addiction prevention • 46 states welcome a $206 billion settlement with the tobacco industry over health care costs for treating sick smokers
	1999	**1999**
	• Star Tobacco introduces new cigarettes with low-nitrosamine tobacco and activated-charcoal filter, but is unable to make health claims	• Justice Department sues the tobacco industry to recover billions of government dollars spent on smoking-related health care, accusing cigarette makers of a "coordinated campaign of fraud and deceit"

Science	Products	Policies or Regulation
		• Government and tobacco industry lawyers present oral argument to the Supreme Court over whether the Food and Drug Administration (FDA) can regulate tobacco as a drug and crack down on cigarette sales to minors
2000	**2000**	**2000**
• National Institute for Environmental Health Sciences publishes ninth report on Carcinogens including ETS as one of the known human carcinogens • Surgeon General's Report, *Reducing Tobacco Use,* published	• Phillip Morris introduces cigarettes with "safer" paper • Star Tobacco—new cigarettes produced in Virginia and Kentucky with low-nitrosamine tobacco, but without health claims	• Tobacco farmers sue cigarette makers for $69 billion, claiming they conspired to undo the federal system that regulates tobacco prices • Supreme Court rules, 5-4, that FDA lacks authority to regulate tobacco as an addictive drug • New York state imposes fire-safety standards on cigarettes

REFERENCES

Associated Press. 2000. Major events in fight vs. tobacco. *The New York Times,* March 21. 2000.

Feder BJ. 1996. A safer smoke or just another smokescreen? Reynolds courts support for a new product. *The New York Times.* April 12, 1996:(Business Day);D1.

Gorrod JW, Jacob P III. 1999. *Analytical Determination of Nicotine and Related Compounds and Their Metabolites.* PAF2. Amsterland, Netherlands: Elsevier Science B.V.

Hoffmann D, Hoffmann I. 1997. The changing cigarette, 1950-1995. *J Toxicol Environ Health* 50(4):307-364.

NIH (National Institute of Health). 1996. *The FTC Cigarette Test Method for Determining Tar, Nicotine, and Carbon Monoxide Yields of U.S. Cigarettes. Monograph 7.* NCI Smoking and Tobacco Control. Bethesda, MD: National Institutes of Health, National Cancer Institute.

R.J. Reynolds Tobacco Company. 1988. *Chemical and Biological Studies: New Cigarette Prototypes That Heat Instead of Burn.* Winston-Salem, NC: R.J. Reynolds Tobacco Company.

U.S. DHHS (U.S. Department of Health and Human Services). 1964. *Smoking and Health: Report of the Advisory Committee to the Surgeon General of the Public Health Service.* Atlanta, GA: U.S. Public Health Service.

U.S. DHHS (U.S. Department of Health and Human Services). 1973. *The Health Consequences of Smoking: A Report of the Surgeon General.* Atlanta, GA: U.S. Department of Health and Human Services.

U.S. DHHS (U.S. Department of Health and Human Services). 1982. *The Health Consequences of Smoking: Cancer; A Report of the Surgeon General.* Atlanta, GA: U.S. Department of Health and Human Services, Centers for Disease Control and Prevention.

U.S. DHHS (U.S. Department of Health and Human Services). 1986. *The Health Consequences of Involuntary Smoking: A Report of the Surgeon General.* Atlanta, GA: U.S. Department of Health and Human Services, Centers for Disease Control and Prevention.

Wynder EL, Hoffman D. 1979. Tobacco and health: a societal change. *NEJM* 300(16):894-903.

Index